35th Ethylene Producers Conference 2023

AA002830

Topical Conference at the 2023 AIChE Spring Meeting and 19th Global Congress on Process Safety

Houston, Texas, USA
12-16 March 2023

ISBN: 978-1-7138-7933-6

Printed from e-media with permission by:

Curran Associates, Inc.
57 Morehouse Lane
Red Hook, NY 12571

proceedings.com

Some format issues inherent in the e-media version may also appear in this print version.

Printed with permission by Curran Associates, Inc. (2023)

For permission requests, please contact AIChE
at the address below.

AIChE
120 Wall Street, FL 23
New York, NY 10005-4020

Phone: (800) 242-4363
Fax: (203) 775-5177

www.aiche.org

Additional copies of this publication are available from:

Curran Associates, Inc.
57 Morehouse Lane
Red Hook, NY 12571 USA
Phone: 845-758-0400
Fax: 845-758-2633
Email: curran@proceedings.com
Web: www.proceedings.com

TABLE OF CONTENTS

CIRCULAR ECONOMY/DECARBONIZATION

CAUSTIC SYSTEM TUTORIAL

CONTAMINANTS AND IMPURITIES IN ETHYLENE PLANTS

ETHYLENE PLANT ENVIRONMENTAL AND SUSTAINABILITY

ETHYLENE PLANT MAINTENANCE AND RELIABILITY

ETHYLENE PLANT OPERATIONS

ETHYLENE PLANT ROTATING EQUIPMENT

ETHYLENE PLANT SAFETY

ETHYLENE PLANT TECHNOLOGY FUNDAMENTALS

INDUSTRY 4.0 IN ETHYLENE PRODUCTION

HEATERS/H2 FIRING – DECARBONIZATION OPTIONS

EPC KEYNOTE: DEVELOPING A ROADMAP TO NET ZERO

155b Low Carbon Intensity Ethylene Overview

Steven Slome

Sustainability has become increasingly important in the chemical and energy industries in recent years. Now, due to pledges of carbon neutrality (many by 2050), the chemical industry is searching for options to reduce the carbon intensity of their products (primarily committed to as neutrality based upon scope 1 and 2 emissions, i.e., those directly emitted and those indirectly emitted for utilities), and this need is further supported by the increased chance of expanding carbon taxes and cross border adjustment mechanisms for carbon intensity into key markets. Ethylene is especially important as it is both responsible for a significant fraction of global emissions from the chemical industry as well as being an important feedstock for further chemistry (e.g., MEG, EO, PE., etc.). Because of this, ethylene as well as methanol and ammonia are currently under increased pressure to decarbonize and are seeing an increase in activity focused on sustainability in these sectors (methanol and ammonia are the other "big two" from an emissions perspective: combined these two are, according to IEA estimates, responsible for almost 75 percent of direct emissions from the chemical industry). Methanol decarbonization is especially interesting as it is a potential feedstock for ethylene as well.

NexantECA will discuss the different options for ethylene carbon intensity reduction from a cost and carbon intensity perspective, with insights and data coming from NexantECA's recently published multi-subscriber report Low Carbon Intensity Ethylene: A Technoeconomic and Carbon Intensity Study. Current approaches for decarbonization with high interest include cracker focused approaches such as, alternative cracker feedstocks, carbon capture, changes to the cracker (e.g., electrification and/or hydrogen co-firing), using decarbonized methanol (including very low CI e-methanol) as a feedstock for MTO, as well as alternative chemistries (e.g., ethanol to ethylene). There are also some very early stage developments and recent announcements of technologies that may hold promise such as CO2 to ethylene, however these are far from commercial and have significant hurdles to overcome to be considered commercially viable.

The reality of achieving global net zero emissions for ethylene will likely be an uphill battle and require a combination of multiple approaches with the likelihood high currently that we will fall short by the deadline—but should it ultimately fall to zero by 2050 or some later date, it is most probably that the death of the carbon footprint of ethylene will not be one swift blow by shutdown economics of a superior technology, rather it will be one of a thousand paper cuts by incremental improvements and the confluence many small steps.

AICHE PAPER NUMBER: 155C

ENERGY EFFICIENT OLEFIN-PARAFFIN SEPARATIONS WITH OPTIPERM™ MEMBRANE SYSTEM

Robert Somich

Process Technology Engineering

Braskem

Jacob Wargo

Membrane Development Engineer

Compact Membrane Systems

Kenneth Pennisi

Sr Process Design Engineer

Compact Membrane Systems

Prepared for Presentation at the Year 2023 Spring National Meeting

Houston, Texas, March 12-16, 2023

ENERGY EFFICIENT OLEFIN-PARAFFIN SEPARATIONS WITH OPTIPERM™ MEMBRANE SYSTEM

Robert Somich

Process Technology Engineering

Braskem

Jacob Wargo

Membrane Development Engineer

Compact Membrane Systems

Kenneth Pennisi

Sr Process Design Engineer

Compact Membrane Systems

Abstract:

Compact Membrane Systems has developed the novel Optiperm™ membrane platform for decarbonizing today's industry across two applications: olefin paraffin separation and carbon capture. The separation of light olefins from paraffins is one of the most energy intensive in the world. The state of the art separation method is distillation, a technology that has remained unchanged over decades. Membranes hold the promise of a more modular, energy efficient and capital efficient method for separation. However, they have historically been hampered by low performance and stability. CMS has developed a transformational membrane technology that is being demonstrated at commercial scale with Braskem.

The first Optiperm™ membrane utilizes a silver embedded polymeric material to selectively transport olefins. This material has demonstrated over 3 years of longevity in laboratory operation and has shown high selectivity and permeance under realistic operating conditions. Additionally, Optiperm™ membranes were first trialed at the Delaware City Refining Company in over 300 days of field operations validating the robustness and capability of the technology in an operating plant environment.

In continuing the commercialization of Optiperm™ Membranes, CMS' has partnered with Braskem to jointly develop and fabricate a multistage pilot unit to test commercial scale spiral wound membranes in a mixed C_3 process stream. This talk will discuss results from the pilot unit that was deployed in a Braskem Polymer facility in January 2022 and was brought on stream in May 2022. This is the first commercial scale operation of Optiperm™ membrane modules under field conditions using actual process streams. While the membrane as the technical unlock, the engineering system allows this novel technology to cross the chasm from the lab to commercial applications. We will describe the pilot unit design which incorporates all the process equipment of a commercial scale unit including an advanced process control system that enables the unit to simultaneously produce two high quality product streams.

Introduction

Background

Olefins are the basic building block of all polyolefin polymers. In order to efficiently produce these polymers a high purity olefin feed stock is required; this is produced by removing nonreactive paraffins from a mixed olefin/paraffin feed. The separation of light olefins and paraffins is one of the world's most energy intensive industrial applications, consuming up to 0.5% of global energy use[1]. The energy costs associated with the production of olefins make them a relatively expensive feed stock and a significant component of the manufacturing cost of derivative products. Consequently, there is a significant economic benefit to maximizing process yield, the fraction of feed that is converted into product. Frequently a challenge to maximizing process yield is losses of olefins with off-gas or effluent streams. Addressing this loss is a major value process intensification opportunity.

Yield in reactors may be limited by reaction kinetics, reaction equilibrium, the need to limit the extent of reaction to avoid undesired byproducts, or the need to remove non-reactive contaminants that build up in the reactor as the reaction progresses. These effluent streams contain valuable olefins from the reactor. If un-recovered, the olefins in these streams can represent a substantial cost to the process, increased emissions, and missed opportunity to increase product volume. Ideally these olefins are captured, purified, and returned to the process rather than lost.

Today, there are few options for processing these olefins. Some options result in the loss of the olefins completely: Venting of off-gas streams to the atmosphere results in a complete loss of the valuable feedstock and is often prohibited by environmental regulations. Venting to a flare or fuel gas stream, while preferred environmentally, also results in loss of valuable olefins. In some cases, olefins may be recovered by distillation. However, these separations can be difficult and energy intensive often requiring columns having multiple high efficiency trays, making the process capital intensive as well, particularly at the relatively small scales required for a column dedicated to a single reactor. In addition to high energy consumption and related GHG emissions, polyolefin catalysts advancements have increased reactor throughput over the years resulting in the splitter becoming the process bottleneck and the plant capacity limiter. Due to these challenges, a technology that could bolt on to existing plants and increase capacity, capture and upgrade waste streams, and debottleneck the plant offers substantial value in today's environment of declining feedstock and price volatility.

Membranes hold the promise to be that solution – dramatically reducing energy consumption, capturing and upgrading waste streams, and expanding capacity. A membrane that can efficiently separate the olefins from reaction mixtures and has good long-term performance stability can provide substantial economic benefits to a wide range of chemical manufacturing processes. Historically, membranes have not met the stringent demands of operating in a real-world environment within a petrochemical plant. Despite these historical challenges, CMS has demonstrated the efficacy of the Optiperm™ membrane for recovery of propylene from a process gas mixture.

Optiperm™ Facilitated Transport Membrane Technology

CMS has developed a silver-containing fluoropolymer membrane known as Optiperm™. The technology exploits a mass-transfer mechanism known as facilitated transport. Silver is incorporated into the membrane polymer and acts as a binding site for the double bond in the olefins. Paraffins and other chemicals, lacking a double bond, do not have the same interaction. This difference in affinity allows for olefins to pass through the membrane at a significantly higher rate than paraffins and other molecules which rely on the solution diffusion mechanism to pass through the membrane.

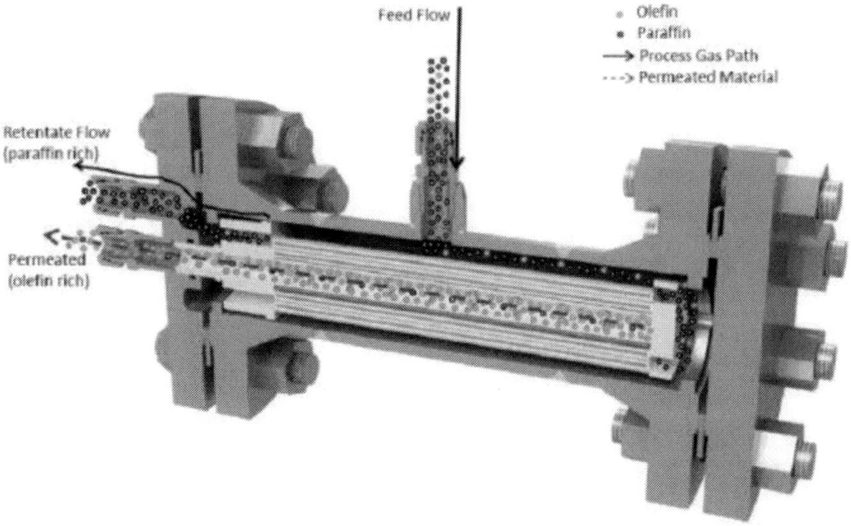

Illustration of Optiperm™ membrane module.

Although various silver bearing facilitated transport membranes for olefin/paraffin separation have been studied for many years, the Optiperm™ membrane is unique in that it exhibits stable performance over a much greater time period as compared to earlier membranes. The improvement in membrane stability is made possible by incorporating the silver ion into the backbone of the fluoropolymer which makes up the bulk of the membrane structure. As with most fluoropolymers, the fluoropolymer used in the Optiperm™ membrane is very resistant to chemical attack. It is this property that has been successfully utilized to create a membrane that provides stable separation performance in olefin/paraffin separations and recovery of olefins from mixed gases. CMS has performed extensive testing in the laboratory to verify the stability of the membrane.

Optiperm™ operates at a wide range of industrially relevant temperatures, pressures, and compositions. CMS has demonstrated olefin separation at temperatures up to 70°C (158°F) and pressures up to 250 psig. The membrane can remove olefin in both concentrated and dilute streams for compositions ranging from 5-95% by volume of olefin. The membrane is limited to C2, C3, or C4 olefins gas mixtures because it requires vapor phase operation. As with all facilitated transport membranes, Optiperm™ requires humidification of the process stream. CMS has developed a proprietary humidification system within the module to ensure optimal performance.

The graph below illustrates Optiperm™ stable performance profile when separating a 20% propylene/80% propane mixture at 60 psig and room temperature. The membrane exhibits stable performance for over 2 years in the laboratory. We estimate membrane life at 2 years in most streams and up to 3 years in a high-quality feed.

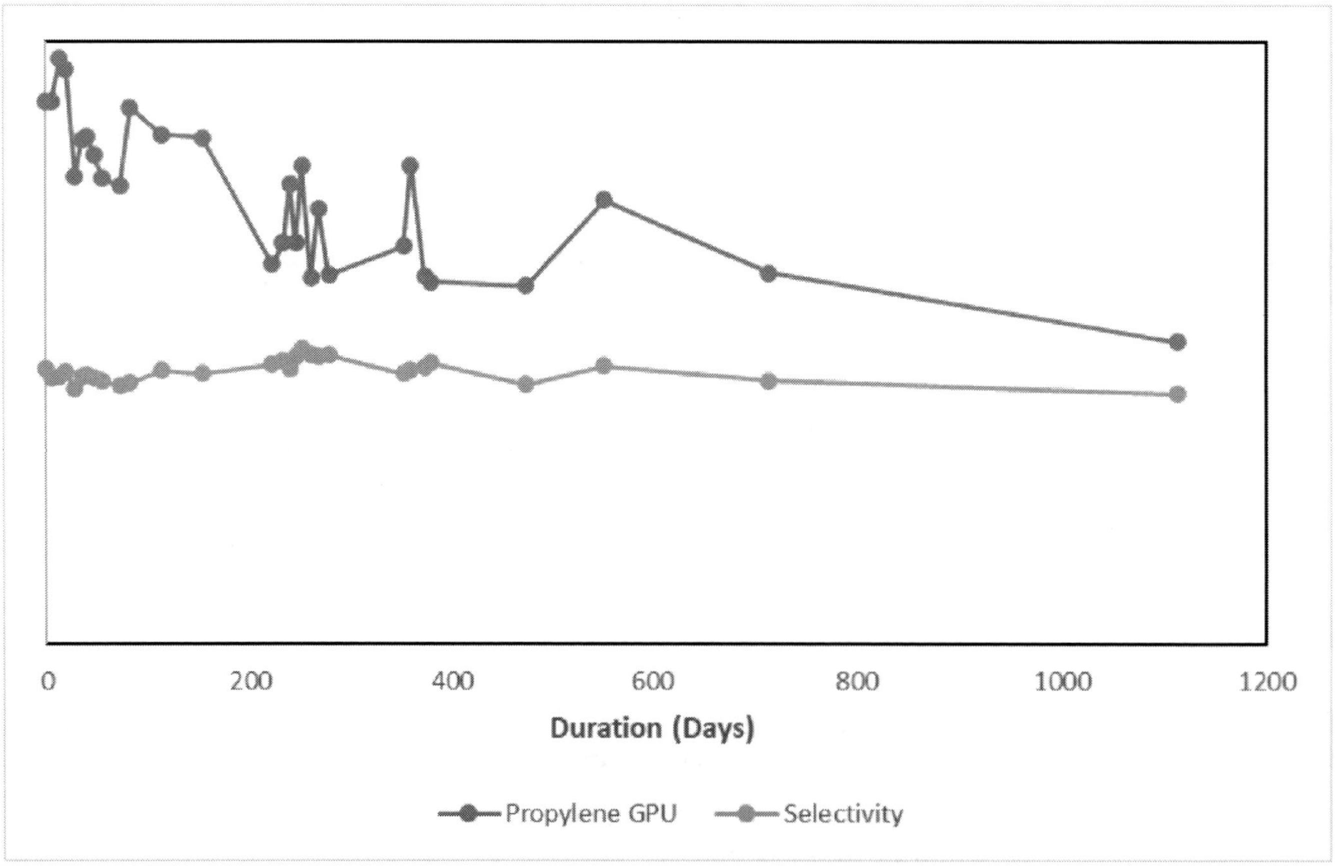

Optiperm™ membrane long term aging study.

Optiperm™ has also undergone long term continuous testing at conditions representative of a full-scale application. Its viability in the field was first tested at the Delaware City Refining Corporation where the membrane was separating a real process stream for over 300 days. CMS also built a test rig capable of continuously operating with full length membrane modules at pressures ranging from 90-200 psig and temperatures up to 70°C within the laboratory. On this test rig Optiperm™ demonstrated stable performance for over 70 days under realistic operating conditions. The results from these experiments reinforced Optiperm's™ stability and chemical resistance along with its viability for use in full scale Olefin/Paraffin separations. Currently Optiperm™ membranes are in Pilot testing at one of Braskem's production sites. This is CMS first deployment of a commercial scale membrane operating in a plant environment. The results of this testing have further demonstrated that Optiperm™ is an energy efficient and economical technology for Olefin/Paraffin separations.

Methodology

The effluent at Braskem's facility is currently separated by a distillation column (a C3 Splitter) that produces polymer grade propylene (PGP – 99.5% propylene) and HD5 propane (95% propane). The C3 splitter is operating at full capacity which precludes any increase production. It is considered impractical to replace the C3 splitter or add an additional splitter due to the high capital cost. However, elimination of one of the feed streams to the column would unload the column and enable a small capacity gain as well. An economic and engineering study of the application indicated the Optiperm™ membrane process to be a cost-effective solution to the problem. A pilot test unit was installed at the site to demonstrate technical feasibility. Specifically, the purpose of the pilot unit is to:

1) Demonstrate a process that separates a propylene/propane mixture into a salable HD5 propane product and 90+% propylene product,
2) Maintain production of on-spec product streams as feed composition varies,
3) Show durability and stability of the membranes over months of operation, and
4) Test the impact of varying operating conditions, particularly pressure, on membrane separation performance.

The Pilot Test Unit

The pilot unit at Braskem's facility is a two-stage membrane process. A two-stage membrane design was selected because a key objective of the pilot test is to simultaneously demonstrate recovery of HD5 propane as well as 90+% propylene. Independent specification of two product stream purities with a widely varying feed stream composition requires more than a simple single stage membrane process. Moreover, recoveries of both propylene and propane exceed 90% at some feed compositions. High recoveries are more efficiently achieved with multistage membrane processes. A two-stage membrane process was considered suitable for a pilot unit due to its relative simplicity. A process flow diagram of the system is presented below. The humidification water is supplied to the membranes by CMS' patented humidification process. In the two-stage process, the retentate from the first stage membrane is the propane rich product and the permeate from the second stage membrane is the propylene rich product. The permeate from the first stage membrane is compressed and fed to the second stage membrane. The retentate from the second stage membrane is recycled to the first stage membrane

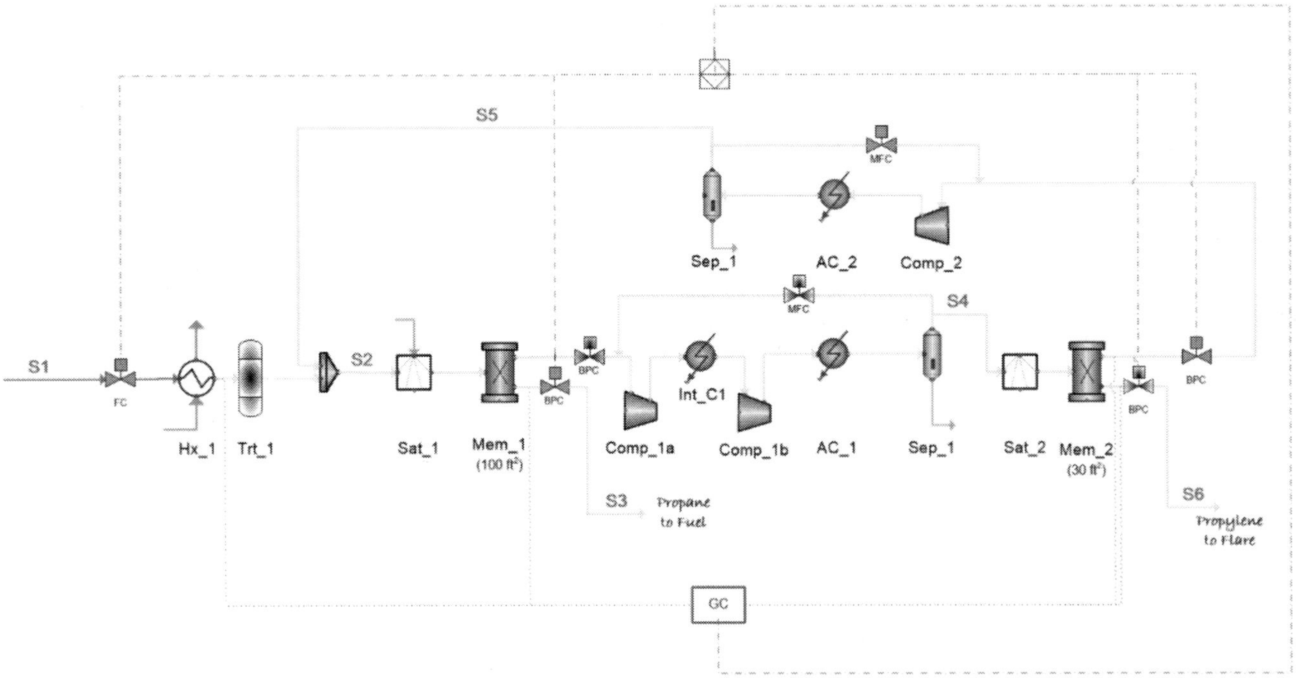

Diagram of the Optiperm™ pilot test unit.

The feed flow rate to the system and various pressures are controlled by automated valves. Stream compositions are monitored by a gas chromatograph. The photos below provide a visual tour of the Pilot Unit.

Photos of Optiperm™ pilot unit. Key pieces of equipment are labelled.

Results

At the time of writing, the process has been operating more than 90 days, 45 of them continuously without interruption. System utilization is 85%, excluding the initial system shake-out test period. After routine unit operations were established, commercial scale Optiperm™ membranes were installed in both membrane stages and have now been in service more than 60 days. The first stage membrane has an average feed flow of 26 lbs/hr and the second stage is processing 24 lbs/hr. The charts below show the changes in membrane properties (propylene permeance and propylene/propane selectivity) over time, as calculated from the separation data. Apart from two periods where operating parameters were significantly altered, the properties of both membranes are stable with no significant performance decay noted in the membrane properties charts. Additionally, as indicated in the 1st stage membrane properties chart, as the stage cut was adjusted to the design flow condition the membrane performance properties showed a distinct improvement in both permeance and selectivity.

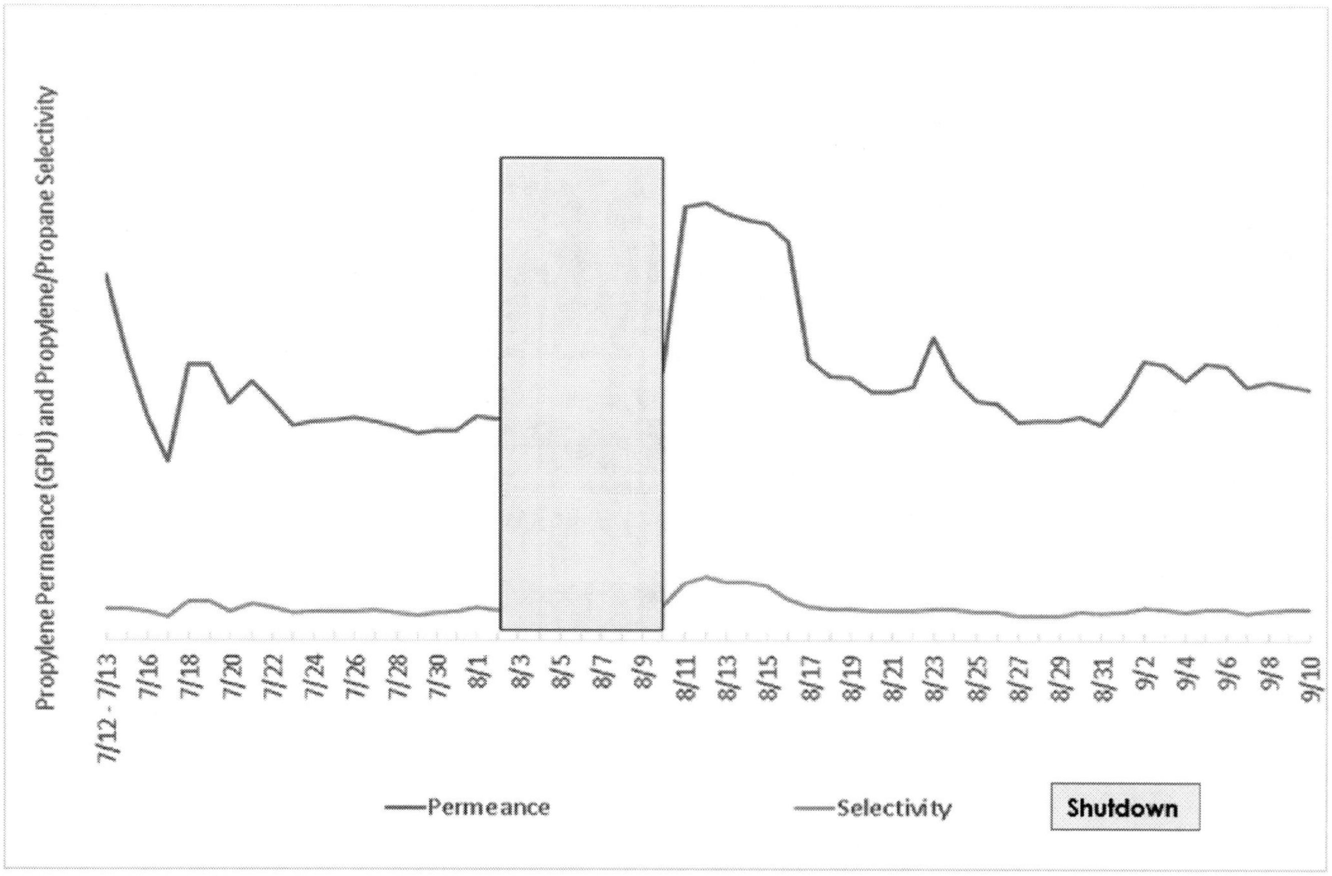

Average stage 1 membrane properties

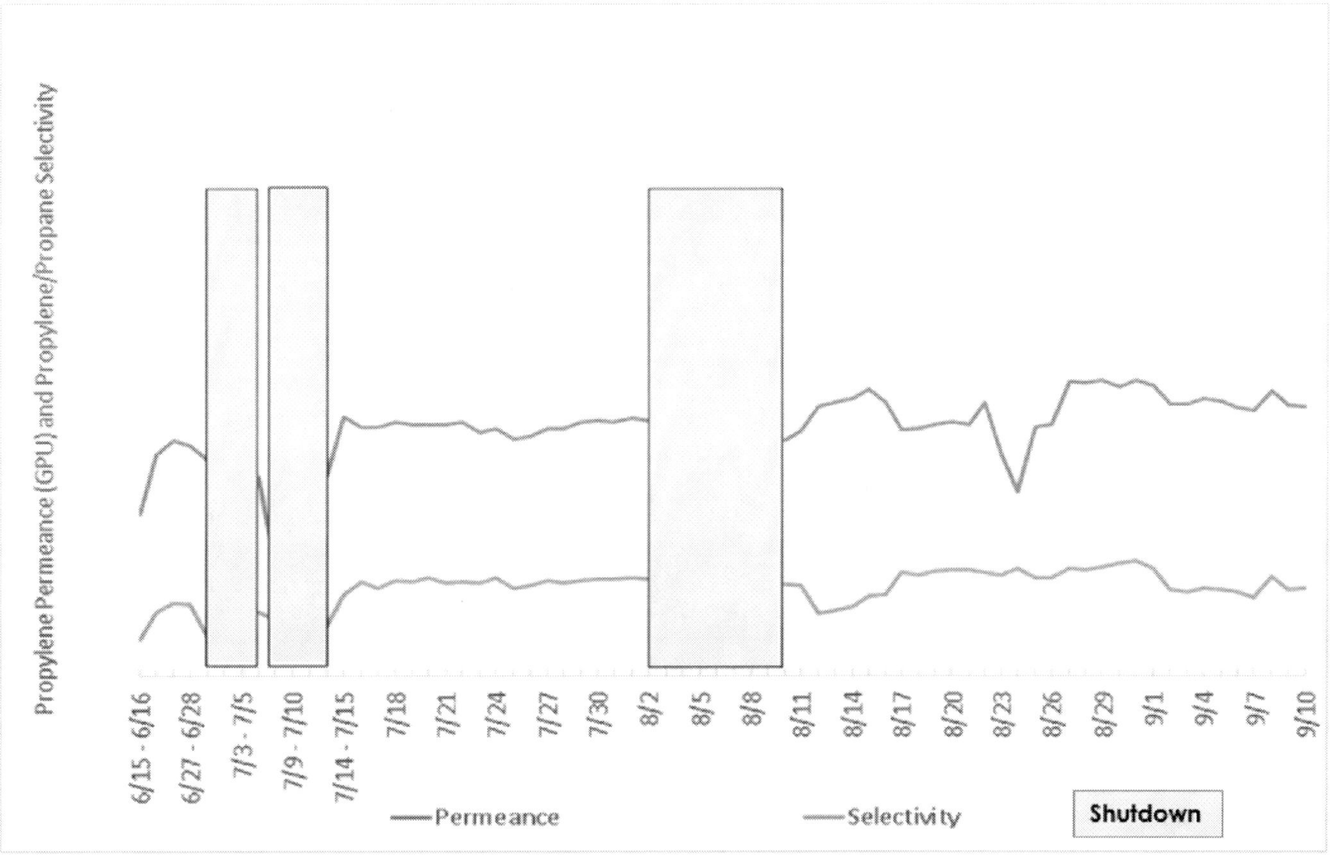

Average stage 2 membrane properties

The chart below shows the feed and product compositions during a period of operation when the objective was to produce both product streams at purities exceeding the minimum specification. After adjusting operating conditions early in the test segment, both product streams were maintained within the required specification ranges for about five consecutive days. Although the feed composition varied during this time period, product compositions were maintained above the required minimum purity.

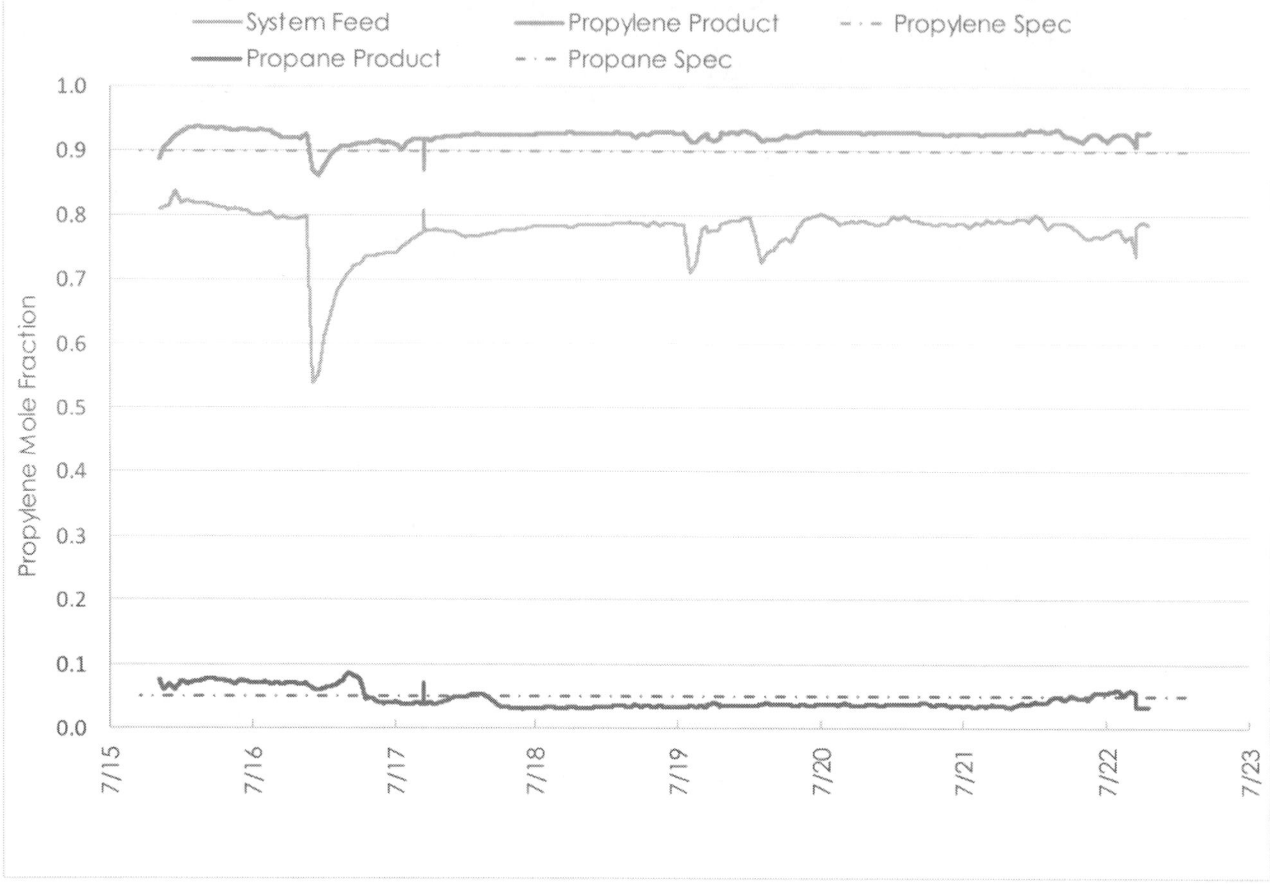

Average feed and product stream compositions.

Economic Analysis of a Commercial Scale Process

An economic analysis of recovery of propylene from a propylene/propane stream projects a significant cost advantage of an Optiperm™ membrane process over distillation. The feed stream to the separation process was assumed to contain 90% propylene and the assumed flow rate was 2300 pounds per hour. Two recovery processes were evaluated, a distillation column, and a two-step Optiperm™ membrane process. Both the distillation column and membrane processes were designed to recover chemical grade propylene at 95% purity for use as reactor feed, as is possible in a polypropylene reactor purge case, while leaving a salable HD5 grade propane stream. Greater than 99% propylene recovery is required to attain this result.

The results of the economic analysis of the two processes are presented in the Table below. All costs in the table are in year 2020 dollars. The reported CAPEX of the distillation process and membrane process represent the installed cost of each of the technologies. Of the two propylene recovery processes, the membrane process has a substantially lower recovery cost, 0.065 $/lb vs. 0.16 $/lb, a 58% cost savings. The cost savings is primarily a result of lower

CAPEX and lower utilities costs (cooling water, electricity, and steam). The membrane process offers an 84% savings in CAPEX and a 47% savings in utilities costs compared to distillation. The energy demand, which is reduced by 68%, also translates into a 48% reduction in greenhouse gas emissions compared to the distillation process. The maintenance costs of the membrane process are greater than the maintenance costs of the distillation process. This is primarily a result of membrane replacement costs. The membranes are projected to have an average life of three years in this application.

Table 1.

	Distillation	Membrane
Steam Power Input (kW)	444	52
Electrical Power Input (kW)	0	85
Total Energy Input (steam & electric) (kW)	444	137
GHG Emissions (t CO_2/yr)	993	520
CAPEX including auxilliary equipment	$5,487,600	$1,594,700
Utilities Costs ($/yr)	$121,400	$64,700
Maintenance Costs ($/yr)	$109,800	$204,400
OPEX (utilities + maintenance $/yr)	$231,200	$269,100
Total Annualized Cost ($/yr)	$1,625,800	$678,800
Olefin Recovery Cost ($/lb)	$0.16	$0.065
Recovery Cost Savings	-	58%

Conclusions

The Optiperm™ Pilot Unit has demonstrated ease of startup and consistent stable operation. Furthermore, it has shown that Optiperm™ membranes have maintained stable separation performance for over 60 days of operation and is also capable of continuously producing in spec product, HD5 grade propane and 90+% propylene, despite variation in feed composition. Furthermore, the current pilot unit performance data demonstrates potential for significant economic benefits compared to traditional separation at full scale operation. Moving forward CMS and Braskem plan to further demonstrate stable separation performance while producing on spec product for 1000 continuous hours along with demonstrating that Optiperm™ is capable of full scale operation at a wide range of pressure and flow rates.

References

1. Sholl, D.S., Lively, R.P., Nature, **532**, 435-437, 28 April 2016

AIChE Paper Number: 155d

OCM & Steam Cracker Integration

Laurin, Marc

Technology Manager, Ethylene Business Group

Lummus Technology LLC

Prepared for Presentation at the 2023 Spring National Meeting
Houston, TX, March 16, 2023

OCM & Steam Cracker Integration

Laurin, Marc

Technology Manager, Ethylene Business Group

Lummus Technology LLC

Abstract: Oxidative Coupling of Methane (OCM) offers a synergistic integration opportunity with Steam Crackers. Steam cracking is a highly endothermic reaction that fires large amounts of fuel gas, thereby emitting large amounts of CO2. OCM is a highly exothermic reaction that can offset a portion of the energy needs for the Steam Cracker, thereby reducing firing requirements and hence, CO2 emissions. Normally, Steam Crackers produce excess fuel gas recovered from the cracker effluent The excess fuel gas can be converted to valuable chemicals in OCM, thereby improving feedstock yield while simultaneously lowering the CO2 foot of the Steam Cracker. Some of the salient features of this technology will be discussed in this paper.

Introduction

Ethylene is produced by thermal cracking of hydrocarbons mixed with dilution steam at high temperatures (750-900°C) in fired heaters. The reaction is highly endothermic. During thermal cracking there is fuel gas, a mixture of methane and hydrogen, produced as a by-product. The fuel gas is normally fired in the cracking heaters to supply the heat of reaction with excess energy being supplied to generate steam, which is used in the recovery section. In naphtha cracking, there is remaining fuel gas by-product that is exported from the ethylene plant.

There are many concepts being developed in the industry to decarbonize ethylene plants. Some concepts, for example air-preheating, will reduce firing and focus the heater duty more on providing the heat of reaction. Others will shift the heater fuel away from fuel gas by-product to hydrogen, ammonia or electricity to reduce CO_2 emissions. All of these will result in an increased export of fuel gas by-product.

Oxidative Coupling of Methane (OCM) provides a way to convert excess methane to valuable olefin products while reducing the specific CO_2 footprint (kg CO_2/kg ethylene) of the plant instead of sending out the fuel gas for combustion. OCM is a highly exothermic process and the available heat can be used to crack ethane, replacing the recycle heater in an ethylene plant, further reducing specific CO_2 footprint. A portion of reaction heat can be recovered as steam that can supplement the ethylene plant or other units in the complex.

OCM Process

Siluria® OCM Technology has been previously presented [1, 2, 3, 4] and is now owned by Lummus Technology. Methane is an extremely stable molecule chemically with strong C-H bonds and it does not contain any functional group or polar distribution to facilitate chemical activation. As a result, the splitting of the C-H bonds requires high temperatures (>900°C), and/or the use of oxidation agents to complete conversion. Catalysts can be used to reduce the extreme conditions required to activate the C-H bond provided they do so without forming a carbon oxygen bond (CO_2).

Siluria OCM is a direct and efficient route to ethylene unlike multi-step routes through synthesis gas, a mixture of H_2 and CO, to methanol/olefins or directly to olefins. The hydrocarbon products that result from synthesis gas routes often have broad product distribution and require significant additional refining/separations leading to further cost, energy input and lower yield for the feed.

In Siluria OCM, methane and oxygen react over a catalyst exothermically to form ethane and ethylene (C2) and small amounts of higher hydrocarbons (C3+ products).

$$2\ CH_4 + O2 \rightarrow C_2H_4 + 2H_2O + \text{heat}$$

The key catalyst performance parameter is to minimize non-selective reactions that result in CO_2 and CO and to maximize the coupling reaction.

Lummus Technology has currently two branded main olefin producing processes; Gemini™ Process (Ethylene production via OCM process) and the Orion™ Process (Propylene production via OCM). The input stream for both processes is low value methane-containing streams such as, natural gas, off-gas from various petrochemical facilities, propane de-hydrogenation units, and refineries. This paper will focus on the steam cracker off-gas to produce ethylene.

OCM Reactor

At the core of the Gemini process design is the OCM adiabatic reactor that consists of two main sections: the OCM Reaction Section, which contains the OCM catalyst bed, followed by an adiabatic chamber designed to perform Post Bed Cracking (PBC) of ethane and propane.

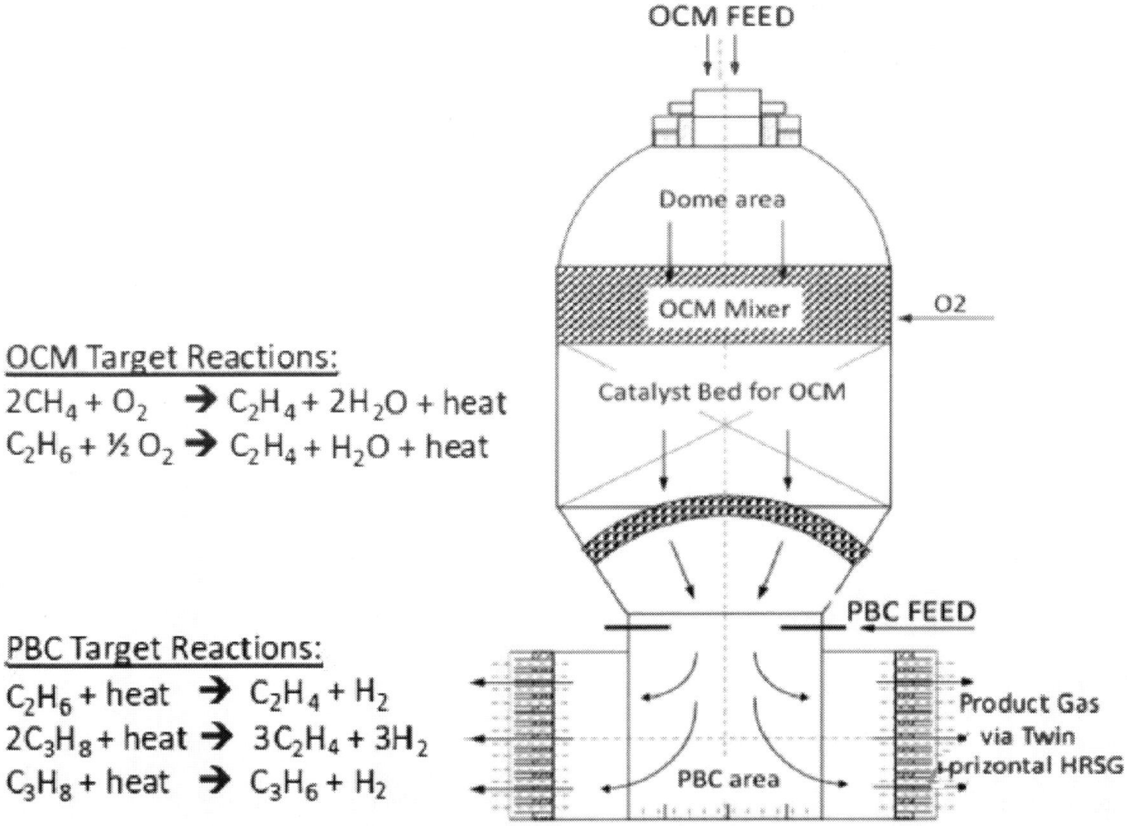

Figure 1 Core of Siluria Process: OCM Reactor

The OCM feed to the reactor, primarily consisting of methane with minor amounts of ethane, is mixed with oxygen in the OCM mixer and sent to the catalyst bed. OCM reactions alongside other side reactions begin to occur. The main reaction that takes place in the catalyst bed is shown in the following reaction 1:

$2\ CH_4 + O2 \rightarrow C_2H_4 + 2H_2O$ + heat **Reaction 1**

In addition to methane oxidation, Siluria's OCM catalyst simultaneously promotes the oxy de-hydrogenation of the ethane present in the feed according to the following reaction 2:

$C_2H_4 + 1/2O_2 \rightarrow C_2H_4 + H_2O$ + heat **Reaction 2**

Following the reaction section is the PCB. Ethane and propane can be injected into the PCB to utilize the available heat to crack the alkanes according to the following reactions 3, 4, 5:

$C_2H_6 + heat \rightarrow C_2H_4 + H_2$	**Reaction 3**
$2C_3H_8 + heat \rightarrow 3C_2H_4 + 3H_2$	**Reaction 4**
$C_3H_8 + heat \rightarrow C_3H_6 + H_2$	**Reaction 5**

OCM Process Main Blocks

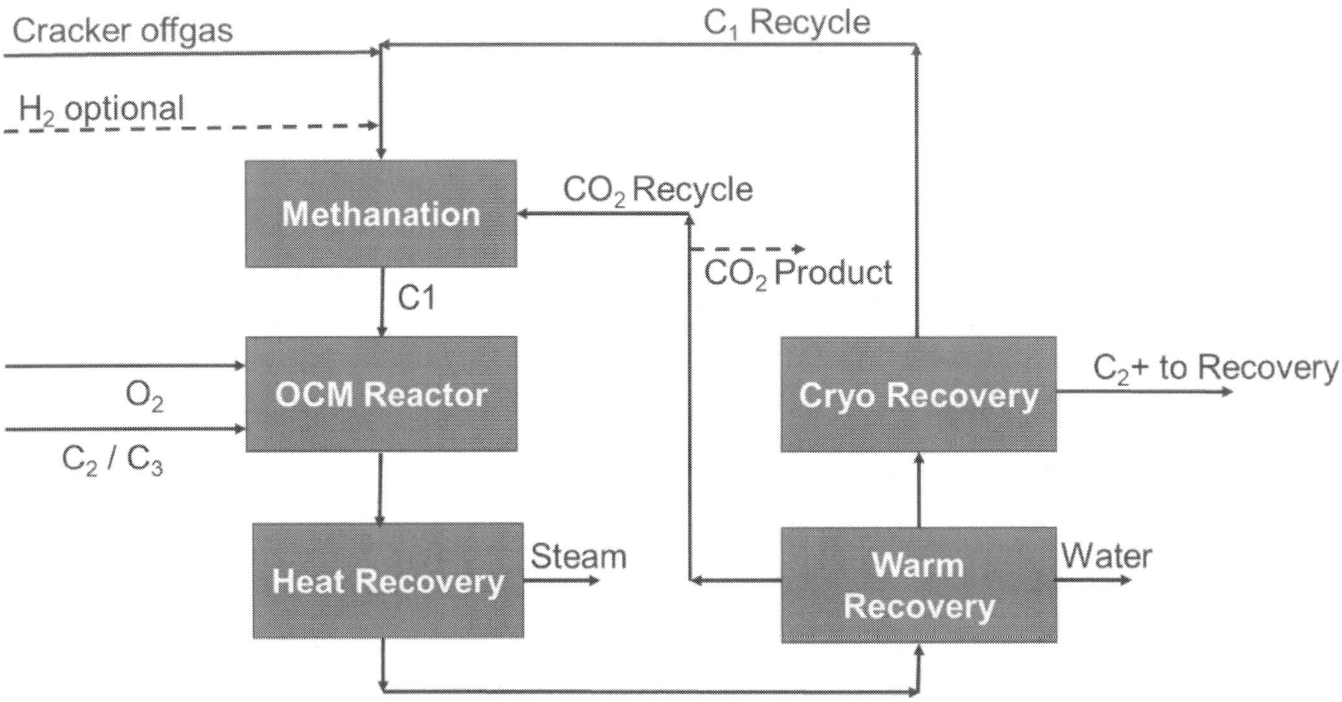

Figure 2 Gemini™ Block Flow Digram

The Gemini process consist of two main sections:

1. The Hot Section which consist of the methanation, the OCM reactor and the Heat Recovery
2. The Cold Section which consists of the compression for separations and fractionation

Hot section

The excess fuel gas feed from the steam cracker plus the recycle stream from the Cold Section - consisting of CH_4, H_2 and CO is mixed with a specified amount of CO_2 from the CO_2 removal unit and fed to the methanator. The CO and CO_2 are converted to methane, to consume the available H_2. Oxygen and the methanator effluent are then mixed in a proprietary mixer within the OCM reactor before being sent to the catalyst bed. Post the OCM reactions on the catalyst bed, ethane and even propane, are cracked to additional olefins in the post bed cracking section as described previously. The heat from the OCM reactor is then recovered for feed pre-heating and by producing high pressure steam.

Cold Section

Once the process effluent is cooled, it is compressed to about 25 to 40 barg and CO_2 is removed by amine and caustic wash. The effluent is dried, chilled and sent to a demethanizer. The demethanizer utilizes refrigeration shared with the ethylene plant, with the operating conditions aligned for ease of integration. The C2+ stream from the bottom of the demethanizer is sent to the steam cracker for recovery of olefins and unconverted ethane.

The OCM reaction does produce some CO_2 by product as well as H_2. The CO_2 is recovered by the amine wash and the hydrogen stays with the unconverted methane and recycled to the methanator. The fuel gas by-product from the steam cracker is a mix of methane, hydrogen and some small amount of CO. Depending on the level of residual hydrogen in the fuel gas by-product a CO_2 product may be produced, a small hydrogen import can be used to balance the CO_2 production so that there is no CO_2 product.

OCM Integration with a Steam Cracker

As an example of the impact of integrating OCM with a Steam Cracker consider a 1500 kta naphtha fed ethylene unit with a 0.65 P/E. Olefins Conversion Technology (OCT) has been included to increase the yield to High Value (HV) products from the feed naphtha. OCT is not required for the integration of the OCM unit with the steam cracker.

All values in kta	SCU w/ OCT
Naphtha	**4,630**
Hydrogen + Fuel Gas	**787**
PG Ethylene (E)	**1,500**
PG Propylene (P)	**972**
1,3 Butadiene (BD)	**246**
Isobutene product	**116**
Benzene (BZ)	**304**
Aromatic rich C7-C8	**382**
C_9^++PGO+PFO	**323**
Total HV E+P+ BD+BZ	**3,022**
Total Heater Effluent	**5,319**
CO_2 Emission - heaters	**2,046**
CO_2 Emission – export penalty	**70**
Specific CO_2 kg/kg ethylene	**1.41**

Table 1 Base naphtha fed steam cracker balance

The block flow is shown below.

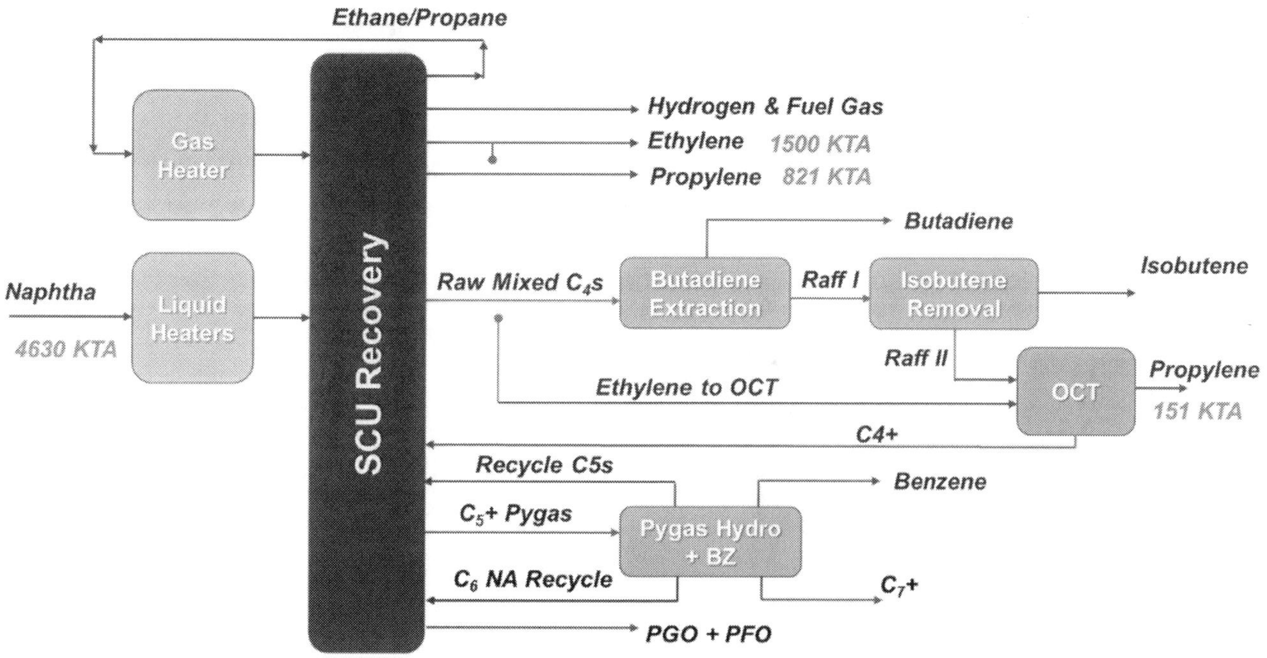

Figure 3 Based naphtha cracker block flow

Most of the fuel gas stream is fired in the cracking heaters, there is a resulting export of a mixed fuel gas by-product to the off sites. The fuel gas by-product is a loss in yield to HV products and a CO_2 source within the complex since it will end up as fuel and be burned. The addition of an OCM unit will convert this stream to ethylene. Refrigeration is supplied from the steam cracker for the OCM demethanizer and steam is generated in the OCM unit and sent to the steam cracker.

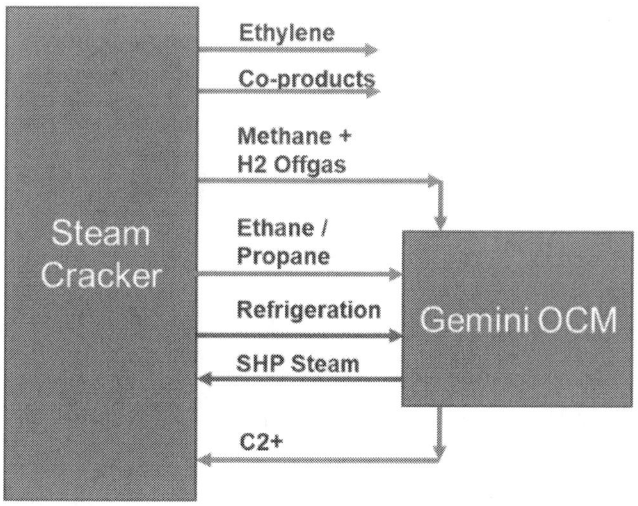

Figure 4 Integration of steam cracker and OCM

The recycle ethane and propane recovered in the steam cracker can be cracked in the OCM reactor instead of a cracking heater, reducing total fuel firing requirement, freeing up additional mixed fuel gas by-product. The block flow for the modified flow sheet follows:

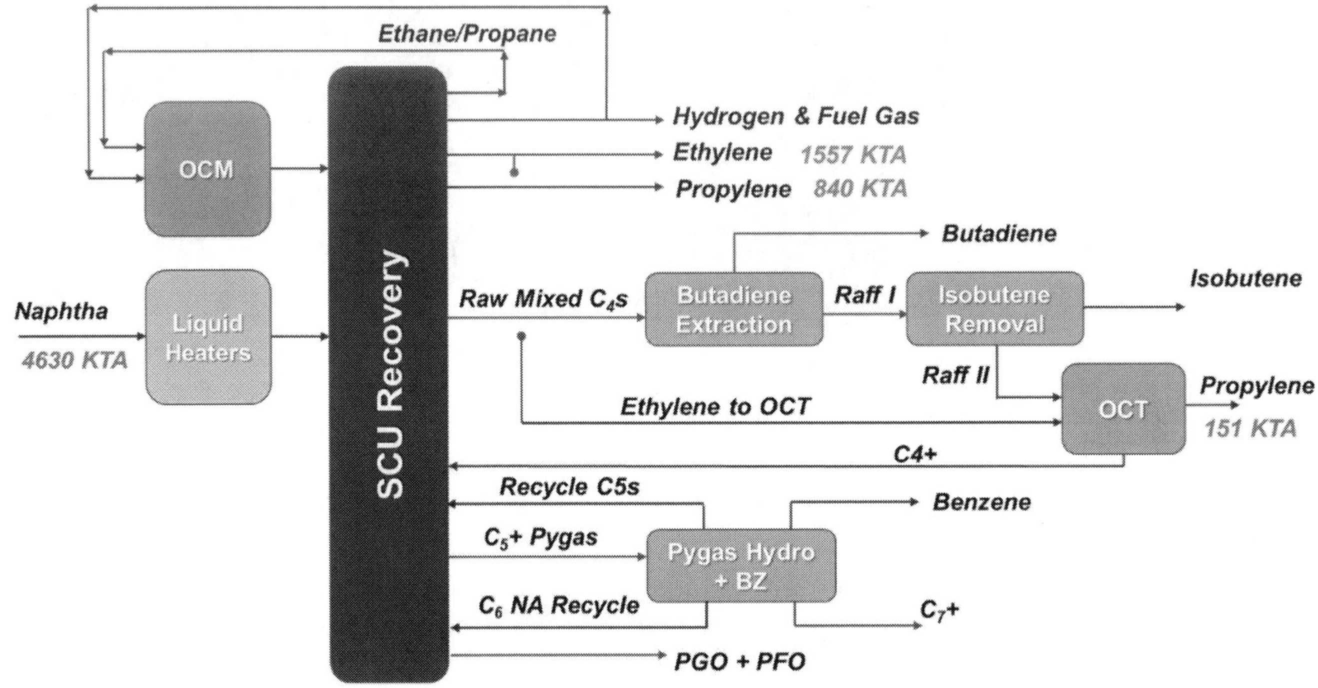

Figure 5 OCM integrated block flow diagram

All values in kta	SCU w/ OCT	Addition of OCM
Naphtha	**4,630**	**4,630**
Oxygen		**371**
Hydrogen + Fuel Gas	**787**	**661**
PG Ethylene (E)	**1,500**	**1,557**
PG Propylene (P)	**972**	**991**
1,3 Butadiene (BD)	**246**	**246**
Isobutene product	**116**	**116**
Benzene (BZ)	**304**	**305**
Aromatic rich C7-C8	**382**	**384**
C_9^++PGO+PFO	**323**	**324**
Water	**-**	**417**
Total HV E+P+ BD+BZ	**3,022**	**3,099**
CO_2 Emission - heaters	**2,046**	**1,943**
CO_2 Emission – export penalty	**70**	**-**
Specific CO_2 kg/kg ethylene	**1.41**	**1.25**

Table 2 OCM integration balance

There is an improvement in the feed stock yield. There is a 3.8% increase in the ethylene product and an overall increase in high value products of 2.5% by not exporting fuel gad by-product. There is a simultaneous reduction in CO_2 emission from the combined SCU with OCM. The source of the major reduction is from not sending the recycle ethane and propane to a cracking heater but instead using the heat from the OCM reactor to crack the recycles streams. The overall yield pattern of high value products is maintained, there is no tradeoff between the HV products.

OCM expansion of a Steam Cracker

The addition of an OCM unit to an existing 1500 kta naphtha fed steam cracking unit can be used to expand the capacity of the plant while lowering the specific CO_2 emission. The OCM unit is designed to process fuel gas by-product, and ethane recycle from the steam cracking unit. Since the steam cracker recycle gas heater will be under loaded, imported fresh propane feedstock is supplied to fill out the recycle gas cracking heater. The OCM unit demethanizer bottoms is routed to the acetylene converter in the SCU for recovery of products. The fuel gas stream is reduced to what is required operate the cracking heaters.

All values in kta	**SCU**	**SCU & OCM**
Naphtha	**4,494**	**4,494**
Propane		**363**
Oxygen		**395**
Hydrogen + Fuel Gas	**771**	**730**
PG Ethylene (E)	**1,500**	**1,725**
PG Propylene (P)	**832**	**904**
1,3 Butadiene (BD)	**253**	**272**
Isobutene product	**119**	**120**
Benzene (BZ)	**293**	**307**
Aromatic rich C7-C8	**398**	**405**
C_9^++PGO+PFO	**329**	**339**
Water		**443**
Total HV E+P+ BD+BZ	**2,878**	**3,208**
Total Heater Effluent	**5,335**	**5,474**
CO_2 Emission - heaters	**2,026**	**2,026**
CO_2 Emission – export penalty	**78**	**-**
Specific CO_2 kg/kg ethylene	**1.40**	**1.17**

Table 3 Expansion material balance

The capacity factor impact on major equipment sections is indicated in the table below:

	SCU	SCU & OCM
Heater Effluent	1.00	1.02
Gasoline Fractionator	1.00	1.01
Quench Tower	1.00	1.01
Dil Steam Drum	1.00	1.02
Charge Gas Compressor	1.00	1.01
Chill Train & Demethanizer	1.00	1.02
Deethanizer System	1.00	1.17
Ethylene Fractionation	1.00	1.18
Propylene Fractionation	1.00	1.10
Charge Gas Compressor Power	1.00	1.01
Propylene Refrigeration Compressor Power	1.00	1.19
Binary Refrigeration	1.00	1.22

Table 4 Expansion area factors

In the expansion illustrated above there is a 15% expansion in ethylene capacity with a reduction in the specific CO_2 production. The total CO_2 emission remains about the same. The expansion is made without impacting the hot section or the heaters of the steam cracker. The heater effluent mass rate did increase due to replacing the recycle ethane with fresh propane but the volume feed to the heaters remained the same. The expansion, of the SCU is confined to the recovery area downstream of the Demethanizer, avoiding complicated cold box expansions.

Conclusion

There is great interest in the industry to decarbonize the production of base olefins. The major concepts of heater modifications to reduce firing or changing the source of heat, such as hydrogen/ammonia firing or electrical heater cracking will result in additional fuel gas by-product available. The additional fuel gas by-product can be converted into high value products with no increase in CO_2 generated; in fact, the product yield from feedstock will increase. Siluria® OCM Technology offers a path to reduce CO_2 emissions now and in future.

References:

1. Radaelli, Guido, Iyer, Rahul, Rafique, Humera. (2017) "Building with Natural Gas," presented at the 2017 AIChE Spring Meeting, San Antonio, TX
2. Radaelli, Guido, Rafique, Humera. (2019) "Conversion of Low-cost Feedstock to Petrochemical Products," presented at the 2019 AIChE Spring Meeting, New Orleans, LA
3. Radaelli, Guido, Kemper, Rainer. (2015) " Siluria and Linde OCM: Turning A Breakthrough Into Commercial Practice," presented at the 2015 AIChE Spring Meeting, Austin, TX
4. Weinberger , Sam, Scher, Erik, Iyer, Rahul. (2012) " Natural Gas to Ethylene in One Step Siluria Technologies OCM (Oxidative Coupling of Methane)," presented at the 2012 AIChE Spring Meeting, Houston, TX

AIChE Paper Number: 155e

Full Machine Electrification of a Grass Roots Ethylene Plant

Hendrik Reyneke
Lead Process Engineer
Linde GmbH, Engineering Division

Mark Whitney
Principle Process Engineer
Linde Engineering Americas

Martin Kamann
Rotating Equipment Engineer
Linde GmbH, Engineering Division

Peter Brehm
Electrical Engineer
Linde GmbH, Engineering Division

Prepared for Presentation at the
2023 Spring Meeting and 19th Global Congress on Process Safety
Houston, Texas, March 16, 2022

Full Machine Electrification of a Grass Roots Ethylene Plant

Hendrik Reyneke
Lead Process Engineer
Linde GmbH, Engineering Division

Mark Whitney
Principle Process Engineer
Linde Engineering Americas

Martin Kamann
Rotating Equipment Engineer
Linde GmbH, Engineering Division

Peter Brehm
Electrical Engineer
Linde GmbH, Engineering Division

Abstract

Traditionally, large compressors in ethylene plants are driven by steam turbines which are fed by steam generated in conventional cracking furnaces as well as auxiliary boilers. In future low-emission olefin plant designs, significantly reduced generation of waste heat carriers such as steam from the furnaces is targeted. Accordingly, reduction of steam consumption in the separation section will play a key role in holistic low-emission plant designs.

In a recent grass roots ethylene plant design, Linde Engineering has electrified the drive trains of all large compressors - Cracked Gas, C2 Refrigerant and C3 Refrigerant Compressors.

The full machine electrification step was a key element to achieve a substantial CO2 emission reduction and an increased energy efficiency.

Due to the required shaft power and speed range of the cracked gas compressor and the lack of as-built references for this particular combination of shaft power versus speed, specific tailormade process and electrical solutions were created by smart and innovative process and electrical design.

In this paper, the authors will take a detailed look at the process solutions applied for this specific grass roots ethylene plant with its unique constraints and explain the selected electrical drive concept with its advantages. The paper will also compare a conventional steam turbine drive solution with the chosen electrical drive solution from a reliability and failure rate point of view highlighting surprising advantages from the selected electrification concept.

1. Introduction

Linde performed a Front End Engineering Design (FEED) for a mega grass roots ethylene plant in China. The ethylene plant is part of a new chemical complex for a world player in the chemical industry. The chemical complex and ethylene plant are currently in the detail engineering phase and start-up is planned for the end of 2024.

A very important Basis of Design (BOD) requirement of the ethylene plant was the full machine electrification which means full electrification of the large machines cracked gas compressor (CGC), C2 refrigerant compressor (C2C) and C3 refrigerant compressor (C3C).

For all new grass roots plants in China, an Environment Impact Assessment (EIA) has to be performed for authority approval with the requirement of reducing emissions. With full machine electrification a substantial CO_2 emission reduction was achieved.

New ethylene plants in China also have to fulfill a very stringent energy consumption guarantee (with reference to specific Chinese GB-codes) and the goal posts are moved from year to year to achieve lower energy numbers. With this particular plant, an even lower energy consumption figure could be achieved and guaranteed compared to the 2022 GB-code ethylene plant energy requirements.

2. Relevant constraints for electrification and impact on other design aspects of the plant

Electrical motors for the large machines

<u>Power and Speed requirements for the three large compressors</u>

In assessing which electrical motor is suitable as a compressor drive, shaft power AND speed of the compressor have to be considered.

In the next figure the following can be seen:

- Optimal compressor speed with associated power for process conditions – green line
- Maximum synchronous motor speed with associated power – blue line
- Compressor design speed including a 5% API margin – purple line

From **Figure 1** it can be seen that a potential 10 to 20 % reduction in optimal compressor speed can occur when using an electric drive directly as a compressor drive.

The figure also indicates that a gearbox between electrical motor and compressor could be an advantage to design the compressor at its optimal speed.

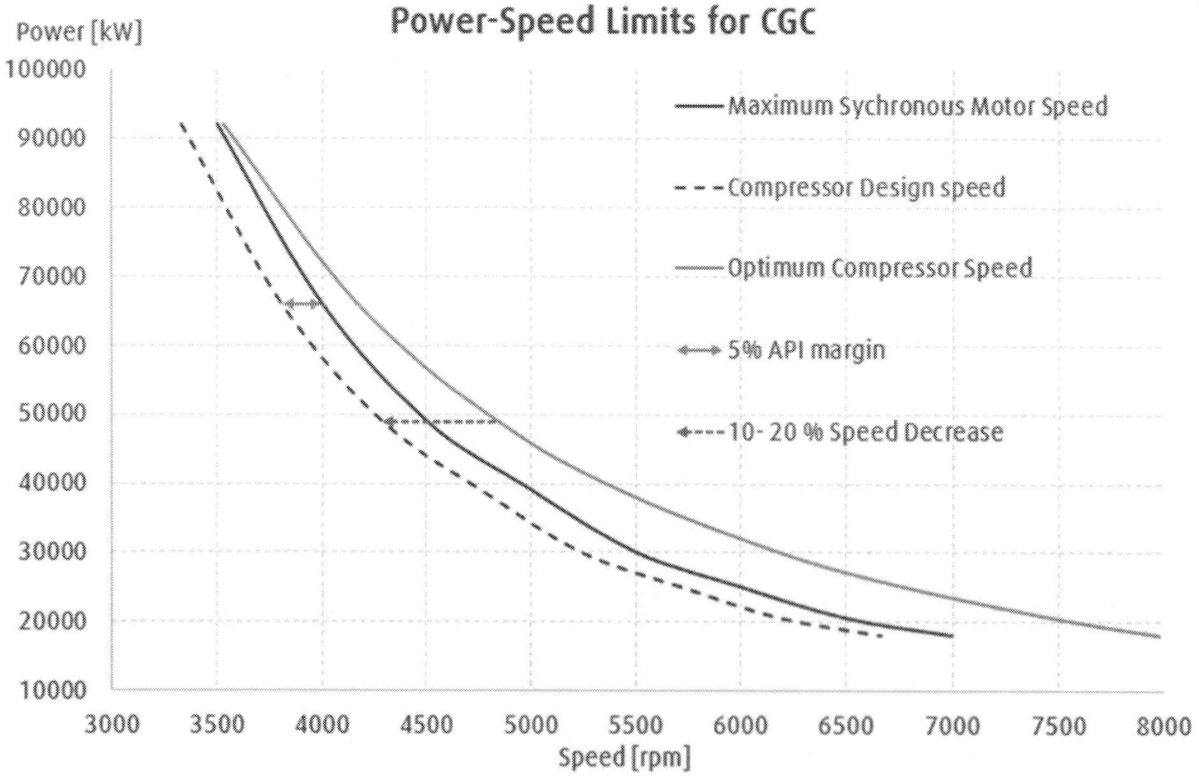

Figure 1: *Power vs Speed limits for compressors and electrical motors*

Ethylene Plant large compressor shaft power vs speed requirements

Based on the process configuration and requirements of the ethylene plant, the following shaft power vs speed requirements were originally established in the described project.

Table 1: *Shaft Power vs Speed requirements for large compressors in the ethylene plant*

Name of Compressor	Shaft Power [MW]	Speed [RPM]
Cracked Gas Compressor	55	4000
C2 Refrigerant Compressor	27.9	5650
C3 Refrigerant Compressor	23.2	3600

Conclusions on the Power and Speed requirements for the three large compressors

During the technology qualification and technical risk assessment steps of the project the following became clear:

- In the required shaft power and speed range, an adequate number of electrical motor references were available for the C2C and C3C driver applications
- A very limited number of electrical motor references were available for the driver application of the CGC in the required shaft power and speed range
- Very limited Mean Time Before Failure (MTBF) and Mean Time to Repair (MTTR) information were available on electrical systems with a power rating > 35 MW
- It became clear that in case of full electrification of all large machines, the CGC required to be split into two compressors to be in a power/speed range with sufficient references and MTBF /MTTR information.

Impact on steam system of the ethylene plant

Very High Pressure (VHP) steam is generated in the furnaces and functions as the main thermal energy carrier in the conventional ethylene plant. Steam is consumed on different pressure and temperature levels. Furthermore, the 3 main compressors in the ethylene plant are major consumers via steam turbine drives, which typically also extract the required amounts of steam on lower pressure levels, controlling the supply and demand of steam for thermal use. Another aspect of the steam turbine design is that considerable amounts of steam are condensed against vacuum conditions consuming huge amounts of cooling water. The so-called condensing part of the turbines is required to balance the shaft power for the large machines. The described conventional scheme is however characterized by a poor thermodynamic efficiency, which can be pinned down to the use of the condensing parts of the turbine. Looking in isolation at the condensing part of the turbine, typically only about 25-30 % of the introduced fuel heat to the furnaces and/or boilers is recovered as shaft power, which is a very poor efficiency compared to modern steam or combined cycle power plants.

As a consequence, there is typically a large demand for steam within the steam system of a cracker, which is partly sufficed by conventionally designed cracking furnaces generating considerable amounts of VHP steam and – if required – further steam import from boilers.

In the particular example of this project, the overall steam configuration of the complex allowed for more degrees of freedom within the cracker steam system:

- Export of excess VHP steam to ethylene plant OSBL. This enables the OSBL utilization of the VHP steam for higher efficiency mechanical power and thermal purposes
- Import of HP, MP and LP steam from OSBL for ethylene plant consumers
- No need of primary steam header control inside battery limits of the plant

The above-mentioned constraints allow for replacing the steam turbines with electrical motors, which causes an excess of steam within the boundaries of the conventional ethylene plant, possibly exceeding the export requirements. The reduction of VHP steam production is therefore a typical target when the large machine drivers are electrified.

Reduction of furnace steam production is predominantly achieved by air pre-heating. Air preheating can be implemented with various set-ups and has the potential to reduce fuel consumption by 15-25 % using simple concepts. Up to 35 % and more can be achieved by using advanced concepts. Keeping the main purpose – conducing the cracking reaction – such fuel savings result in a significant reduction of steam generation.

In the described project, a rather simple, referenced and robust air preheating concept against waste heat and flue gas was implemented (

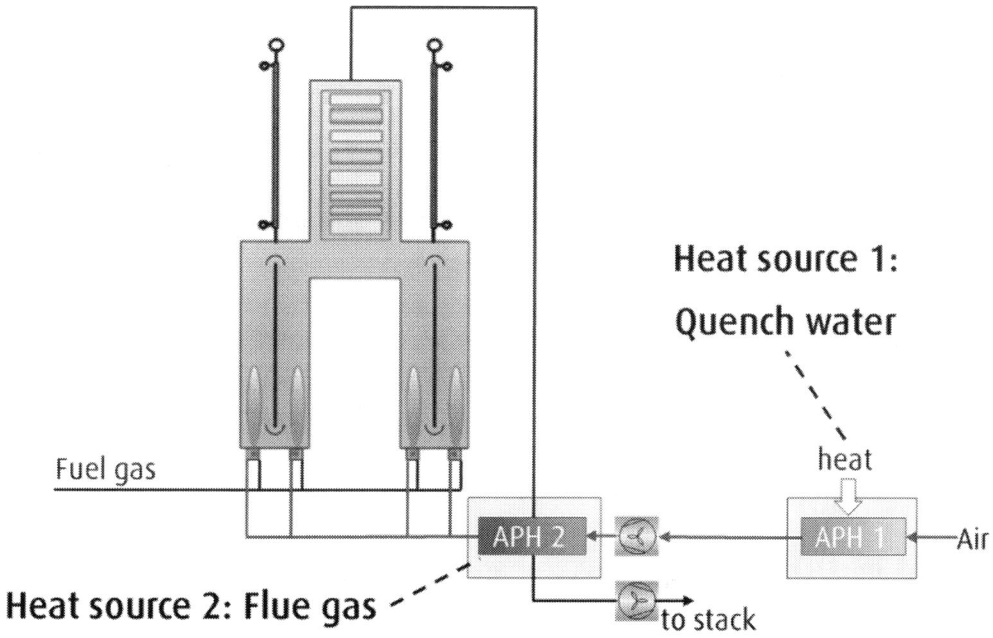

Figure 2), saving approximately 15% fuel. The resulting furnace design has maximum energy efficiency by minimum loss of heat to atmosphere and noticeably reduces the generation of the secondary heat carrier steam.

Depending on the detailed boundary constraints regarding steam export and import, the concept can be adjusted and – if required – more sophisticated air preheating concepts can be implemented.

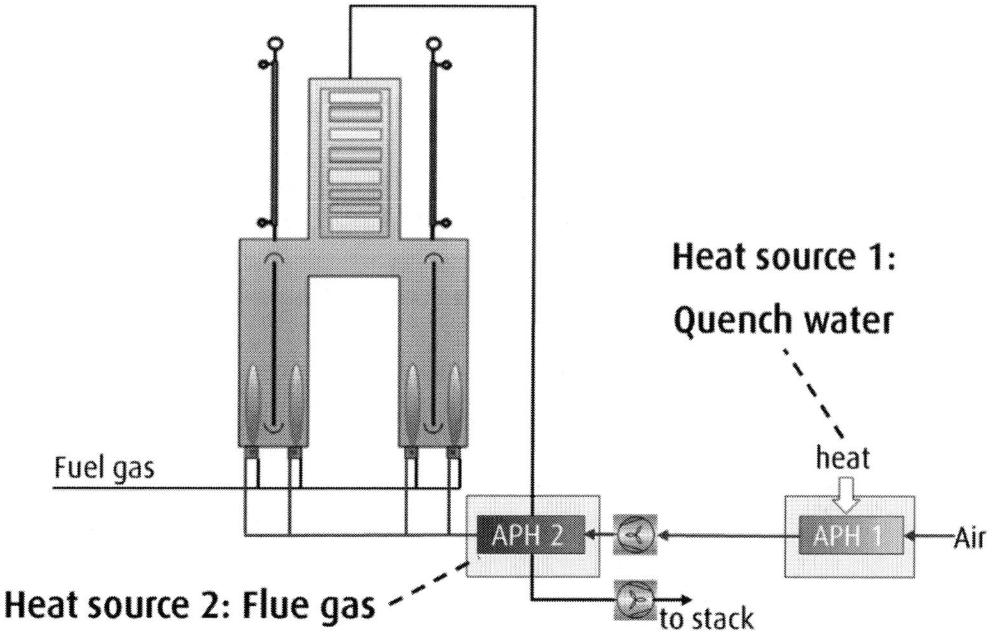

Figure 2: *Simple air preheating concept utilizing external waste heat and flue gas.*

3. Design solutions and features of the electrification concept

Splitting of cracked gas compressor

It was decided to split the cracked gas compressor vertically into a low pressure (LP) section and a high pressure (HP) section. This enabled the selection of two electrical motors as drivers for the CGC that are better referenced and also have well referenced reliability and availability data.

Stages 1 and 2 of the conventional single-shaft 5 stage CGC were combined into one single shaft LP section with 2 casings and stages 3, 4 and 5 were combined into one single shaft HP section with 2 casings.

The reasons for splitting the CGC vertically and not horizontally were the following:

- Splitting the machine vertically saves equipment count. In a horizontal split, equipment is smaller but requires to be doubled in case of two parallel 50 % CGC trains
- Less plot space is required for a vertical-split CGC
- Minimizing potential control issues with a vertical split. No effort is required in matching the suction and the discharge pressures of 2 compressors in parallel. With the vertical split, the only interface between the two machines is the LP section discharge/HP section suction pressure. The two machines can also run completely independent.
- Optimizing the speed of the LP and HP sections independently, thereby achieving a higher overall efficiency compared to a conventional single shaft compressor

Process optimization of cracked gas compressor

The aim of the optimization of the CGC process parameters was to achieve an approximate 50/50 shaft power split for the LP and HP sections. This enabled the selection of two electrical drivers with the same shaft power for the CGC saving overall CAPEX.

In a 5 stage single shaft CGC, the shaft power of the individual stages is approximately one fifth of the overall shaft power of the CGC with proportionally more shaft power on stages 1 and 2 compared to stages 3,4 and 5. This is due to the higher condensate discharge and higher efficiency of the first 2 stages (speed is optimized for the first 2 stages).

This means less shaft power has to be shifted from the last 3 stages to the first 2 stages to achieve an approximately 50/50 power distribution for the LP and HP sections.

This shift of power was done by optimizing the process parameters of the CGC subject to the normal objectives of increased individual stage efficiencies and minimizing overall fouling potential in the CGC.

Shaft Power vs Speed requirements after vertical CGC split and optimization

In the following table the shaft power and speed requirements for the 4 large compressors in the ethylene plant following the CGC vertical split and optimization can be seen.

Table 2: *Power vs Speed requirements for large compressors after CGC Split*

Name of Compressor	Shaft Power [MW]	Speed [RPM]
LP Cracked Gas Compressor	27.1	3900
HP Cracked Gas Compressor	27.3	4100
C2 Refrigerant Compressor	27.9	5650
C3 Refrigerant Compressor	23.2	3600

From the table above it is apparent that

- All four large compressors have approximately the same shaft power
- All four large compressors have four different speeds
- If an electrical driver solution can be found where 4 identical electrical motors can be used with the same speed and shaft power, a standardized electrical motor approach can be followed saving significant CAPEX
- Such a solution is possible by installing 4 different gearboxes for the 4 equivalent and standardized motors accommodating the 4 different compressor speeds.
- It enables the selection of 4 identical and standardized control systems for the 4 identical electrical motors.

Process Control of the 4 large compressors

The process control of the large 4 compressors can be seen in the table below:

Table 3: *Process Control Parameters for large compressors*

Name of Compressor	**Control Parameters**
LP Cracked Gas Compressor	1st Stage Suction Pressure via motor/compressor variable speed
HP Cracked Gas Compressor	3rd Stage Suction Pressure via motor/compressor variable speed
C2 Refrigerant Compressor	C2 Splitter Overhead Pressure via motor/compressor variable speed
C3 Refrigerant Compressor	1st Stage Suction Pressure via motor/compressor variable speed

For all 4 large compressors, a standardized control set-up as per **Figure 3** was chosen.

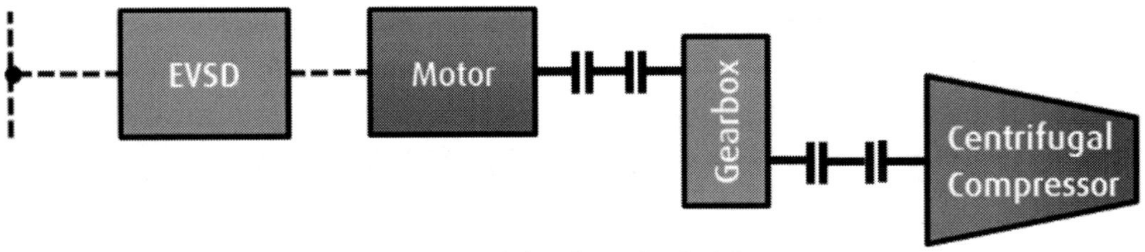

Figure 3: *Compressor Set-up with electrical drives*

For Electronic Variable Speed Drives (EVSDs) two options exist:

- Load Commutated Inverter (LCI)
- Voltage Source Inverter (VSI)

For the project, VSIs were chosen for the following reasons:

- VSIs were found to be commercially more attractive than LCIs
- Less harmonics in the supply network
- Less harmonics and inter-harmonics in the compressor and electrical motor shaft system
- No issues with avoiding natural frequencies of the compressor and electrical motor shaft system (VSI rated speed of 1800 rpm vs. LCI rated speed of 1200rpm)

Anti-Surge Control of the 2 CGC solution compared to a single CGC

A consequence of splitting the cracked gas compressor is that the anti-surge control has to be updated.

In the following diagrams a comparison of the two CGC anti-surge control configurations can be seen.

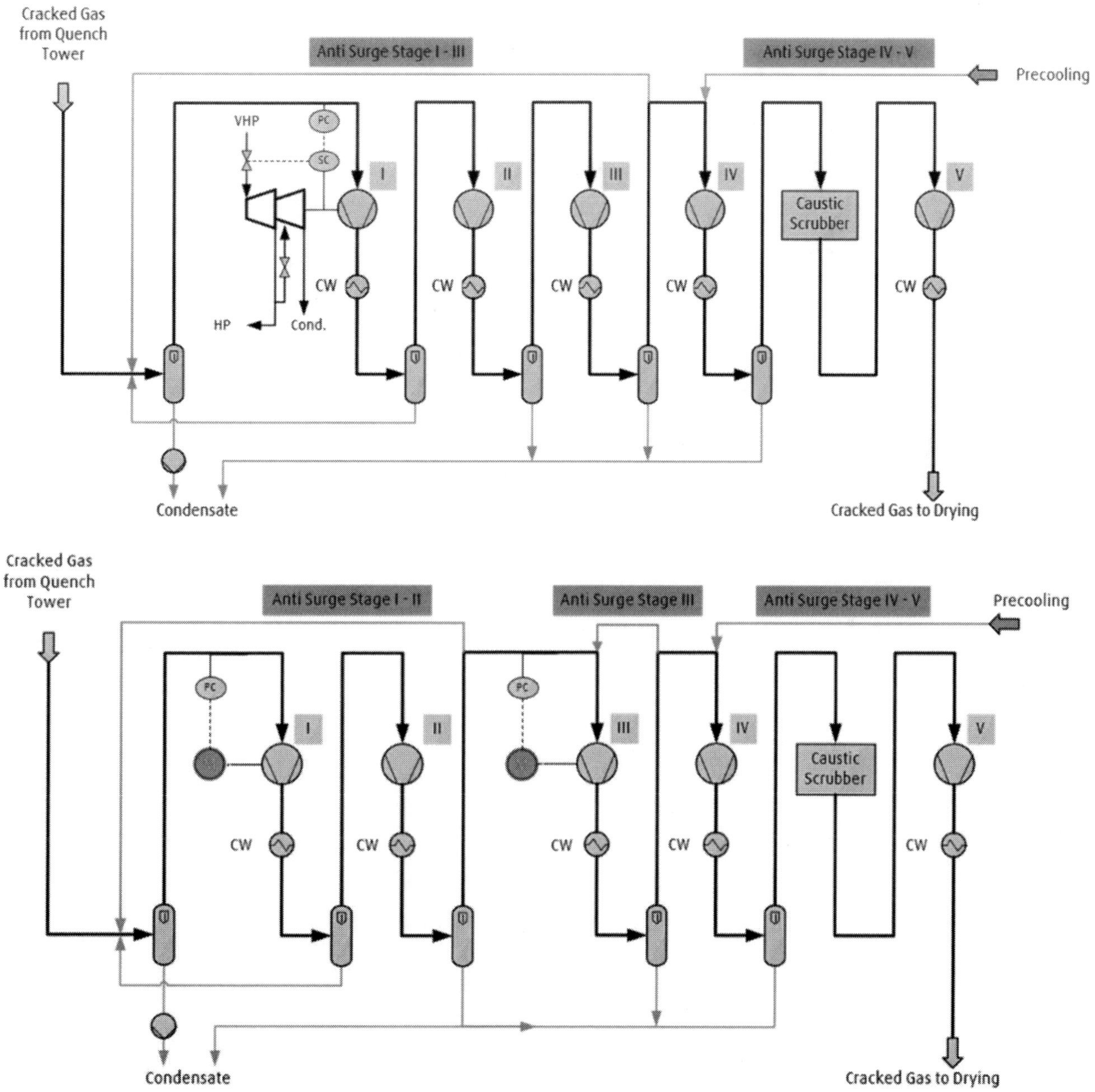

Figure 4: *Control of single shaft CGC with conventional steam turbine (upper section) and Control of vertically split CGC with electrical motors (lower section)*

In the vertically split CGC compressor, the interstage pressure between the LP and HP section can be controlled adding another degree of freedom.

Configuration of the 4 large compressor for full electrification

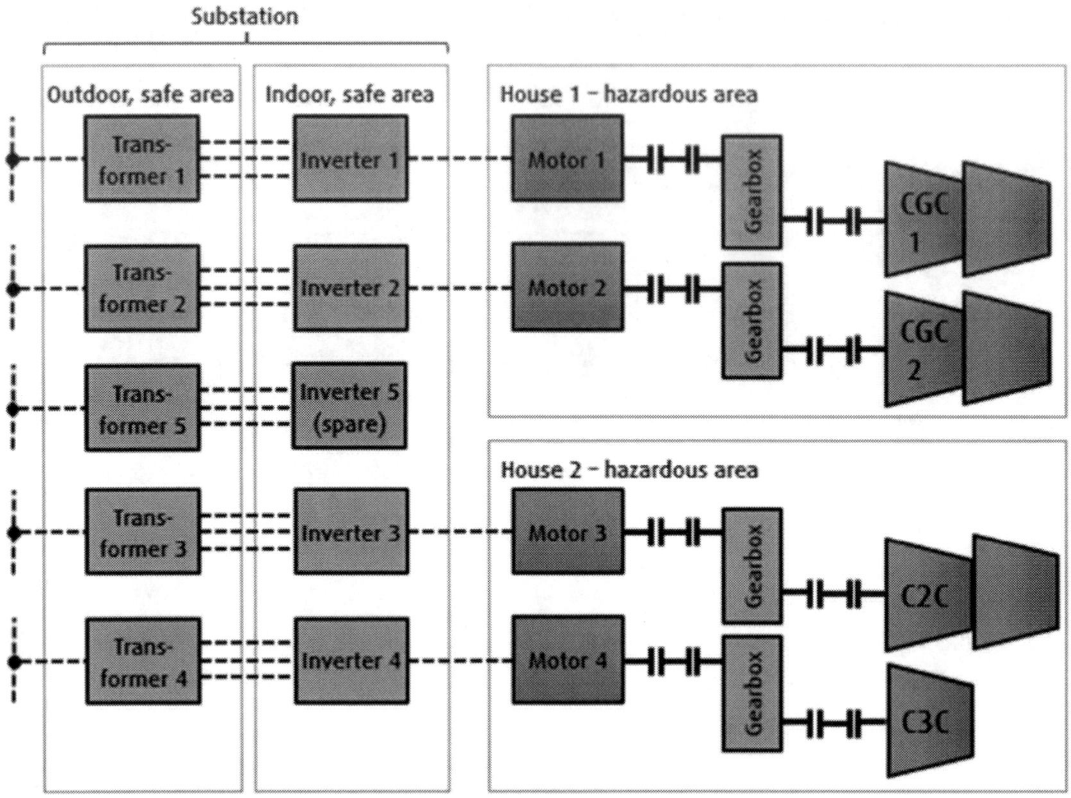

Figure 5: *Configuration of large machine electrification*

The following four identical electrical motors were selected:

- Type: Synchronous 4-pole
- Speed: 1800 RPM
- Rated Power: 33 MW

A spare 5th VSI is also installed as a common hot spare for the 4 VSIs with an energizing time of less than 500 milliseconds to increase the reliability and availability of the system.

Layout of the compressors for full electrification

The four machines can for example be arranged within two housings.

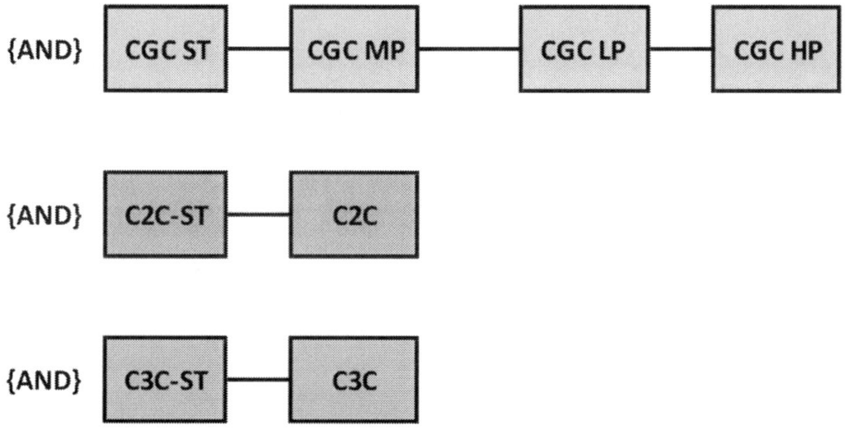

Figure 7 shows an arrangement with the two cracked gas compressors in one building and the two refrigerant compressors in a 2nd building.

Figure 6: *Electrification of large machinery – machinery house layouts*

4. Reliability and Failure Rate of the large machines after full electrification

The reliability and failure rates of the following three drive set-ups were evaluated:

- Conventional steam turbine solution
- Single shaft CGC with a 3 LCI control solution
- Split CGC with 4+1 VSI control solution as implemented for the project

The reliability was evaluated using the equation:

$$R_{iso} = (PH - FOH)/PH$$

Where R_{iso} = Reliability according to ISO 3977-9
PH = Period Hours
FOH = Forced Outage hours

Remark: The forced outage hours do not include the time to restart the plant.

Reliability of a conventional steam turbine solution

The reliability of the conventional set-up as per

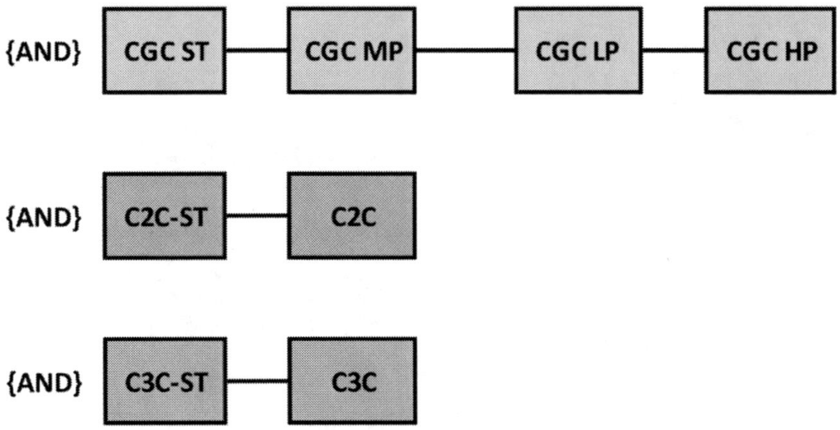

Figure 7 with steam turbines (ST) was evaluated based on reported failure rates from existing Linde plants.

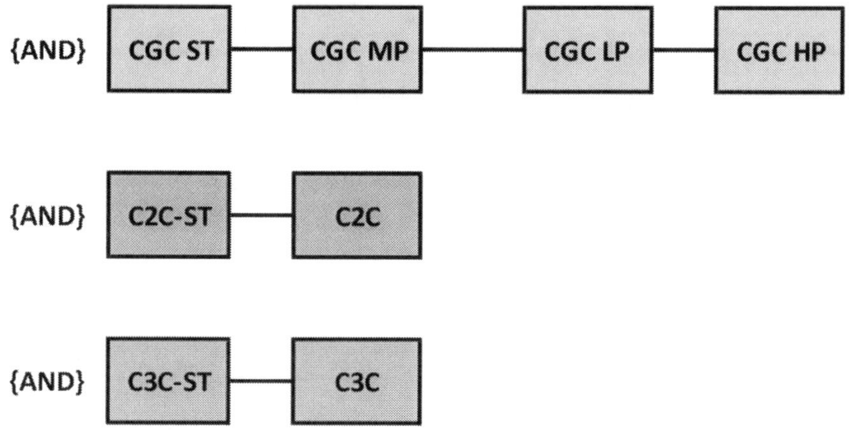

Figure 7: *Large machine set-up with steam turbines*

- Reliability = 99.53 %
- Failure Rate of drive system (all turbine drives) = 0.03/year

The reliability of the turbine drive concept is already excellent with the failure rate extremely low.

Reliability of a Single shaft CGC with a 3 LCI control solution with no spare

The LCI drive concept is indicated in

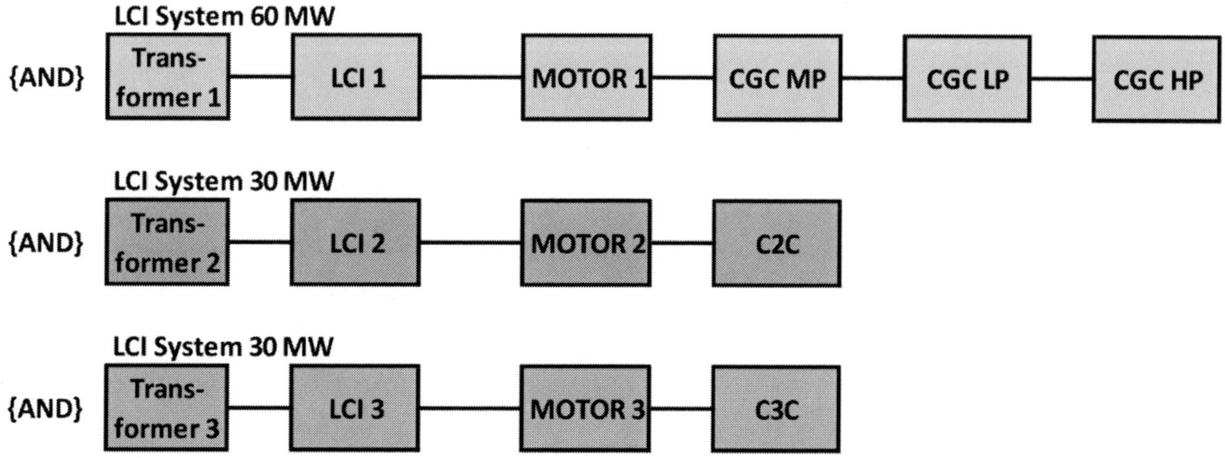

Figure 8.

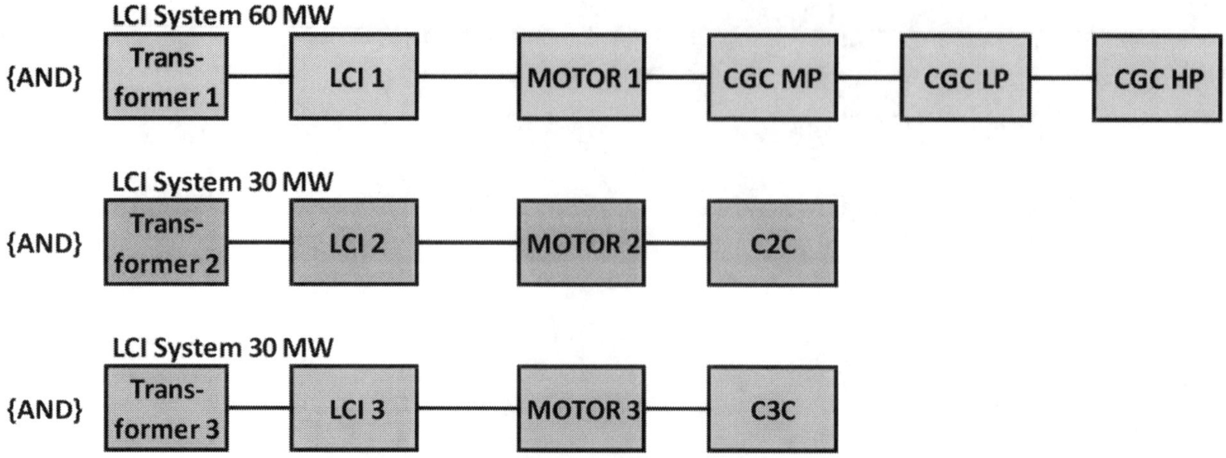

Figure 8: *Large machine set-up with electrical motors and LCIs*

- Reliability = 99.40 %
- Failure Rate of drive system (electrical system + motors) = 0.43/year

The reliability of the LCI drive concept is reduced mainly due to more frequent failures of the LCI systems without any spare. The reliability still implies to be acceptable, which comes from rather short expected mean times to repair of the respective electrical systems. However, the very high expected failure rate of the unspared electrical drive system would have a further detrimental impact on the production unit in terms of rather frequent upset and restart requirements (which is not fully covered by the calculation approach of the reliability). This unacceptably high failure rate thus disqualifies the system.

Reliability of a Split CGC with 4+1 VSI control solution as implemented for the project

The solution as implemented for this project is shown in **Figure 9**.

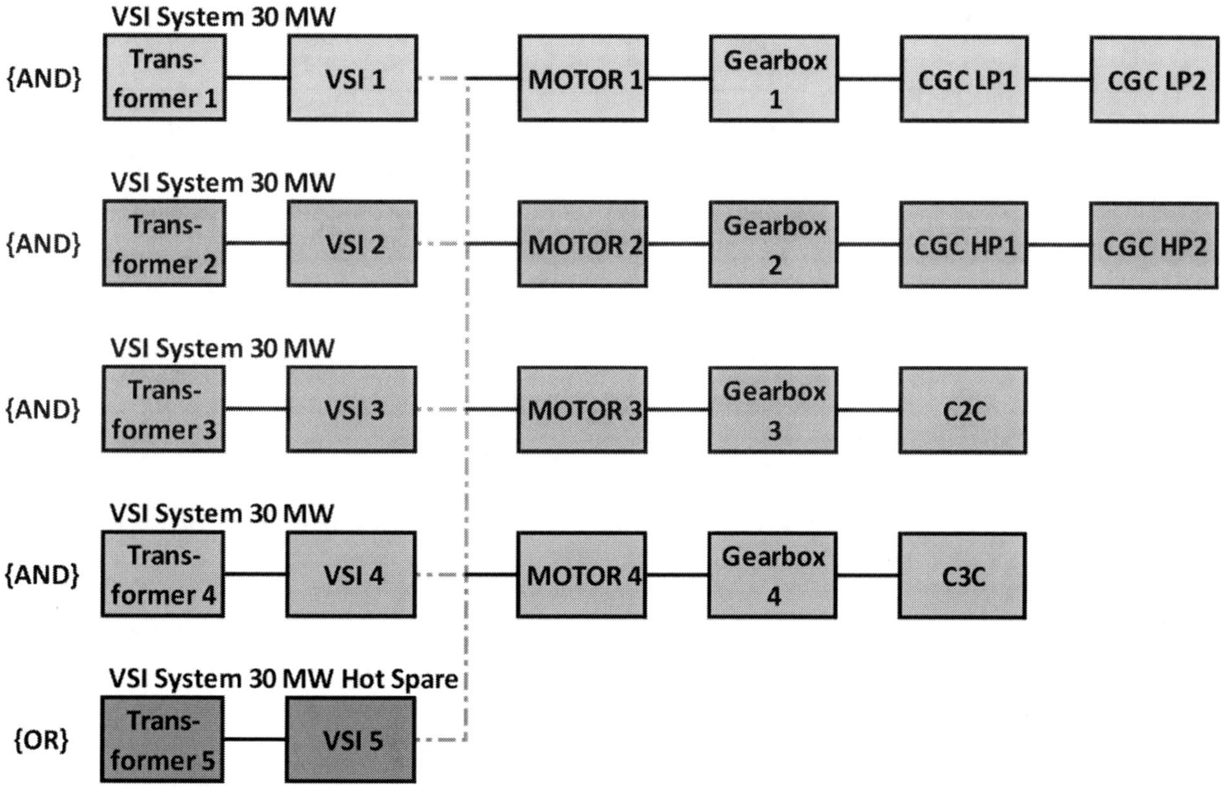

Figure 9: *Large machine set-up with electrical motors and spared VSIs*

- Reliability = 99.57 %
- Failure Rate of drive system (electrical system + motors) = 0.07/year

The reliability of the VSI drive concept including a spare unit is the best in the forecasted reliability and the failure rate of the drive system is also acceptable (once in approx. 15 years).

5. Summary and Conclusions

Full large machinery electrification of a grass roots ethylene plant can significantly reduce the steam consumption within the steam system of the cracker and avoids energy loss by vacuum steam condensation inside battery limits. Such an electrification approach can be ideally combined with furnace designs consuming considerably reduced fuel and generating less steam, e.g. by applying air preheating. Fuel and emission savings inside battery limits of 15% as described for the given project can be achieved using simple, reliable and proven technology elements. Furthermore, the concept allows for further significantly increased fuel and emission savings by applying respective air preheating concepts, depending on the required target.

The configuration of the overall chemical complex addressed herein allows for the export of excess VHP steam and import of steam on the respectively required pressure levels. This constraint allows that the VHP steam is mechanically and thermally utilized outside battery limits with higher efficiency compared to the conventional use in cracker steam systems, where considerable amounts of heat is typically lost during vacuum steam condensing. The concept benefits thus from significantly increased overall efficiency and reduced CO_2 emissions, even if the required extra electrical power would be generated with natural gas fueled power plants. Moreover, the option to import larger amounts of renewable and/or low-carbon power would further contribute to the reduction of the scope 1+2 emissions.

For the C2 and C3 refrigerant machines, sufficient electrical motor references for the shaft power/speed combination including MTBF/MTTR values were found to replace the steam turbines with single electrical motors.

For the cracked gas compressor these references were not sufficiently available which led to a "vertical split" of the machine, further enabling a higher overall compressor efficiency compared to a single shaft compressor.

A standardized electrical drive train solution could then be applied with 4 identical motors and 4 identical electronic control systems. This was achieved by the introduction of 4 gearboxes to cater for the 4 different compressor speeds.

An installed standby electronic control system was added to achieve sufficient reliability figures, which are expected to be even slightly better compared to the already very high reliable conventional steam turbine drive solution.

AIChE Paper Number: 155f

Practical Flue Gas Carbon Capture at Steam Crackers

Torben Hoefel
Product Manager Sustainable Olefin Technologies
Linde Engineering

Alexander Laugwitz
Team Lead CO_2 Plants
Linde Engineering

Mohamed Anas
OASE® Gas Treating Excellence
BASF

Prepared for Presentation at the
2023 Spring Meeting and 19th Global Congress on Process Safety
Houston, Texas, March 16, 2022

Practical Flue Gas Carbon Capture at Steam Crackers

Torben Hoefel
Product Manager Sustainable Olefin Technologies
Linde Engineering

Alexander Laugwitz
Team Lead CO_2 Plants
Linde Engineering

Mohamed Anas
OASE® Gas Treating Excellence
BASF

Abstract

"Post-combustion CO_2 capture" is a widely used technology for decarbonizing plants in various sectors, including chemical manufacturing, power generating, and industrial facilities. In the context of steam cracking plants, "Post-combustion CO_2 capture" can be understood as CO_2 capture from flue gas mainly originating from cracking furnaces and auxiliary steam boilers. In principle the respective flue gases are routed through a column with circulating liquid solvent, which absorbs the CO_2 molecules. A heating medium such as steam enables the adjacent regeneration column to release the highly concentrated CO_2 from the solvent, which is sent to a processing unit for further upgrade (e.g. drying, compression, liquefaction) for storage or utilization.

This paper evaluates steam cracker flue gas carbon capture using the example of mainly methane-fired cracking furnaces. The case of a world-scale liquid cracker decarbonization is presented, where the CO_2 from the flue gases is captured utilizing the BASF OASE® blue process, which is mutually developed by BASF and Linde. More than 95% of the direct continuous CO_2 emissions from the cracking furnace and boiler stacks can be efficiently removed.

The paper further outlines how existing cracker assets can be retrofitted with appropriate ducting to collect and route the individual flue gases to the capture facility. Despite being an elegant "add-on" solution for retrofitting existing facilities with relatively low impact to the existing fired capital equipment, flue gas capture units consume a noticeable amount of low-temperature heat. Energy-integration potentials between cracker and the flue gas capture unit are highlighted, mitigating or even avoiding extra thermal energy demand and associated additional CO_2 generation.

1. Background

Heavy industries including cement, iron and steel, oil refining, and petrochemicals are collectively responsible for about 22% of global CO_2 emissions [1] and among these, petrochemicals contribute roughly one fourth, accounting for about 5% of global CO_2 emissions [2].

One possible solution for the reduction of these emissions is carbon capture, utilization and/or storage (CCUS). Methods to do so are typically distinguished into three possibilities: "Post-combustion CO_2 capture", "Pre-combustion CO_2 capture" and "Oxy-fuel combustion CO_2 capture" [3,4,5].

"Post-combustion CO_2 capture" is a widely used technology for eliminating the CO_2 emissions of industrial plants in various sectors, including the heavy industries mentioned above. In the context of petrochemical plants, "Post-combustion CO_2 capture" can be understood as CO_2 capture from flue gas mainly originating from cracking furnaces and auxiliary steam boilers [5]. In the commercially ready solutions, the respective flue gases are routed through a column with circulating liquid amine solvent, which absorbs the CO_2 molecules. A heating medium such as steam enables the adjacent regeneration column to release the highly concentrated CO_2 from the solvent, and the CO_2 is sent to a processing unit for further upgrade (e.g. drying, compression, liquefaction) for storage or utilization. Considering the CO_2 quantities of a steam cracker storage / sequestration will be the prevailing scenario.

2. BASF OASE® blue flue gas carbon capture

The BASF OASE® blue flue gas carbon capture technology is jointly developed by Linde and BASF using a novel propriety solvent [6,7,8]. The solvent is key when it comes to an efficient and sustainable absorption of the CO_2. The technology is typically embedded into a holistic solution for the whole CO_2 "value chain", including downstream conditioning, compression, logistics and applications [7,8]. A typical process scheme of the actual capture process is shown in **Figure 1**.

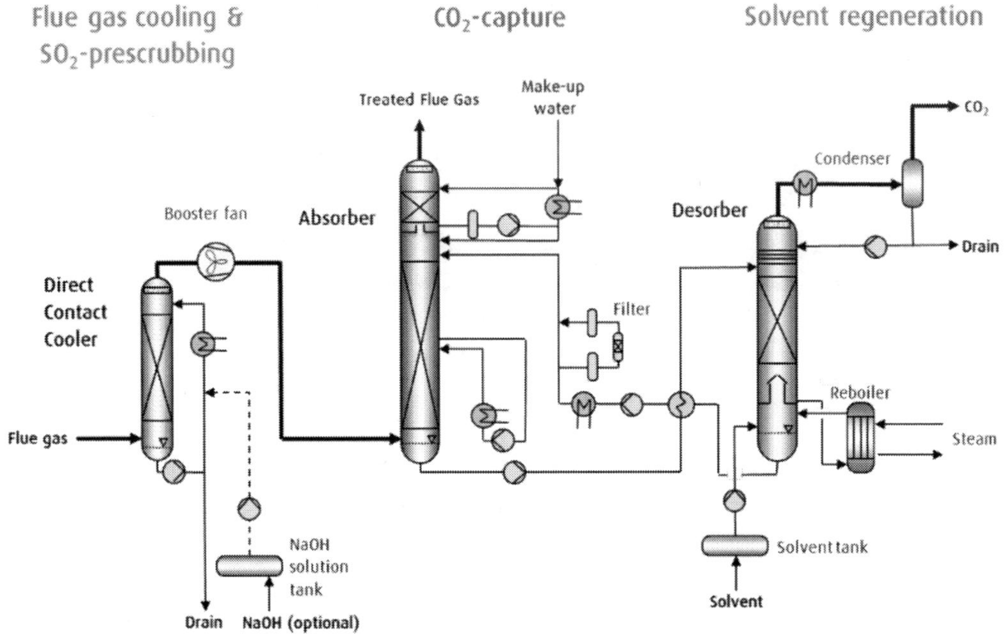

Figure 1: *BASF OASE® blue flue gas carbon capture plant in detail.*

With the OASE® (registered trademark of BASF SE) blue technology, CO_2 is removed from flue gases through chemical scrubbing with an aqueous amine-based solvent. For existing assets, it can be implemented without interfering with upstream processes. Moreover, advanced plant integration concepts have been developed for new assets to further improve total costs of ownership. The BASF/Linde joint development of OASE® blue technology leverages BASF´s capabilities in high-performance gas treatment technologies and Linde´s strength and proven track record in design and delivery of turn-key industrial plants. The result is an optimal interplay of solvent, process design, equipment and plant integration.

OASE® blue technology can be applied to flue gases from 4 to 25 vol% CO_2 content, hence covering a broad spectrum of CO_2 sources such as different types of power plants, gas motors, cement and lime kiln plants, furnaces, reformers and crackers. The technology allows for CO_2 capture rates higher than 95%. Without additional treatment, CO_2 product purity of 99.9 vol% (dry) is achieved, allowing for direct use in many downstream applications. There are more than 500 OASE® gas treatment plants in operation world-wide for different applications. The OASE® blue process has been operated for more than 100,000 hours in reference plants.

3. Large single-train designs suitable for world scale olefin plants

Mainly methane-fired world-scale steam crackers with CO_2 concentrations in the wet flue gas around 7-9 mol% are an interesting application case for flue gas carbon capture. With total CO_2 quantities of 150-200 t/h, large-capacity single train designs can be built with absorber column diameters ranging from 12 to 14 m. This enables a compact and cost-efficient single train design with significant economy-of-scale savings, reaching a CO_2 recovery rate of 95 %. With the above-mentioned wet flue gas CO_2 concentrations, thermal energy requirement for the desorber as low as 2.5 GJ/ton of captured CO_2 can be reached (without heat integration). Cooling duty can be provided by making use of cooling water and/or air cooling as required by the specific conditions of the facility. Typically, a heat-integration study is conducted to reduce the requirement for external supply of thermal energy. Despite the relatively demanding flue gas conditions originating from steam cracker furnaces containing noticeable amounts of oxygen, the excellent properties of the tailored OASE® blue solvent avoids excessive solvent deterioration, keeping the make-up below 0.4 kg amine per ton of captured CO_2.

A typical simplified layout of a BASF OASE® blue flue gas carbon capture plant is shown in **Figure 2**, with the main equipment displayed in the back: The direct contact cooler is located on the left, the large absorber tower in the middle and the regenerator on the right.

It is vital that sufficient plot space is available in close proximity to the primary CO_2 emitters. This avoids long-distance, and large diameter ducting for the volumetrically large flue gas flows. If that condition is met, the flue gas carbon capture unit can be designed for a compact plot, despite the large single train equipment. For above mentioned capacities, a plot space of approximately 5000 m^2 would be sufficient. If required, an even smaller footprint in a more compact layout is possible.

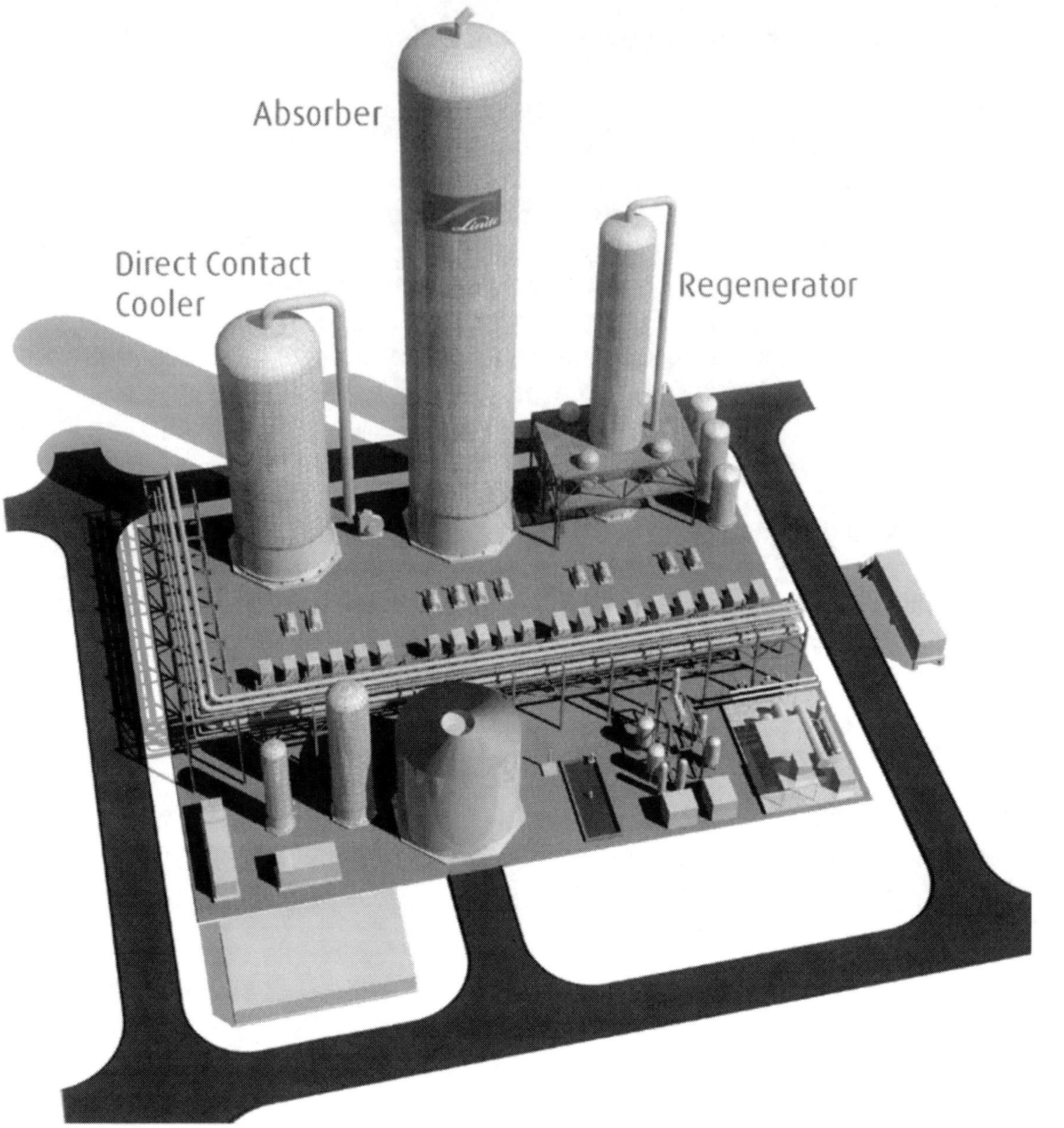

Figure 2: Typical simplified layout of a *BASF OASE® blue flue gas carbon capture plant.*

4. Process integration into new and existing steam cracker assets

From a process integration point of view, the flue gas carbon capture unit is an elegant add-on solution with relatively low impact to the steam cracker and its fired capital equipment. One typical embodiment is described herein.

The flue gas is typically cooled down in the furnace convection section to temperatures around 100 to 140 °C. In modern steam cracker designs, the furnaces are equipped with individual flue gas fans and separate stacks. This enables flexible operation with and without the flue gas carbon capture unit. Alternatively, and as often found in very old plants, the flue gas is discharged from the furnaces at higher temperatures, which is then routed to a common heat recovery unit with a single stack. The routing to flue gas carbon capture will branch off downstream of the flue gas fans and upstream of the stack(s). Mechanically or logically linked dampers in stack and CCS duct branch ensures an adequate and safe control of the flue gas.

The individual furnaces flue gas fans will not necessarily require significant additional power versus operation without the flue gas capture unit. Installation of a separate flue gas fan in the common flue gas line inside the flue gas capture unit is instead considered, which is located downstream of the direct contact cooler and upstream of the CO_2 wash tower. Ambient temperatures and reduced flue gas flow allow efficient operation of the fan at this location. More importantly, this design allows greater flexibility and fully independent furnace operation with existing appropriately sized furnace flue gas fans.

The described design is suitable for new and existing steam cracking plants. An example from a study for a retrofitted flue gas duct on an existing steam cracking plant is shown in **Figure 3**. Careful review of the existing mechanical and static design, as well as operation and safety is required for each project. This also ensures the long-term high unit availability.

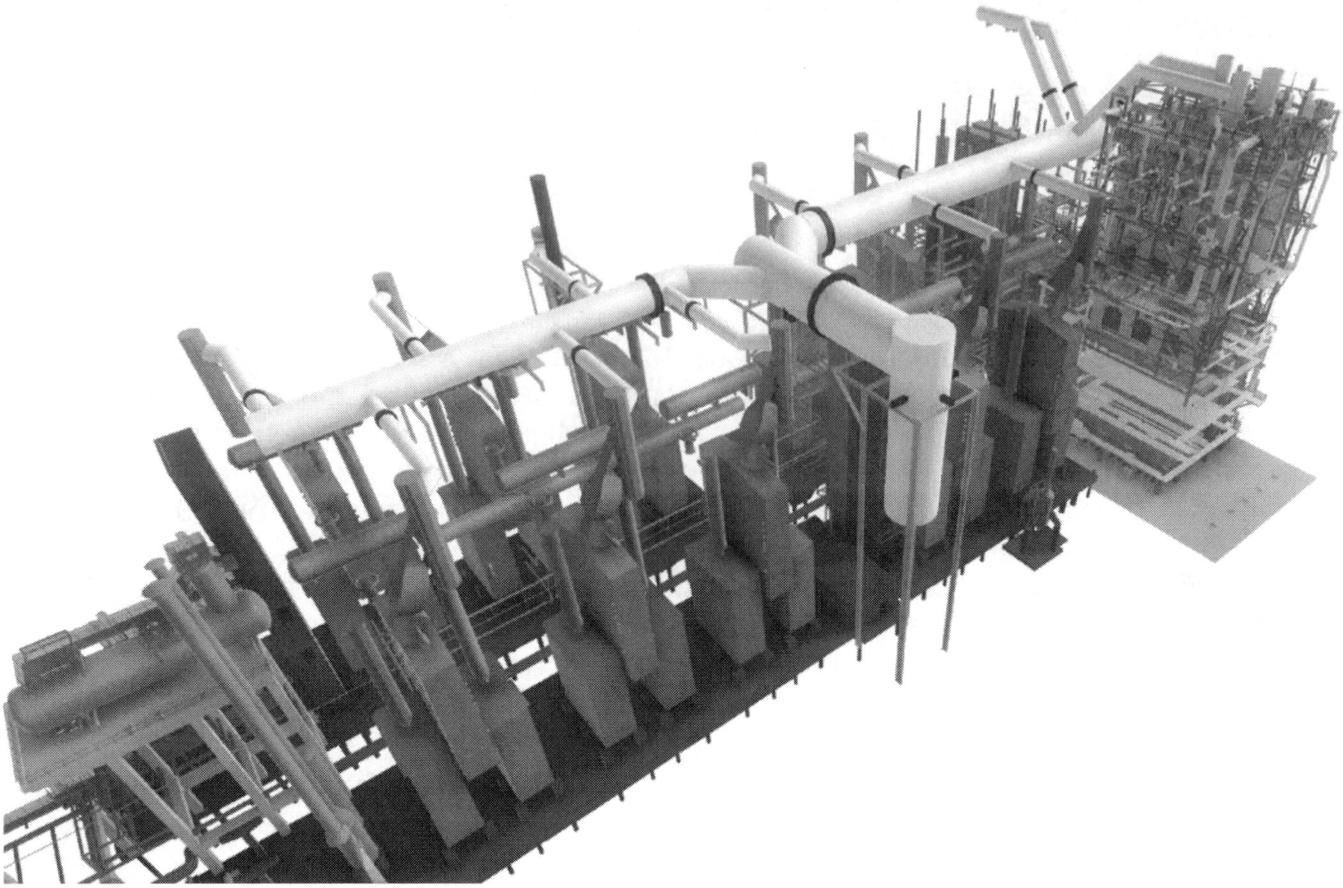

Figure 3: *Flue gas duct retrofitted into an existing steam cracking plant.*

5. Relevance on designing energy-integrated systems

It is often correctly mentioned that despite being optimized for flue gases with dilute CO_2 concentrations flue gas carbon capture units require noticeable amounts of low-temperature thermal heat in the form of low-pressure (LP) steam[4]. Recognize, however, that the impact of such requirements on the overall energy efficiency is drastically influenced by the chosen (energy)-integration approach. This aspect will be further explained with the following generic examples, in which a flue gas carbon capture unit shall be installed for an existing world-scale steam cracking plant processing mainly naphtha and generating >1000 kta ethylene and >500 kta propylene. It is assumed that the flue gas carbon capture unit utilizes the BASF-Linde OASE ® blue process and a CO_2 compression and drying section delivering the dry CO_2 at approximately 35 bar a.

A first and the basis case is shown in **Figure 4**. A steam cracker is displayed on the right side, where 720 MW of CH_4-rich gas (natural gas and/or cracker CH_4 fraction) is utilized for the operation of the cracking furnaces. Very high-pressure (VHP) steam (about 400 t/h) is generated, which is used for the operation of three large steam turbines driving the cracked gas compressor (CGC), C3 refrigerant compressor (C3C) and C2 refrigerant compressor (C2C). The turbines extract some steam to high-pressure (HP), medium-pressure (MP) and low-pressure (LP) conditions, where it is utilized as heat carrier within the cracker as well as adjacent units (steam export). Noticeable amounts of steam is expanded to vacuum conditions in order to balance the shaft power demand. A boiler is operated to balance VHP steam demand, consuming 90 MW fuel. A total of approximately 157 t/h CO_2 is generated.

If a flue gas CO_2 capture unit and its energy supply is planned as an isolated facility (**Figure 4**, left side), it will require the generation of LP steam within its battery limits by burning CH_4-rich gas or natural gas within another steam boiler generating e.g. HP steam. Conveniently, the required CO_2 compressor could be driven by a back-pressure turbine, using the high-pressure steam generated within the boiler. On the downside however, the total CO_2 generated increases from 157 t/h to 190 t/h, which is a remarkable increase of 21 %. The increase is leveraged by an unfortunate "windup" effect, as the LP steam demand requires the generation of CO_2 within the separate boiler, which in turn again increases the LP steam demand due to the increasing CO_2 capacity. This example shall highlight that flue gas carbon capture in general suffers from "extra thermal energy demand", increasing the total CO_2 to be captured. The extra total CO_2 depends on the flue gas properties and unintegrated solutions may cause extra CO_2 generation, for storage or utilization, in the range of 10 to even 30 % (in most unfavorable conditions). Furthermore, flue gas carbon capture units require the removal of considerable amounts of low temperature heat caused by the condensation of water from the flue gas (among others). If such heat is removed by water cooling, the total cooling water demand of cracker and flue gas capture unit can noticeably increase.

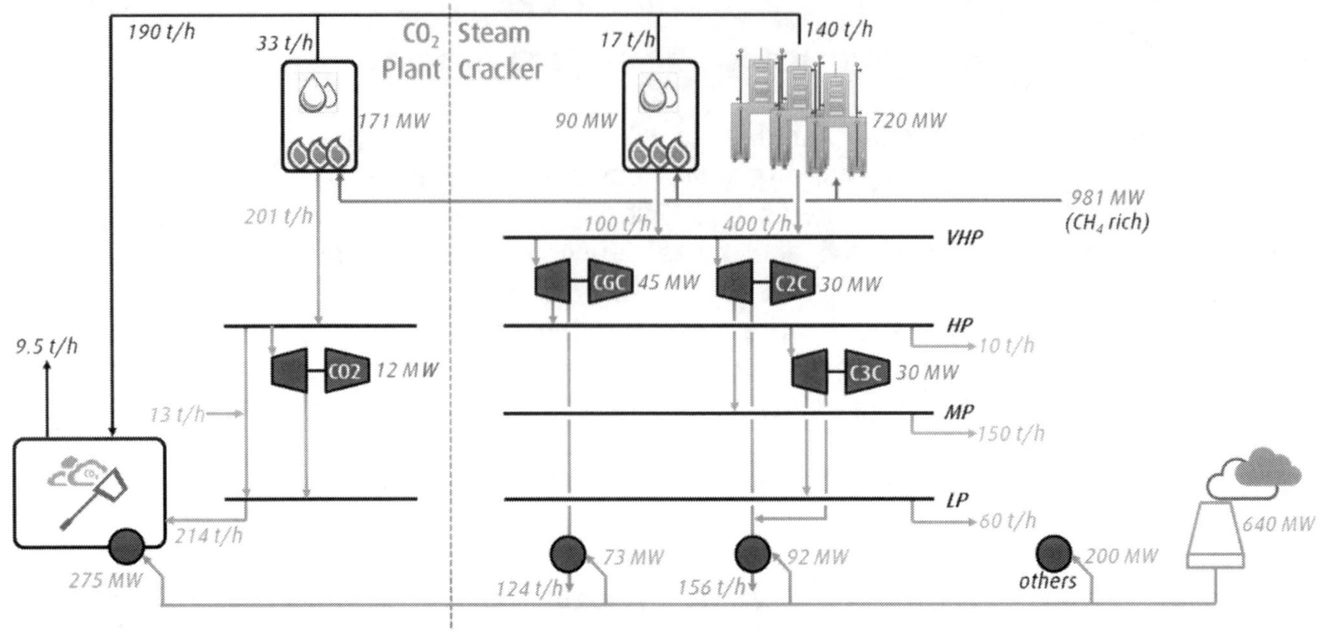

Figure 4: *Unintegrated heat supply to cracker flue gas carbon capture.*

6. Embodiments of energy-integrated systems

Considering however that only low-temperature thermal heat is required by flue gas carbon capture units, simple and promising heat integration opportunities together with steam cracker units can be identified. Conventional steam cracking furnaces are only able to supply a fraction of the consumed heat into the actual endothermal reaction (25-35 % of the supplied heat). The remaining heat must be used for other "low-temperature" purposes, e.g. for feed preheating and eventually for the generation of secondary heat carriers like steam. Consequently, steam crackers traditionally come along with a sophisticated steam system, where the large machinery required for the separation section is driven by steam turbines. It is also often observed that only a fraction of the steam is used in the context of "heat and power", utilizing the depressurized steam as heat carrier. Another large fraction of the steam is processed to vacuum conditions just for the generation of shaft power, not leveraging on the efficient concept of "heat and power".

A first integrated solution is shown in **Figure 5**, where LP steam from the steam cracker is transferred to the carbon capture unit. Here, the concept of "heat and power" is maximized within the steam system of the cracker and fewer steam is sent to vacuum conditions. As a result, the additional steam required from boilers is significantly reduced and a total of 172 t/h CO_2 is generated, which is an increase of only 9 % compared to the design without flue gas carbon capture.

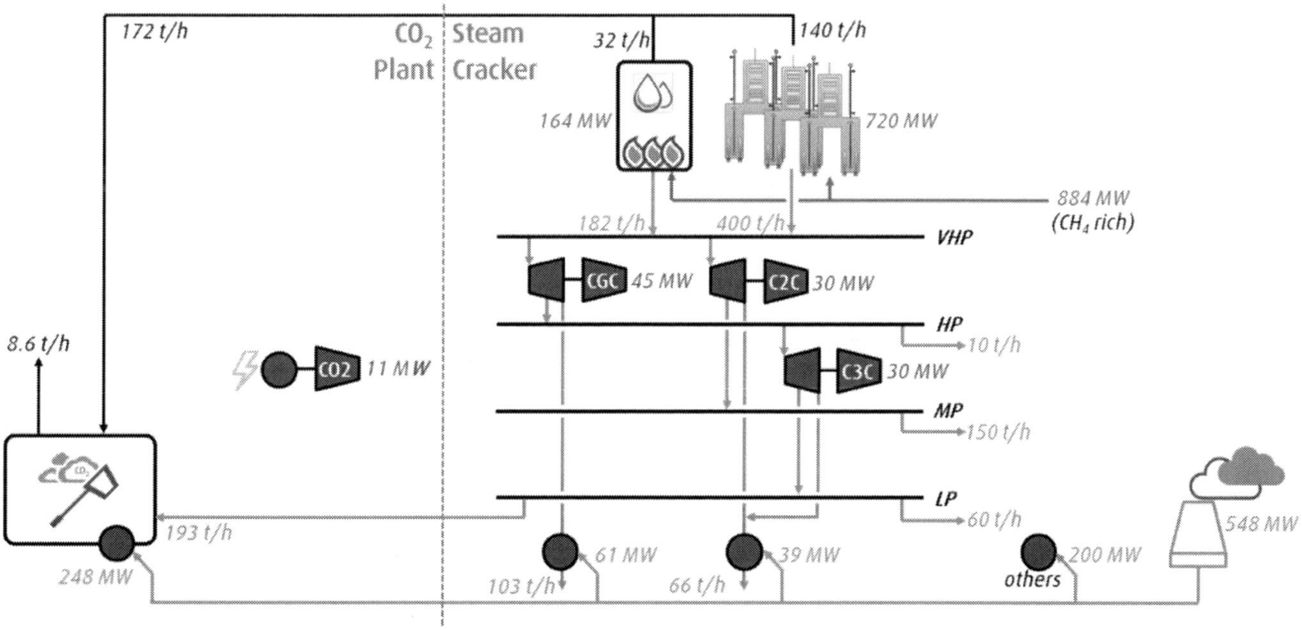

Figure 5: *Integrated heat supply to cracker flue gas carbon capture, increasing LP steam generation in the cracker.*

Of course, it needs to be mentioned that such concept likely requires some modifications within the steam system of the cracker: In the shown example, the CGC and C3C turbines would require significantly increased steam extraction, which might require a retrofit or replacement of the respective turbines. Also, the installed boiler capacity needs to be reviewed with respect to the increased steam generation requirements.

The example shown in **Figure 6** further minimizes on vacuum steam generation by electrification of a refrigerant compressor, in this case the C3 compressor. With a well-thought-out concept, such electrification is often feasible even in brownfield conditions and can be executed by an experienced licensor. Furthermore, the CGC compressor turbine is redesigned/replaced for LP steam extraction instead of HP steam extraction, as the HP steam demand is now minimal. As a result, the total CO_2 generation even slightly decreases by 3% compared to the design without flue gas carbon capture, further reducing the LP steam demand of the capture unit as well as the power demand of the CO_2 compressor. As further advantage, the heat demand can now be delivered by the existing boiler capacity. Also, utilities like cooling water require less capacity expansion or even none, as the integrated cases described in **Figure 5** and **Figure 6** drastically reduce the cooling water consumption within the steam cracker. Furthermore, the electrification of the compressor drives enables the introduction of larger volumes of renewable power and contributes to any company's sustainability roadmap.

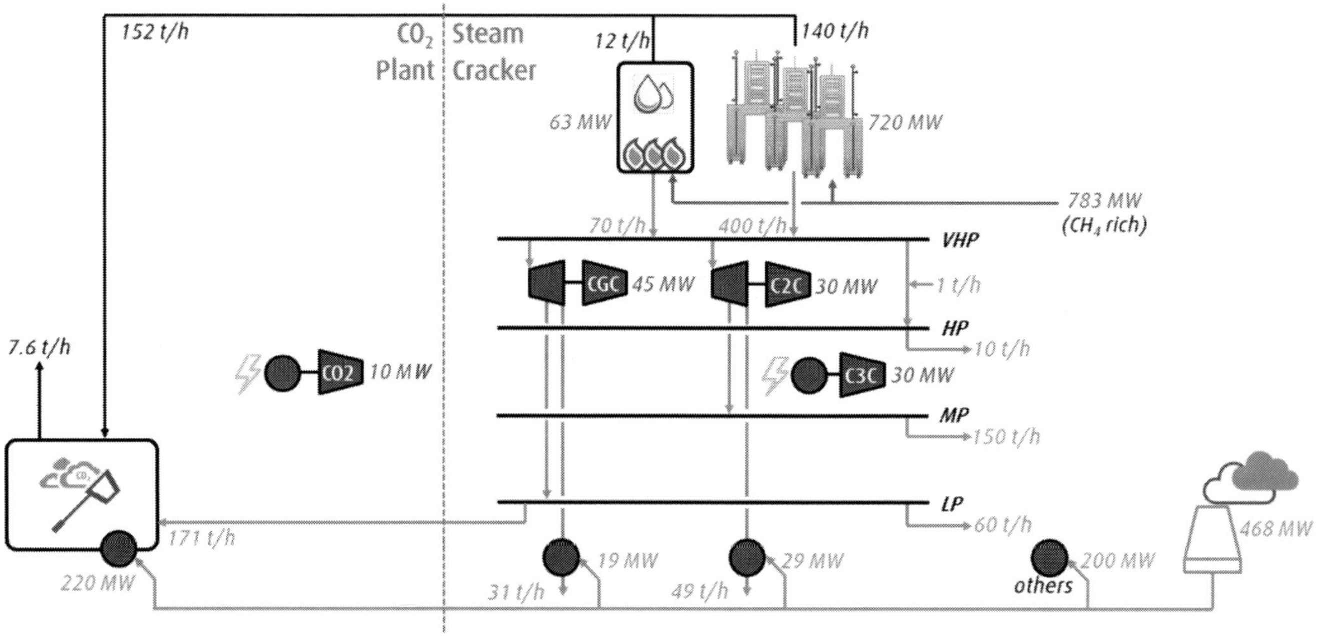

Figure 6: *Integrated heat supply to cracker flue gas carbon capture, increasing LP steam generation in the cracker and further reducing on vacuum steam condensation by machine electrification.*

7. Impact on key performance indicators

Such energy-integration opportunities can have a major impact on the economic assessment, as the level of integration can positively influence multiple parameters like total CO_2 to be sent to sequestration, size/CAPEX of the capture unit, total primary fuel demand and other utility demands like cooling water. Some high-level performance indicators are provided in **Table 1** to highlight the differences.

Table 1: *High-level performance indicators for flue gas carbon capture units integrated into steam crackers (world scale steam cracking plant as described in the text).*

Case		1	2	3
Description		Unintegrated	Integrated	Integrated, C3C electrified
Extra fired duty	MW	171	74	-27
Extra electrical energy (flue gas capture unit + CO_2 compressor + C3C)	MW	9.9	20.5	49.8
Total extra energy @ efficiency factor of 37.5 % for electrical energy	MW	197.4	128.8	105.7
Extra CO_2 from fired duty (CH_4-rich gas)	t/h	33.2	14.4	-5.2
Total extra CO_2 from fired duty + electrical energy import (combined cycle, 433 kg CO_2 / MWh)	t/h	37.5	23.3	16.3
CO_2 eliminated from base case[1)]	t/h	147.9	148.9	149.9
Required CO_2 premium[2)] to compensate cost from extra fired duty (0.044 USD/kWh) and extra electrical energy (0.128 USD/kWh), scenario 1	USD/t	60	40	35
Required CO_2 premium[2)] to compensate cost from extra fired duty (0.105 USD/kWh) and extra electrical energy (0.178 USD/kWh), scenario 2	USD/t	133	77	40

1) As the total CO_2 generation increases for case 1 and 2, the eliminated CO_2 compared to the case without carbon capture slightly reduces at same capture efficiency of 95%.
2) Scenario 1 marks a "medium/high energy cost" scenario with medium fuel/electricity spread. Scenario 2 marks a "high energy cost" scenario with low fuel/electricity spread.

It should be noted that the total additional energy consumption with a typically applied efficiency factor for electricity of 37.5 % is lowest for case 3, despite the high import of electrical energy. A similar trend is observed if one cumulates the extra CO_2 generation caused from the extra firing and the extra electrical energy import: Even if fossil fuel would be used for electricity generation (e.g. combined cycle, 433 kg CO_2 / MWh), extra scope 1+2 CO_2 generation would be lowest for case 3. It needs to be mentioned that case 3 would further benefit from application of low-carbon power. The last two rows show the fraction of the CO_2 premium, which would be required to compensate for the additional cost of fuel and electrical energy, depending on two cost scenarios. In high energy cost situations as they were present in the EU and USA in 2022, such energy costs can heavily influence the business case. It shall be highlighted that in both cost scenarios, case 3 is able to greatly buffer the cost impact caused by the additional energy demand. The effect for the scenario 2 is more significant because the cost spread between electrical energy and fuel is lower (1.7) compared to scenario 1 (2.9).

Of course, the complete economic assessment and the total required CO_2 tax/premium depends on further factors, including CAPEX and others, as it is evaluated on for other sectors in other studies [9]. Regarding CAPEX for example, each of the mentioned cases has somewhat different cost-driving characteristics: While case 3 requires the modification of the cracked gas turbine as well as the electrification of the C3 compressor, it benefits from a reduced size of the flue gas capture unit and does not require extra boiler capacity.
Roughly summarized, one cannot make a clear general statement in favor or against any of the described solutions when it comes to CAPEX and the main difference originates from the cost drivers as highlighted in **Table 1**. In today's environment of increasing prices for energy and CO_2 (or respective incentives for CO_2 avoidance) it is however anticipated that the most economic solutions will shift towards "life-cycle oriented" designs with an increasing weighting of low-energy solutions. Considering this, the described integrated cases are expected to yield significant advantages.

8. Takeaway

The given generic examples are intended to highlight how an often-mentioned disadvantage of flue gas capture units (the large low-temperature heat demand) can be mitigated or even avoided in the context of steam cracker applications, without requiring sophisticated and costly methods like e.g. high-temperature heat pump technology. Although these examples were intentionally designed in favor of the integrated solutions, the underlying correlation with respect to energy efficiency increase have been observed in real projects. Operating companies can clearly benefit here from a licensor with experience in steam cracking technology as well as flue gas carbon capture units.

9. References

1. Güleç, F., W. Meredith and C. E. Snape, "Progress in the CO_2 Capture Technologies for Fluid Catalytic Cracking (FCC) Units—A Review", *Frontiers in Energy Research*, 2020
2. Lange, J. P., "Towards circular carbo-chemicals – the metamorphosis of petrochemicals", *Energy & Environmental Science*, 2021
3. Santos, M. P. S. and D. P. Hanak, "Carbon capture for decarbonisation of energy-intensive industries: a comparative review of techno-economic feasibility of solid looping cycles", *Frontiers of Chemical Science and Engineering*, 2022
4. Jha, A., V. Ravuru, M. Yadav, S. Mandal and A. K. Das, "A critical analysis of CO_2 capture technologies", *Hydrocarbon Processing*, June 2021
5. Buehler, J., "Overview of decarbonization pathways for the oil and gas and petrochemical industries—Part 2", *Hydrocarbon Processing*, June 2022
6. Online: https://www.basf.com/global/en/media/news-releases/2016/07/p-IR-160719.html
7. Online: https://www.linde-engineering.com/en/process-plants/co2-plants/carbon-capture/post-combustion-capture/index.html
8. Mirza, N. and D. Kearns, "State of the art: CCS Technologies 2022", page 138, *Global CCS Institute*, 2022
9. Panja, P., B. McPherson and M. Deo, "Techno-Economic Analysis of Amine-based CO_2 Capture Technology: Hunter Plant Case Study", *Carbon Capture Science & Technology*, 2022

95a Caustic Tower Design Fundamentals

Miguel Maldonado, Min Wang, Mark Loeffler

The Caustic Tower in an Ethylene Unit removes the acid gases (H2S and CO2) generated during the cracking process. This is required to meet Polymer Grade Ethylene product specifications and to protect downstream systems.

This presentation will provide an overview of the Caustic Tower, the basic reactions, and other design fundamentals and concepts. Caustic Tower design features for fouling prevention will also be covered, including potential impact of recycles introduced into the cracked gas system.

95b The Fundamentals of Caustic Tower Fouling

David Hood, Dwight Hines

The tutorial presentation reviews the fundamentals of process-side fouling in an ethylene unit's caustic tower. The general mechanisms of fouling are first discussed, with a particular focus on the precursors and differences between liquid- and gas-heavy feed slates. Common techniques for anticipating and monitoring the fouling of equipment are then presented. Finally, an assortment of proven ways to mitigate fouling such as process treatment, operational adjustments, and maintenance practices are discussed.

95c Fouling in the Caustic Tower – Mechanism and Mitigation

Dave Smith

In the ethylene plant, by products of the cracking process include acid gases such as CO2, CO, and H2S and carbonyls. The acid gases and carbonyls are contaminants for the olefine products, as well as a corrosion risk in processing equipment. To remove these contaminants, ethylene plants employ an acid gas removal treatment operation. This operation is typically performed via scrubbing the process gas with caustic to neutralize the acid gases via contact in a Caustic Tower. The spent caustic is then treated or disposed of in a wet air oxidation unit, as wastewater or other means.

Certain of the contaminants can undergo polymerization to create polymer that can then lay down on trays or packing in the Caustic Tower, causing a variety of operational issues. These issues include increased differential pressure, reduced scrubbing efficiency, foaming, and ultimately can result in a throughput limitation and even premature shutdown if the Caustic Tower fouling is severe enough. The foulant in the caustic tower has several various names in the olefines production vernacular, such as “red oil” or “red tide”, as well as others that are more descriptive about the physical description of the foulant material.

The purpose of this tutorial is to delineate the various mechanisms for fouling in the Caustic Tower, their impacts on plant operation, and then the various methods that can be employed to mitigate this fouling. Mechanical, operational, and chemical methods will be included.

95d Caustic Tower Red Oil – How to Effectively Mitigate the Oil Soluble Aldol Polymers

Joice Boll, Mikhail Knis, Bishwanath Mondal

The formation of "red oil," (long chain aldol polymers) in the caustic tower continues to be a recurring problem that can impact the efficiency of acid gas removal and lead to fouling and emulsions in the spent caustic recovery system. A key reason this is a recurring problem is the inherent limitations of controlling the aldol polymerization reactions using inhibitors that work in the aqueous phase with no activity in higher molecular weight polymers that are insoluble in caustic. Over time, even small upsets in matching the water phase inhibitor with demand can cumulatively lead to deposit accumulation of red oil over the long run-length of best-in-class ethylene plants. This can impact efficiency and ease of operation. Red oil deposits can accumulate and potentially increase the risk of not being able to meet efficiency and run-length targets. This paper reviews the chemistry of aldol polymerization and explains the limitation of using chemistry that works only in the aqueous phase. Case histories are presented to demonstrate the power of inhibitor chemistry that works in both oil and water phases. These concepts are timely as plants increase run-length targets and for those trying to recover from upsets in control using chemistry that is only active in the water phase.

95e Aldol Polymerization Inhibitor Evaluation Using Turbiscan

Bhaskar Reddy Aluri, Ian Robson, Geoff Chappell

In a steam cracker, caustic tower is located between later stages of compression, which serves the purpose of removal of acid gases (H_2S and CO_2) through absorption using dilute caustic soda. Similar to other process equipment, the caustic tower is prone to fouling resulting in reduced caustic circulation rate, foaming, and pressure drop. A major contribution to fouling in caustic tower is due to base catalyzed aldol polymerization of aldehydes. The rate of polymerization depends on caustic tower operating temperature and strength of caustic. In order to prevent/minimize aldol polymerization various inhibitors are available in the market from different companies. Often, it is challenging for plant personal to choose the most suitable inhibitor among available options for his unit. Traditional laboratory flask test[1-3], followed by gravimetric separation of formed polymer is a method used to evaluate different inhibitors. Also, other innovative methods such as UV spectroscopy measurement or Evaporative Light Scattering Detector (ELSD)[4] for quantification of aldol polymer as a function of time is reported.

In the current paper, we will describe an easy and efficient method of evaluating caustic inhibitors at reducing aldol polymerisation as a function of time, temperature and caustic strength using a Turbiscan™ LAB stability analyzer. The extent of polymerisation observed as an increase in destabilization behaviour of the reactant mixture, measurements derived from back scattering and transmission signals from the test solution. This method for evaluating aldol inhibitors is considered an improvement on methods described earlier.

References:

(1) McClain Robert D, Patel Natu R, Glath Raymond M, Reducing undesired polymerization in hydrocarbon cracking processes, WO 02/36715A1.

(2) Mahesh Subramaniyam, Banker Prashant, Method for prevention of fouling in basic solution by inhibiting polymerization and solubilizing deposits using amino acids, US6986839B2

(3) Hiren M. Bhajiwala, Harshad R. Patil, Virendrakumar Gupta, Studies of amino acids for inhibition of aldol condensation and dissolution of polymeric product of aldehyde in alkaline media, Appl Petrochem Res (2013) 3:17–23.

(4) Satheesh Sambandam, Bassam Alnasleh, Debby Rossana, Study of aldol polymerization in ethylene plant Caustic towers, Presentation at the 2018 Spring National Meeting Orlando, Florida, April, 23, 2018.

95f Management of Caustic Tower Fouling

Alexander Fritz, Betty Latosinsky, Grace Kim

Ethylene plants utilize caustic towers for the removal of acid gas (H2S/CO2) from the cracked gas downstream of the furnaces. Typically, caustic towers experience base-catalyzed polymeric fouling due to the inherent presence of condensed hydrocarbon in contact with sodium hydroxide solution. This fouling can become severe if left unchecked, so the summary herein outlines strategies for the management of caustic tower fouling. At the root, caustic fouling mechanisms include primarily aldol condensation but also free-radical and Diels-Alder reactions. Fouling progression is a function of many parameters, including the plant feedstock being cracked and associated impurities, the tower operating temperature, the location of the tower within the process, fouling precursors from recycle streams, and any chemistries added to the tower. Additionally, impurities in the fresh caustic solution can also play a role in the overall fouling propensity. All these factors offer levers to control system performance and reduce foaming or flooding impacts. To aid in the discussion, analytical methods for foulant characterization are outlined. Ultimately, a comprehensive understanding of the fouling within caustic systems leads to enhanced plant efficiency and product on-spec for H2S/CO2 concentration.

AIChE Paper Number: 60a

Towards high-quality petrochemical feedstocks from plastic waste pyrolysis oils: Contaminant removal via depths filtration and its impact on steam cracking

Marvin Kusenberg
Ghent University, Ghent, Belgium

Martijn Roosen
Ghent University, Ghent, Belgium

Andreas Eschenbacher
Ghent University, Ghent, Belgium

Emmanuelle Biadi
Pall Corporation, France

Nicolas Laudou
Pall Corporation, France

Leonor Casado
Pall Corporation, Madrid, Spain

Astrid Doktor
Pall corporation, Bad Kreuznach Germany

Ali Arshad
Pall Corporation, Houston, USA

Steven De Meester
Ghent University, Ghent, Belgium

Kevin M. Van Geem
Ghent University, Ghent, Belgium

Prepared for Presentation at the 2023 Spring National Meeting
Houston, Texas, March 12 – March 16, 2023

Abstract:

Advanced recycling of plastic waste has the potential to overcome current hurdles of post-consumer plastic waste recycling as it is believed to be more robust towards contaminants and mixed feedstocks compared to mechanical recycling. Converting waste polymers into their chemical building blocks via pyrolysis followed by steam cracking enables the production of virgin-grade polymer products. However, advanced recycling still faces several hurdles which are related to the high variety of different polymers and additives present in plastic waste that lead to contamination of the corresponding pyrolysis oils. In fact, it has been shown that contaminants are the main factor that decides if plastic waste pyrolysis oils are acceptable for industrial steam crackers. Particularly, metal contaminants are highly problematic in terms of crucial aspects such as fouling, coke formation or downstream catalyst poisoning. In this study, the potential of depth filtration as an effective upgrading step to remove solids and metal contaminants from pyrolysis oil is investigated. In order to assess the effect of the filtration step on the steam cracking performance, continuous experiments have been performed at bench-scale including the detailed on-line analysis of the entire product range and the assessment of the radiant coil coke formation. Based on the detailed analysis of the filtrates using ICP-OES, it was found that a substantial reduction of metals was reached with the tested filter media, however, with different removal efficiencies based on the fineness of the filter media. By means of the total metal concentration in the filtered pyrolysis oils, between 20.9% and 81.8% of the metals were removed. Steam cracking tests using the respective filtrates blended with naphtha in a weight-based 1:3 ratio show that a substantial reduction in radiant coil coke formation of 40-50% could be achieved compared to the unfiltered pyrolysis oil.

Introduction

The production of plastics exceeded 60 million tons in Europe alone in 2018 and almost 60% of the produced plastics are used for packaging applications [1]. Mechanical recycling, although more energy efficient, is limited in terms of mixed and contaminated waste streams. As a consequence, recycled products often have lower quality compared to virgin plastic products leading to downcycling. Chemical or advanced recycling of plastic packaging waste that mostly consists of polyolefins (i.e., polyethylene (PE), polypropylene (PP)), is believed to be more tolerant towards contaminants and mixed feedstocks compared to mechanical recycling. Converting waste polymers into base chemicals such as ethylene and propylene enables the production of virgin-grade polymers [2-4]. Furthermore, by avoiding waste incineration and the (partial) replacement of fossil feedstocks for steam crackers, a substantial amount of greenhouse gases can be prevented [5, 6]. The current trend in electrification of petrochemical processes is another driver to substantially reduce emissions from end-of-life plastics management [7, 8].
Challenges in advanced recycling are related to the diverse mix of different polymers in plastic waste and the extensive use of additives in the plastics formulation [9-11]. Typically, metal-containing additives are used as inert fillers, (e.g. SiO_x, $CaCO_3$), pigments (e.g., TiO_2, iron oxide) and stabilizers [11-14]. The resulting contamination of pyrolysis oils is the key reason why advanced recycling has not yet been established on an industrial scale [15]. This is mainly due to operational risks in large-scale steam crackers, such as corrosion, increased coke formation or downstream catalyst poisoning [9, 16]. This was found in our recent study, where steam cracking of crude

untreated plastic waste pyrolysis oils blended with fossil naphtha was investigated [17]. Substantial coke formation was observed which could be related on the one hand to the higher olefin and/or aromatic concentrations of the pyrolysis oil and on the other hand to high amounts of metal contaminants such as iron or sodium present in the pyrolysis oils.
In this study, the potential of depth filtration as an effective upgrading step to remove solids and metal contaminants from a crude plastic waste pyrolysis oil was investigated. Three filter media with different fineness were tested and the pyrolysis oils were characterized before and after filtration. Analytic techniques used were inductively coupled plasma – optical emission spectrometry (ICP-OES) and comprehensive two-dimensional gas chromatography (GC × GC) coupled to flame ionization detection (FID). Furthermore, the effect of depth filtration on the steam cracking performance was investigated in bench-scale steam cracking experiments including the detailed on-line analysis of the entire product range and the quantitative assessment of the radiant coil coke formation.

Materials and Methods

Feedstocks: The pyrolysis oil used in this study was produced in a continuous pyrolysis unit consisting of a single-screw extruder (LabTech, Thailand) which is connected to a 5.7 L stirred-tank reactor [18-20]. Pyrolysis was performed at 450 °C and atmospheric pressure [17]. The starting material was a representative post-consumer mixed polyolefin (MPO) fraction collected at curb-side and sorted in a typical waste sorting and separation plant in Belgium. The material consisted mostly of PE (~76 wt.%) and PP (~16 wt.%) and ~8 wt.% of other polymers. A typical light fossil naphtha feedstock was used for blending and as a reference feedstock for the steam cracking experiments.

Experimental: Flowrate filtration tests were performed using depth filter media of different fineness. Main applications for the filter media are clarifying filtration of viscous and highly viscous oils (F1), catalyst retention (F2) and the retention of activated carbon (F3). Filtration tests were performed at 80 °C due to the highly viscous nature of the pyrolysis oil using a peristaltic pump (Watson-Marlow 520S, UK) with a continuous flow rate set at 50 mL/min. 3 L of pyrolysis oil was processed per filter medium. The respective filter media were installed in stainless steel discs with pressure sensors. The differential pressure (Δp) was monitored over time. The observed Δp during filtration increased from 0 to 100 mbar for F1 and to 200 mbar for F2. The Δp of F3 increased to 500 mbar which is related to the different fineness. Steam cracking experiments were performed using a continuous bench-scale steam cracking unit (BSSC) as described in detail elsewhere [17, 21-24]. The feed section controls the supply of the feedstock and the water to the reactor coil. The flow of liquids is regulated by Coriolis flow-meter (CORI-FLOW) controlled pumps (Bronkhorst, The Netherlands). Steam and hydrocarbon feedstock are pre-heated and mixed prior to entering the reactor. The mixture enters the 1.475 m long reactor (Incoloy 800HT Ni: 30-35 wt%, Cr: 19-23 wt%, Fe: > 39.5 wt%) which is placed vertically in a rectangular electrically heated furnace. The furnace is divided into eight separate sections which can be controlled independently to set a specific temperature profile. Two pressure transducers, placed at the inlet and outlet of the reactor, indicate the coil inlet (CIP) and the coil outlet pressure (COP), respectively. The BSSC comprises a dedicated on-line

analysis section as consisting of two analytical setups. On the so-called Refinery Gas Analyzer (RGA) the light products (i.e. C_{4-}) are analyzed. The concentrations of H_2, CO_2, CO, N_2, methane, ethane, ethylene and acetylene are detected via thermal conductivity detection (TCD) and methane, ethane, ethylene, propane, propylene, acetylene, propadiene, iso-butane, n-butane, trans-2-butene, 1-butene, cis-2-butene, methylacetylene and 1,3-butadiene are analyzed using FID. Quantification of the compounds is possible using a well-defined continuous flow of nitrogen as internal standard. The response factors of each component were determined using a well-defined calibration gas mixture. The methane concentration determined on the RGA is used as a secondary internal standard to quantify the entire reactor effluent composition via on-line GC × GC-FID [25]. The response factors of the C_{5+} compounds are determined with the effective carbon number approach as described elsewhere [21, 22, 25-27].
In order to facilitate feeding, the highly viscous pyrolysis oils were blended with fossil naphtha in a weight-based 1:3 ratio. Moreover, a heating mantle was used to keep the feedstock bottle at ~60 °C which was sufficient to avoid solidification of the pyrolysis oil/naphtha blends. The steam cracking conditions are shown in **Table 1**. Two cracking severities were tested according to the coil outlet temperature (COT). The coil inlet temperature (CIT) was kept at 550 °C for every experiment and the evaporation and mixing section was operated at 500 °C to ensure complete evaporation of the feedstocks. The coil outlet pressure was kept at 1.7 bar for all experiments and a pressure drop was negligible. The hydrocarbon flow rate was set at 150 g/h with a steam dilution rate of 0.5 wt_{H2O}/wt_{HC}, which corresponds to an average coil residence time of around 0.5 s.

Table 1. Steam cracking process conditions.

Temperature profile 1 [°C]	550-590-630-670-700-740-780-820
Temperature profile 2 [°C]	550-590-630-680-720-770-810-850
Coil outlet pressure [bar]	1.7
Liquid hydrocarbon flow rate [g/h]	150
Steam flow rate [g/h]	75
Average coil residence time [s]	~0.5

The amount of coke that was deposited in the reactor during a 6-hour experiment was decoked afterwards in a controlled experiment. Using an infrared (IR) analyzer (Fuji Electric, Japan), the CO and CO_2 concentrations in the reactor effluent during the decoking experiment were measured [26]. Using a volumetric gas flow meter (Ritter, Germany), the volumetric concentrations of the reactor effluent were measured. Combining the volumetric flow rate with the volumetric concentrations of CO and CO_2 measured using the IR-meter, the coke that was formed and deposited in the reactor during the steam cracking experiments was quantified.

Analytics: The total suspended solids (TSS) present in the respective samples were determined by a gravimetric technique: 100 mL of heated sample was drawn through a 47 mm diameter pre-weighted 0.8 µm analysis membrane which was afterwards rinsed with a combination of hexane, acetone and xylene. Subsequently, the membrane was dried for 30 min at 80 °C and desiccated. The amount of solids on the test membrane was then calculated by deducting the clean membrane weight from the final membrane weight. The TSS present in the sample were calculated in mg/L. Furthermore, the particles found per 1 mL of sample were counted individually and sorted according to the particle sizes: >5 µm, >10 µm, >20 µm, >50 µm and >100 µm. To assess the stability of the filtrates over time, the TSS analysis was repeated after 2 days of storage. Moreover, all samples have been characterized using comprehensive two-dimensional gas chromatography (GC × GC) coupled to a flame ionization detector (FID) and inductively coupled plasma – optical emission spectrometry (ICP-OES) [17, 19].

Results and Discussion

Total suspended solids (TSS) analysis: In Fig. 1. microphotographs of the TSS test membranes of the analyzed samples (i.e., unfiltered pyrolysis oil and the three filtrates) are shown.

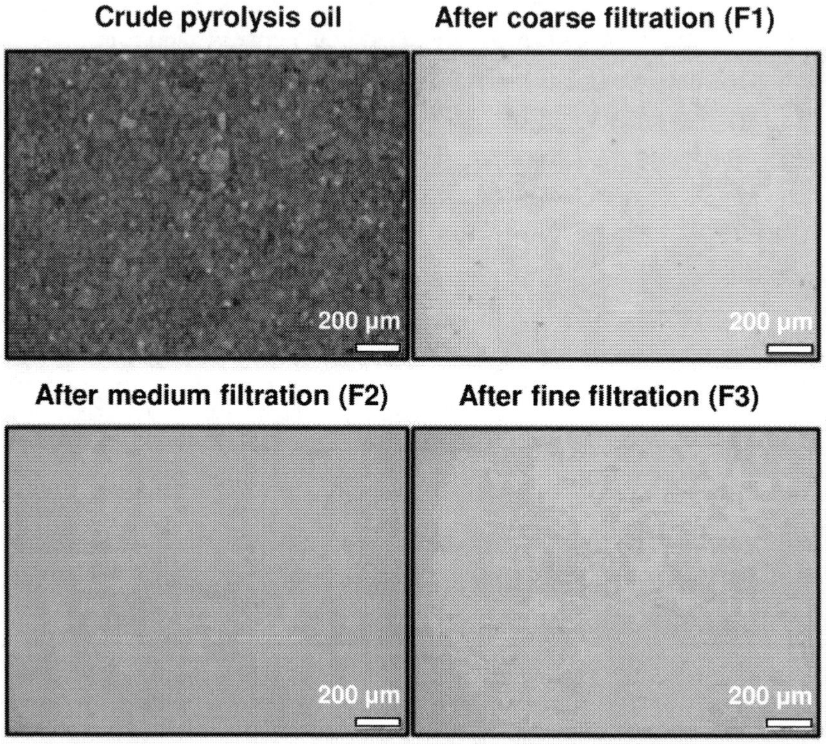

Fig. 1. Photographs (magnification ×100) of the TSS test filter membranes (0.8 µm, nylon) immediately after filtration.

The total particle count per mL of sample was 2595, mostly present as fine particles in the ranges >5 µm (2595), >10 µm (1420) and >20 µm (599) with smaller amounts of larger particles in the ranges >50 µm (91) and >100 µm (14). The average particle size was smaller than 10 µm. The detailed TSS and particle counts per mL of sample are listed in Table 2. It can be seen that filter media were effective in reducing the total suspended solids to <1 mg/L (limit of detection) when finer (F3) and medium (F2) filters were used. The TSS level dropped to 2 mg/L with the coarser (F1) filter.

Table 2. Particle count per 1 mL of sample and TSS of the crude oil and filtrates.

	Particle count per 1 mL of sample					
Tested medium	>5 µm	>10 µm	>20 µm	>50 µm	>100 µm	TSS [mg/L]
Unfiltered pyrolysis oil	2595	1420	599	91	14	69
Filtrate of coarse filter medium (F1)	7	4	1	0.8	0	2
Filtrate of F1 after 2 days storage	18	7	2	2	0.9	7
Filtrate of medium filter medium (F2)	5	3	1	0.4	0	<1
Filtrate of F2 after 2 days storage	17	6	1	1.5	3	4
Filtrate of fine filter medium (F3)	1	0.9	0.5	0.1	0	<1
Filtrate of F3 after 2 days storage	3	1	0.5	0	0	2

The three media had an overall solid removal efficiency > 97 %, and > 99% on particles greater than 5 µm, leaving only the finest particles in the filtrates. Furthermore, it was found that after 2 days, the TSS of the F1 (coarse) filtrate increased from 2 to 7 mg/L, of the F2 (medium) filtrate from <1 mg/L to 4 mg/L, and of the F3 (fine) filtrate from <1 mg/L to 2 mg/L. This observation suggests that re-agglomeration of finer particles occurred during storage [28, 29]. It is therefore advisable to perform filtration immediately prior to the final use of the purified pyrolysis oil. Furthermore, SEM-EDX analysis confirmed the re-agglomeration phenomena as illustrated in Fig. 2, which shows a SEM image of the TSS membranes immediately after filtration with the fine filter medium F3 and after two days storage. It can be clearly seen in the images that a larger quantity and bigger sized particles were retained on the TSS membrane two days after filtration using the fine medium F3.

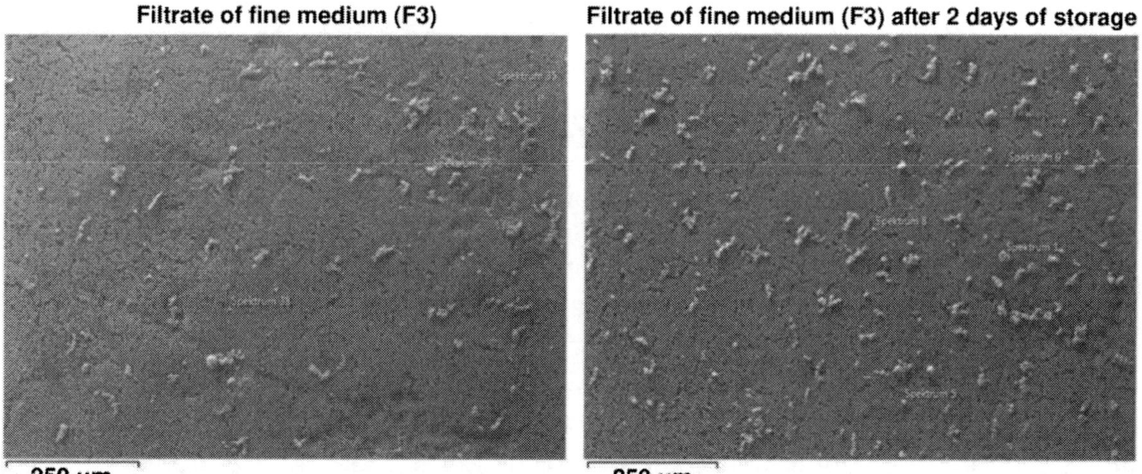

Fig. 2. SEM images of the TSS membranes of filtrate F3 directly after filtration and after 2 days storage.

Metal contaminants: Metal contaminants in steam crackers are unwanted for several reasons, however, the two most important reasons are coke formation/fouling and downstream catalyst poisoning. In terms of coke formation, iron plays an important role due to its catalytic activity facilitating coke formation [16, 30]. ICP-OES analyses were performed to detect the absolute contaminant concentrations in the respective filtrates (see Table 3).

Table 3. Absolute metal concentrations (in ppmw) in the unfiltered pyrolysis oil and the respective filtrates measured via ICP-OES.

Element	Unfiltered oil	Coarse filtrate (F1)	Medium filtrate (F2)	Fine filtrate (F3)
Na	18.83	10.29	6.98	1.72
Ca	16.75	13.25	5.07	2.08
Zn	8.54	7.19	8.54	1.31
Al	12.37	12.37	8.41	2.56
Fe	6.57	5.81	3.54	1.42
Cu	3.73	4.87	1.57	0.87
Si	2	1.38	<0.04	0.59
As	0.31	0.24	0.22	0.12
Sb	2.1	0.5	1.62	0.83
Mg	0.56	0.63	0.84	0.24
Ti	0.24	0.16	0.18	0.14
Co	0.11	0.35	0.78	0.09
Cr	1.81	1.89	2.06	1.65
Cd	0.17	0.15	0.15	<0.01
Mn	0.06	0.03	0.03	<0.01
Ni	0.15	<0.01	<0.01	<0.01
Pb	0.11	<0.01	<0.01	<0.01
Se	0.05	<0.01	<0.01	<0.01
Sr	0.11	0.13	0.04	<0.02
Tl	0.17	<0.01	<0.01	<0.01
V	0.1	<0.01	<0.01	<0.01
SUM	74.84	59.24	40.03	13.62

As indicated by the TSS analysis, the finer grade filter medium F3 performed best with a total metal removal efficiency of 81.8% compared to a removal efficiency of 46.5% (medium – F2) and 20.9% (coarse – F1). In terms of individual metals, it can be seen that the removal is not proportional for every element but that some metals are removed more efficiently than others. This shows that the metals can be present in different chemical forms. Results further show that with the fine filter medium (F3) calcium, iron and sodium were reduced to trace levels while the respective concentrations are almost unchanged when passing through the coarser filter medium (F1). Furthermore, most metals which were present in low concentrations in the unfiltered sample were reduced to below the detection level (cadmium, manganese, nickel, lead, selenium, strontium, thallium and vanadium) which is in the low ppb-range.
In terms of steam cracking it was found that the concentrations of some of the metal contaminants in filtrate of F3 (fine) still exceeded the industrial thresholds for steam cracker feedstocks; These are namely arsenic (0.12 vs. 0.005 ppm), calcium (2.08 vs. 0.5 ppm), iron

(1.42 vs. 1 ppm), and sodium (1.72 vs. 0.125 ppm). In terms of iron, mild blending of filtrate F3 with fossil steam cracker feedstocks such as naphtha would be sufficient to comply with feedstock specifications. For the other filtrates, blending with naphtha at a slightly higher mixing ratio would also suffice to comply with industrial specifications. Another important coke inducer is sodium which is also present in all filtrates at higher concentrations than the threshold value of 0.125 ppm [16, 31]. On the other hand, the coke promotor nickel was removed below the limit of detection. Calcium, which contributes to fouling to a great extent was reduced by all filter media substantially, however, the threshold value (0.5 ppm) is still (substantially) exceeded in all filtrates [16]. The concentration of silicon in the respective filtrates is close to the threshold value of 0.5-1 ppm in the cases of F3 (0.59 ppm) and F2 (<0.04 ppm), while the filtrate of F1 (1.38 ppm) slightly exceeds the specification. To further lower the contaminant concentrations, additional steps such as dissolving the pyrolysis oils to extract or precipitate out the extra fine contaminants could be performed.
In conclusion, the metal reduction shown in Table 3, while significant, is still not sufficient to fully comply with the maximum feedstock specifications for all elements. However, considering the likely scenario of co-feeding of pyrolysis oils with fossil feedstocks on an industrial scale, the most severe impacts by the metals still present could be leveled out by appropriate mixing ratios. Hence, filtration proves to be an efficient technique for reducing the content of metal contaminants in plastic waste derived pyrolysis oils.

Hydrocarbon composition: Table 4 shows the hydrocarbon composition of the respective filtrates in comparison with the crude pyrolysis oil as published in [17]. The results were obtained using comprehensive two-dimensional GC × GC coupled to FID. As it was expected, the results indicate that the filtration step did not have a considerable effect on the hydrocarbon matrix of the respective samples.

Table 4. Hydrocarbon composition of the respective filtrates and the crude pyrolysis oil (taken from [17]).

	Unfiltered pyrolysis oil [17]	Coarse filtrate (F1)	Medium filtrate (F2)	Fine filtrate (F3)
n-paraffins	31.97	31.71	32.20	32.38
Iso-paraffins	6.25	6.08	5.71	6.36
α-olefins	24.00	24.26	23.88	23.85
Iso-olefins	18.51	18.82	18.79	18.30
Linear dienes	3.53	3.27	3.20	3.28
Branched dienes	1.32	1.07	1.25	1.19
Naphthenes	12.34	12.36	12.69	12.46
Aromatics	2.06	2.44	2.29	2.18

Steam cracking performance: Fig. 3 depicts the most important steam cracking product yields methane, ethylene, propylene and 1,3-butadiene. As expected based on the hydrocarbon composition, the light product yields of the different filtrates and the unfiltered oil are very similar [17]. It can therefore be concluded that the filtration step had no significant effect on the steam cracking product yields.

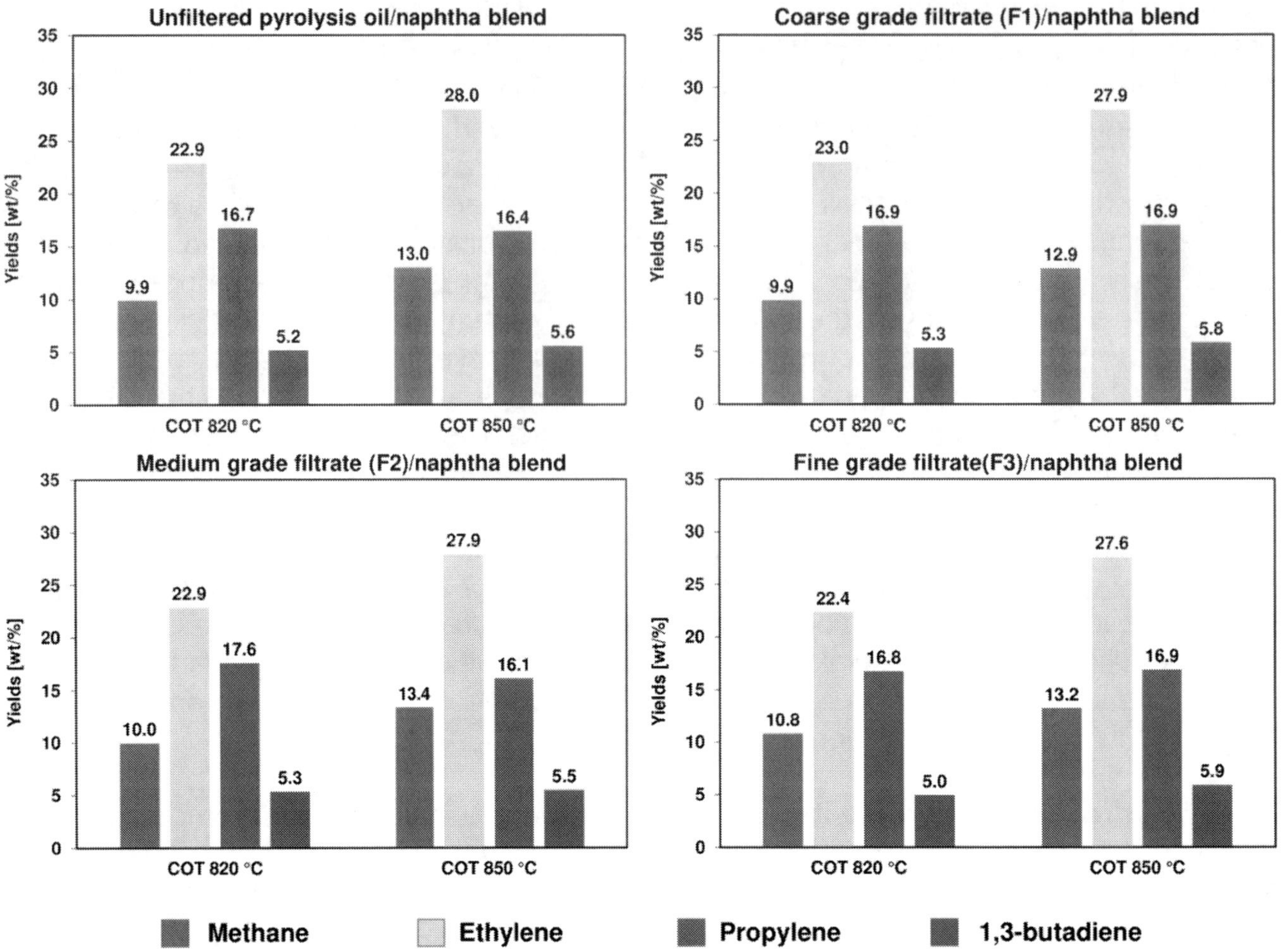

Fig. 3. Steam cracking yields of the most important products methane, ethylene, propylene and 1,3-butadiene measured using on-line GC × GC. Unfiltered pyrolysis oil/naphtha blend was cracked in a previous study [17].

Coke formation is a relevant factor for the time-on-stream or the so-called run-length of a steam cracking furnace. More frequent decoking cycles lead to reduced production capacities and hence economic losses, ultimately making pyrolysis oils unfeasible as steam cracking feedstocks. Especially metal contaminants can have a severe effect on coke formation due to catalytic effects (i.e., iron, nickel, sodium) [9, 32, 33]. It is therefore probable that next to the coking induction by active metal sites located at the inner surface of the reactor coil (material used in this study: Incoloy 800HT Ni: 30-35 wt%, Cr: 19-23 wt%, Fe: > 39.5 wt%), the initial coking phase can be accelerated by metal contaminants present in the feedstocks. Table 5 shows the radiant coil coking rates measured after the respective steam cracking experiments of the coarser and medium grade filtrate. To put the coke formation data into context, coke formation data from experiments performed at identical conditions of pure fossil naphtha and the untreated pyrolysis oil/naphtha blend are shown. Table 5 shows a significant reduction in the radiant coil coking rate after filtration of the pyrolysis oil with the coarse and medium grade filter media. Given the observation that the filtration steps did not influence the hydrocarbon composition of the pyrolysis oils, the metal contaminants can indeed be considered a strong driver for coke formation.

Table 5. Radiant coil coking rate of the filtrates compared to the unfiltered pyrolysis oil and pure fossil naphtha [17]. All pyrolysis oils were blended with fossil naphtha in a 1:3 weight-based ratio.

	Coking rate [mg/h]
Pure fossil naphtha [17]	14.0
Unfiltered pyrolysis oil/naphtha blend [17]	33.5
Coarse grade filtrate (F1)/naphtha blend	20.8
Medium grade filtrate (F2)/naphtha blend	16.5

Furthermore, it can be seen that the coking tendency of the medium grade filtrate (F2) is lower than that of the coarse filtrate (F1) and almost in the same ballpark as the pure fossil naphtha. Nevertheless, there is still a higher coking rate when using 25 wt.% of pyrolysis oil regardless of the contamination. However, an experimental uncertainty of 10-20% in terms of the measured coking rate needs to be considered. It is important to note that the coke formation was observed in the initial coking phase in experiments of limited duration and no long-term studies were performed. Hence, additional long-term studies at a larger scale are required to conclusively evaluate the steam cracking feasibility of the filtered products. Furthermore, fouling issues in the convection and transfer line exchanger sections are likely to occur when cracking pyrolysis oils due to their heavy nature and crucial contaminants such as calcium or silicon. These effects were not investigated in this study. However, the results show that filtration of pyrolysis oils for the removal of particles leads to a considerable reduction in radiant coil coke formation and hence to an improved applicability of pyrolysis oils in the context of advanced recycling.

Conclusions and Outlook

In this work, a detailed depth filtration study of a representative contaminated post-consumer plastic waste pyrolysis oil was performed including the thorough analysis of the filtered products to assess the impact of filtration on the steam cracking performance. ICP-OES analyses confirmed that the used filter media were effective in reducing the particulate contamination of the pyrolysis oil depending on the fineness of the used filter media. While the coarse filter medium led to a reduction of ~21 % of the metal contaminants, the finest medium led to a metal reduction of ~81 %. At the same time, the hydrocarbon composition of the pyrolysis oils was not significantly affected by the filtration step as confirmed by comprehensive two-dimensional GC × GC. The steam cracking performance of the filtered pyrolysis oils blended with naphtha was investigated in a continuous bench-scale steam cracking unit and compared to the unfiltered pyrolysis oil. Very similar steam cracking product yields were obtained for all pyrolysis oil/naphtha blends. This finding was expected based on the almost identical hydrocarbon compositions of the pyrolysis oils before and after filtration. On the contrary, the radiant coil coke formation was reduced by 40-50% compared to the unfiltered pyrolysis oil.

This study shows that depth filtration is an efficient purification technique for contaminated plastic waste pyrolysis oils, substantially increasing the steam cracking feasibility towards closed-loop recycling. Based on the bench-scale investigation, technical filtration systems at a larger scale could be designed using multi-stage filtration and continuous processing of the filtered products.

Acknowledgment

The authors acknowledge financial support by the Catalisti SBO project WATCH (Plastic Waste to Chemicals). Furthermore, Kevin M. Van Geem is holder of the ERC Grant OPTIMA (Process Intensification and Innovation in Olefin Production by Multiscale Analysis and Design) with the grant agreement ID 818607.

References

1. Plastics Europe, *Plastics - The Facts 2020*, 2020.
2. M. Solis, et al., *Waste Manage.*, 2020, **105**, 128-138.
3. K. Ragaert, et al., *Waste Manage.*, 2017, **69**, 24-58.
4. R. Palos, et al., *Energy Fuels*, 2021, **35**, 3529-3557.
5. M. G. Davidson, et al., *J Clean Prod*, 2021, **293**, 126163.
6. H. Jeswani, et al., *Sci Total Environ*, 2021, **769**, 144483.
7. K. M. Van Geem, et al., *Science*, 2019, **364**, 734.
8. K. M. Van Geem, et al., *MRS Bull.*, 2022, DOI: 10.1557/s43577-021-00247-5.
9. M. Kusenberg, et al., *Waste Manage.*, 2022, **138**, 83-115.
10. M. Roosen, et al., *Environ. Sci. Technol.*, 2020, **54**, 13282-13293.
11. S. Ügdüler, et al., *Waste Manage.*, 2020, **104**, 148-182.
12. J. Murphy, *Additives for Plastics Handbook*, Elsevier Science, 2 edn., 2001.
13. A. Turner, et al., *Environ Int*, 2021, **156**, 106622.
14. G. M. Zhang, et al., *Adv. Mater. Res.*, 2011, **174**, 486-489.
15. M. Kusenberg, et al., *Fuel Process. Technol.*, 2022, **238**, 107474.
16. K. M. Sundaram, et al., presented in part at the AIChE 2018 Spring National Meeting, Orlando, Florida, 22-26 April 2018.
17. M. Kusenberg, et al., *Waste Manage.*, 2022, **141**, 104-114.
18. A. Zayoud, et al., *Waste Manage.*, 2022, **139**, 85-95.
19. M. Kusenberg, et al., *Fuel Process. Technol.*, 2022, **227**, 107090.
20. M. S. Abbas-Abadi, et al., *J. Anal. Appl. Pyrolysis*, 2022, **166**, 105614.
21. M. R. Harper, et al., *Combust. Flame*, 2011, **158**, 16-41.
22. R. De Bruycker, et al., *Ind Eng Chem Res*, 2014, **53**, 18430-18442.
23. K. M. Van Geem, et al., *Ind Eng Chem Res*, 2005, **44**, 3402-3411.
24. K. M. Van Geem, et al., *Chem. Eng. J.*, 2007, **134**, 3-10.
25. S. P. Pyl, et al., *J. Chromatogr. A*, 2011, **1218**, 3217-3223.
26. S. P. Pyl, et al., *Chem. Eng. J.*, 2011, **176-177**, 178-187.
27. K. M. Van Geem, et al., *J. Chromatogr. A*, 2010, **1217**, 6623-6633.
28. J. Fritzsche, et al., *Adv. Eng. Mater.*, 2013, **15**, 1299-1306.
29. S. Pischinger, et al., presented in part at the Internal Combustion Engine Division Spring Technical Conference (ICES), 2003.

30. R. Orriss, presented in part at the 8th Annual Ethylene Producers' Conference, New Orleans, Louisiana, 1996.
31. M. Brayden, et al., presented in part at the AIChE Spring National Meeting, Ethylene Producers Conference, Orlando, Florida, April 23 – 27, 2006.
32. F. D. Kopinke, et al., *Carbon*, 1988, **26**, 117-124.
33. M. J. Bennett, et al., *J Mater Sci*, 1981, **16**, 170-188.

60b Impact of Recycled Plastic Composition on Honeywell's Upcycle Process Pyrolysis Oil for Steam Cracking Units

Chris Anderle

To increase the percentage of recycled plastic used in consumer packaging, additional methods of recycling beyond mechanical recycling are required. Mechanical recycling is effective, but often results in product downgrading and typically only applies to rigid HDPE, PP, and PET. To increase recycling rates, we need to find solutions for flexible and film-based plastics. Depolymerization or pyrolysis is one of those solutions. UOP's pyrolysis process for recycled plastics is called the UpCycle process.

For the implementation of the UpCycle process, we face many challenges – from the feedstock acquisition all the way to the product disposition. The available feedstocks need to be sampled, sorted, cleaned and sized for the pyrolysis process. The pyrolysis process needs to be robust enough to handle feedstock impurities and still provide the desired product – Recycled Polymer Feedstock (RPF) - for the downstream steam crackers. The desired RPF quality is highly dependent on the plastic feedstock sorting and cleaning steps. Our UpCycle process is inclusive of product clean up and fractionation, if necessary, to meet the customer boiling range and key product specifications. The light gases generated from the pyrolysis reactor are routed to the unit's fired heater and used as fuel.

AICHE PAPER NUMBER 60C

THE CHALLENGES BEHIND CARBON DISULFIDE REMOVAL IN NAPHTHA-DERIVED CUTS

Henrique Silva
Regional Engineer - Latin America
Veolia Water Technologies & Solutions

Tatiana Barbosa
Industry Subject Matter Expert
Veolia Water Technologies & Solutions

Prepared for Presentation at the 2023 Spring Meeting,
March 12-16, 2023

THE CHALLENGES BEHIND CARBON DISULFIDE REMOVAL IN NAPHTHA-DERIVED CUTS

Henrique Silva
Regional Engineer - Latin America
Veolia Water Technologies & Solutions

Tatiana Barbosa
Industry Subject Matter Expert
Veolia Water Technologies & Solutions

Abstract: A known contaminant in naphtha feedstock to olefins' crackers is carbon disulfide (CS_2). Sometimes, this sulfur compound is imperceptible to ethylene units and its harmful effects are not often discussed by producers. However, for downstream cuts, particularly to C5 processing, where CS_2 tends to aggressively concentrate, it may be a very difficult challenge to handle. Over the past decade, Veolia has been working on how to protect operation assets from carbon disulfide byproducts' corrosion and fouling mechanisms. Although there are known technologies in the market targeting carbon disulfide decontamination, several factors must be considered such as the cost of solution as well as its compatibility to the process. This paper shares the experience to date, reviewing the effects of high carbon disulfide concentration in hydrocarbon medium, the challenges in dealing with this, and important operational events observed along this period.

Introduction

The global scenario affecting the petrochemical industry has been challenging producers to respond to a huge variety of feedstock quality in ethylene and its downstream plants which recovers olefins' byproducts. One of the concerning features about feedstock composition is sulfur concentration and its different species. Amid sulfide compounds, carbon disulfide (CS_2) is often dissolved in liquid feeds and, although its usual concentration is a bit low in naphthas, when the cracked gas becomes purified, CS_2 tends to concentrate in the pygas fractions, especially the C5 cut because its volatility is close to this cut's compounds. Highly concentrated, CS_2 can promote undesirable reactions, particularly to basic-nitrogen compounds, which can lead to fouling mechanisms as well as corrosion.

In some experiences, Veolia has helped to make the production of C5 derivates almost free from CS_2. Not only is CS_2 a problem in itself; there are also consequential issues due to its decontamination, particularly using a chemical scavenger, that should be considered if CS_2 is required to be eliminated from the process. Veolia has used a combination of different technologies to remove CS_2 in various units. Although they can enable plants to run at full capacity, operational issues have been experienced in some units over the past decade and more, making the desired run-length difficult to achieve.

This paper discusses how techniques were employed to remove CS_2, focusing on chemical scavenger injection and what problems are still being faced given that not all questions have been answered. This paper also aims to share information about carbon disulfide with the petrochemical community and stimulate discussions to better understand the challenges it presents as a contaminant.

Although processing C5 cuts is naturally more sensitive to carbon disulfide concentration, this paper will discuss the decontamination mechanism, with special attention to chemical scavengers, and its collateral effects that should be handled and considered regardless of the production process.

How carbon disulfide can be removed

Different technologies can be used to decontaminate a hydrocarbon stream from CS_2. Mainly, three technologies are often employed to remove it at different stages in the unit:

- Cationic resin
- CS_2 Adsorption
- Chemical injection of CS_2 scavenger

Due to efficiency, practicality, and cost, a CS2 scavenger is usually the preferred technology to eliminate most of the weight of carbon disulfide from hydrocarbon medium.

Since the early 2010s, Veolia has been developing and studying CS_2 scavengers to allow CS_2 decontamination. Chemical scavenging easily removes CS_2 if properly dosed and given enough residence time, the expected effect can be immediate. Then again, without precautions, impacts can be catastrophic.

The injection of CS_2 scavenger has the purpose to turn CS_2 into an organic sulfur-nitrogenous salt (SNS) that can be water washed and discarded in the aqueous effluent. In a common design, the scavenger is injected into a hydrocarbon rich in CS_2, passing through a mixer which enables enough residence time to scavenge CS_2, turning it in a salt. Demineralized water is used to wash this salt out from the system and a lean CS_2 stream can go to downstream applications, meeting specification. The CS_2 scavenger has a dosage controlled by the target concentration of CS_2 in the outlet hydrocarbon. The water fed has the property to wash salt produced by scavenged CS_2 and leaves through the bottom of the column, and some light compounds can be recovered by following distillation (Frac Column). Figure 1 shows a simplified sketch of the application described.

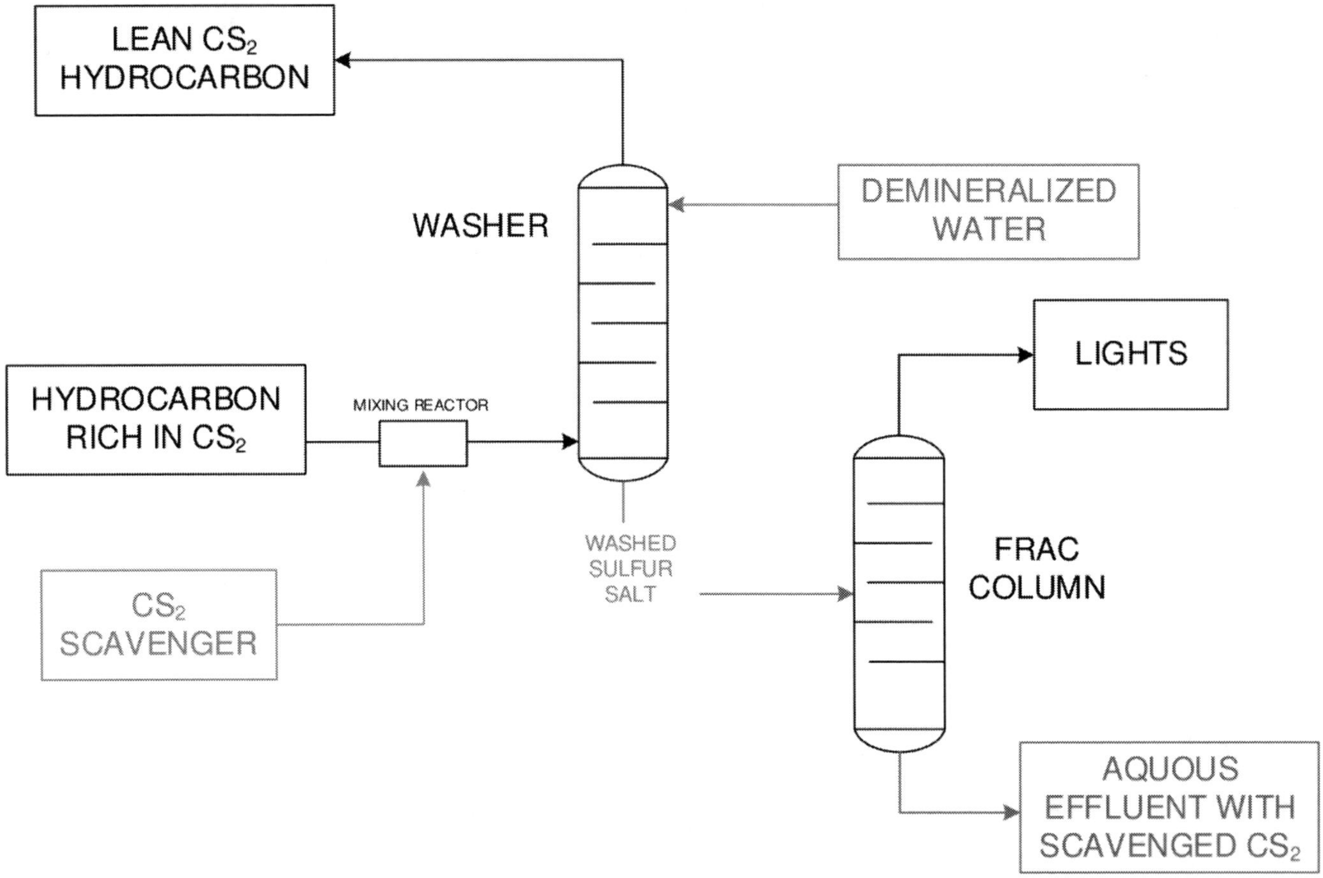

Figure 1. Simplified model sketch of the CS_2 scavenger application

The harmful corrosion effects of the scavenging application

In C5 cuts, for example, it is common to deal with corrosive species intrinsic to the design, such as carbamates, which is a derived compound from nitrogen-bases and carbon dioxide

(CO_2). Similar to this, there are thiocarbamates, which derive from CS_2 instead of CO_2. A strong base, such as caustic, is often used to minimize corrosion from those species, by controlling pH, and the success rate is often acceptable.

Caustic is the most common acid-salt neutralizer. It can be injected into the solution to react with the acid-salt to form a sodium salt. At low concentration, caustic provides a very cost-effective neutralization.

What was observed in Veolia's experience was that the caustic used into the feed to the frac column was not enough to meet specified corrosion rate limits. Historically, pH control is often successful in preventing carbamates corrosion, but not always. As a result, the first symptom noticed is the column pressure drop decreasing to values unnaturally low (Figure 2), which could be linked to corrosion of the column's internals. Also, total iron analysis in frac column bottom samples could range from 20 to 45 w ppm; too high in comparison with the desired < 5 w ppm in this case.

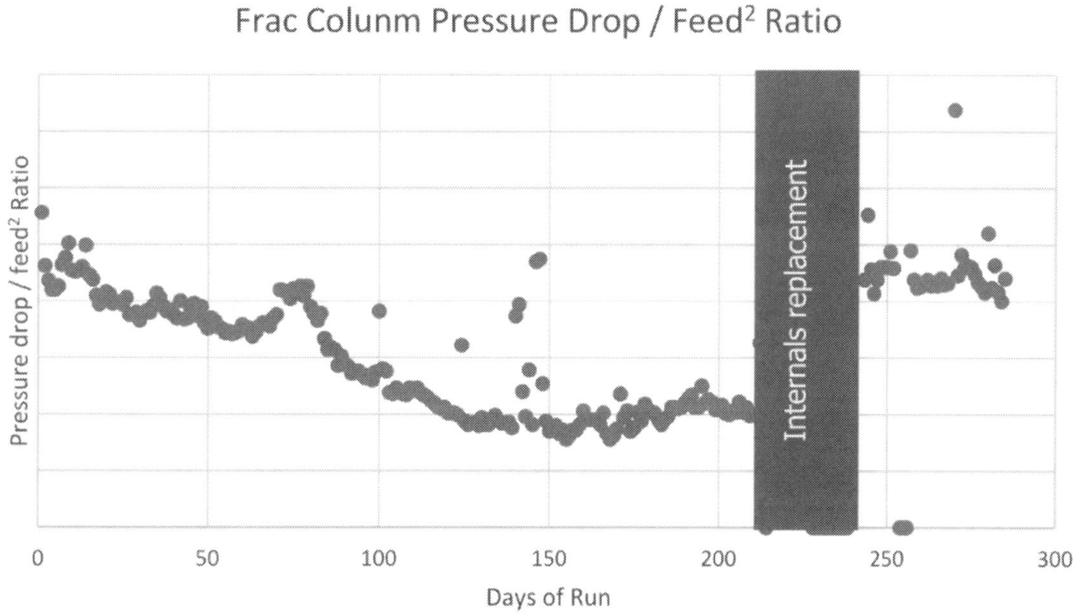

Figure 2. Corrosion impact on the frac column's pressure drop.

Internal inspections would then reveal extreme corrosion over the frac column's internals. The mechanism would be clearly limited to the liquid phase below the feed tray: trays above the feed are seen intact, while the trays of the bottom section are totally gone. The corrosion is severe enough to dissolve tray metal material into the aqueous medium.

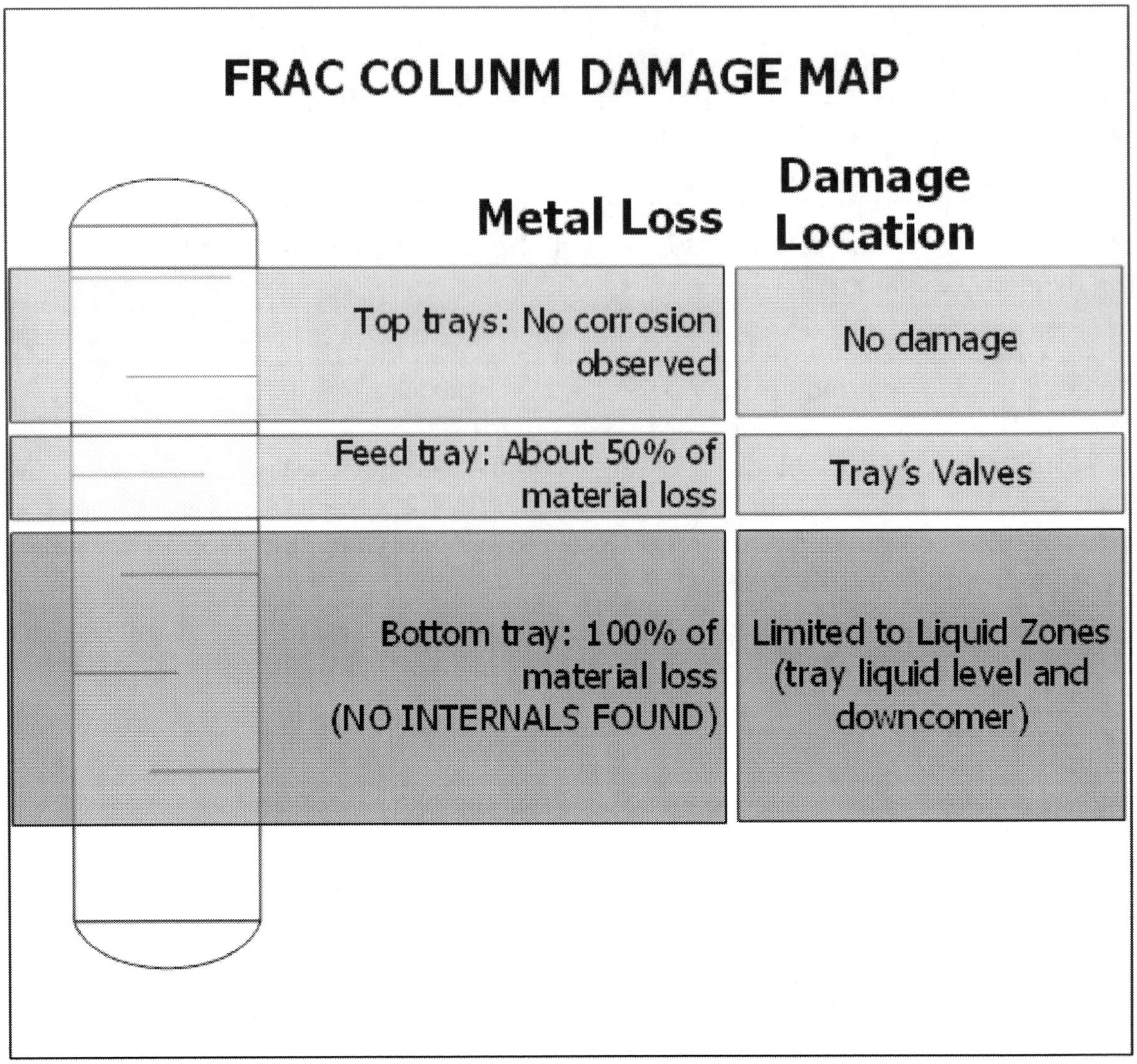

Figure 3. Corrosion severity over frac column.

Within only 200 days of run, affected plants would shut down to replace internals and carbon steel coupons were installed on the inlet and outlet streams. Uniform corrosion rates pointed values above 20 mpy in the bottom stream, but when the real rate was estimated from loss of thickness of corroded internals from the column, values reached much higher levels, up to ten times the measured value.

Several mechanisms were discussed to describe which phenomena could be so aggressive as to disintegrate so much of the column's internals in just a few months. Coupons and corrosion probes in the feed and the bottom streams presented black and adherent deposit coating over the surface, but with distinct corrosion rates (Figure 7).

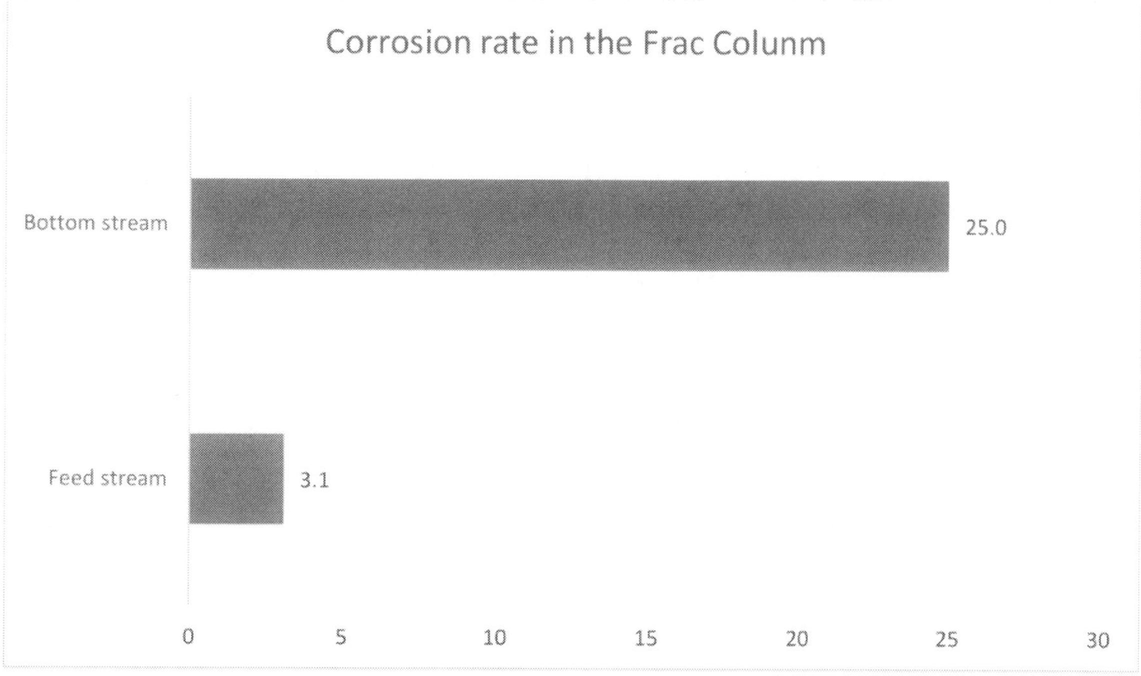

Figure 4. Typical corrosion rates from the frac column.

The main difference between the washer column conditions and the downstream process in these problematic applications was the operating temperature. In the frac column, the bottom reaches a temperature over 100°C, which is much higher than the temperature under washing conditions.

Usually, the scavenged forms of CS_2 are thiocarbamic acid derivatives like salts or even thiocarbamates. Some literature suggests that both forms, when submitted to heat, can be thermally decomposed, releasing hydrogen sulfide (H_2S) as shown in Figure 5. However, no gas phase showed corrosion issues along the tower and pH was previously controlled with caustic to neutralize any H_2S formation. Therefore, a clear explanation to the mechanism of corrosion in these systems remain elusive. Nonetheless, it has a high potential of being related to a thermally induced process.

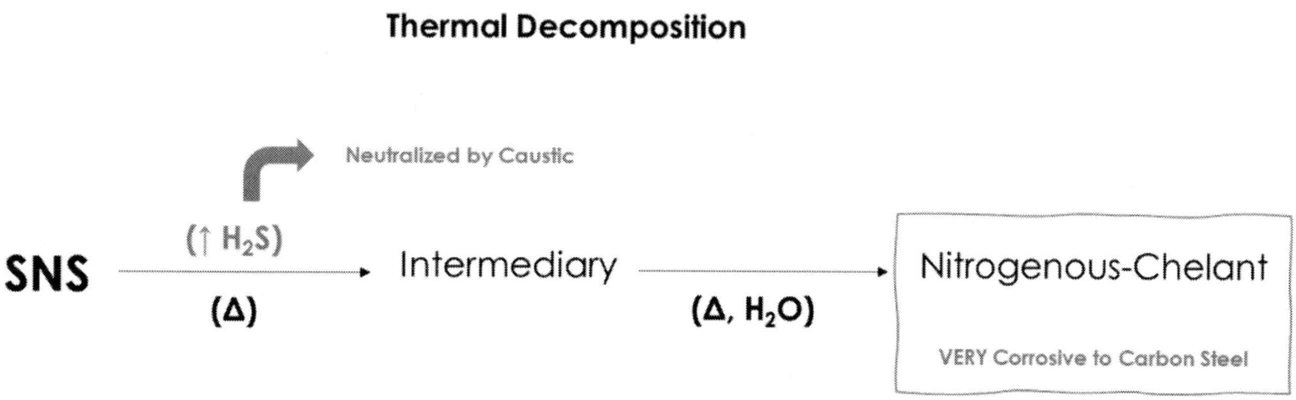

Figure 5. Thermal decomposition of SNS

Another hypothesis was that a chelant-family chemical could be formed under operation conditions. Chelants are strong sequestrants of metals, making carbon steel alloy vulnerable. In aqueous solution with excess of nitrogenous scavenger and under high temperature, thiocarbamate can become a nitrogenous-complex compound with strong chelation properties.

Apparently, stainless steel can respond better than carbon steel and no relevant reaction have been observed in stainless steel alloy.

The organosulfur reaction produces fouling

By the time corrosion was noticed in these systems, black deposit fouled the corroded areas of the frac column, as well as the upstream system, particularly in the bottom of the washer column. Although the deposition in the column did not provoke corrosion, its volume was great enough to necessitate cleanings during turnarounds (Figure 6). Veolia introduced a second generation of CS_2 scavenger, which formed lower molecular weight, thus more soluble, organosulfur reaction products. This reduced the corrosion and fouling tendencies, without sacrificing the efficiency of the scavenger reaction.

The typical deposit from this mechanism is mainly composed of an organosulfur chain. Fouling tendency is mostly related to the CS_2 concentration, the excess amount of the scavenger, and the water/hydrocarbon ratio. The greater the amount of water available to solubilize the reaction products, the lower will be the deposition rate. Also, the targeted CS_2 specification will determine how much scavenger will be needed, and hence, the amount of produced reaction organosulfur salts as well.

Figure 6. A fouled exchanger from scavenging CS_2 reaction products.

Recent years' inspections, however, have shown that water wash was not enough to prevent deposition in the column. Quite the contrary, fouling has been seen to intensify on the water/hydrocarbon interface at the bottom of the column, where water concentrates. This would suggest that deposits are not behaving as a soluble product that can be washed in water.

Recent detailed deposit analysis revealed the salt's properties have changed, making it insoluble in water, strengthening the theory of an oxidizing mechanism occurring in the washer, decomposing the salt just like the thermal decomposition does, shown in Figure 7.

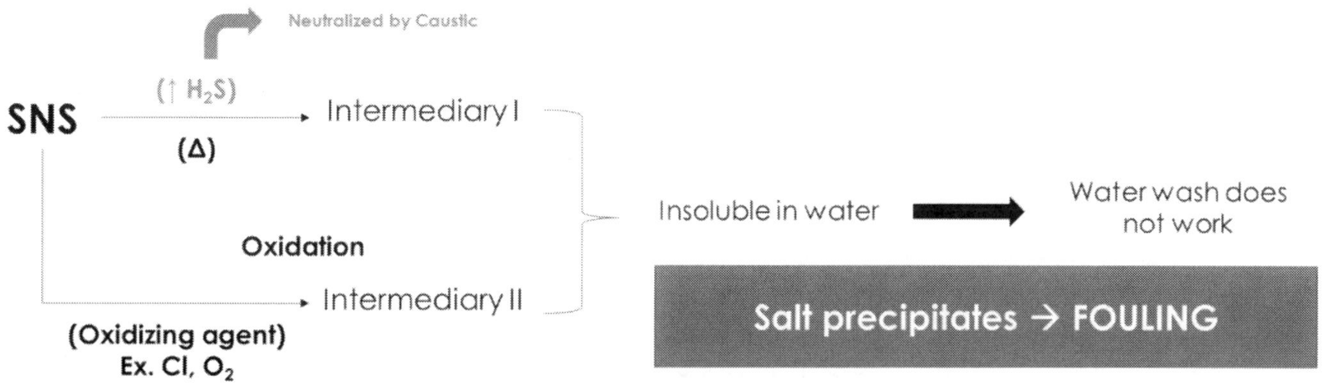

Figure 7. Mechanisms to form insoluble residues.

Free-radical polymerization of dienes was suspected but gas-chromatography of volatilized deposit samples found no evidence of polyolefinic material.

There are some hypotheses as to how an oxidizing mechanism, or perhaps even a secondary reaction, could occur, enabling changes in the salt's solubility in water such that it cannot be extracted from the washer column. An understanding of the wash water and hydrocarbon chemistry could well be the key to avoiding undesirable reactions beyond scavenging carbon disulfide.

A summary of scavenging process issues and challenges

Figure 8 summarizes the points of concern when a scavenging technology is used to control CS_2 decontamination.

Washing the salt is mandatory to prevent fouling. The salt must also be discarded in some way that avoids decomposition so that corrosion of incompatible metallurgy does not occur.

Optimizations of wash water feed rate and scavenger dosage are needed to ensure that salt production is minimized and washing potential is maximized.

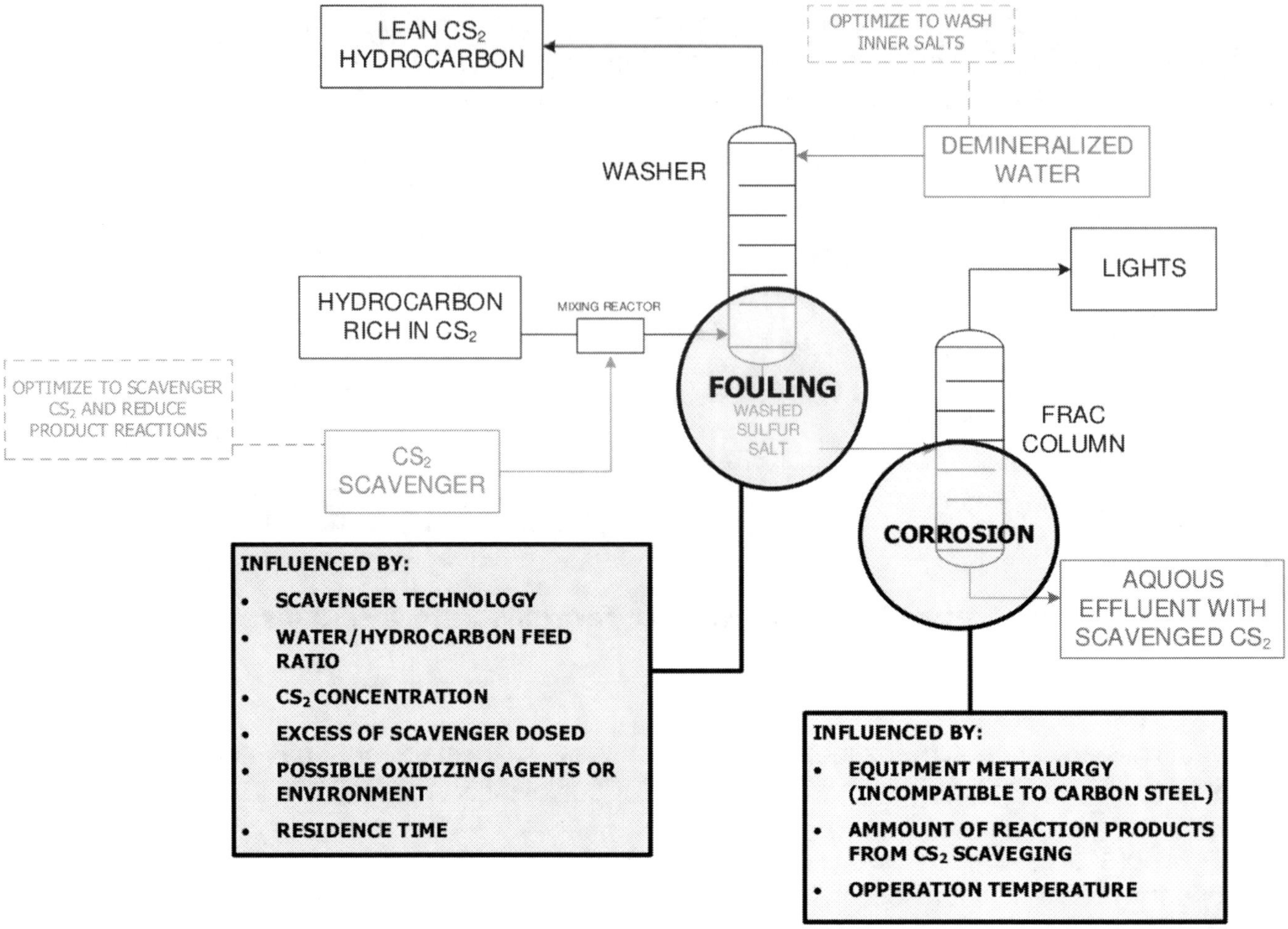

Figure 8. Main issues from scavenging consequences

Conclusion

There is still much work to be done to better understand all the mechanisms for the fouling and corrosion issues derived from a CS_2 scavenger. So far, what is known is that a CS_2 scavenger is very useful, but it must be applied with great caution. The scavenger chemical selection must consider not only the carbon disulfide removal efficiency, but also the reaction products' properties and their consequences to the process. It is strongly recommended that the system avoids thermal decomposition and an oxidizing environment to enable the scavenger to operate with better performance, reducing or mitigating corrosion and fouling issues.

References

1. Kirk-Othmer 5th ed. "Encyclopedia of Chemical Techonolgy". Vol 4. pages 365 to 384
2. Robert A. Donia et al. "Reactions of mono- and di-amines with carbon disulfide. I. N,n'-dialkylethylenediamine-carbon disulfide reactions". Jan. 1948
3. Weiss, Noah. "Investigation of Isoprene Process Deposits". May 2021

60c The Challenges behind Carbon Disulfide Removal in Naphtha-Derived Cuts

Henrique Silva, Tatiana Barbosa

A known contaminant in naphtha feedstock to olefins crackers is carbon dissulfide. Sometimes, this sulfur compound is imperceptible to ethylene units and its harmful effects are not often discussed by producers. However, for downstream cuts, particularly to C5 processing, where CS_2 tends to agressively concentrate, it may be a very difficult trouble to handle. Over the past decade, Braskem & Veolia has been working in how to protect operation assets from carbon dissulfide byproducts mechanisms related to corrosion and fouling. Although there are known technologies in the market aiming carbon dissulfide decontamination, several consequences must be considered as the cost of solution as well as its compatibility to the process. This parper has the purpose of share both companies experience lived to date, reviewing the effects of high carbon dissulfide concentration in hydrocarbon medium, the challenges to deal with this and important operational events observed along this period.

SULFUR CHEMICALS FOR ETHYLENE PRODUCTION

Jonathan W. VanScoyoc
Americas Account Manager
Chevron Phillips Chemical Company

Kevin Kabell
Senior Olefins Technical Support Engineer
Chevron Phillips Chemical Company

Prepared for Presentation at the 2023 Spring National Meeting
Houston, TX, March 12–16, 2023

SULFUR CHEMICALS FOR ETHYLENE PRODUCTION

Jonathan W. VanScoyoc
Americas Account Manager
Chevron Phillips Chemical Company
Kevin Kabell
Senior Olefins Technical Support Engineer
Chevron Phillips Chemical Company

Abstract:

Sulfur chemicals, primarily dimethyl sulfide (DMS) and dimethyl disulfide (DMDS), have long been reliably and safely used as sulfiding agents to generate the needed H_2S utilized in the production of ethylene. The addition of sulfur chemicals in the feedstock being cracked stabilizes the cracking furnace radiant coil inner diameter surfaces which reduces the formation of coke and CO that adversely impacts furnace operational run lengths and some acetylene removal operations. Sulfur chemicals allow operators to maximize furnace utilization rates and profitability for their overall production unit.

In North America, both DMS and DMDS are produced domestically as either by-products or as intermediates within other dependent chemical processes, or they are imported from Europe or Asia. In the last 5 years, plentiful advantaged feedstocks and increased global demand for polyethylene have driven significant investments that have expanded ethylene production in North America. At the same time, changes in unrelated markets and technology are impacting the traditional supply balance of the by-product or intermediate sulfur chemicals. In addition, challenges with global supply chains also continue to occur. This paper discusses the dynamics driving uncertainties in the supply of these critical sulfur chemicals as well as Chevron Phillips Chemical's effort to evaluate the sulfur chemical dimethyl sulfide (DES) as a potential candidate.

Biography

Jonathan VanScoyoc is currently the Americas Account Manager for Chevron Phillips Chemical supporting their Sulfides, Fuels, and Sulfolane/ene products. He has over 25 years of experience in the industry providing technical, developmental, supply chain, and sales support for sulfur chemicals used in both the ethylene and refinery industry. Jonathan has a B.S. Degree in Business Logistics from Penn State University and a Finance MBA from Temple University.

Kevin Kabell is currently a Senior Olefins Technical Support Engineer for Chevron Phillips Chemical. His support focus is in olefins plant support, capital project development and support, and in process safety enhancement. He has over 26 years of technical and operational experience in the ethylene industry as part of Chevron and Chevron Phillips Chemical companies in multiple domestic and international locations. Kevin has a B.S. Degree in Chemical Engineering from Texas A&M University.

Introduction: Sulfiding Applications

The use of sulfur chemicals as sulfiding agents has long been established as critical for steam crackers in the production of ethylene. At the conceptual core, a stable metal–oxide layer formed upon startup or after decoke is one of the critical elements of a reliable furnace cracking operation. There are numerous past available papers amongst the EPC literature and other technical forums that can be referenced with greater details on their application usage. A basic background is included below to provide a foundation and some basic background for the application. A crucial factor that affects pyrolysis furnace performance is the deposition of carbon in the hydrocarbon boundary layer in a typical tube/coil wall (Figure 1). In a clean coil, coke forms as a result of surface reactions initiated kinetically by the interior wall temperature. A stable metal–oxide layer (e.g., chrome oxide, aluminum oxide) helps to stabilize the formation of catalytic coke against the metal wall of the coil tube. Sulfiding agents promote a thicker and even initial layer of controlled coke to minimize the catalytic coking impact with a more rapid shift to thermal coking. Many steam cracker plants continue to add sulfur chemicals after an initial sulfiding to maintain the optimal amount of sulfur required to inhibit radical induced coke; however, some plants discontinue sulfiding after the initial layer is established.

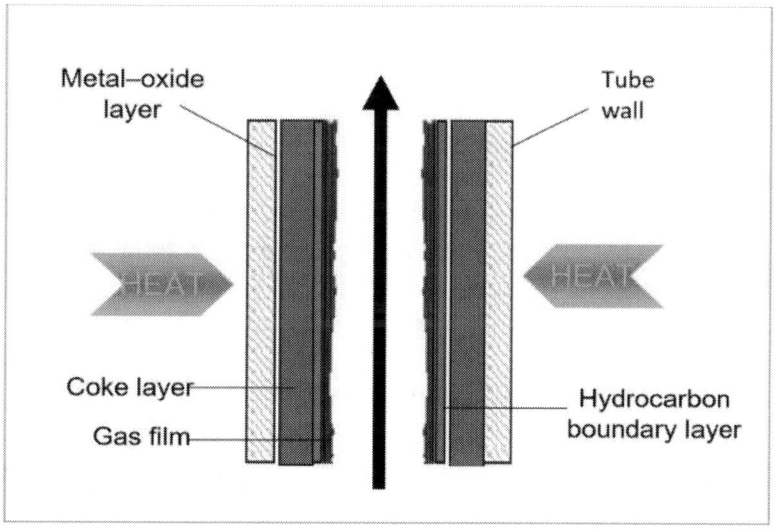

Figure 1. Process Gas Flow.

Sulfiding applications can vary with the life of the tube as chrome or other oxide-forming metals are lost to boundary layer decoking cycles as the coil ages. Coking stability yields multiple benefits to an operation, including the reduction of spalling risk on older aged furnace coils, longer furnace cycles between decokes, a more ratable carbon monoxide (CO) concentration in furnace effluent, and, perhaps, even higher charge rates or conversion to a furnace.

Sulfur Chemicals in North America

The four primary sulfur chemicals available in North America used as sulfiding agents for ethylene production are dimethyl disulfide (DMDS), dimethyl sulfide (DMS), ethyl mercaptan (EtSH), and sulfurized olefins (TBPS 454); see Table 1 for key properties of each chemical. Hydrogen sulfide (H_2S) is the one consistent building block for all of these sulfur chemicals, providing some circular usage. In the ethylene and refinery market, DMDS, DMS, EtSH, and TBPS 454 were developed to safely provide clean reliable H_2S into those processes for use, especially in cases where sour feeds are not directly available. Chevron Phillips Chemical is unique because it not only produces ethylene, but also currently or previously manufactures all four of these sulfur chemicals.

Table 1. Sulfiding Agent Properties

Property	**DMDS**	**DMS**	**EtSH**	**TBPS 454**
Sulfur % (weight)	68%	52%	47%	54%
Density (pounds/gallon)	8.9	7.11	7.03	9.0
Typical Purity	99.0%	99.0%	99.0%	N/A
Boiling Point (°F)	229°F	99°F	95°F	350°F
Flash Point (°F)	59°F	-36°F	-55°F	217°F
Vapor Pressure at 70°F (PSI)	0.45	8.1	8.5	0.01
Decomposition Temperature (°F)	1150	1470	1300	625
Viscosity at 70°F (mPas)	0.62	0.285	0.32	12.8

DMDS, dimethyl disulfide; DMS, dimethyl sulfide; EtSH, ethyl mercaptan; H_2, hydrogen; mPas, millipascal seconds; PSI, pounds per square inch; TBPS 454, sulfurized olefins (polysulfides).

When selecting a sulfur chemical for ethylene production, four factors are typically considered:

1. Ease of use for storage, handling, and transporting the chemical
2. Safety implications and community requirements
3. Effectiveness in the furnace coil, particularly its use in the production of ethylene across the full coil life
4. Business dynamics regarding availability and cost

If we want to follow the principle that the market gets it correct, then the ease of use should also properly align with that as well. In the following sections, DMDS, DMS, EtSH, and TBPS 454 are ranked according to these factors. It is important to note that the rankings are specific to the North American market. The rankings would be different in Europe and Asia, for example, where DMDS is the predominant sulfur chemical used due to its widespread availability.

There is no perfect sulfur chemical for use. There have been several opinions by experts in the industry on why one is better than another. There have been informative papers and trials at specific plants demonstrating potential better performance of one chemical versus another, but it is critical to remember every plant is different and individual trials should be completed.

#1: Dimethyl Sulfide

DMS is the most popular sulfur chemical in the North American sulfiding market. We have it ranked as 4/5 on the ease-of-use scale. It is estimated that approximately 55% of ethylene production units in North America use DMS as a sulfiding agent. The industry is experienced at handling DMS safely, and its unique properties enable it to perform well, particularly for ethane feedstock crackers. With its low boiling point and lighter density, it mixes well with a feedstock stream. This generally allows DMS to be injected at one point as it disperses well with the feedstock. DMS decomposes completely to H_2S at 1470 °F, which makes it a good match for a standard cracking furnace. Historically, DMS has been readily available and economical.

DMS does have several disadvantages. Due to its flammable and high vapor pressure properties, DMS must be transported in MC-331 pressure trailers and stored in pressure tanks under a nitrogen blanket. It also has a very strong odor. Finally, there is the well-known issue of carbon disulfide (CS_2) impurities not decomposing, which will be discussed in further detail later.

#2: Dimethyl Disulfide

DMDS is the second most popular sulfur chemical in North America, and we have it ranked 3/5 on the ease-of-use scale. It is estimated that 40% of the sulfiding market uses DMDS as its sulfiding agent. The primary advantage of DMDS is its high sulfur content (68%), the highest of the four sulfur chemicals; therefore, it is more efficient and less of it is required. DMDS decomposes at a lower temperature (1150 °F), making it suitable for use in a standard cracking furnace, and has less of an odor compared to DMS. DMDS does not have the CS_2 impurity issues that occur with DMS, although CS_2 is generated during decomposition. DMDS, however, does share some of the same disadvantages. Most notably, it is also rated as flammable and must be stored in a pressure vessel under nitrogen. Although not required, transportation in MC-331 pressure trailers is recommended. In addition, both DMDS and DMS will attack rubber and nitrile elastomers, so it is critical that proper gaskets and other materials are compatible.

The main challenge with DMDS is its higher boiling point and heavier density, which inhibits it from potentially dispersing well with a feedstock, particularly ethane. Generally, it needs to be properly injected at a point closer to the furnace, so it is better dispersed in the feedstock. Most plants are traditionally designed for DMS, so this may lead to different engineering requirements. DMDS has historically been at a higher cost compared to DMS, although it is important that any comparison be done on a sulfur content basis. Both DMS and DMDS use methanol as a key raw material. With the growing potential of green methanol, both products could improve their sustainability product cycle and lower their environmental impact in the future.

#3: Sulfurized Olefins (Polysulfides)

Sulfurized olefins (TBPS 454), or polysulfides, are less widely used in North America. There are several sulfur content–level polysulfides on the sulfiding market; the focus here is on the 54% sulfur polysulfide, the highest sulfur polysulfide on the sulfiding market and the most suitable for this application. At the time of writing, there is just one cracker plant in North America that uses polysulfides. Polysulfides are not classified as flammable and have a lower US Department of Transportation (DOT) regulation rating; therefore, they can be transported and stored with standard equipment. They also a have low odor.

Polysulfides have several disadvantages when it comes to use. Like DMDS, they have a heavy density and thus do not potentially mix as well with feedstocks depending on the location of injection point and equipment. Polysulfides are viscous at low temperatures, which may make them difficult to use in colder climate conditions. Since they are mixtures and not a single molecule, they can also be unstable, and, over time, sulfur may drop out and crystalize. Finally, polysulfides generally cost more than DMS and DMDS.

#4: Ethyl Mercaptan

Like sulfurized olefins, ethyl mercaptan (EtSH) is less widely used in North America than DMS and DMDS and generally more costly. At the time of writing, it is used at just 2 to 3 cracker plants in North America. With a thermal decomposition temperature of approximately 1300 °F and a light density that mixes well with feedstocks, it performs well. The biggest challenge with EtSH is handling and storage. Like DMS and DMDS, it must be stored in a pressure vessel under nitrogen and utilize specific compatible friendly materials. EtSH also requires transport in MC-331 pressure trailers due to its flammability rating and extremely high vapor pressure. It is best handled by well-trained professionals.

EtSH is known for its strong and pungent odor; so strong that just a small vapor release or leak can be detected by smell miles away. EtSH is used to odorize propane; thus, EtSH leaks are often mistaken for propane gas leaks.

Sulfur Chemical Market Analysis

Over the past 10 years, demand for sulfur chemicals in North America has approximately doubled, which has resulted in a supply–demand imbalance. This has been driven by two markets: ethylene and renewable refineries.

Ethylene Market

It is estimated that the sulfur chemical market in North America for the ethylene market is 12-15M MT (metric ton) annually. In the past 10 years, ethylene production has increased approximately 50% in North America, primarily in the Gulf Coast of the United States (Figure 2).

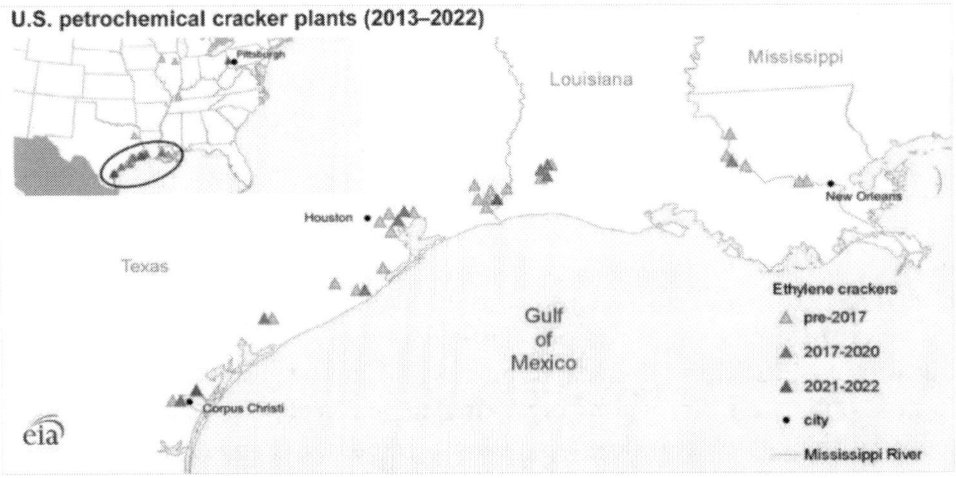

Figure 2. Growth in US Petrochemical Cracker Plants (2013–2022).
Source: US Energy Information Administration, based on corporate filings and trade press; 2021.

This growth has been driven by abundant availability of ethane from shale fracking in the United States. Current capacity on ethylene production in North America is estimated at approximately 50M MT based on a recent Dow Jones analysis. The expansion has, in turn, driven a similar increase in demand for sulfur chemicals. This is, however, a difficult number or correlation to exactly determine because what is unknown it is actual output that will drive sulfur chemical demand. It can fluctuate at any time depending on market conditions and unplanned outages. Any expansion has potentially cannibalized other production units. Traditionally production expansion comes in waves, and then takes the market demand time to then catch up to the expanded available capacity.

The fact that not every unit continues to inject a sulfur chemical continuously after the initial sulfiding is another complication. Each world-scale size unit expansion on average requires the equivalent of approximately 225K MT annually of sulfur. This paper assumes that future production planned increases will result in a similar correlation for demand in sulfur chemicals. The same Dow Jones analysis estimated that production capacity will further increase by 5% in the next 10 years, so demand for sulfur chemicals could increase accordingly.

Renewable Refineries Market

Recent changes in the North America refinery market have led to sulfur chemical demand almost tripling over the past 10 years to an estimated 12-15M MT annually. Over the past 5 years, new refineries have been built to process renewable fuels as drop-in replacements for traditional crude oil–based fuels, primarily diesels and jet fuels. In addition, some refineries, such as the well-publicized Phillips 66 refinery in Rodeo, California, have been converted to use renewable feedstocks instead of traditional crude oil. The two new predominate renewable refinery technologies available are UOP's Ecofining™ process and Haldor Topsoe's HydroFlex™ process. These processes utilize plant-derived oils, animals fats, and used cooking oils instead of traditional crude petroleum oil. As a result, they burn cleaner, release less carbon, and potentially reduce CO_2 emissions by up to 100% compared to traditional petroleum-derived fuels according to these two technology providers.

Unlike crude oil, renewable feedstocks do not contain any natural sulfur; therefore, the addition of a continuous injection of a sulfur chemical is required to prevent coking throughout the process, which could harm catalysts and lead to other issues. A traditional petroleum refinery requires a sulfur chemical for only a 12- to 48-hour period depending on the size and capability of the unit when activating a new hydrotreating catalyst after a change-out and does not require continuous injection outside very small reformer units. Hydrotreating catalysts are used in the refinery process to remove sulfur from fuels to meet sulfur-content regulations to lower SO_2 emissions, which has been critical in reducing environmental impacts for traditional fuels. To reiterate, a sulfur chemical is used to activate the catalyst, which ultimately reduces the sulfur content in that hydrocarbon stream. An example of sulfur chemical usage of a traditional refinery versus a renewable refinery is as follows: a traditional large diesel unit processing petroleum crude oil would require a catalyst change and activation once every 3 to 4 years and approximately 150 MT of a sulfiding agent. A renewables diesel unit with an equal size output requires now approximately 650 MT of a sulfur chemical per year to operate, which is an order of magnitude higher than a traditional refinery over a 3 to 4 year period.

There is a great deal of uncertainty on the long-term viability and demand for renewable fuels, which drives uncertainty in supporting industries and materials such as sulfur chemicals. At present, renewable refineries are economical solely due to significant government subsidies. The current technology is in its infancy; it is possible that technological advances could reduce the operational need for such high levels of sulfur chemicals. To date 11 new or converted plants are producing renewable fuels and several more are planned, raising questions about the long-term demand of sulfur chemicals for this application. In addition, the current push to electrify transportation and move away from internal combustible engine (ICE)-powered vehicles may adversely impact the demand for sulfur chemicals for the refineries market.

Complexities in Meeting Sulfur Chemical Demand

As previously noted, both ethylene and sustainable refining markets are growing and evolving such that forecasting sulfur chemical demand is challenging; however, one thing is clear—more chemicals are needed. So why not just make more product? That is a complex question as the four sulfur chemicals addressed in this paper are tied to other products and markets. The complexities and dynamics surrounding each of the chemicals are addressed in the following sections.

Dimethyl Sulfide

DMS, the workhorse in North America, is produced as either an intermediate or co-product. Prior to the market changes noted previously, this product experienced a stable balance between demand and production. When DMS is produced as a co-product, it is tied to the larger production of methyl mercaptan (MeSH). This is the case for Chevron Phillips Chemical. MeSH is used to produce methionine, an essential supplement used primarily for poultry livestock. The amount of DMS produced is dependent upon demand for MeSH, process recycling capabilities, and economics. In North America, there has been a recent move to integrate current independent MeSH production into the production chain of methionine to eliminate MeSH transportation challenges. MeSH is a toxic and extremely flammable colorless gas, which makes it challenging to handle. This is the case for the methionine production chains in Europe and could potentially complicate DMS availability in the future in North America.

When DMS is produced as an intermediate, availability is dependent on demand and economics for the final end product—dimethyl sulfoxide (DMSO). DMSO is a multiple purpose solvent used in growing markets such as pharmaceuticals and electronics, as well as the agricultural market, which can be seasonal. Because DMS production is tied to these other potentially higher value products and markets, meeting ethylene market demand can be challenging.

Dimethyl Disulfide

DMDS, the dominant sulfur chemical used in the refinery market, is almost exclusively dependent upon imports from Europe, primarily, and some supply from Asia. This speaks to the popularity of DMDS in the European and Asian markets, as well as the dearth of DMS in those markets. While DMDS is produced to support both refinery and ethylene markets, North American supply is susceptible to supply chain disruptions such as those experienced in 2020/2021 and prioritization to local needs and regulations. DMDS production in North America is limited and also tied to MeSH production and other complicating unrelated market dynamics. It is estimated that supply can support just approximately 15% to 20% of the current DMDS market demand.

Sulfurized Olefins (Polysulfides)

TBPS 454 is produced in Europe, with no known current domestic production. TBPS 454, as with all polysulfides, is primarily used in metal working as a lubricant additive due to its superior lubricity, high flash point, good mixing capabilities with base oils and fluids, and lower odor. TBPS 454 makes an excellent lube additive in various gear oils, which is also seeing growth in higher value emerging technology markets such as automotive and energy generation. Renewable refineries prefer TBPS 454 versus DMDS due to its ease of handling and storage. This is especially true for refineries in California and other environmentally sensitive areas.

Ethyl Mercaptan

With the closing of the Occidental Chemical plant in Baytown, Texas, approximately 15 years ago, there is now just one known active domestic producer of EtSH. Chevron Phillips Chemical produces EtSH in Europe. EtSH has traditionally been produced to support the propane odorant market, as well as the agricultural market as a pesticide/fungicide.

Potential Solutions for Meeting Sulfur Chemical Demand

Dimethyl Sulfide with CS_2

As previously noted, one of the potential negatives of DMS is the presence of CS_2 as an impurity in some supplies. It is naturally found in the production process for DMS; thus, further processing, such as treating or distillation, is required to remove/reduce. CS_2 will also break down and decompose to H_2S, but at a higher temperature than DMS (approximately 1560°F). Those pesky double bonds are more difficult to break. Because of this and depending on an operating temperature and length and time of the furnace, CS_2 impurities can pass through before decomposing to H_2S. CS_2 then tends to concentrate in the C5 streams, attributing to the total volume sulfur in streams such as the C5 diolefin product, among others, causing it to be off-specifications and/or lose value.

Experience has shown that when evaluating CS_2 content in various available DMS options, results can range from <20 parts per million (PPM) to 2000 PPM. High-purity DMS is defined as having CS_2 content ≤20 PPM. Based on Chevron Phillips Chemical's studies, standard-grade DMS typically averages 500 to 600 PPM. An analysis shows that an estimated 5% of ethylene units are currently using high-purity DMS by request. The remaining units use either standard DMS, using high-purity DMS for other reasons, or other sulfur chemicals.

As demand for sulfur chemicals grows, trials may be used to determine if current CS_2 still continues to impact heavy olefin products, and economic evaluations of the aforementioned may identify opportunities that can be captured. The following case studies demonstrate recent independent studies and trials completed by ethylene plants to study and evaluate the impact of CS_2 content in DMS.

<u>Case Study 1: Financial Study</u>

A plant that traditionally uses high-purity DMS evaluates the use of standard DMS based on economics. After field testing, CS_2 passes through the process and into the C5 stream. A financial/specification impact study was then conducted on the sulfur impact on the C5 stream versus the costs associated with using standard DMS.

For this plant, it was determined that a maximum specification of 320 PPM CS_2 content was required. This was based on average flow rate and end-product TVS specifications of 60 PPM. Next, a financial study was completed to compare the market premium of the C5 stream meeting this specification versus the higher cost of high-purity DMS over a 3-year period. The plant decided to keep utilizing high-purity DMS based on the negative financial impact.

Case Study 2: New Plant Design Study

A new unit is designed to evaluate how to use standard DMS without CS_2 impurities impacting the value of the C5 stream. Engineering solutions, treatments, or distillation or other, could be pursued. The plant decided that the small capital investment required for the engineering solution was warranted to enable the use of standard DMS with higher CS_2 content.

Case Study 3: Availability Issues

A plant that traditionally used high-purity DMS but could not find it available and has little choice but to quickly trial standard DMS. After 3 to 4 months of use of standard DMS, the plant did not experience any impact to their C5 stream and required no operational changes to the process.

Case Study 4: Supply Reliability

A plant that traditionally used standard DMS was experiencing issues with supply reliability and sought to reduce the risk of a supply disruption. They developed a solution to increase supply flexibility by setting up their operations to use either DMS or DMDS by investing in an engineering solution that would allow either chemical or a mix to be used successfully if needed.

Di-Ethyl Sulfide Study at a Chevron Phillips Chemical Ethylene Unit

In 2022, Chevron Phillips Chemical underwent an evaluation of an available sulfur chemical di-ethyl sulfide (DES) as a potential sulfiding agent in the ethylene industry. See Table 2 for key properties.

Table 2. Diethyl Sulfide Properties

Property	**DES**
Sulfur % (weight)	36%
Density (pounds/gallon)	7.0
Boiling Point (°F)	198°F
Flash Point (°F)	14°F
Vapor Pressure at 70°F (PSI)	1.45
Decomposition Temperature (°F)	1250
Viscosity at 70°F (mPas)	.422

A partial ISO container of DES was shipped to one of the Texas facilities and used for a full tank cycle instead of the normal sulfiding agent. One of the focus areas centered on contained "other" sulfurs in the 97% pure DES received. As noted with CS_2, there was a need to understand how the sulfurs were reacted, diluted, or tolerated in downstream processes. One advantage of DES is that primarily decomposes to H_2S and ethylene. Samples of raw gasoline and debutanized aromatic concentrate (DAC), as well as several heavy olefins product streams were sampled prior and during the trial for total sulfur content. Results are still currently being tabulated, but no adverse effects were noted as of the end of the trial in late 2022. As results are published and if acceptable, this may offer an additional option for a tight sulfur chemical supply in need of alternatives for variable growth.

Refinery Sulfur Contained Feeds (General)

Many facilities adjoining refinery operations, already take advantage of sulfur contained feedstocks for crackers at affordable pricing. The contained sulfur in the feedstocks eliminates or reduces the sulfiding requirement for those furnaces processing those streams (sour propane, etc.). However, where sulfur chemical supplies are tight, facilities looking for new options may be able to find solutions previously untapped at their refining neighbors.

All plants are different and will see different results. The primary purpose here is to demonstrate that the best way to determine solutions is with trials and the exploration of new opportunities. As plants evolve over the years, prior experience may need to be challenged when seeking new alternatives.

Conclusion

Looking for sulfiding supply solutions within increasing tight sulfur chemical supply has become more prominent in recent years. Choosing a sulfur chemical is critical for any ethylene plant. It is important for every plant to evaluate their own situation and complete an analysis on which sulfur chemical is best for the operation. When possible, it is best to establish a flexible system to utilize a diversified chemical supply. Chevron Phillips Chemical, as a leading ethylene, polyethylene, and sulfur chemical producer in North America, is committed to continuing to research, performing trial runs, and developing solutions for the industry. With efforts like these from Chevron Phillips Chemical and others, the needs and challenges of an ever-growing petrochemicals and sulfur chemicals demand can be successfully met.

Advances in Purification of Monomers for Petrochemical and Polymerization Applications

Stephen Caskey, Jay Gorawara, Allan Hatami

Petrochemical and polymer producers are evaluating opportunities to improve operational expenditures (OPEX) and increase flexibility of their operation. One way producers can achieve this is by utilizing lower-cost feedstocks. Unfortunately, these feedstocks typically have higher levels of contaminants, requiring a more robust contaminant removal solution to purify them before processing. If not properly treated, even trace levels of these contaminants can permanently deactivate a costly process unit catalyst, requiring a plant shutdown to reload. This results in added catalyst cost as well reduced revenue due to lost production time. To avoid these costs, producers need a robust contaminant removal solution that has long life and robustly removes contaminants.

Honeywell UOP has developed a new multifunctional purification adsorbent AZ-500 for monomer pretreatment to remove acid gases as well as oxygenates, organic nitrogen and organic sulfur species. A typical propylene producer customer operating a metathesis process unit using AZ-500 could see up to a 30% reduction in regeneration frequency, resulting in lower utility consumption and a substantial reduction in regenerant consumption and cost. In addition, AZ-500 offers an increase in adsorbent life, resulting in reduction in costs due to less frequent turnarounds. Finally, AZ-500's increased performance allows producers more flexibility in feedstock selection, which could result additional savings from cheaper feeds. A new state of the art hybrid alumina-zeolite composite adsorbent is ready for commercial use.

AIChE Paper Number: 60e

Impact of excessive sulfur dosing to coil material deterioration and ageing

Dr. Dietlinde Jakobi
Schmidt + Clemens GmbH + Co. KG
51789 LINDLAR/GERMANY

Lukas Halbach
Schmidt + Clemens GmbH + Co. KG
51789 LINDLAR/GERMANY

Towje Kirchner
Schmidt + Clemens GmbH + Co. KG
51789 LINDLAR/GERMANY

Prepared for Presentation at the 2023 Spring National Meeting
Houston, TX, March 14, 2023

Impact of excessive sulfur dosing to coil material deterioration and ageing

Dr. Dietlinde Jakobi

Schmidt + Clemens GmbH + Co. KG

51789 LINDLAR/GERMANY

Lukas Halbach

Schmidt + Clemens GmbH + Co. KG

51789 LINDLAR/GERMANY

Towje Kirchner

Schmidt + Clemens GmbH + Co. KG

51789 LINDLAR/GERMANY

Abstract:

The design and the process robustness of radiant sections in steam cracking furnaces play an important role in improving ethylene plant efficiency as well as achieving their sustainability goals – both of which being strongly influenced by the physical and chemical stability of the installed tube materials.

Radiant coils are operated under severe conditions at temperatures even beyond 1100 °C (2012 °F) and alternating reducing and oxidizing conditions that represent an enormous challenge for the materials – especially related to their corrosion resistance and mechanical properties.

An oxide barrier layer, generated during furnace start-up on the inner surface of high-alloyed radiant coils, protects the material initially against hazardous atmospheric conditions. With progressive ageing, corrosive attack can deteriorate the protectiveness of such oxide-barrier layers. Further impact of temperature and aggressive atmosphere can also strongly impair the mechanical properties of the bulk material by internal oxidation, sulfidation and by carburization. As a further consequence, the inner tube surface can become strongly susceptible to catalytic coke deposition. During the aging of materials installed in these very harsh conditions, the originally formed oxide layer is degraded and catalytically active elements like nickel and iron get access to the surface leading to catalytic reactions with the process gas (coking).

Coke deposition on the inner surface of components is a limiting factor when it comes to furnace run lengths, thus affecting e.g. the furnace availability and the maintenance costs, but the need to increase firing rates due to coking also affects the efficiency of the furnace as well as the CO_2 footprint.

Normally, small amounts of sulfur-containing components are added to the feedstock to de-activate catalytically active sites present on the inner surface of components in order to reduce the CO formation in the cracked gas. When such concentration limits are exceeded, e.g. by accident, reactions taking place on the inner surface may lead to extensive and irreversible damage to the coil material induced by sulfidation.

This paper will reflect the strong corrosive attack of centrifugally cast tubes in a several years old radiant tube section resulting from an unintentional, excessive sulfur dosage occurred during presulfidation.

Results of metallurgical investigations will be presented. These are focused on the degradation mechanisms leading to the service life reduction of the entire radiant section assembly under attack of sulfur.

The influence on the operating parameters will also be addressed, since the coking rates suddenly increased after the event, leading to strongly reduced run-lengths.

In this special case study, coke formation was so strong and furnace efficiency so severely decreased, that the operation of the radiant coils was no longer economically justifiable.

Introduction

Presulfidation of radiant coils is a commonly used technique to control carbon monoxide levels in the effluent of a cracking unit, but excessive sulfiding can lead to deterioration and ageing of the coil material:

The carbon monoxide (CO) levels in the effluent during the feed-in phase after start-up can pose a significant risk if they reach critical levels. This is due to the water-gas shift reaction that takes place within the inner tube of the cracking unit. The catalytic active sites, composed of elements such as iron and nickel, promote the formation of carbon monoxide.

High concentrations of CO can harm downstream processes and applications, as it inhibits catalysts in the selective hydrogenation of alkenes. This makes it essential to control CO levels in order to ensure the safe and efficient operation of the cracking unit.

One common approach to poison the active sites is presulfidation of the radiant coils before the start of a cracking run. Only a monoatomic layer of sulfur is necessary in order to reduce the formation of CO.

While sulfiding has been shown to have some positive effects, such as reducing coking rates and increasing run lengths, in some literature [1] it continues to be a controversial topic in the industry. Some studies have reported negative effects, such as accelerated coke formation [2]. It is important to note that excessive sulfiding can harm the metal surface of the radiant tubes, leading to increased coking rates and decreased efficiency.

In conclusion, controlling CO levels is crucial for the safe and efficient operation of the cracking unit. While presulfidation is a commonly used method, it is important to carefully consider its effects and ensure that the appropriate amount is used.

It is well known from literature that sulfur shows significantly higher reaction rates with metals (Ni, Co, Fe and Cr) compared to oxygen [3]. Very important is also the penetration of sulfur through the oxide scales. Stroosnijder et al. [4] showed that oxide scales (chromia and alumina) are in general stable against conversion to sulfides in sulfur containing, oxidizing atmospheres. At 1000 °C, the sulfur does not react with the oxide scales [3] but it penetrates e.g. through chromia scales (Cr_2O_3) into the bulk material, while sulfidation trials with alumina scale forming alloys do not show any signs of sulfur penetration through the alumina scale (Al_2O_3). Stroosnijder et al. [5] found that at 1200 °C the sulfur diffusion coefficient is "four orders of magnitude larger" for Cr_2O_3 than for Al_2O_3 protective scales.

In combination with other contaminants, especially sodium (Na), which forms in steam cracking conditions Na_2SO_4, sulfur accelerates corrosion attack also known as hot corrosion. This type of corrosion takes place at temperatures between 550 °C and 950 °C and is induced by the formation of liquid salts at the corrosive site.

This paper addresses as Case 1 the unforeseen consequences of an excessive amount of sulfur dosing in a steam cracker, which was introduced inadvertently as part of the presulfidation process along with steam. The unintended high dose caused severe corrosion attack and significantly increased coking rates. After the accident, the steam cracker was shut down and affected tube sections were cut out for investigation at Schmidt + Clemens GmbH & Co KG.

Samples were appropriately prepared and investigated using optical microscopy (OM), scanning electron microscopy using back scattered electrons (SEM/BSE) and electron beam microanalysis (SEM/EDX), optical emission spectroscopy (OES) as well as combustion analysis.

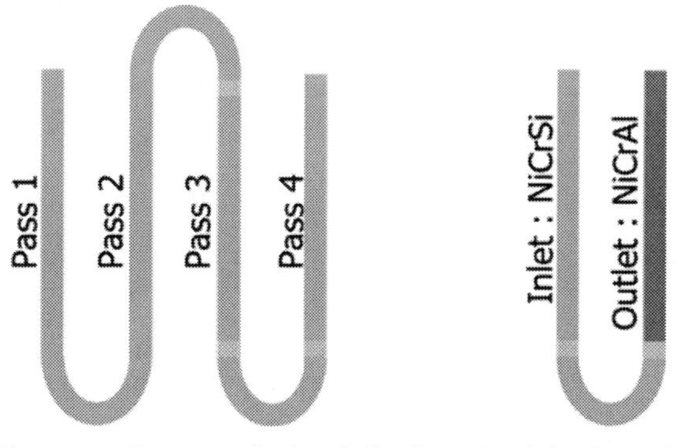

Figure 1: Furnace design left: Case 1; right Case 2

The furnace, equipped with four-pass design coils (see Figure 1), was in service since 2014 using butane and recycle-ethane as feedstocks. The investigated tube sections were taken from pass three (bottom and top) and pass four (bottom only). All tubes consist of the chromia-forming alloy Centralloy® ET 45 Micro (NiCrSi).

Additionally, Case 2 was investigated - dealing with a two-pass furnace: In this case, the inlet sections were consisting of NiCrSi, while the outlet section was equipped with the alumina-forming alloy Centralloy® HT E (NiCrAl).

Results

In **Case 1**, three radiant coil sections made of alloy NiCrSi were under investigation. The examinations were performed in the certified S+C in-house laboratory. This included the analysis of the chemical composition and metallurgical investigations in different levels of detail (optical microscopy to scanning electron microscopy).

The optical emission spectroscopy, performed on the core of the tube wall, revealed that samples were consistent with the typical chemical composition of NiCrSi. The analysis of sulfur, carbon, oxygen, nitrogen and hydrogen was carried out by combustion analysis performed on metal chips, which were drilled from different layers across the thickness of the tube wall (see Figure 2).

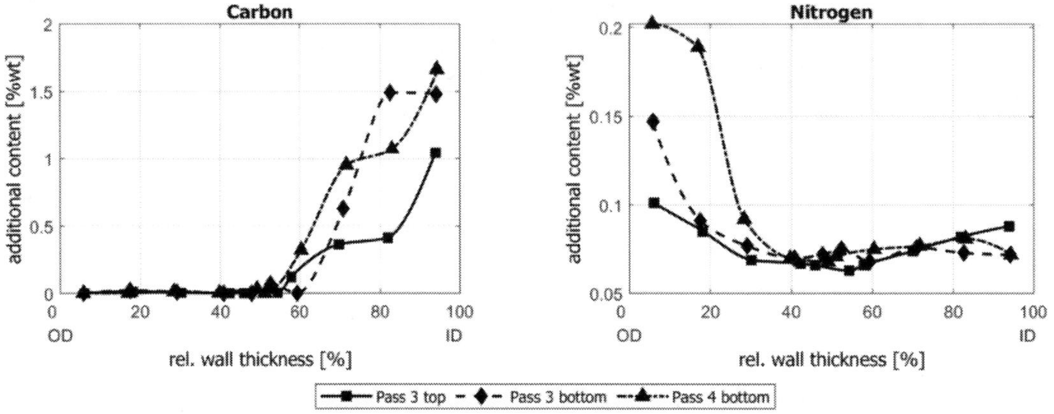

Figure 2: Results of combustion analysis for carbon (left) and nitrogen (right) content in the tube wall ("0" corresponds to outer surface, OD/"100" to inner tube surface, ID)

Figure 2 (left) clearly demonstrates that all tubes are significantly carburized at the inner surface towards the tube core. The increase in nitrogen levels indicate a noticeable nitridation of the tubes from the outer surface. Both phenomena are diffusion controlled. From the atmosphere inside the tube, carbon is diffusing into the alloy (carburization). At the outer tube surface, nitrogen from the burner atmosphere is diffusing into the alloy (nitridation). Diffusion is controlled by temperature and time. In this case, diffusion is solely influenced by temperature, as all tubes were installed at the same time. Hence, higher amounts of carbon and nitrogen typically indicate a higher tube metal temperature. As the process gas is heated in the steam cracker, the gas as well as the tube metal temperature along the unraveled coil path raise. This makes pass 4 the hottest pass. The detected carbon and nitrogen levels correlate with this theoretical consideration.

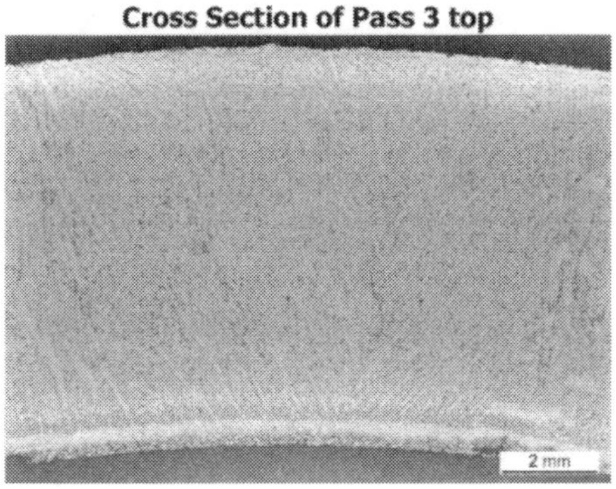

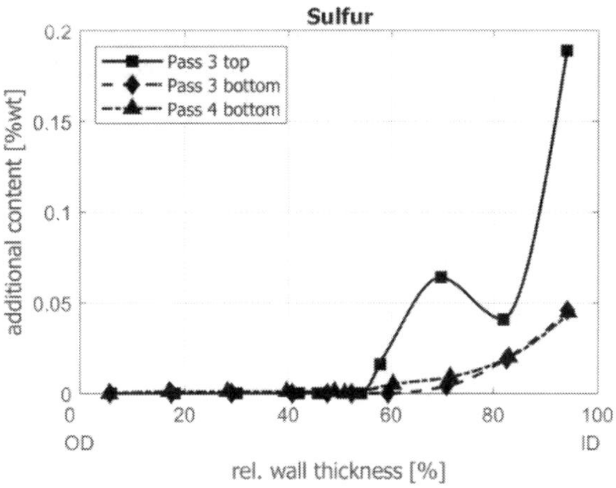

Figure 3: Left: Optical micrograph of etched cross section (pass 3 top) Right: Results of combustion analysis reflecting sulfur content in the tube

Figure 3 shows a micrograph of the corresponding tube section, and the diagram depicts the sulfur content measured across the tube wall. While at the tube wall center none of the tubes show an increased sulfur content, the areas near the inner tube surface show a massive increase in sulfur content – especially for pass 3 top section.

To investigate how sulfur was interacting with the base metal, cross sections from the tube sections were prepared and examined.

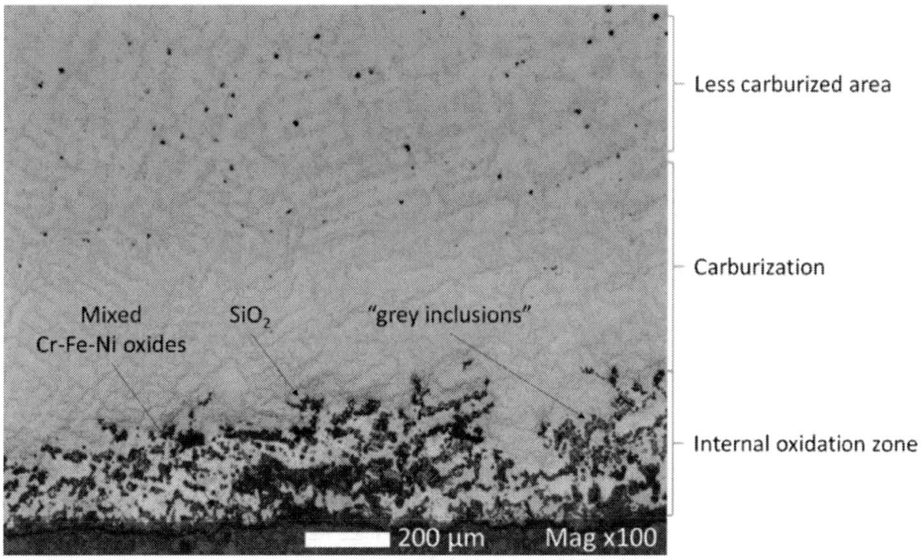

Figure 4: Polished cross section of sample pass 3 top

Figure 4 shows a polished cross section, which reveals progressed corrosion at the inner tube surface. There is no continuous oxide scale present and the oxide forming elements have been precipitated as SiO_2 and mixed Cr-Fe-Ni oxides in the subsurface area forming an internal non-protective oxidation zone. Carbon ingress through this zone has led to formation of a bulky metal carbide network (carburization). Due to the chromium depletion of the subsurface area (caused by internal oxidation & carburization), a self-healing of the material is not possible.

The investigation also revealed unknown phases near the inner surface. These "grey inclusions" can be observed in Figure 4. These phases can be detected near the phase boundary metal-atmosphere as well as between the internal oxidation zone and the carburized base metal.

The samples were prepared for investigation with SEM (EDX) to identify the phases.

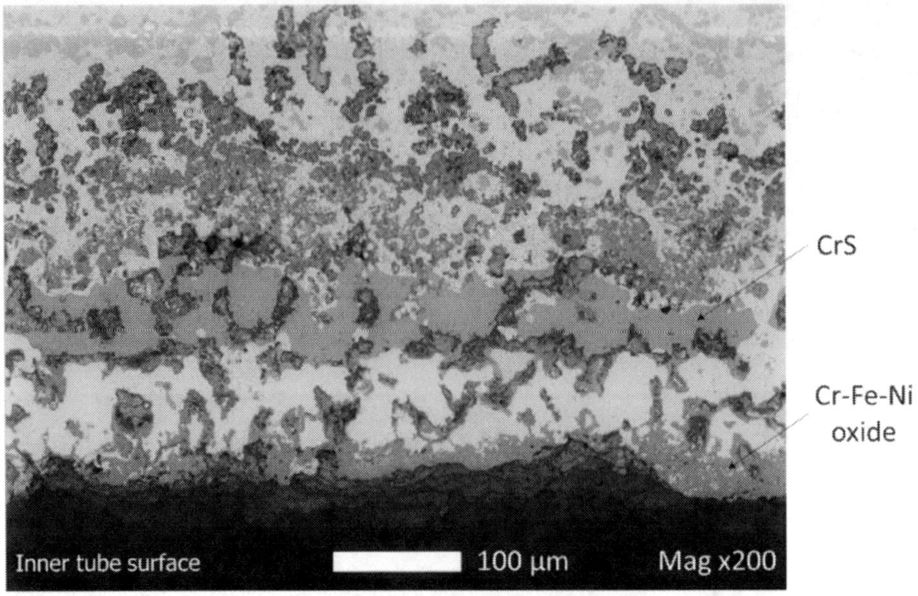

Figure 5: SEM BSE of ID surface of pass 3 top sample showing internal corrosion

In the SEM BSE contrast (Figure 5), the internal corrosion zone with a depth of 400 µm from the inner tube surface is visible and the EDX mapping reflects the different compounds/phases in the corroded area (see Figure 6).

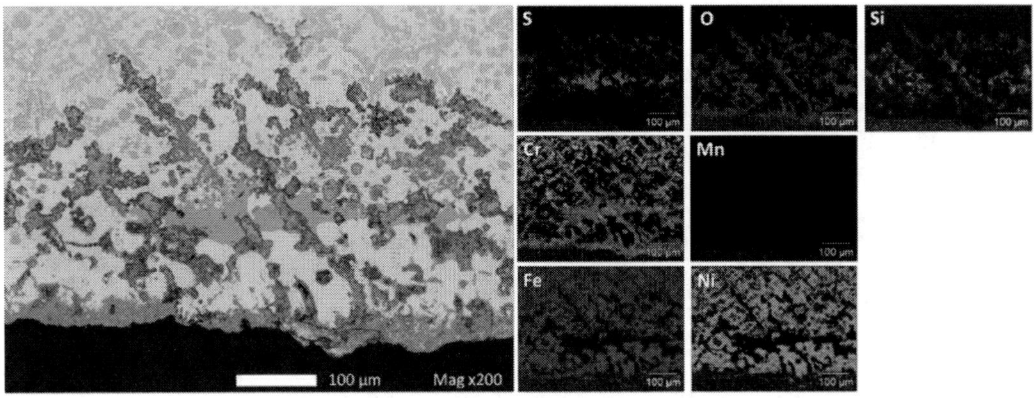

Figure 6: EDX mapping of sample pass 3, top section

In the EDX mapping (Figure 6) the internal corrosion of Cr-oxide and Si-oxide can be seen (brighter color reflects higher element content). Additionally, a locally concentrated very high amount of Cr-sulfides is detected. These are mainly located at the transition between internal oxidation towards the carburized matrix, but also distributed within the entire internal oxidation zone.

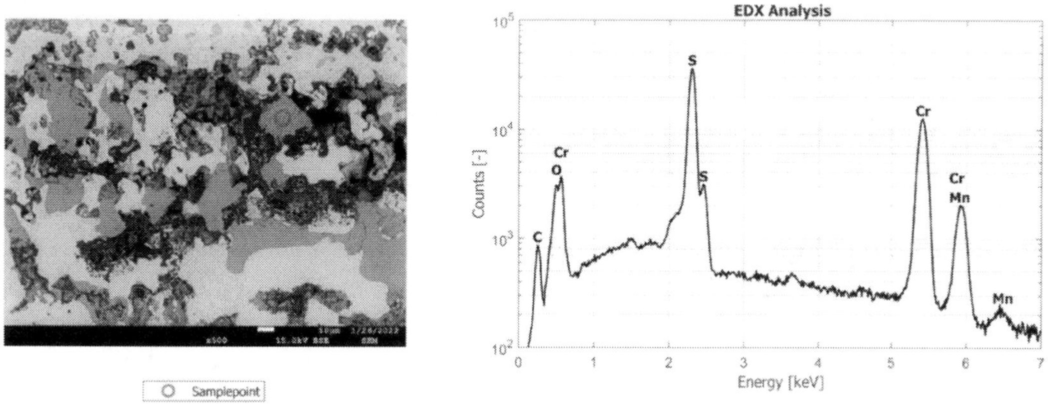

Figure 7: EDX analysis of Cr-sulfide phase

An EDX analysis (Figure 7) revealed the following chemical composition of the Cr-sulfides (see sample point marked in Figure 7):

At-%	Cr	S
Sample point*	48%	48%

*remains: Mn, Ni, Si

This phase can therefore be clearly identified as CrS.

In **Case 2** (as mentioned earlier), the investigated inlet tube was made of alloy NiCrSi, while the outlet tube was made of NiCrAl material. Both tubes were connected through a manifold. The tube sections removed for investigation were cut close to the manifold to ensure similar operating conditions (see Figure 1).

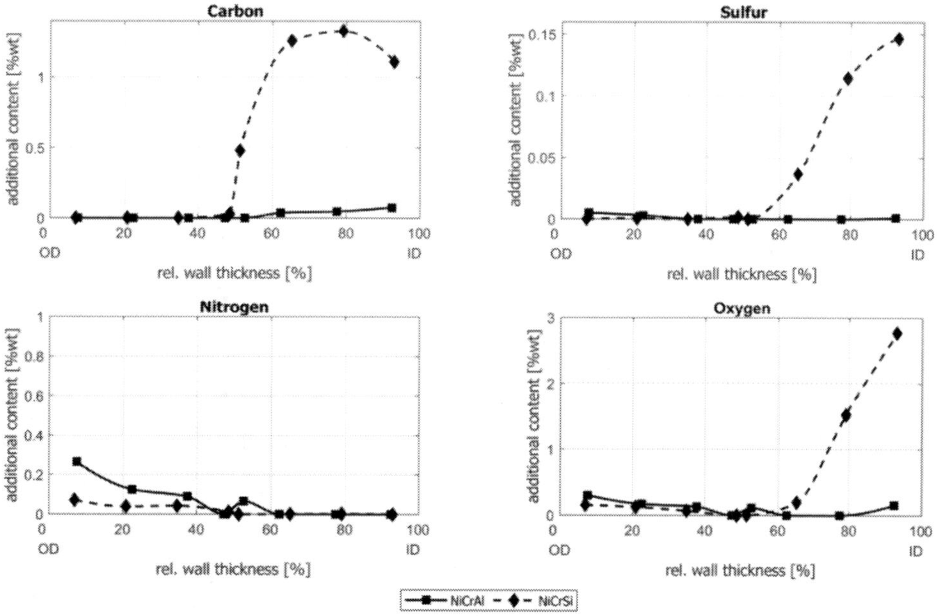

Figure 8: Results of combustion analysis for S, C, N and O content over tube-wall thickness for NiCrSi (dashed line) & NiCrAl (continuous line) materials

The NiCrSi tube section shows internal oxidation, carburization and sulfidation at the inner tube surface. The alloy NiCrAl tube connected to the ET 45 Micro tube section via the manifold shows no signs of internal corrosion.

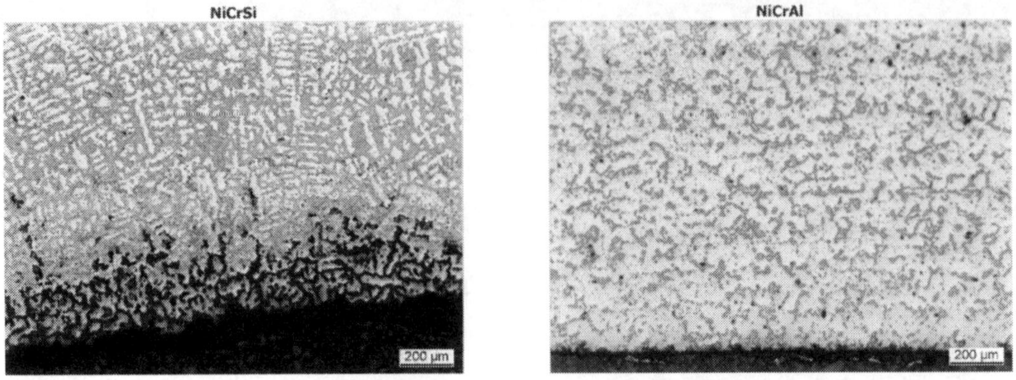

Figure 9: Optical micrographs /cross sections of the inner subsurface area reflecting left: NiCrSi; right: NiCrAl material.

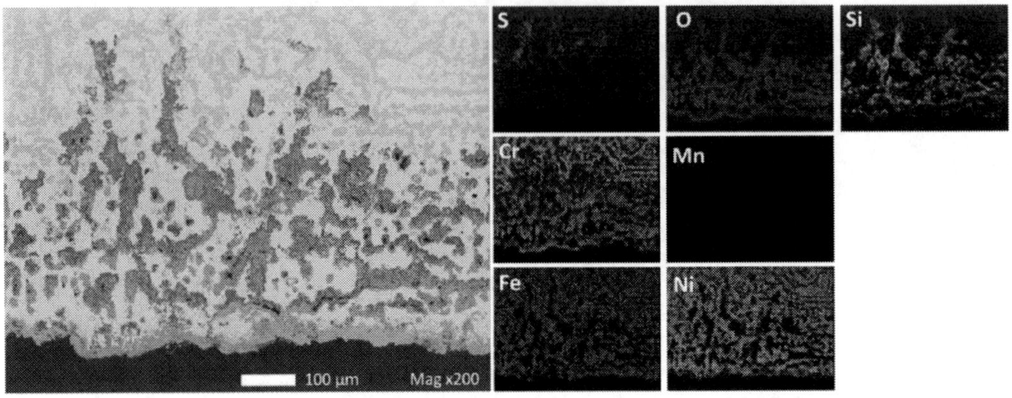

Figure 10: SEM EDX element mapping of NiCrSi with grey phases behind internal corrosion front

Figure 9 reflects the hazardous corrosion on the inner surface of the alloy NiCrSi sample. In the transition zone between the internal oxidation zone (black precipitates) and the carburized base material (blue color), grey phases have been detected. Figure 10 shows the EDX measurement with enriched S content and internal oxidation.

A continuous and protective alumina scale (see Figure 11) is present on the inner tube surface of NiCrAl material. It therefore does not show any corrosion attack.

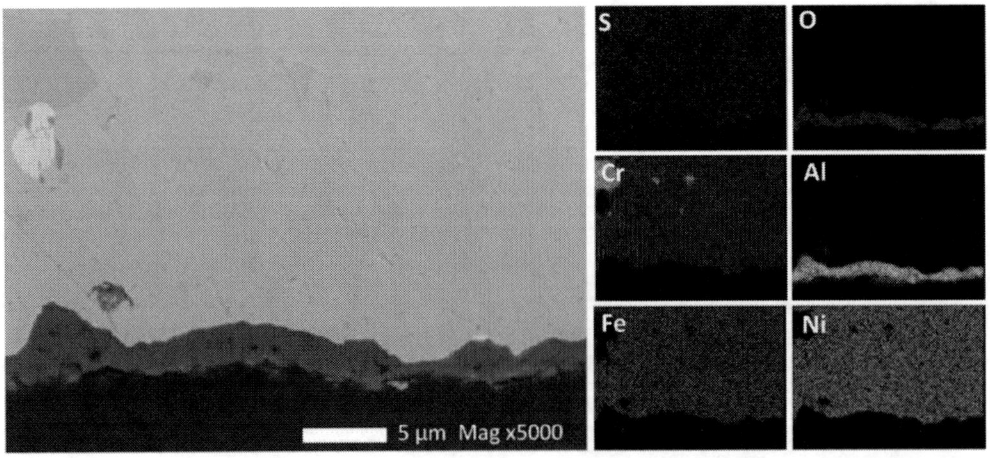

Figure 11: SEM EDX measurement of subsurface scale of alloy NiCrAl

Discussion

The use of sulfur-containing compounds as additives for the thermal cracking of hydrocarbons is unavoidable for some stream cracker operators. Such additives are fed-in continuously in lower concentrations with the feedstock to the coils, or, as a presulfidation step in higher concentrations before taking the coils into operation.

This is carried out due to the beneficial effect of sulfur compounds to the reduction of the CO-peak and in general, the de-activation of catalytically active sites present on metallic surfaces of radiant and convection section assemblies.

In this investigation, it could be shown that an excessive addition of sulfur-containing feedstock can lead to strong corrosion phenomena, and therefore, fast material degradation. Essential during this contamination event is whether the alloy is able to form and maintain a protective oxide scale or not.

Sulfidation of chromia forming alloys (35Cr-45Ni)

In carburizing atmospheres like in the steam cracking process, chromia forming alloys like 35Cr-45Ni suffer from various corrosion phenomena. Firstly, the transformation of chromia scales into chromium carbides in reducing conditions during the cracking phase and secondly, the re-oxidation of compounds under oxidative conditions at temperatures above 950 °C (1742 °F) during decoking. These alternating reducing and oxidizing conditions lead to the failure of the surface protection [6].

Therefore, these alloy classes are not able to maintain the protective oxide scales but are affected and impaired by corrosion over time when exposed to steam cracking conditions.

As known from literature, chromia scales do not act as diffusion barriers for sulfur diffusion. If sulfur is able to diffuse into the alloy and form chromium sulfides, the corrosion rate will be increased, because less chromium (main oxide scale forming element) is available to form a continuous and dense oxide scale. Chromium depletion in the near-surface area leads to the remaining material components (mainly nickel and iron) being indirectly enriched at the metal surface.

Especially iron and nickel, known to act as active sites for catalytic coke formation, are therefore increasing the coke rate and reducing run length of the cracker unit.

The more iron and nickel is present on the surface, the more catalytic coke will be formed, the shorter the run lengths will be and the less efficient the steam cracker will operate.

Unlikely the scale will heal itself, because sulfidation increases subsequent oxidation. The oxidized sulfides lead to a release of sulfur that is diffusing deeper into the alloy to form new sulfides [3].

The impact of sulfur-induced corrosion at high temperature can accelerate even further, if other contaminants are present in the steam cracking atmosphere. At temperatures up to 950 °C e.g. in combination with sodium (Na), hot corrosion can take place. Liquid phases are formed with low melting points, which attack the metallic material and lead to accelerated corrosion and premature failure of components. Also at temperatures higher than 950 °C, contaminants can have a negative impact in impairing or preventing the formation of a dense and protective oxide layer.

Once a part has been sulfidized and corroded, it cannot heal itself. Consequently, accelerated corrosion will reduce the remaining lifetime of the component. As a consequence, the corrosion products on the inner surface will lead to increased coke formation, reduced run lengths and as a result to an inefficient furnace operation. A replacement of corroded parts is the only possibility to regain the performance of the furnace.

Sulfidation for alumina forming alloys (NiCrAl)

Alumina forming alloys like Centralloy® HT E are able to form and maintain a protective alumina scale over several years – also proven on sections examined after 10 years in service. This unique corrosion-resistance depends on the ability to self-heal the protective oxide, which may spall off, e.g. due to erosion by carbon particles. The alumina itself on the surface not only forms a thinner and denser oxide scale, but also it is more stable in steam cracking as well as in decoking conditions. In the investigated "Case 2", the Centralloy® ET 45 Micro tube showed progressed corrosion, while the Centralloy® HT E sample did not show any corrosion, even though it was processed in the same operating conditions.

The investigation results confirmed the findings from the literature. The alumina scale formed on the alloy NiCrAl was able to protect the alloy from sulfur-induced and accelerated corrosion. The chromia scale formed on NiCrSi does not hinder sulfur ingress. This leads to progressed corrosive attack and to reduced plant performance as well as reduced mechanical integrity of the coils.

References:

[1] Froment, Marie-Francoise S. G. Reyniers and Gilbert F. (1995), Influence of Metal Surface and Sulfur Addition on Coke Deposition in the Thermal Cracking of Hydrocarbons, Industrial & Engineering Chemistry Research, pp. 773-785

[2] Manjunath, Patil S. A.-F. (2020), Catalytic Effect of Dimethyl Disulfide on Coke Formation on High-Temperature Alloys: Myth or Reality? *Industrial & Engineering Chemistry Research*, p. 59 (34).

[3] Kofstad, Per (1988), High Temperature Corrosion, Elsevier, London, p. 429-458

[4] M. F. Stroosnijder & W. J. Quadakkers (1986), Review of high temperature corrosion of metals and alloys in sulphidizing/oxidizing environments I. Corrosion of metals, High Temperature Technology, 4:2, p. 83-96

[5] M.F. Stroosnijder & W.J. Quadakkers (1986-2), Review of high temperature corrosion of metals and alloys in sulphidizing/oxidizing environments II. Corrosion of alloys, High Temperature Technology, 4:3, p. 141-151

[6] Jakobi, Dietlinde & Karduck, Peter & Richthofen, A.F. (2013), The high-Temperature corrosion resistance of spun-cast materials for steam-cracker furnaces - A comparative study of alumina- and chromia-forming alloys. NACE - International Corrosion Conference Series.

SULSAFE™ Sulfiding Agent – Enhanced Performance for a Sustainable Operation

Joice Boll
Global Technical Support, DORF KETAL, Houston - TX

Mikhail Knis
North America Sales Manager, DORF KETAL, Houston - TX

Sudarshan Vijayaraghavan,
SEA Sales Manager, DORF KETAL, Singapore

Behzad Parvizi
Dr. ir.
Laboratory for Chemical Technology, Ghent University, Ghent, Belgium

Kevin M. Van Geem
Prof. Dr. ir.
Laboratory for Chemical Technology, Ghent University, Ghent, Belgium

Prepared for Presentation at the 2023 Spring National Meeting
Houston, Texas, March, 12-16, 2023

SULSAFE™ Sulfiding Agent – Enhanced Performance for a Sustainable Operation

Joice Boll
Global Technical Support, DORF KETAL, Houston - TX

Mikhail Knis
North America Sales Manager, DORF KETAL, Houston - TX

Sudarshan Vijayaraghavan,
SEA Sales Manager, DORF KETAL, Singapore

Kevin M. Van Geem
Prof. Dr. ir.
Laboratory for Chemical Technology, Ghent University, Ghent, Belgium

Behzad Parvizi
Dr. ir.
Laboratory for Chemical Technology, Ghent University, Ghent, Belgium

Abstract: The use of sulfiding agents in steam cracker furnaces is standard practice in the Olefins industry. Sulfiding agents thermally decompose in the furnace to generate H_2S that reacts with the metal oxide to form a passivating metal sulfide. DMS and DMDS (di-methyl-sulfide and di-methyl-di-sulfide) have the highest market share despite their toxicity and bad odor. Sustainability and supply chain issues are motivating ethylene producers to look for other options. Dorf Ketal has developed a new generation of Sulfiding Agents, presented to the market as SULSAFE™. In this paper, we will share laboratory and field data comparing DMDS with the new polysulfide.

The laboratory data is from the pilot ethylene furnace at Ghent University. Compared to DMDS, the new polysulfide is shown in the lab to potentially improve the yield of ethylene, propylene, and butadiene with a reduction of coke precursors and coke. CO control is improved with less sulfur. Decomposition studies are presented to show that DMDS generates CS_2, meaningfully impacting the efficiency of conversion of sulfur to H_2S while adding a harmful contaminant to the system. Polysulfide thermal decomposition data is presented to show that it does not generate CS_2.

Dorf Ketal field experience confirms the lab data, providing valuable benefits in terms of furnace run length, CO control, and yields. The new polysulfide offers excellent HSE characteristics and superior performance, contributing to the sustainability goals of current & future ethylene producers. For new units, this technology can reduce CAPEX for the injection systems as it is not flammable and does not require nitrogen blanketing. Odor problems with DMS/DMDS are also eliminated.

Introduction

Olefin producers have an increasing need for improved carbon efficiency and sustainability. One area of investigation is to find ways to improve run-length and reliability in the furnace section of Ethylene Plants. In the furnace process, improved coking control plays a crucial role in productivity and reliability. Improved coking control can reduce maintenance costs, increase production, and reduce risk of unplanned shutdowns.

The furnace is comprised of two sections, convection section and radiant section, depending upon the mode of heat transfer. In the radiant section coils, a major challenge to meeting targeted run-length is coke generation and deposition. Coking is the deposition of carbonaceous substances which form a layer on the inner surface of the coils. Coke formation leads to an increase in tube metal temperature, coil pressure ratio, and fuel consumption to maintain the same severity of cracking.

Coke in the furnace is usually identified as catalytic or pyrolytic, with both having the free radical mechanism to grow the molecular weight of the carbon deposit. Catalytic coke is generated due to the catalytic effect of metals present in the coil alloy (with different levels of nickel, chromium, and iron) in the hydrocarbon dehydrogenation process [1], and hence are predominant in the start of run when the metal surface is more exposed. Once the metal surface is coated by either coke and/or sulfiding agents, the pyrolytic coke, plays the major role, being governed by the high temperatures in the radiant section and feedstock composition – heavier molecular weight feeds tend to increase coking rates [2].

Apart from coke, ethylene producers need to have good control of CO generation, which is also a result of metal catalysis (especially nickel) in the steam reforming reaction [3]. CO is a poison to acetylene reactors and a contaminant in the final ethylene. Just as with catalytic coke, CO formation peaks in the beginning of the furnace run. The CO content in the furnace effluent can be used as an indirect way to verify the quality of the passivation layer formed inside the coils throughout campaign, due to natural coke deposition and/or the use of chemicals.

To reduce the catalytic coke and control CO generation, ethylene producers commonly use sulfiding agents like DMDS or DMS as a source of sulfur, to generate H_2S which reacts with the metal surface to control the catalytic effect. The motivation to use DMDS/DMS is usually related to price point and market availability, even though the compounds are well known to carry a number of health and environmental hazards. What is not yet well known is the real availability of the H_2S generated and the byproducts created. This article describes the a new polysulfide technology in comparison with DMDS for improved CO and coke control. SULSAFE™ is the brand name used by Dorf Ketal to market a new polysulfide agent technology.

A study was carried out by Ghent University in a pilot plant to corroborate the use of the new polysulfide on the commercial scale. The description of the pilot plant and the findings of the study carried out at Ghent University will be further described in this paper. The new polysulfide reduced CO generation and coking rate, and also positively impacted the furnace lights yields. Case studies presented in this paper demonstrate how the chemical use on a commercial scale is consistent with pilot plant results measured Ghent University.

Furnace operation and coking mechanism

Cracking of feedstock to produce ethylene is an endothermic process. The heat of reaction is supplied by combustion of fuel gas in the furnace. Unstable carbon free radicals are generated in the process, they tend to react with other free radicals to generate higher molecular weight coke precursors. The coke precursors can eventually build enough molecular weight to deposit on the inner surface of the coils as coke [2, 4].

This high molecular weight carbon compounds called coke act as an insulation to reduce heat transfer efficiency. As heat transfer efficiency declines, the tube metal temperature (TMT) is increased to maintain the desired coil outlet temperature (COT). The accumulation of coke also increases the pressure drop across the coils, resulting in lower ethylene yields. Nickel (Ni) in the metal surface acts as a catalyst and promotes the catalytical coke formation. Modern furnaces use the so-called micro alloys with nickel (>30 wt. %), chromium (>25 wt. %) with the balance being iron and a few additives (Si, C, W, Mn). To minimize catalytic coking, it is necessary to passivate the metal surface with a protective layer.

CO and coke suppression method

For Olefin plant furnaces, sulfur compounds that generate H_2S are found to be most effective technology to passivate the metal surface of the radiant coils, with DMDS and DMS having the highest market share. The passivating action of sulfiding agents serves to control CO formation and reduce the catalytic coke. However, the interaction these compounds can have in the termination of free radical reactions, and hence in total coke generation, is still discussed by academia and industry [3 - 6].

All sulfiding agents available in the market dissociate in presence of temperature, steam, and hydrocarbon to generate H_2S. This H_2S comes in contact with the Ni in the metal surface to form nickel sulfide. Once nickel sulfide is formed, the passivation layer will prevent metal contact with the steam and hydrocarbon suppressing CO and catalytic coke formation [5]. It has become standard practice in the industry to differentiate the cost efficiency of different sulfiding agent based on "cost per kg of sulfur." We are learning that different sulfiding agents are not equally efficient in preventing coke formation and this difference can also impact furnace efficiency and carbon footprint. For example, all of the sulfur in DMDS does not convert directly to H_2S, [5, 7], and the temperature range (600-650°C) where this decomposition process occurs can also affect the performance as a nickel passivator [7]. In order to select the most cost-effective sulfiding agent, Ethylene producers can benefit by choosing the sulfur containing compound with the lowest total cost of operation including the value of improved carbon efficiency, rather than a simple "cost per kg of sulfur" evaluation.

Apart from the total cost of operation evaluation, it is also important to consider the HSE aspects related to the use of sulfiding agents in the furnaces. DMDS and DMS, the most popular chemicals amongst the producers, are known to be highly toxic and difficult to handle due to low flash point and toxicity. In contrast, polysulfides like TBPS and the new material developed are recognized as better options to meet the best-in-class HSE standards, in line with the United Nation (UN) and companies' sustainability goals.

Shortcoming of DMDS and improvements with new polysulfide

The new polysulfide development was done with the objective of offering to the market a sulfiding agent option which is aligned with global HSE sustainability goals, and also to offer improved performance and carbon efficiency compared to DMDS. DMDS was used as comparison point as this compound has the highest sulfur concentration amongst the regular agents and is the most preferred sulfiding agent as measured by market share.

The HSE aspects related to polysulfide and DMDS are described in the Figure 1. The flash point of DMDS is lower than the ambient atmospheric temperature, it gets vaporized at room temperature and it is highly flammable, which forces the ethylene producers to keep it under nitrogen blanketing. Also, DMDS has very low odor threshold and is extremely dangerous and toxic to people and the environment, which makes it necessary for the ethylene manufacturer to spend extra CAPEX on the special injection system to use DMDS. By comparison, the polysulfide is environmentally friendly, has a mild odor and a flash point high enough to not require a nitrogen blanket. Injection system of the polysulfide is advantageous to ethylene producers. Also, storage and transportation of polysulfides reduce the risk of an accident compared to DMDS.

POLYSULFIDE	
Flash Point (°C)	62
Vapor Pressure (psi 25°C)	0.1 – 0.2
Eye Irritation	No
Respiratory Irritation	No
Oral Toxicity Limit (LD50 ppm)	7150
Handling & Storage	Easy
N_2 Blanket (Pressurized Vessel)	No
Transport Danger	Low
Odor Description	Light Diesel

DMDS	
Flash Point (°C)	15
Vapor Pressure (psi 25°C)	0.55
Eye Irritation	Yes
Respiratory Irritation	Yes | Toxic
Oral Toxicity Limit (LD50 ppm)	300 to 500
Handling & Storage	Difficult
N_2 Blanket (Pressurized Vessel)	Yes
Transport Danger	Medium
Odor Description	Pungent

Figure 1 – Overall comparison of DMDS and polysulfide HSE properties.

To compare the sulfiding agents performance, several studies were conducted by Ghent University, including a decomposition profile in a bench scale reactor and a cracking test in the furnace pilot unit.

Decomposition Study – by Ghent University

The decomposition products of DMDS, the new polysulfide and TBPS were studied in the Bench Scale Steam Cracker (BSSC) setup at the Laboratory for Chemical Technology (LCT) at Ghent University in Ghent, Belgium. The decomposition runs were carried out an incoloy metal reactor, to have insight on the wall effect on the decomposition products. The conversions and selectivity for the sulfiding agents were evaluated in three different reactor temperatures, from 500 °C to 700 °C. The coil outlet pressure (COP) has been set at 1.7 bar to have a similar pressure as in the steam cracking runs in the pilot unit. The mass flow of hydrocarbons was fixed to 100 g/h for all of the samples. Furthermore, the sulfur content of the samples has been adjusted to around 500 ppm by addition of the respective sulfur agent in Hexane. Nitrogen was fed as carrier gas to keep the pressure of the reactor with a fixed mass flow of 70 g/h. For quantification of the sulfur compounds, a flow of 30 g/hr of 3-chlorothiophene in Hexane with concentration of 100 pm has also been fed to reactor effluent line as an internal standard.

The decomposition profile of DMDS is significantly different than TBPS and the new polysulfide (Figure 2.) DMDS decomposes at a lower temperature range. This permits the DMDS decomposition to occur above the cross-over point (600-650°C), potentially losing H_2S to areas where the passivation is not required in the furnace. the new polysulfide decomposes at the widest range of temperatures, even when compared with another polysulfide, in this case TBPS. This assures formation of H_2S throughout the furnace coils.

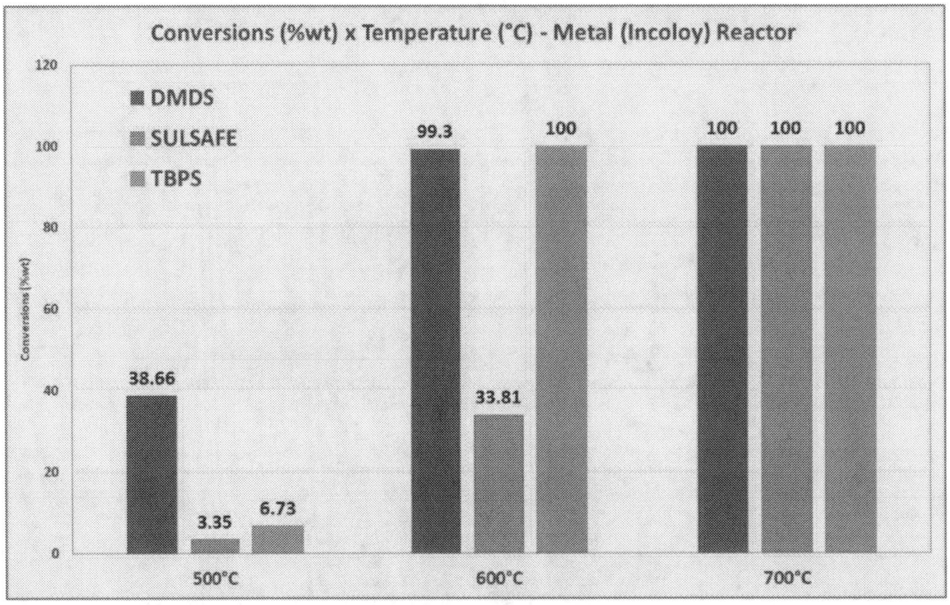

Figure 2 – Sulfiding Agents dissociation with respect to temperature.

The selectivity towards H_2S was evaluated in the reactor outlet, and can be seen in Figure 3. Apart from H_2S, CS_2 was the only other sulfur specie generated, and its selectivity for the different sulfiding agents can be seen in Figure 4.

It is clear that DMDS decomposition starts with the production of CS_2 only, and with the increase in the temperature the reaction finally generates the H_2S, which is actually the compound required for furnace passivation. Even at 700°C, when completely decomposed, DMDS still has only 80% selectivity to H_2S in the conditions tested. The polysulfides don't produce CS_2 at any temperature, having 100% selectivity towards H_2S.

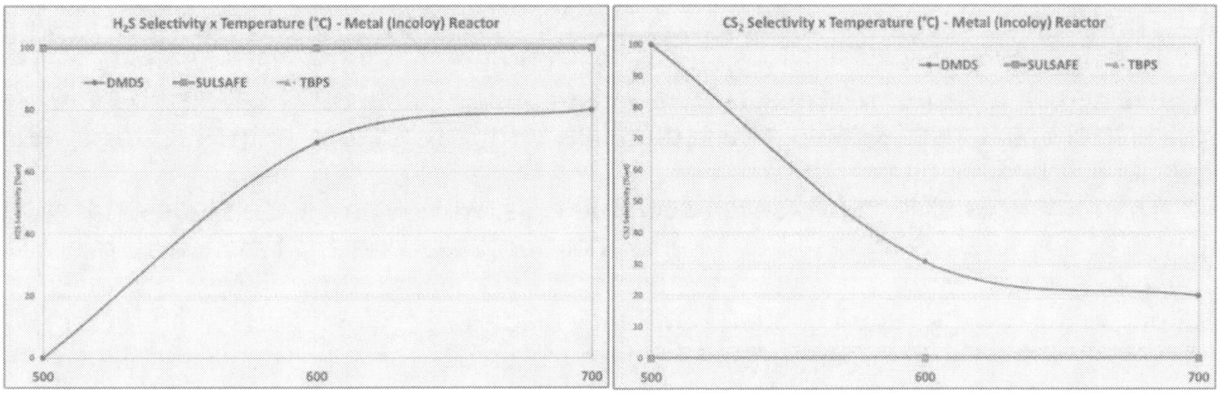

Figure 3 (left) and 4 (right) – H_2S and CS_2 selectivity at different temperatures, respectively.

The CS_2 present in the feed and generated in the furnace gets carried over with the py-gas to the GHU (Gasoline Hydrogenation Unit). CS_2 is a well known feed contaminant and acts as a strong inhibitor for palladium-based catalyst in the GHU unit. Many py-gas units operating with a nickel-based catalyst have observed that the catalyst is poisoned in the presence of CS_2 [8]. CS_2 is also known to create corrosion issues in the Depentanizer overhead and Isoprene units, increasing the risk of reliability issues. Many Ethylene producers with liquid feedstocks have limited the content of CS_2 to something in the range of less than 3 to 10 ppm in the feed.

The bench scale reactor study has demonstrated that the decomposition profile and H_2S selectivity are different between the chemicals, indicating that the need for sulfur when using the new polysulfide molecule can be lower than with DMDS. Efficiency of DMDS is lost due to the formation of contaminant CS_2 with DMDS, and also from the H_2S being produced at lower temperatures zones than the radiant section.

Pilot Plant Study – by Ghent University

The experimental work was carried out in the pilot plant set-up for steam cracking of hydrocarbons at the Laboratory for Chemical Technology (LCT) at Ghent University in Ghent, Belgium. Since the main parts of this unit, the analytical equipment and the calibration procedure have been described elsewhere [9, 10], only a brief description, based on Wang et al., Dhuyvetter et al. [11] and Van Geem et al. [12] will be given. Special attention will be paid to the analytical section because this section has been improved recently [12].

The feed section controls the supply of the different feedstocks to the reactor coil. The flow is regulated by the pumping frequency of a coriflow pump. The mass flow rate of all feeds are measured continuously to avoid inaccuracies due to volume dependence on temperature and pressure. The measurement of the mass flow rate of liquid hydrocarbons and water is carried out using an electronic balance on which an intermediate barrel filled with the feedstock is placed. Every minute, the weight indicated by the balance is sent to the computer. The flow rate can be easily calculated from the weight reduction per time interval. This value is compared with the value obtained from the CORI-FLOW mass flow controller, which measures the flow each second. Bronkhorst designed the CORI-FLOW series mass flow metering instruments to provide high accuracy: ±0.2%. CORI-FLOW utilizes a patented, advanced Coriolis type mass flow sensor to achieve unsurpassed performance, even with changing operating conditions in pressure, temperature, density, conductivity and viscosity. The instruments are offered as separate flow meters or with a pump. The pumping frequency is automatically adjusted by the controller of the CORI-FLOW. When the fluid level in the intermediate barrel is too low, the barrel is automatically refilled with feed from the storage barrels or gas cylinders. Gaseous hydrocarbons are fed to the reactor by means of a separate CORI-FLOW.

The furnace, built of silica/alumina brick (Li23), is about 4m long, 0.7m wide and 2.6m high. It is fired by means of ninety premixed gas burners, mounted with automatic fire checks and arranged on the side walls in such a way as to provide a uniform heat distribution. The fuel supply system comprises a combustion controller, for the regulation of the fuel to air ratio, and the usual safety devices.

The furnace is divided into seven separate cells, which can be fired independently to set any type of temperature profile [12]. Twenty thermocouples and eight manometers are located along the reactor coil to measure the temperature and pressure of the reacting gas. The reaction section of the tube is about 12m long, made of Incoloy 800HT, and has an internal diameter of 9mm. These dimensions are chosen to achieve turbulent flow conditions in the coil with reasonable feed flow rates (Re>7000).

The cooling section consists of a heat exchanger, a so-called Transfer Line Exchanger (TLE), and is made of Incoloy 800HT. It consists of two concentric tubes: the reactor effluent flows through the inner tube while air, providing cooling of the effluent, flows co-currently through the outer tube. A schematic representation of the reactor and the cooling section is given in Figure 5. The process gas is further cooled to 150°C by means of cooling oil.

At the reactor outlet, the injection of nitrogen provides an internal standard and contributes to a certain extent to the quenching. Nitrogen is used as the internal standard for hydrocarbon fractions of the effluent, while sulfur standard (3-cholorothiophene), which is also injected at the reactor outlet (see Figure 5), serves as an internal standard for sulfur containing compounds present in the reactor effluent. Before the cooling of the effluent in a TLE, a sample is taken for the on-line C_5^+ analysis.

The liquid and the tar are separated from the cooler exit flow by means of a knock-out vessel and a cyclone. The pressure at the exit of the reactor is controlled by a reduction valve. A fraction of the product gas is then withdrawn for on-line C_4^- analysis, while the rest of the effluent is sent directly to the flare.

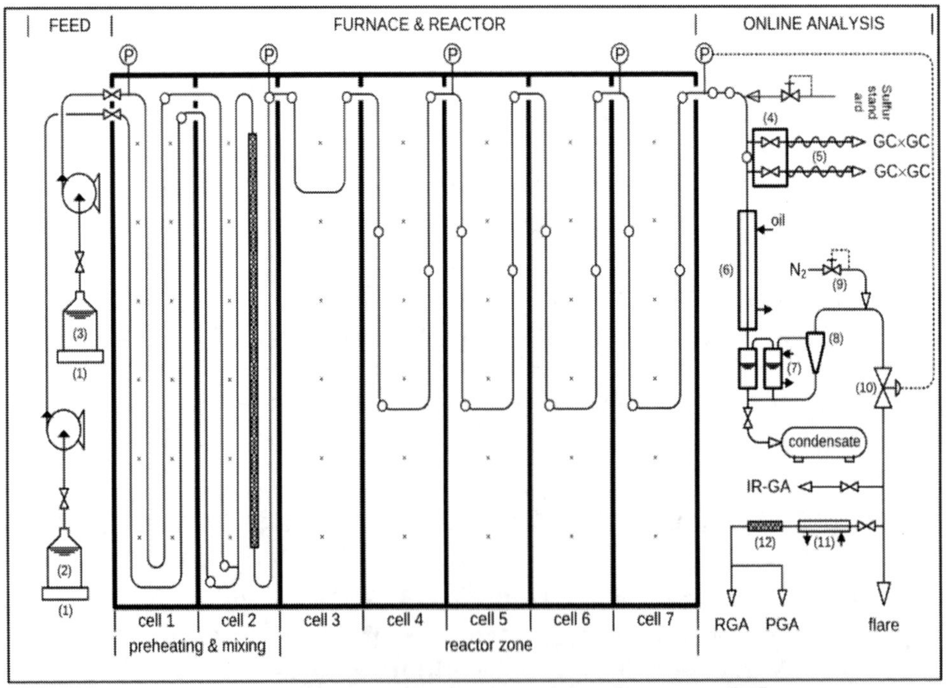

Figure 5 – Schematic of pilot plant furnace [7].

The furnace effluent yields comparing DMDS and the new polysulfide runs are presented in the Figures 6, 7, 8, and 9. The sulfiding agents were fed at 100 ppm in ethane feed, as product base (not sulfur base). Being a polysulfide with higher thermal stability, the new polysulfide molecule has lower weight percentage of sulfur than DMDS. By running the comparison tests at the same product dosage, the user can easily compare line-item chemical costs.

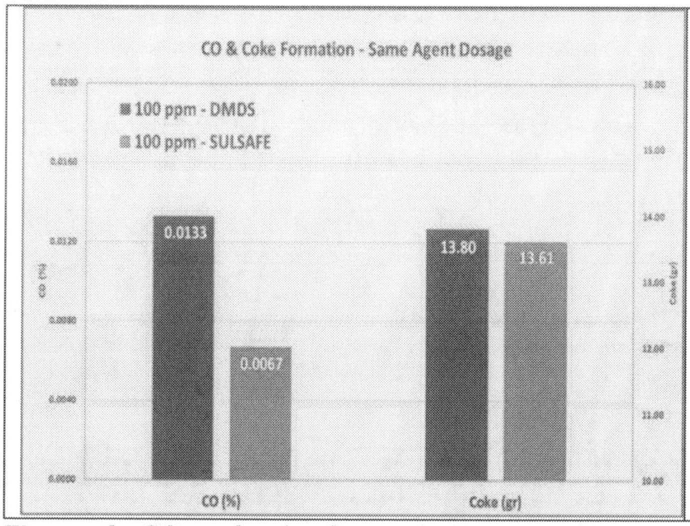

Figure 6- CO and coke formation comparison.

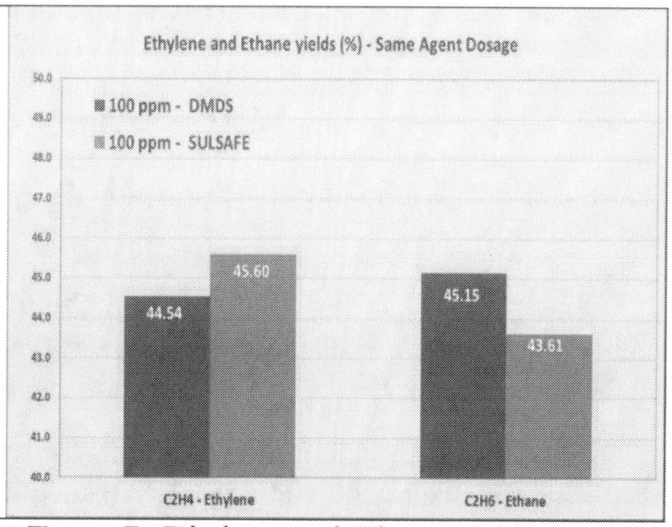

Figure 7- Ethylene and Ethane yields data.

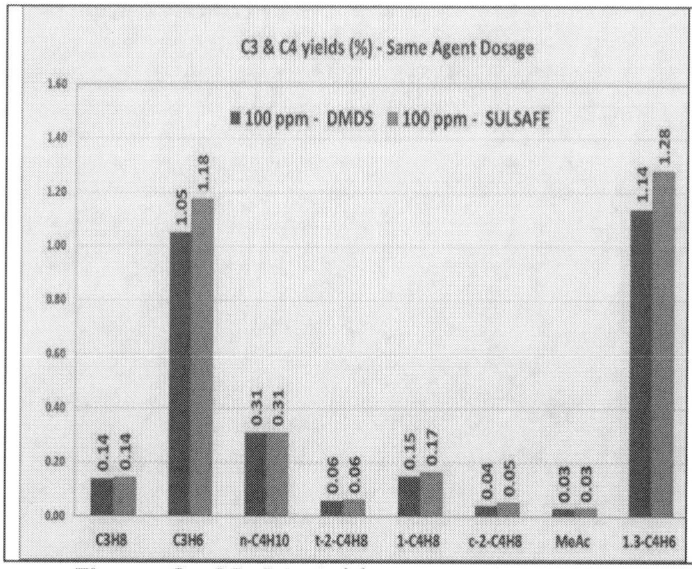

Figure 8- C3-C4 yields comparison.

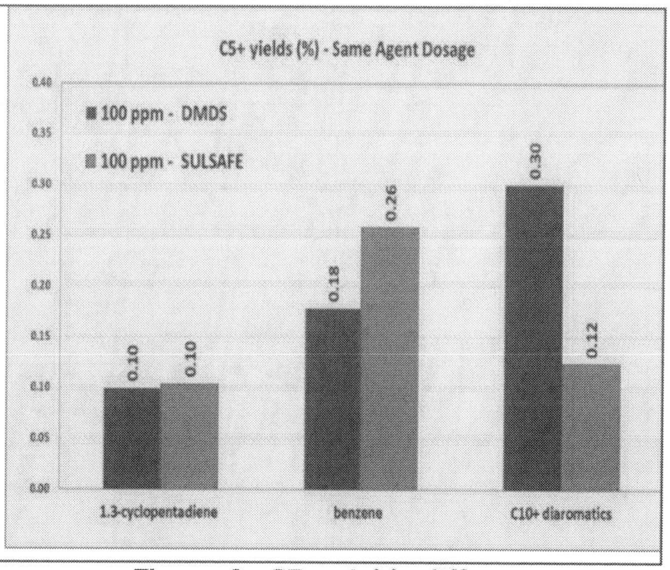

Figure 9- C5+ yields difference.

There are several interesting observations from the pilot tests:

1) The amount of CO generated with new polysulfide is much lower than with DMDS (less than half). As stated earlier, CO will act as a catalyst poison in the acetylene converter, and is a very good indicator of the effectiveness of the passivation layer created by the sulfiding agents.
2) Coke generation was measured at the end of the procedure, through carbon mass

balance during the decoke. Although the quantity in grams of coke generated with the new polysulfide is only slightly lower than the numbers seen with DMDS, the C10+ diaromatics are significantly reduced in the polysulfide run. Diaromatics are known to be coke precursors, especially in the outlet of the coils and TLEs.

3) Ethane conversion was around 3% higher during the new polysulfide runs, which can be translated to less ethane recycle, a significant energy saving in the field application.
4) The new polysulfide resulted in higher Ethylene & unsaturated C3/C4 yields, corroborating with the lower loss to heavies and higher conversion of ethane seen.

The differences in the ethane conversion and unsaturated molecules observed, suggest that the new polysulfide interaction with the free radical terminations are less intense than the ones with DMDS. Relative to DMDS, the new polysulfide has improved furnace yield, reduced heavies generation, lower total cost of operation and superior sustainability. The benefits seen in pilot study conducted by Ghent are corroborated by field applications, and some of them were captured in the form of case study as follows.

Case Studies

Case study 1

An Ethylene plant in Asia Pacific started a trial with the new polysulfide as part of a DMDS replacement plan. The trial was taken in both recycle and naphtha furnaces. The Ethylene producer was able to observe a reduction in CO content measured in the acetylene reactor inlet (Figure 10), and a better coke control (Figure 11). The improvement in the coke control was translated into increase in furnace run-length, since the TMT was the limiting factor for the decoking process.

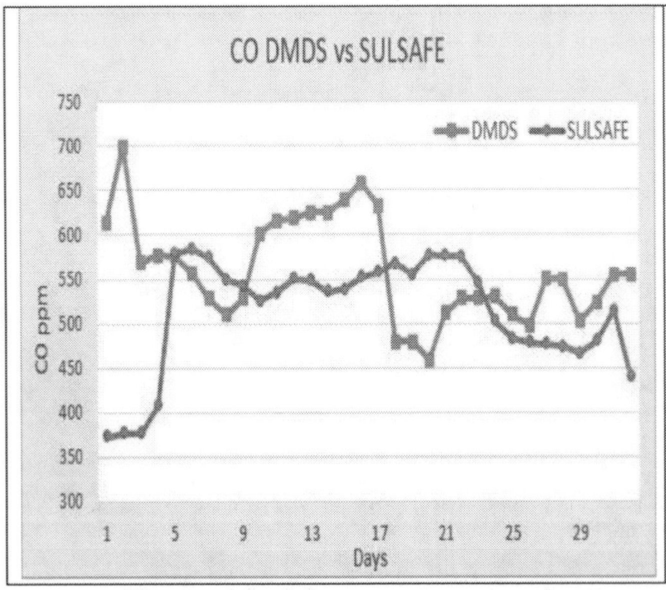

Figure 10- CO content comparison.

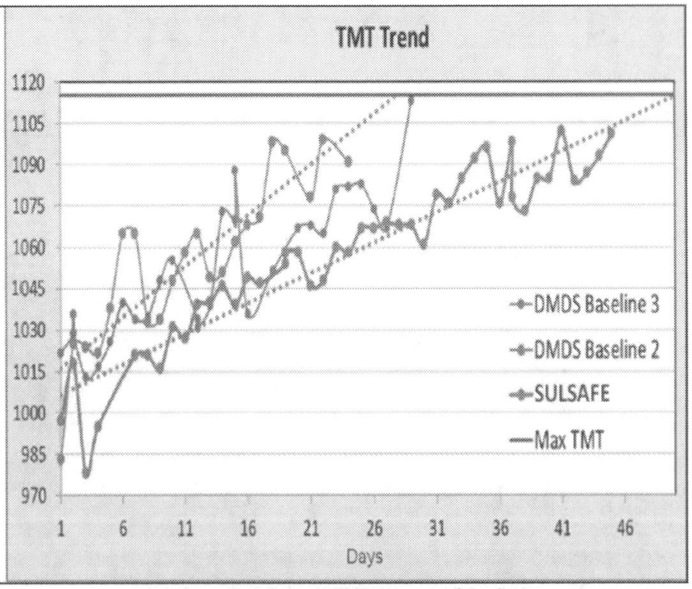

Figure 11- TMT trends data.

Apart from the immediate benefits, after a few runs it was possible to start the comparison of the naphtha furnace yields using DMDS with the yields when using the new polysulfide, for similar operating conditions. The naphtha furnace analyzers are located in the individual outlets.

As can be seen in the Table 1, for 4 different operating regimes, the delta of yield found with the new polysulfide runs was positive to Ethylene (ranges from 0.3 up to 1 mol%) and Propylene (ranges from 0.2 up to 0.4 mol%), when compared to average DMDS runs. Since the furnace operational conditions and feedstock quality are very dynamic, it is definitely more challenging measure the yield improvement at these levels, but the combination with the lower coke rates seen corroborates with the positive impact. Based on the performance, the new polysulfide has been established in this plant as continuous sulfiding agent.

Furnace Condition	Hydrogen (mol%)	Methane (mol%)	Ethylene (mol%)	Ethane (mol%)	Propylene (mol%)	Propane (mol%)
1	-1.740	-0.050	1.070	0.255	0.410	-0.077
2	-1.360	-0.070	0.782	0.230	0.200	-0.003
3	-1.011	-0.039	0.300	0.050	0.190	0.015
4	-0.330	-0.089	0.396	-0.130	0.290	0.015

Table 1 – Delta of SULSAFE™ yields with similar furnace condition.

Case study 2

A plant in Asia trialed with the new polysulfide as a replacement to DMDS. In the trial, the new polysulfide outperformed DMDS, with an improvement in TMT slope (Figure 12). The feed for both (DMDS and polysulfide) runs was C2/C3 recycle and all the other operating conditions were kept constant.

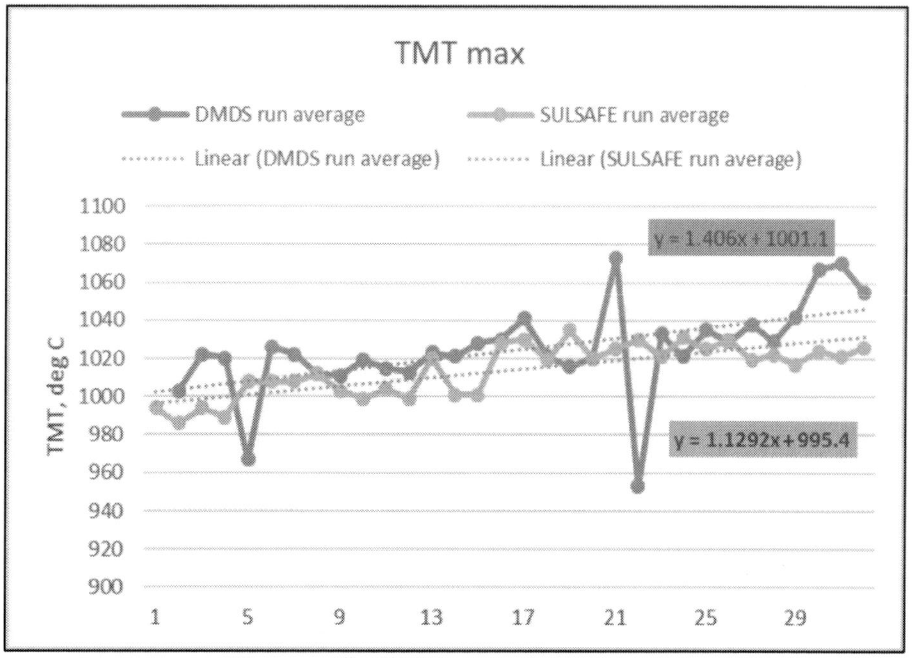

Figure 12- TMT max. trend.

In the above graph we can see a relatively significant amount of drop in slope of TMT rise while using the new polysulfide as compared with DMDS. The rate of change of TMT is one of the indicators used by Ethylene producers to check the performance of a particular run. This difference can be used to calculate fuel savings.

Conclusion

The use of sulfiding agents in ethylene furnaces is necessary to passivate the metal surface and reduce CO generation and catalytic coking intensity. The most common agent used (DMDS) does not convert completely to H_2S, losing passivation performance and generating CS_2, a feedstock contaminant when carried over to GHU section. With less complete passivation, higher amounts of CO and coke are formed. Apart from the performance drawbacks, DMDS is toxic and difficult to handle, adding HSE related hazards.

The studies conducted by Ghent University indicate that the sulfiding agents can have different performance in the furnace, depending on the decomposition profile and H_2S selectivity. Compared to DMDS, the new polysulfide developed reduces CO and coke formation and does not produce any CS_2. With this new option, Ethylene producers will have an interesting alternative to reduce the risk associated with CO spikes, unplanned shutdowns due to coke formation, and increase Plant profitability and carbon efficiency with the improved run-length and light yields. HSE risk is also reduced. Collectively, these observations make the case that producers should reevaluate the selection of sulfiding agents based on total cost of operation, carbon efficiency, and HSE sustainability. This is important paradigm shift in the established mindset of procurement of sulfiding agents as the cost of sulfur per kg is no longer the primary criteria for product selection.

References

1) J. Towfighi, M. Sadrameli, A. Niaei, Coke Formation Mechanisms and Coke Inhibiting Methods in Pyrolysis Furnaces, Journal of Chemical Engineering of Japan, Vol. 35, No. 10, pp. 923-937, 2002.
2) M. B. Jambor, E. Hájeková, Formation of Coke Deposits and Coke Inhibition Methods During Steam Cracking, Petroleum & Coal 57(2) 143-153, 2015.
3) M.R. Djokic, K.M. Van Geem, R. Van de Vijver, C.A.R. Pappijn, N. Olahova, G.B. Marin, Thermal Decomposition of Sulfur Compounds and their Role in Coke Formation During Steam Cracking of Hydrocarbons, Proceedings of the AIChE EPC 2018.
4) H. Zimmermann, R. Walzl, Ethylene, in: Ullmann's Encyclopedia of Industrial Chemistry, Wiley-VCH Verlag GmbH & Co. KGaA, 2000.
5) M.-F.S.G. Reyniers, G.F. Froment, Influence of Metal Surface and Sulfur Addition on Coke Deposition in the Thermal Cracking of Hydrocarbons, Industrial & Engineering Chemistry Research, 34 (1995) 773-785. 5
6) N. Rahimi, R. Karimzadeh, S.M. Jazayeri, K.D. Nia, An empirical investigation of the influence of sulfur additives on the catalytic rate of coke deposition and CO formation in the steam cracking of LPG over Incoloy 600 and stainless steel, Chemical Engineering Journal, 238 (2014) 210-218. 6
7) B. Parvizi, K.M. Van Geem, Influence of DA2628 and DMDS addition on Ethane steam cracking, 2022.
8) M. Bhargava, A. Patil, N. Ahmad, Challenges with Carbon Disulphide removal in petrochemical naphtha, 2000.
9) Vandamme, P. S.; Froment, G. F., Putting computers to work - Thermal cracking computer control in pilot plants. Chemical Engineering Progress 1982, 78, (9).
10) Dierickx, J. L.; Plehiers, P. M.; Froment, G. F., ONLINE GAS-CHROMATOGRAPHIC ANALYSIS OF HYDROCARBON EFFLUENTS - CALIBRATION FACTORS AND THEIR CORRELATION. Journal of Chromatography 1986, 362, (2), 155-174.
11) Dhuyvetter, I.; Reyniers, M. F.; Froment, G. F.; Marin, G. B.; Viennet, D., The influence of dimethyl disulfide on naphtha steam cracking. Ind. Eng. Chem. Res. 2001, 40, (20), 4353-4362.
12) Van Geem, K. M.; Pyl, S. P.; Reyniers, M. F.; Vercammen, J.; Beens, J.; Marin, G. B., On-line analysis of complex hydrocarbon mixtures using comprehensive two-dimensional gas chromatography. Journal of Chromatography A 2010, 1217, (43), 6623-6633.

DMS/ DMDS Overview

Debby Rossana, Sherri Elder, Lisheng Xu, Andre Bernard

The cracking furnace in an Ethylene plant is prone to coke formation and fouling. Organosulfur compounds have been shown to help reduce coke formation in the furnace tubes. Dimethyl Sulfide (DMS) and Dimethyl Disulfide (DMDS) are the two most commonly used organosulfur compounds. There are drawbacks to these materials and new metal passivating materials are always being developed . The use of sulfur species in the Ethylene plant has been covered extensively in the EPC proceedings. This paper will look at the mechanism of DMS and DMSO in the furnace and provide a summary of the current literature available in the EPC proceedings.

Advances in Purification of Refinery Off-Gas Streams - Including a New Lead-Free Arsine Adsorbent

Stephen Caskey, Jay Gorawara, Allan Hatami

Refinery off-gas streams derived primarily from fluid catalytic cracking units (FCCU's) and coker units have historically been sent to fuel gas systems, but these light hydrocarbon rich streams can be a cost advantaged feedstock for steam crackers (ethylene plants). Prior to being fed to the fractionation section of a steam cracker, removal of a variety of contaminants is required to meet ethylene plant product specifications or result in downstream deactivation of selective hydrogenation or polymerization catalysts.

Adsorbents are cost effective and efficient means of removing many of these problematic contaminants including arsine, mercury, chloride, oxygenates, sulfur compounds and more. The adsorption systems can be operated in either regenerative or non-regenerative mode. Honeywell UOP with its extensive adsorption technology expertise, field service capabilities, broad adsorbent product portfolio and commercial experience in refinery off-gas treatment can offer solutions to these purification demands.

Honeywell UOP has developed new materials for application in refinery-off gas streams. A lead-free arsine removal adsorbent has been patented. The new arsine removal adsorbent, free of environmentally problematic lead oxide, has been field tested at a major olefins producer in Asia and is now available for use.

In this paper we will discuss:

- Contaminants of concern in refinery off-gas streams fed to steam crackers.
- Appropriate adsorbents for each purification service with case studies and resultant attainable contaminant effluent specification.
- Optimum treater operating guidance.
- Honeywell UOP's adsorption unit design and field service capabilities.

AIChE Paper Number: 123a

The U.S. Approach to Regulating GHG Emissions– A Practical Summary

Everard Ashworth
Principal
Ashworth Leininger Group, Camarillo, CA

Prepared for Presentation at the 2023 Spring National Meeting
Houston, Texas, March 23, 2023

The U.S. Approach to Regulating GHG Emissions– A Practical Summary

Everard Ashworth
Principal
Ashworth Leininger Group, Camarillo, CA

Abstract:

In more than a decade since the Supreme Court held that GHGs are "air pollutants" under the Clean Air Act, EPA has wrestled with how to regulate GHG emissions from existing, new, and modified sources. In June 2022, the Supreme Court held in *West Virginia v. Environmental Protection Agency* that EPA lacked authority to issue regulations restricting GHG emissions from power plants, finding that Congressional authorization was needed for EPA to do so. Sixteen states and Puerto Rico have adopted their own GHG emission reduction requirements independent of federal law; these are unaffected by the Supreme Court decision. Other states and the District of Columbia have leveraged administrative and executive action to quantify GHG emissions and to commit to GHG reduction. This paper and related presentation provide a practical review of how GHG emissions are regulated in the U.S. at the federal and state levels.

INTRODUCTION

GHGs are a class of compounds that trap heat in earth's atmosphere. Under U.S. Environmental Protection Agency (EPA) programs, GHGs are defined to include carbon dioxide (CO_2), methane, nitrogen oxides, and fluorinated gases, each of which has a unique Global Warming Potential (GWP) expressed as CO_2 equivalent (CO_2e).[1] This paper and related presentation provide a practical review of how GHG emissions are regulated in the U.S. at the federal and state levels.[2] The first section of this paper identifies the seminal actions taken by the judicial, executive, and legislative branches of the U.S. government to control GHG emissions. The second section provides a broad overview of the regulatory mechanisms that state governments have fashioned to inventory and control GHG emissions from major sectors and sources within their jurisdictions.

SECTION 1 – FEDERAL GHG REGULATORY SCHEMES

EPA regulation of GHG at the federal level has focused on mobile sources, large stationary sources, and power generation, which accounted for 27%, 24%, and 25% respectively of U.S. GHG emissions in 2020.[3] Seminal Supreme Court decisions have upheld EPA's authority to regulate GHG emissions; however, the *West Virginia* decision effectively limited the extent to which EPA may regulate GHG emissions from existing "major stationary sources" under EPA's established permitting and emission control programs. In response to various court decisions, recent federal legislative efforts and EPA proposed regulatory actions seek to add GHG emissions to EPA's so-called "command and control" rules and to offer significant federal funding and tax advantages to new and emerging technologies that reduce GHG emissions. Additionally, regulations proposed by the Securities and Exchange Commission (SEC), when adopted, will require disclosure and transparency for corporate GHG inventories as part of routine financial investment reporting. This practical summary of federal GHG regulations first considers current rules in force affecting mobile and stationary sources, respectively, followed by focused summaries of pending regulatory actions. Relative to ethylene producers, Section 1.3-HFC Emissions, Section 1.4.1 – GHG Emissions in the Permitting Context, Section 1.4.2-GHG Reporting, Section 1.5-Inflation Reduction Act, and Section 1.6-Security and Exchange Commission Proposed GHG Disclosure Scheme will be of most interest.

---

[1] See generally https://www.epa.gov/ghgemissions/overview-greenhouse-gases Recent proposed SEC regulations and state GHG programs include other compounds such as black carbon, sulfur hexafluoride and other gases.

[2] This paper does not consider voluntary commitments under the various Conference of Parties (COP) held to advance the United Nations Framework Convention on Climate Change.

[3] See generally https://www.epa.gov/ghgemissions/sources-greenhouse-gas-emissions

1.1 Motor Vehicle GHG Emissions[4]

The Clean Air Act requires the EPA Administrator set emission standards for "any air pollutant" emitted from motor vehicles "which in his judgement "cause[s] or contribute[s] to air pollution which may reasonably be anticipated to endanger public health or welfare." (42 USC §7415(a)) In 2003, under the George W. Bush Administration, EPA made two determinations relative to GHG emissions from motor vehicles: first, that it lacked the authority to regulate GHGs for climate change purposes and second, that even if it did have such authority, it would decline to set GHG emission standards for motor vehicles. The State of Massachusetts together with a broad spectrum of states, cities, environmental groups, and other non-government organizations sued EPA. In *Massachusetts v. EPA*, a divided panel of the D.C. Circuit denied petitions for review of EPA's action. A number of states, local governments, and environmental organizations sought further review in the Supreme Court. The Supreme Court first held that Massachusetts had standing to seek review of EPA's denial in light of the harms global warming could cause that State, including submersion of State-owned property on the coast due to increase in ocean levels.

The Court then held that "greenhouse gases fit well within the Clean Air Act's capacious definition of 'air pollutant.'" The Court reversed EPA's denial of the petition, finding that EPA hadn't given a reasoned explanation for its refusal to decide whether GHGs cause or contribute to climate change. The Court remanded the matter to EPA, but expressly didn't reach the question of whether EPA must make a finding that GHGs endanger public health or welfare, or whether policy concerns may inform EPA's actions in the event that it makes such a finding.

Using the authority affirmed by the Court in *Massachusetts v. EPA*, subsequent Obama Administration EPA Administrator Lisa Jackson found that GHG emissions are air pollutants that endanger public health or welfare, and that motor vehicles cause or contribute to that pollution. Jackson issued her "endangerment and cause-or-contribute findings" in December 2009.[5] In the period 2010-2016, EPA established engine performance standards to limit CO_2 emissions from light duty and medium and heavy duty engines with phased compliance based on the year of manufacture beginning in 2012 through 2025. EPA estimated that when the rules are fully implemented GHG emissions from light duty cars and trucks will be reduced by half and medium/heavy duty truck GHG emissions will decline by a third. EPA's technical analyses recognized that the rules would increase the cost of internal combustion engines but demonstrated that these costs would be recouped through fuel savings accruing over a vehicle's useful life.

---

[4] See generally: EPA regulation of light duty cars and trucks, https://www.epa.gov/regulations-emissions-vehicles-and-engines/regulations-greenhouse-gas-emissions-passenger-cars-and; EPA regulation of commercial trucks, https://www.epa.gov/regulations-emissions-vehicles-and-engines/regulations-greenhouse-gas-emissions-commercial-trucks; EPA regulation of aircraft engine exhaust, https://www.epa.gov/regulations-emissions-vehicles-and-engines/regulations-greenhouse-gas-emissions-aircraft.

[5] These findings were later upheld by the Supreme Court in *Util. Air Regulatory Group v. EPA* (2014).

On April 30, 2020, the Trump Administration issued its proposed Safer Affordable Fuel-Efficient (SAFE) Vehicles Rule for Model Years 2021–2026 Passenger Cars and Light Trucks, which rolled back both the rate of reduction in CO_2 emissions and the Consumer Label Fuel Economy standards. Concurrent with its proposed rulemaking, the Trump Administration sought to harmonize motor vehicle emission standards nationwide by eliminating a waiver that allowed the California Air Resources Board's to set separate motor vehicle regulations, which were more stringent than the rollbacks finalized by EPA. Interestingly, automotive manufacturers were divided on this approach, some siding with the Trump Administration, while others expressed their intention of manufacturing engines to meet California's stricter requirements, which have been adopted by twelve other states, representing 30% of the domestic motor vehicle market.

With the transition to President Biden's Administration, EPA re-visited the Trump motor vehicle regulatory rollbacks and in April 2021 withdrew the Trump Administration's SAFE Vehicle Rule. In August 2021 EPA proposed to reinstate Obama-era standards for light duty vehicles but at more stringent levels. The rule was finalized in December 2021. Table 1 presents a summary of the Obama, Trump, and current Biden Administration light duty vehicle rules.

Table 1: Rulemaking Summary, EPA Light Duty Vehicle Model Year (MY) Fleet Average Standards and Projections

Date, Rule Status, and MY Deadline	EPA Projected Fleet-wide CO_2 Emissions Standards	EPA CO_2 Standards expressed as "MPG equivalent"**	EPA Estimated Real World Label Value***
2021 *Final Rule* MY 2026	161 grams/mile	55 mpg	40 mpg
2021 *Proposed Rule* MY 2026	173 grams/mile	52 mpg	38 mpg
2020 *Final Rule* (SAFE2) MY 2026	208 grams/mile	43 mpg	32 mpg
2012 *Final Rule* MY 2025 Standards	180 grams/mile	50 mpg	36 mpg

*All values calculated using the final rule updated fleet mix of 47% cars and 53% trucks in MY2026.
**MPG equivalent is the MPG assuming the GHG standards are met exclusively by reducing tailpipe CO_2.
***This is a value that would be comparable to what a consumer would see on a fuel economy label and reflects real-world impacts on GHG emissions and fuel economy that are not captured by the compliance tests, including high speed driving, air conditioning usage, and cold temperature operation.

EPA analysis shows that the cost of the cleaner, more efficient engines is offset by the attendant fuel savings, which in some cases can be recouped in less than 2 years. However, the practical difficulty of realizing significant reductions in GHG emissions through improved GHG emission performance turns on the median useful life of the vehicle. Based on U.S. Bureau of Transportation Statistics, the average age of passenger cars and light duty trucks has increased steadily from 5 years in 1969 to over 12 years in 2021[6]; the average age of heavy duty trucks is even greater. New cars today are expected to last for 18 years or more. Therefore, to spur integration of new, efficient technology some states have adopted fleet average performance standards, which require accelerated retirement of older vehicles; this is discussed further in Section II.

Concurrent with its light duty vehicle rule, the Biden Administration laid out its plans to regulate commercial trucks and buses and heavy duty vehicles (the so-called Clean Trucks Plan). The regulatory actions that make up the Clean Trucks Plan are:

- Setting stronger nitrogen oxide (NO_x) standards for heavy duty trucks beginning in model year (MY) 2027 and "Phase 2" tightening of GHG emissions standards for MY 2027 and beyond;
- Setting stronger emissions standards for medium-duty commercial vehicles for MY 2027 and later (These revised standards will be proposed in combination with new standards for light-duty vehicles for MY 2027 and beyond.); and
- "Phase 3" GHG standards for heavy-duty vehicles beginning as soon as MY 2030 that are significantly stronger than the Phase 2 MY 2027 GHG standards.

In the final rule, signed December 20, 2022, EPA implemented these three actions. Additional forthcoming aspects of the Clean Trucks Plan are discussed in the future regulations section, below.

EPA did not take final action on the portion of the proposed rule regarding changes to Phase 2 heavy-duty GHG standards. EPA intends to undertake a separate rulemaking regarding more stringent heavy-duty GHG standards in the future (i.e., heavy-duty GHG Phase 3 standards), as part of the Clean Trucks Plan.

<u>1.2 Aircraft GHG Emissions</u>

In 2016, under Clean Air Act Section 231(a), EPA made findings that GHG emissions from certain classes of engines used in aircraft contribute to the air pollution that causes climate change, endangering public health and welfare. These findings triggered a requirement for EPA to promulgate standards addressing GHG emissions from engines on certain aircraft. On January 11, 2021, EPA adopted aircraft engine CO_2 emission standards that match those adopted by the International Civil Aviation Organization (ICAO) in 2017. EPA has announced that it will press for more stringent CO_2 emission controls, including use of low carbon/carbon neutral aviation fuels, as part of the ongoing ICAO GHG engine standards review.

[6] See: https://www.bts.gov/content/average-age-automobiles-and-trucks-operation-united-states

1.3 HFC Emissions[7]

Hydrofluorocarbons (HFCs) are potent greenhouse gases intentionally developed as replacements for chlorofluorocarbons (CFCs), identified as significant ozone-depleting substances (ODS) used in the refrigeration, air conditioning, aerosols, fire suppression, and foam blowing sectors. As noted in the introduction, the global warming potential (GWP) of a GHG measures its relative climate impact. In the case of HFCs, GWP can be hundreds to thousands of times greater than CO_2. HFC use is growing worldwide due to the phaseout of ODS and increasing use of refrigeration and air-conditioning equipment globally.

1.3.1 Existing HFC Requirements

The U.S. Congress sought to reduce HFC GHG emissions through a cap and trade program that applies to manufacturers, importers, and end users of HFCs through the American Innovation and Manufacturing (AIM) Act, enacted on December 27, 2020[8]. The AIM Act aligns with the international goal of the Kigali Amendment to the Montreal Protocol and directs EPA to phase down both production and consumption of HFCs in the U.S. by 85%. This reduction is to be secured in three ways:

- First, by phasing down production and consumption of HFCs;
- Second, by maximizing reclamation and minimizing releases of HFCs from equipment; and
- Third, by facilitating the transition to next-generation technologies through sector-based restrictions.

The AIM Act identifies eighteen individual HFCs to be regulated, defines how the annual production and consumption baselines are to be established among producers and consumers of HFCs, and directs EPA to secure the 85% reduction from baseline levels by 2035. Once fully implemented, EPA expects the HFCs control program will prevent 171 million metric tonnes of CO_2 equivalent ($MMTCO_2e$) emissions annually– roughly equal to the annual GHG emissions from one seventh of the passenger vehicles registered in the United States.

EPA's initial phasedown regulation, referred to as the Framework Rule, was finalized on September 23, 2021. EPA's initial rulemaking established the methodology used for baseline calculations, allocating allowances to both producers and consumers based on the three highest non-consecutive years between 2011 and 2019 and sought a 30% reduction from baseline levels by 2024, as summarized in Table 2.

---

[7] See generally: https://www.epa.gov/ods-phaseout/phaseout-class-ii-ozone-depleting-substances

[8] https://www.epa.gov/climate-hfcs-reduction/aim-act

Table 2: Summary of HFC Phasedown

Baseline Consumption: 303.89 MMTEVe*
Baseline Production: 382.55 MMTEVe

HFC Phasedown Schedule by Year Interval	Consumption & Production Allowance Caps as a Percentage of Baseline	Estimated Consumption and Production Allowance Caps in MMTEVe	
2020–2023	90%	Consumption:	273.5
		Production:	344.3
2024–2028	60%	Consumption:	182.3
		Production:	229.5
2029–2033	30%	Consumption:	91.2
		Production:	114.8
2034–2035	20%	Consumption:	60.8
		Production:	76.5
2036 & after	15%	Consumption:	45.6
		Production:	57.4

* Million Metric Tonnes of Exchange Value Equivalent, which is numerically equivalent to the 100-year GWP of HFCs listed in the AIM Act . Under the AIM Act, the production baseline is calculated by adding: (i) the average annual quantity of all regulated substances produced in the U.S., from January 1, 2011, through December 31, 2013, and (ii) 15% of the production level of HCFCs in calendar year (CY) 1989, and (iii) 0.42% of the production level of CFCs in CY 1989. Similarly, the consumption baseline is calculated by adding: (i) the average annual quantity of all regulated substances consumed in the U.S. from January 1, 2011, through December 31, 2013, and (ii) 15% of the consumption level of HCFCs in CY 1989, and (iii) 0.42% of the consumption level of CFCs in CY 1989.

The Framework Rule also establishes how allocations can be traded among participants, defines the monitoring, recordkeeping, and reporting required of participants, and provides "administrative consequences" for non-compliance, principally retiring existing allowances or withholding future allowances.

1.3.2 Forthcoming HFC Requirements

On November 3, 2022, EPA issued a proposed rule to establish the methodology for allocating HFC production and consumption allowances in the second phase that spans years 2024 to 2028. The rule, which will be finalized in early 2023, continues applying a similar methodology to allocate general pool production and consumption allowances as the Framework Rule did for calendar years 2022 and 2023 while incorporating former new market entrants from an earlier set-aside pool as general pool allowance-holders. While EPA did not propose any changes to the methodology for allocating "application-specific" allowances (issued to entities using HFCs in specific applications), it corrected consumption baseline data and made other adjustments to the HFC phasedown implementation, compliance, and enforcement provisions. From 2024 to 2028, HFC production and consumption allowances would be capped at 40% below the baseline (i.e., historic levels).

EPA has yet to regulate equipment that processes or reclaims HFCs, although it has issued a notice requesting data on the appropriate practices, processes, or activities regarding the servicing, repair, disposal, or installation of equipment, for purposes of maximizing the reclamation and minimizing the release of certain HFCs and substitutes from equipment and ensuring the safety of technicians and consumers. EPA likely will fashion control rules similar to those in place for the repair and maintenance of air conditioning systems containing HFCs established under Title IV of the Clean Air Act.

In December 2022, EPA issued a proposed rule that focuses on the third area, the transition to alternatives through sector-based restrictions. Consistent with the AIM Act, EPA is proposing to restrict the use of certain higher-GWP HFCs in aerosols, foams, and refrigeration, air conditioning, and heat pumps. The proposed rule would prohibit the manufacture and import of products containing restricted HFCs by January 1, 2025, in most cases, and would prohibit the sale, distribution, and export of products containing restricted HFCs a year later, which in most cases would be January 1, 2026. EPA intends to finalize the rule in 2023.

EPA has not regulated other GHG compounds with extreme GWP though a similar phase down. However, both California and Massachusetts have adopted regulations to limit, then phase out, sulfur hexafluoride (SF_6)– which has a GWP 23,900 times that of CO_2. These state requirements are discussed in Section II.

1.4 New and Modified Stationary Source GHG Emissions[9]

1.4.1 GHG Emissions in the Permitting Context

EPA's initial GHG Tailoring Rule, finalized in May 2010, sought to integrate GHG emission thresholds into existing EPA permitting schemes: EPA's Prevention of Significant Deterioration (PSD) regulations applicable to certain new and modified "major stationary sources" and EPA's Title V Operating permit program covering operating permits for all "major sources" of GHG emissions, *regardless of the status of the source for conventional pollutants* (i.e., even if the source is considered a "minor source" of oxides of nitrogen and sulfur, carbon monoxide, particulate matter (PM, PM10 and PM2.5) and lead). The new rule phased in requirements: Step 1 for so-called "anyway sources" that triggered applicability, which was in place for a six-month period, and Step 2 in place for a 12-month period, applicable to major sources of GHG. EPA subsequently modified this program with Step 3 of the GHG Tailoring Rule, issued in June 2012. Both the PSD and Title V operating permit requirements would apply to "major stationary sources" as established in Steps 1 and 2 based solely on potential emissions of GHG in excess of 100,000 tons/year. In addition, the Step 3 rule allowed major sources to secure plantwide applicability limitations (PALs) by allowing GHG PALs to be established on carbon dioxide equivalent (CO_2e) emissions, in addition to the already available mass emissions PALs, and to use the CO_2e-based applicability thresholds for GHGs provided in the "subject to regulation" definition in setting the PAL on a CO_2e basis. The rule also revised the PAL regulations to allow a source that emits or has the potential to emit at least 100,000 tons per year of CO_2e but has

[9] See generally: https://www.epa.gov/nsr/clean-air-act-permitting-greenhouse-gases

minor source emissions of all other regulated NSR pollutants, to apply for a GHG PAL while still maintaining its minor source status.

On June 23, 2014, the U.S. Supreme Court issued its decision in *Utility Air Regulatory Group v. EPA* ("UARG Decision"), overturning the Tailoring Rule. The Court held that EPA may not treat GHGs as an air pollutant for purposes of determining whether a source is a major source required to obtain a PSD or Title V permit. The Court also held that PSD permits that are required based on emissions of other pollutants may continue to set limitations on GHG emissions based on the application of Best Available Control Technology (BACT) to the extent GHG emissions exceed a "significance" threshold, which the Court directed EPA to establish. In accordance with the UARG Decision, on April 10, 2015, the D.C. Circuit issued an amended judgment in *Coalition for Responsible Regulation, Inc. v. EPA*, which, among other things, vacated the PSD and Title V regulations under review in that case to the extent that they required a stationary source to obtain a PSD or Title V permit solely because the source emits or has the potential to emit GHGs above the applicable major source thresholds. The D.C. Circuit also directed EPA to consider whether any further revisions to its regulations are appropriate in light of the UARG Decision, and if so, to undertake to make such revisions.

In response to these court decisions, EPA modified its permitting programs to only require "major sources" of conventional pollutants to consider GHG emissions. In the context of its PSD program, only new sources that are considered "major" for a conventional pollutant (i.e., the potential to emit more than 100 tons/year for listed sources or 250 tons/year for non-listed sources), or modifications to major sources that are calculated to result in a "net significant increase" of a conventional pollutant. (See 40 CFR §51.166(b)(23)). Should these PSD thresholds be exceeded, a new or modified major source need only consider GHG emissions where the increase in GHG emissions exceeds zero, and net emissions increases of CO_2e exceeds 75,000 tons/year. If triggered, the new or modified source must apply BACT to reduce emissions of GHG. No ambient impact modeling is required (although many states have modeling requirements that apply to specific GHG compounds). In the context of the Title V program, consideration of GHG is limited to sources that are considered "major sources" independent of potential GHG emissions.

1.4.2 GHG Reporting – Major Stationary Sources

Separate of its NSR/PSD/Title V Operating Permit programs, EPA requires annual GHG reporting under 40 CFR Part 98[10] if:

- GHG emissions from covered sources exceed 25,000 metric tons CO_2e per year;
- Supply of certain products would result in over 25,000 metric tons CO_2e of GHG emissions if those products were released, combusted, or oxidized; or
- The facility receives 25,000 metric tons or more of CO_2 for underground injection.

[10] https://www.ecfr.gov/current/title-40/chapter-I/subchapter-C/part-98?toc=1; see generally https://www.epa.gov/ghgreporting

Notably, EPA's mandatory GHG reporting does not require any control or percent reduction of GHG emissions, and unlike California, Washington, and Oregon GHG reporting requirements, does not require third party attestations as to the accuracy of the reported data. Part 98 requires reporting for over forty different source categories, with more than 8,000 facilities reporting annually. Earlier this year[11], EPA issued a proposed rule that would change the existing program's data collection requirements, expand the collection of data to new industrial sectors, and alter the confidential business information (CBI) protections. Key changes in the proposal include the aggregation of emissions reported for combustion sources (Subpart C) by equipment configuration or type, reporting of GHG emissions associated with a facility's energy consumption, requiring facilities to use specific GHG emissions factors, timing for implementation of these new reporting requirements, and expansion of emission sources in the petroleum oil and gas sector.

Separately, EPA has also sought to regulate methane emissions from oil and gas production facilities under its New Source Performance Standards (NSPS) program. As with the Clean Power Plan, the oil and gas NSPS has been modified in succession from the Obama Administration to the Trump Administration, and most recently, to the Biden Administration, which in November 2022 issued a supplemental proposed rule.[12] The proposal seeks to harmonize NSPS, the Inflation Reduction Act (described below), and proposed GHG reporting requirements, illustrating the complexity of regulating GHG emissions from a specific sector.

1.4.3 Electric Power Generation GHG Emissions

EPA's first effort to control GHG emissions from stationary sources focused on the power industry, which, as noted at the beginning of Section I, in 2020 accounted for approximately 27% of U.S. GHG emissions. While, as discussed below, various court decisions blocked this initial regulation, the energy sector has nevertheless turned away from coal-fired electric power generation. Today there are no pending proposed coal fired plants and between 2000 and 2020 GHG emissions from coal-fired power plants decreased by over 50%, as shown in Figure 1.

[11] See 87 Fed. Reg. 36920, June 21, 2022, and 87 Fed. Reg. 42988, EPA-HQ-OAR-2019-0424

[12] See generally: https://www.epa.gov/controlling-air-pollution-oil-and-natural-gas-industry/actions-and-notices-about-oil-and-natural

Figure 1: U.S. Primary Energy Production, 1950-2021

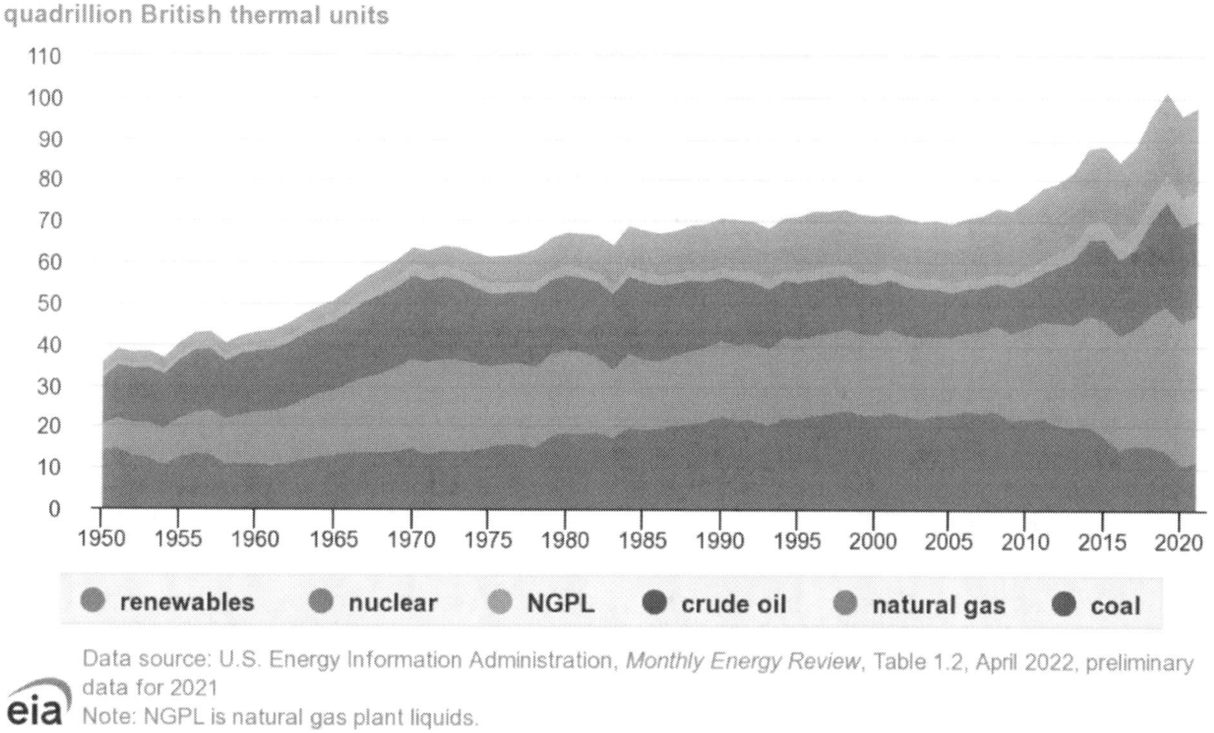

Data source: U.S. Energy Information Administration, *Monthly Energy Review*, Table 1.2, April 2022, preliminary data for 2021
Note: NGPL is natural gas plant liquids.

1.4.3.1 Obama Clean Power Plan

Clean Air Act Section 111(b) directs EPA to identify categories of stationary sources that significantly contribute to dangerous air pollution and to establish emission standards for air pollutants emitted by new sources in the listed categories. In 1971, power plants were listed and regulated under Section 111(b), which directs EPA to establish NSPS maximum emission levels for new and modified major stationary sources. To set the NSPS, EPA determines the best system of emission reduction (BSER) that has been "adequately demonstrated," considering costs and any non-air-quality health and environmental impact and energy requirements. Once EPA promulgates an NSPS under Section 111(b), Section 111(d) requires EPA to promulgate regulations for existing sources. The Section 111(d) rules set procedures for states to submit plans establishing performance standards for existing sources that would be subject to NSPS if they were new, barring an exclusion under Section 111(d).

In August 2015, the Obama Administration EPA promulgated its Power Plant Rule which sought to reduce GHG emissions from the power plant sector by one third over 2005 levels. Leveraging Section 111(d), in the rule EPA directed states to develop customized power plant GHG reduction plans to either establish source-specific requirements for all affected power plants within the state or develop a state measures plan to meet the state's mass-based reduction goal by utilizing a mixture of measures that are not included as federally-enforceable components of the plan, such as renewable energy standards and programs to improve residential energy efficiency.

While this rule ultimately was overturned by the Supreme Court in *West Virginia v. EPA*, it is worth examining the measures EPA instituted in the 2015 Power Plant Rule, as many of the regulatory measures will likely be used in future NSPS regulations as applied to other sectors (e.g., interim targets, mass or rate based performance goals, rate of reduction requirements and backstops). The flexibility in regulatory measures EPA sought to derive from Section 111(d) was rejected by the Supreme Court in *West Virginia v. EPA:*

Existing Units: EPA first established interim and final CO_2 emission performance rates for two subcategories of fossil fuel-fired electric generating units: (1) fossil fuel-fired steam generating units (generally, coal- and oil-fired power plants) and (2) natural gas-fired combined cycle generating units. EPA then offered states three separate metrics to achieve its state-wide interim and final standards:

- A rate-based state goal measured in pounds per megawatt hour (MWh);
- A mass-based state goal measured in total short tons of CO_2; or
- A mass-based state goal with a new source complement measured in total short tons of CO_2.

States were directed to develop and implement plans to ensure that the power plants in their state – either individually, together, or in combination with other measures – achieved the interim CO_2 emissions performance rates over the period of 2022 to 2029 and the final CO_2 emission performance rates – either rate-based goals or mass-based goals – by 2030. The complex rule provided for backstops and coordination among the major electrical grids within the U.S. and among neighboring states.

New Units: EPA promulgated GHG performance standards for new and modified power plants under CAA Section 111(b) concurrent with the 111(d) standards for existing plants (discussed below). The 2015 GHG NSPS for new and modified power plants relied partly on carbon capture and storage (CCS). In 2018, EPA proposed, among other things, to replace the BSER determination for new coal units (i.e., partial CCS) with a determination that BSER for these units would be the most efficient demonstrated steam cycle in combination with best operating practices. Since the 2018 proposal has not been finalized and was not affected by the Supreme Court rulings, the 2015 GHG NSPS for new and modified power plants remains in effect. New natural gas power plants can emit no more than 1,000 pounds of CO_2 /MWh of electricity produced, which is achievable with the latest combined cycle technology. New coal power plants can emit no more than 1,400 pounds CO_2/MWh, which almost certainly requires the use of CCS technology to achieve.

1.4.3.2 Trump Affordable Clean Energy Rule

With Trump's election in fall of 2016, it was obvious that the Clean Power Plan would not be implemented, especially since the rule was stayed by the Supreme Court in February 2016. In March 2017 the new president issued an executive order that laid out a roadmap to Clean Power Plan reversal. Subsequently in October 2017, EPA proposed rescission of the

Clean Power Plan rule, in August 2018 it proposed a substitute rule, and in July 2019 it finalized both proposals as the Affordable Clean Energy (ACE) Rule.[13]

Existing units: The ACE Rule directed states to require coal-fired power plants to use prescribed technologies to improve their heat rate (i.e., increase efficiency), as a BSER for CO_2. It replaced the August 2015 Clean Power Plan's state-specific emission rates and provided various market-based flexible compliance provisions. The ACE Rule was not expected to reduce power sector emissions from a business-as-usual scenario.

New Units: The ACE Rule set BSER for newly constructed large units equivalent to a super-critical coal plant, which has an emissions rate of 1,900 lbs. CO_2/MWh and set BSER for small units at 2,000 lbs. CO_2/MWh. The ACE Rule also sets separate performance standards for newly constructed and reconstructed coal refuse-fired units at an emissions rate at 2,200 lbs. CO_2/MWh. The ACE rule specified that the standards of performance for reconstructed power plants be consistent with the emission rates of newly constructed units.

In its repeal of the Clean Power Plan, EPA stated that it had exceeded its authority under Section 111(d), explaining that the Clean Power Plan had based its regulations on "a shift in the energy generation mix at the grid level." The rulemaking record indicated that EPA determined the interpretive question raised by the Clean Power Plan fell under the "major questions" doctrine, under which a clear statement by Congress is necessary for a court to conclude that Congress intended to delegate authority "of this breadth to regulate a fundamental sector of the economy." In repealing the Clean Power Plan, EPA found it had no such authority. Both rules were immediately challenged by a group of states and environmental organizations, and on the last full day of the Trump administration, two out of three judges on a D.C. Circuit Court of Appeals panel declared the ACE Rule unlawful (*American Lung Association v. EPA)*.

The issue was then taken up by the Supreme Court. On June 30, 2022, the Court issued its decision in *West Virginia v. EPA*, which limited the EPA's ability to regulate CO_2 emissions from power plants by holding that the Clean Power Plan established in 2015 under the Obama administration went beyond EPA's regulatory mandate. Specifically, the court majority found that EPA had exceeded its congressionally delegated authority by pushing utilities to make system-wide moves away from coal-generated power and towards cleaner forms of electricity generation. The issue presented to the Court was whether the U.S. Constitution gives Congress the authority to delegate broad regulatory power to EPA related to GHG emissions. More specifically, the Court framed the issue as whether under Clean Air Act Section 111(d) Congress had authorized EPA to issue rules—including those capable of reshaping the nation's electricity grids and unilaterally decarbonizing virtually any sector of the economy—without any limits on what the agency can require so long as it considers cost, non-air impacts, and energy requirements. The Court held that Congress must provide clear direction to EPA rather than a broad delegation of power for the agency to regulate GHG emissions. The Court's 6-3 decision

[13] See generally https://fas.org/sgp/crs/misc/R46568.pdf

reversed the D.C. Circuit's previous ruling the ACE Rule was unlawful and remanded the case for further proceedings.

EPA has yet to propose a revised NSPS for electrical power generation, but it recently issued a proposed NSPS that seeks to reduce methane emissions from major oil and gas facilities. In light of *West Virginia v EPA*, EPA's ongoing efforts to control GHG emissions from major sources will likely follow the approach in that methane rulemaking, which focuses on controls within the fence line of individual affected sources.

1.5 Inflation Reduction Act Incentives to Reduce GHG Emissions[14]

While not a regulatory control program, the federal Inflation Reduction Act (IRA) enacted in August 2022 provides unprecedented funding through tax credits for a broad spectrum of clean energy and climate programs, representing perhaps the most significant piece of climate legislation adopted by Congress to date. The IRA directs nearly $350 billion in federal funding to clean energy, with the goal of substantially lowering the nation's carbon emissions by the end of this decade. The funds will be delivered through a mix of tax incentives, grants, and loan guarantees. Since the scope and level of engagement by industry to develop and implement clean energy projects to secure these credits is unknown, the final cost of the IRA remains unknown as well. Clean electricity and transmission grid enhancements received the largest allocation ($250 billion), followed by manufacturing ($48 billion), clean transportation, including electric vehicle incentives ($23 billion), agriculture ($21 billion), and water projects ($5 billion). A focused summary of the major credit schemes is provided below.

Section 45Q Carbon Capture and Sequestration Payments and Credits: This new provision of the Internal Revenue Code establishes a two-tiered tax credit scheme for direct air capture facilities with a capacity >1,000 metric tonnes (MT) per year built after January 2023 and electric generation facilities that employ a variety of storage technologies with a design capture capacity >75% of emissions and meet other performance standards. Payments are calculated for a five-tiered system, which provides additional moneys to taxpayers that meet minimum salary, training, and other employee-based standards. Direct payment is available for tax-exempt organizations that include state and tribal governments, electricity cooperatives, and the Tennessee Valley Authority. Otherwise, the taxpayer receives tax credits, which are transferable to a third party.

Wind and Solar Tax Credits: The IRA provides extensions and amendments to existing income tax credits for solar and on-shore and off-shore wind projects that begin construction before January 1, 2025.

Advanced Facilities Deployment Program: The IRA provides $5.82 billion in new funding for high carbon intensity facilities that install, update, or retrofit specific implementing

[14] See generally: https://www.epa.gov/inflation-reduction-act

technologies approved by the Secretary of Energy. Relative to the chemical manufacturing sector these technologies include the following.

- Emissions reductions in medium- and high-temperature heat generation, including
  - Electrification of heating processes,
  - Renewable heat generation technology,
  - Combined heat and power, and
  - Switching to alternative fuels, including hydrogen and nuclear energy;
- Emissions reductions in chemical production processes, including by incorporating, if appropriate and practicable, principles, practices, and methodologies of sustainable chemistry and engineering;
- Leveraging smart manufacturing technologies and principles, digital manufacturing technologies, and advanced data analytics to develop advanced technologies and practices in information, automation, monitoring, computation, sensing, modeling, and networking to
  - Model and simulate manufacturing production lines,
  - Monitor and communicate production line status,
  - Manage and optimize energy productivity and cost throughout production, and
  - Model, simulate, and optimize the energy efficiency of manufacturing processes;
- Leveraging the principles of sustainable manufacturing to minimize the potential negative environmental impacts of manufacturing while conserving energy and resources, including
  - Designing products that enable reuse, refurbishment, remanufacturing, and recycling,
  - Minimizing waste from industrial processes, including through waste energy recovery and combined heat and power, and
  - Increasing resource efficiency;
- Increasing the energy efficiency of industrial processes, including
  - Alternative materials that produce fewer emissions during production and result in fewer emissions during use, including
    - > High-performance lightweight materials and
    - > Substitutions for critical materials and minerals;
  - Development of net-zero emissions liquid and gaseous fuels, including
    - > Emissions reduction in shipping, aviation, and long-distance transportation,
    - > Carbon capture technologies for industrial processes, and
    - > Other technologies that achieve net-zero emissions in nonpower industrial sectors, as determined by the Energy Secretary, in consultation with the SEC.

New Methane Fee for Oil and Gas Industry: The IRA added Section 136 to the Clean Air Act to impose a new emissions fee applicable to oil and gas facilities that emit more than

25,000 MT of CO_2e per year. Beginning in 2024, owners and operators of oil and gas production, processing, gathering, boosting, transmission, storage, and import/export facilities will be required to pay $900 per MT of methane emitted. The fee increases to $1,200 per MT in 2025 and to $1,500 per MT in 2026 and subsequent years. Natural gas distribution facilities are not subject to the charge. Notably, the fee is assessed only on methane emissions that exceed Clean Air Act Section 111 NSPS as proposed by EPA in November 2021, thus encouraging operators to reduce their methane emissions. Further, the fee is applied to CO_2e emissions as reported under EPA's GHG reporting program and directs EPA to amend its GHG reporting rule within the next two years to ensure reporting is based on empirical data that accurately reflect emissions from covered sources. The Congressional Budget Office estimates that when fully implemented, the new methane fee will generate $1.4 billion in net revenue in 2028.

The IRA provides for $850 million in methane reduction-related grants and loan funding, and another $700 million in funding for methane reduction activities at marginal conventional oil and gas wells.

Expanded Scoring of Clean Hydrogen Technologies: The IRA broadens the definition of clean hydrogen projects eligible for funding under the Department of Energy's $8 billion regional clean hydrogen hub programs administered under Section 813 of the Energy Policy Act.

While the practical impact of the IRA will not be known for some time, the Biden administration anticipates this new legislative package will allow the US to reduce GHG emissions by ~40% below 2005 levels in 2030, making the IRA the most significant GHG reduction scheme in the U.S. As discussed below, the Securities and Exchange Commission (SEC) likely will adopt amendments to its financial disclosure regulations to require companies to disclose their GHG inventories. Coupled with the unprecedented funding for clean energy projects, it is reasonable to anticipate that corporate management will direct projects to secure financial incentives, increase market share, and achieve concurrent GHG reductions. The success of these efforts will turn in large part on development and integration of new, innovative technologies that realize sizable GHG reductions as compared to current operations.

1.6 Securities and Exchange Commission Proposed GHG Disclosure Scheme

The Security and Exchange Commission (SEC) requires both domestic and foreign publicly traded companies to register with the agency and provide annual reports of information material to investors' decisions to purchase stock. Since the passage of federal environmental laws in the 1970s, investors have increasingly been interested in the environmental risks posed by a registrant's operations. The proposed rule would codify and standardize disclosure requirements related to climate change-related risks, with many specifics related to GHG emissions. In March 2022, the SEC proposed rule changes that would require registrants to include certain climate-related disclosures in their registration statements and periodic reports, including information about climate-related risks that are reasonably likely to have a material impact on their business, results of operations, or financial condition, and certain climate-related financial statement metrics in a note to their audited financial statements[15]. The required information about climate-related risks also would include disclosure of a registrant's greenhouse gas emissions, which have become a commonly used metric to assess a registrant's exposure to such risks.

In its proposed rule, SEC pointed to the need to provide investors and issuers with consistent, comparable, and decision-useful information that would include GHG emissions data to support investment decisions, as well as the need to provide consistent and clear reporting obligations on GHG emissions for issuers. The agency also pointed out that investors representing literally tens of trillions of dollars support climate-related disclosures recognize that climate risks can pose significant financial risks to companies, and that investors need reliable information about climate risks to make informed investment decisions. In its associated press release, the SEC noted that the disclosures registrants would be required to provide are similar to those that many companies provide under existing disclosure frameworks and standards, such as the Financial Stability Board's Task Force on Climate-related Financial Disclosures (TCFD) and the Greenhouse Gas Protocol.

The proposed rule changes, which are anticipated to be finalized by mid-2023, would require a registrant to disclose information about (1) the registrant's governance of climate-related risks and relevant risk management processes; (2) how any climate-related risks identified by the registrant have had or are likely to have a material impact on its business and consolidated financial statements, which may manifest over the short-, medium-, or long-term; (3) how any identified climate-related risks have affected or are likely to affect the registrant's strategy, business model, and outlook; and (4) the impact of climate-related events (severe weather events and other natural conditions) and transition activities on the line items of a registrant's consolidated financial statements, as well as on the financial estimates and assumptions used in the financial statements.

For registrants that already conduct scenario analysis, have developed transition plans, or publicly set climate-related targets or goals, the proposed amendments would require certain disclosures to enable investors to understand those aspects of the registrants' climate risk management.

The proposed rules also would require a registrant to disclose information about its direct greenhouse gas (GHG) emissions (Scope 1) and indirect emissions from purchased electricity or other forms of energy (Scope 2). In addition, a registrant would be required to disclose GHG emissions from upstream and downstream activities in its value chain (Scope 3) if material or if the registrant has set a GHG emissions target or goal that includes Scope 3 emissions. The SEC stated that these proposals for GHG emissions disclosures would provide investors with decision-useful information to assess a registrant's exposure to, and management of, climate-related risks, and in particular transition risks. The proposed rules would provide a safe harbor for

[15] https://www.sec.gov/news/press-release/2022-46

liability from Scope 3 emissions disclosure and an exemption from the Scope 3 emissions disclosure requirement for smaller reporting companies.

For each required disclosure of a registrant's Scopes 1, 2 and 3 emissions, registrants would be required to disclose the emissions both disaggregated by each constituent GHG and in the aggregate, expressed in terms of CO_2e. Requiring a standard unit of measurement for GHG emissions, rather than different units of measurement for the different GHGs, is intended by the SCE "to simplify the disclosure for investors and enhance its comparability across registrants with different types of GHG emissions." Disclosure of Scopes 1, 2 and 3 emissions would also be required to exclude the impact of any purchased or generated offsets.

Under the proposed rule changes, accelerated filers and large accelerated filers would be required in 2025 and later years to include an attestation report from an independent attestation service provider covering Scopes 1 and 2 emissions disclosures, with a phase-in over time, to promote the reliability of GHG emissions disclosures for investors. The proposed rules would include a phase-in period for all registrants, with the compliance date dependent on the registrant's filer status, and an additional phase-in period for Scope 3 emissions disclosure.

When originally proposed in March 2022, SEC anticipated that the rule would be finalized by year's end. The comment period was extended, and thousands of comments have been submitted to the Commission, such that the rule likely will not be finalized until mid-2023.

SECTION 2 – STATE REGULATION OF GHG EMISSIONS[16]

2.1 State GHG Reduction Targets

Independent of federal GHG regulation and perhaps due to the lack of a comprehensive federal legislative scheme, twenty-four states plus the District of Columbia and Puerto Rico have adopted specific greenhouse gas reduction targets to address climate change. The targets and implementing action plans vary in complexity, with California providing the most comprehensive set of regulatory programs. As summarized below, these state initiatives include GHG reporting and auditing for major sources, carbon pricing, emission limits, renewable portfolio standards, and steps to promote cleaner transportation. Ethylene production within the U.S. is in five states: Texas, Louisiana, Illinois, Kentucky, and Pennsylvania; however, 95% of ethylene production is centered in Texas and Louisiana. Two of these states: Illinois and Louisiana have adopted climate action plans through a mx of executive orders and legislative actions that set out broad objectives to reduce GHG emissions within their state. Kentucky offers tax and production grants for alternative energy but has not adopted a climate action plan. Key initiatives among these three states include a transition from coal/fossil fuels to renewable energy, increased reporting, and auditing of GHG emissions from major stationary sources,

[16] See generally: https://www.c2es.org/category/policy-hub/state/

transition to low- and zero-emitting mobile fleets and improved GHG efficiency for new commercial/residential buildings.

Table 3 – Summary of State GHG Emission Reduction Targets

State or Jurisdiction	Authority Type	Emission Reduction Target (in GHG unless specified)	Target Date
California	Statutory	85% below 1990 anthropogenic levels Carbon neutrality	2045 2045
Colorado	Statutory	26% below 2005 levels 50% below 2005 levels 90% below 2005 levels	2025 2030 2050
Connecticut	Statutory	45% below 2001 levels 80% below 2001 levels	2030 2050
Delaware	Executive	26–28% below 2005 levels	2025
District of Columbia	Executive	50% below 2006 levels 80% below 2006 levels GHG neutrality	2032 2050 2050
Illinois	Statutory	40% renewable power 50% renewable power 100% renewable power	2030 2040 2050
Louisiana	Executive	26–28% below 2005 levels 40–50% below 2005 levels Net-zero emissions	2025 2030 2050
Maine	Statutory	45% below 1990 levels 80% below 1990 levels Net-zero emissions*	2030 2050 2050
Maryland	Statutory	40% below 2006 levels	2030
Massachusetts	Statutory	33% below 1990 levels 50% below 1990 levels 75% below 1990 levels 85% below 1990 levels Net-zero emissions	2025 2030 2040 2050 2050
Michigan	Executive	28% below 2005 levels 52% below 2005 levels Carbon neutrality Net-negative emissions	2025 2030 2050 2050 and thereafter

State or Jurisdiction	Authority Type	Emission Reduction Target (in GHG unless specified)	Target Date
Minnesota	Statutory	30% below 2005 levels 80% below 2005 levels	2025 2050
Montana	Executive	GHG neutrality	2045–50
Nevada	Statutory	28% below 2005 levels 45% below 2005 levels Zero or near-zero	2025 2030 2050
New Jersey	Statutory	1990 levels 50% below 2006 levels* 80% below 2006 levels	2020 2030 2050
New Mexico	Executive	45% below 2005 levels	2030
New York	Statutory	40% below 1990 levels 85% Net-zero emissions	2030 2050 2050
North Carolina	Executive	50% below 2005 levels Net-zero GHG emissions	2030 ASAP but no later than 2050
Oregon	Statutory	10% below 1990 levels 45% below 1990 levels* 75% below 1990 levels 80% below 1990 levels*	2020 2035 2050 2050
Pennsylvania	Executive	26% below 2005 levels 80% below 2005 levels	2025 2050
Puerto Rico	Statutory	25% renewable power 50% renewable power 100% renewable power	2025 2040 2050
Rhode Island	Statutory	10% below 1990 levels 45% below 1990 levels 80% below 1990 levels Net-zero GHG emissions	2020 2035 2040 2050
Vermont	Statutory	26% below 2005 levels 40% below 1990 levels 80% below 1990 levels	2025 2030 2050
Virginia	Statutory	Net-zero GHG emissions	2045

State or Jurisdiction	Authority Type	Emission Reduction Target (in GHG unless specified)	Target Date
Washington	Statutory	45% below 1990 levels 70% below 1990 levels 95% below 1990 levels Net-zero emissions	2030 2040 2050 2050
Wisconsin	Executive	26–28% below 2005 levels	2025

*Target established by executive authority.

2.2 State Strategies for Achieving GHG Emission Reduction Targets

Separate of the state targets, thirty-three states have adopted climate action plans[17] that outline regulatory programs, incentives, and cooperative agreements to secure GHG reductions that vary by sector; including:

2.2.1 Cap-and-Trade Programs

There are three major state-sponsored cap-and-trade programs currently in place: California's broad cap-and-trade program, Massachusetts' cap-and-trade program applicable to specific electrical generating units, and the Regional Greenhouse Gas Initiative (REGGI). It is beyond the scope of this paper to provide a comparison among these GHG reduction schemes; we therefore focus on California's AB32 program to illustrate how the cap and trade programs realize reductions in GHG emissions.

California's cap-and-trade program enacted pursuant to its Global Warming Solutions Act (known as AB32) is authorized through 2030 and the state's Climate Action Plan adopted pursuant to AB32 is essential to the state achieving its interim goal of reducing GHG emissions 40% below 1990 levels by 2030. The cap-and-trade program is comprised of approximately 200 entities that represent some 80% of the state's GHG emissions, including electric power plants, electricity importers, large industrial plants emitting 25,000 metric tons of CO_2 equivalent or more annually, and distributers of transportation fuels and natural gas. The program provides each source with a GHG emissions allocation (i.e., the "cap") that declines over time. One allowance equals one metric ton of carbon dioxide equivalent emissions. Each year, fewer allowances are created and the annual cap among all regulated sources declines. In order to achieve their allocation limit, regulated sources can reduce their emissions or purchase additional credits from the state-run market or from qualified third parties who hold offsets (i.e., the "trade"). Each year, regulated entities must cover their emissions by allowances and a limited quantity of offset credits. The increasing annual auction reserve (or floor) price of allowances, along with a reduction in annual allowances creates a steady and sustained carbon price signal to prompt regulated entities to reduce GHG emissions.

[17] See: https://www.c2es.org/document/climate-action-plans/

AB32 limits the use of third-party offsets to no more than 8% of a company's annual GHG emission totals. These third-party offsets may only be generated through eight approved compliance offset protocols or from three California Air Resources Board-approved credit trading programs: (i) the Low Carbon Fuel Standard (LCFS) establishes tradeable credits for CARB-approved projects that reduce the carbon intensity across transportation fuels consumed in California; (ii) the Renewables Portfolio Standard (RPS) establishes tradeable credits and applies an increasing renewable power standard for each utility's procurement of electricity consumed in California; or (iii) the Zero-Emission Vehicle (ZEV) Program establishes tradeable credits and applies an increasing fleet-wide efficiency standard.

Regulated sources must follow annual mandatory GHG reporting protocols with third-party attestation. If an emitting entity covered by the AB32 cap-and-trade program fails to surrender sufficient compliance instruments (allowances, or allowances plus a limited number of offset credits) to cover its emissions, that entity is automatically assessed an "untimely surrender obligation" of four times whatever is still owed. Failure to surrender this full untimely surrender obligation triggers additional enforcement where each un-surrendered instrument constitutes a separate violation. To date, only one entity has been assessed an untimely surrender obligation, that it then submitted as required. In 2014, California's cap-and-trade program was linked to Quebec's, establishing the first multi-sector cap-and-trade program linkage in North America.

Washington's cap-and-trade program, similar to California's, will go into effect in 2023. Oregon is developing a similar program as well. New York State recently adopted a cap-and-invest program, which differs from the AB32 model in that it encourages investment in GHG emissions reduction by restricting the ability of regulated parties to use carbon offsets by reducing the amount of available emission credits.

Eleven states participate in the Regional Greenhouse Gas Initiative (RGGI), a cooperative, market-based effort among the states of Connecticut, Delaware, Maine, Maryland, Massachusetts, New Hampshire, New Jersey, New York, Rhode Island, Vermont, and Virginia. Established in 2005, RGGI is a regional cap-and-trade program to reduce CO_2 emissions from all fossil fuel-fired power plants with a capacity of 25 MW or higher. The current RGGI emissions reduction target is 30% reduction in CO_2 emissions compared to 2020 levels by 2030.

2.2.2 Industrial and Commercial Sources Reporting

Four states – California, Oregon, Washington, and New York– have adopted or will adopt programs to require annual audited GHG reporting for major sources (including industrial sources, power generation, and fuel sales). These data support participation in the cap-and-trade programs described above that are applicable to large industrial operations. Many states require GHG reporting from industrial sources; however, these programs do not require third-party verification or feed into cap-and-trade control programs.

California and Colorado have adopted control measures to achieve methane reduction from oil and gas, landfills, and other sources, but large chemical and refining operations are not specifically regulated. Relative to commercial operations, states and local governments increasingly are establishing mandatory GHG efficiency standards for new or modified industrial facilities, including commercial; low-cost financing to support construction/integration of GHG reduction capacity.

2.2.3 Industrial Sources Fuel Switching

At a national level, industrial sources contribute approximately 17% of GHG emissions; however, within some states, emissions from the industrial sector can be two or three times this amount; for example: 41% in Texas and 63% in Louisiana. States have focused GHG in from the industrial sector on electrification and fuel switching, both of which require significant capital expenditures. Further complicating this effort is the fact that most industrial sources operate continuously and require power generation in the evenings and overnight. During these periods, solar power is not available and wind power, when available, typically operates a reduced capacity. Therefore, depending on the efficiency of natural gas combustion at the industrial facility, GHG emissions may actually increase when switching from natural gas to electric. Virtually all state climate action plans call for development and installation of alternative fuels, principally hydrogen and bio-based fuels,[18] however market penetration will be limited due to engineering complexities and prohibitive costs of converting the combustion units and providing carbon capture for methane reforming units that generate hydrogen.

2.2.3 Electric Grid

Renewable portfolio standards have been adopted in twenty-nine states and the District of Columbia. These standards require that a certain percentage of a utility's electricity comes from renewable energy sources. Other state initiatives include mandatory conversion from fossil fuels to carbon-neutral generating capacity and providing consumers with the option to purchase electricity from green power plants.

2.3.4 Carbon Storage

At least twenty-one states have adopted legal frameworks governing carbon capture and storage (CCS).[19] California recently adopted legislation and amended its AB32 compliance plan to align with the Governor's goal that state agencies develop regulations to require and provide credit for carbon storage technologies. Such technologies include carbon capture and sequestration, direct air capture, CO_2 removal through natural sinks, and bioenergy with CCS. California expects that use of these technologies can reduce GHG emissions by 20 MT CO_2e/year

---

[18] See https://afdc.energy.gov/laws/state

[19] See: https://www.eli.org/vibrant-environment-blog/carbon-capture-and-storage-history-current-state-and-obstacles-future-part-2

in 2030 and 100 MT CO_2e/year by 2045. Notably, the state has determined that it cannot meet its 2045 carbon neutrality goal without these storage/removal technologies.

2.2.5 Mobile Sources

Thirty-six states have adopted regulations or policies to reduce GHG emissions from the transportation sector. Many initiatives include adopting California GHG standards for new vehicle sold in the state (light duty cars and trucks, medium duty cars and trucks, and heavy duty trucks). However, only five of the seventeen states that previously adopted some or all of the California tailpipe emission standards are expected to follow the state's mandate to eliminate the sale of new gas-powered vehicles by 2035.[20] Other strategies for mobile sources include fleet fuel efficiency standards for government and private/industrial fleets, incentives or funding for the purchase of new, lower GHG-emitting vehicles, and installation of alternative fueling/distribution/transport/charging infrastructure.

2.2.6 Chemical Phase-Out

Both Massachusetts and California have adopted regulations to reduce emissions of SF_6, a GHG with a GWP 23,900 times that of CO_2.[21] Both regulations limit the leak rate of SF_6 from electric switch gear to 1%, with corrective actions required where equipment owners/operators exceed this value. California is seeking to phase out the use of SF_6 after 2025. The eleven states that participate in RGGI also target incremental actions that are aimed towards early retirement and replacement of SF_6-containing electrical equipment. Connecticut provides for SF_6 offset projects to replace SF_6-containing equipment.

Relative to the industrial sector, the dynamic nature and breadth of these state initiatives make it especially difficult to understand the long-term impact of state regulation of GHGs on facility operations. For sources located within states with cap-and-trade or cap-and-invest programs, costs of maintaining and investing in expanded operation will continue to exceed costs for similar facilities in states without such programs. These issues will become even more pronounced as incremental GHG reductions become increasingly expensive and difficult over time. Of equal concern is that re-location of production capacity within the U.S. to avoid local/state regulations likely will have little effect on total GHG emissions with the U.S. Further, state GHG reduction programs may not have the financial resources needed to invest in new/emerging technologies that offer a significant change in GHG emissions from external combustion sources and industrial process equipment. Nevertheless, absent a comprehensive federal GHG program, these state initiatives will continue and likely expand with time.

[20] See: https://www.repairerdrivennews.com/2022/09/07/17-states-mull-californias-ev-2035-mandate/

[21] See: https://www.epa.gov/eps-partnership/state-and-regional-regulations-related-sf6-emissions-electric-transmission-and

CONCLUSION

The federal Clean Air Act does not directly regulate CO_2 or methane, two compounds that constitute 90% of GHG emissions to the atmosphere. However, under *Massachusetts v. EPA,* the Supreme Court held that CO_2 is an "air pollutant" and therefore subject to regulation under the Act. Over the past 15 years, EPA's GHG control efforts have focused on regulating GHG emissions from power generation, industrial sources, and mobile sources, which taken together contribute over 75% of the GHG emissions within the U.S. The results of EPA's efforts are mixed:

> To date, EPA has yet to control the GHG emissions from the **power sector**, despite multiple rulemakings over a 15-year period that have generated millions of comments. While EPA has proposed new NSPS standards for both the power sector and oil and gas production (which comprise ~38% of GHG emissions within the U.S.), these NSPS have yet to be finalized.
>
> With the exception of GHG reporting requirements, GHG emissions from existing industrial **stationary sources** remain largely unregulated at the federal level. Only sources that manufacture or use HFCs have been regulated through EPA's HFC phasedown regulations. By September 30, 2023, annual U.S. manufacture of HFCs will be 10% below a baseline of 303.9 MMTCO2e.
>
> Mobile source regulations applicable to new **light duty vehicles** and **heavy duty trucks** are now in place to reduce GHG and ozone precursors that will realize significant reductions. However, the median useful life of both cars and trucks is such that it will take decades for these new cleaner vehicles to meaningfully penetrate mobile fleets.

Absent clear legislation and direction at the federal level, twenty-four states have adopted GHG reduction targets, the most ambitious of which seek to achieve carbon neutrality by 2045. The most aggressive of these state programs rely on a combination of cap-and-trade or cap-and-investment, low carbon fuel standards, renewable portfolio standards, electrification of mobile sources, and carbon capture/storage technologies to achieve ambitious goals. For example, state climate action plans in California, Washington, Oregon, and New York provide blueprints for how each state will achieve carbon neutrality by 2045 and 2050, respectively, which aligns with the Biden Administration's net zero goal set for 2050.

Within the existing federal and state regulatory schemes, GHG emissions within the U.S. have declined an estimated 15% from 2005 levels, with the largest reduction evidenced in the power sector, where GHG emissions from coal have been reduced by half from 2000 levels, with displaced coal capacity taken up by natural gas and renewable energy. Since 2021 – and for the first time in 60 years— power generated from renewables exceeded that from coal.[22] Notably, these reductions in the power sector were realized absent federal regulation. GHG

[22] Prior to the 1960's hydropower exceeded coal-fired power generation.

emissions from the industrial sector have remained relatively flat, despite a broad spectrum of clean energy projects, including CCS, LCFS, and energy efficiency projects.

GHG emissions from the transportation sector also have remained relatively flat, illustrating the fact that if fleet turnover is limited to median useful vehicle life, it will take decades to achieve significant reductions from the mobile fleet. This has been recognized by California and other states, which have adopted fleet efficiency standards that require accelerated turnover to meet stringent composite fleet performance standards.

Separate of these federal and state regulatory programs, last year Congress adopted the Inflation Reduction Act (IRA) that provides an estimated $350 billion in federal funding for a broad spectrum of clean energy initiatives which have the potential to reduce GHG emissions ~50% below 2005 levels by 2030.[23] These expected reductions are far greater than would be achieved by any proposed or final GHG regulations issued to date by EPA, and could be significantly higher than the previous U.S. pledge to cut GHG emissions 26-28% by 2025. Rather than prescribe emission control plans or mandatory performance standards, IRA provides financial incentives via tax benefits and funding with the goal of substantially lowering the nation's GHG carbon emissions by 2030. Additionally, IRA will support more stringent EPA regulation of mobile and stationary sources as costs to comply with EPA NSPS, and mobile source provisions will be offset by favorable tax benefits. Coupled with the SEC's proposed GHG disclosure regulation expected to be finalized later this year, it is reasonable to anticipate that corporate management will direct projects to secure financial incentives, increase market share, and achieve concurrent GHG reductions. The success of these efforts will turn in large part on development and integration of new, innovative technologies that realize sizable GHG reductions as compared to current operations.

[23] See Energy Innovation Policy and Technology LLC, https://energyinnovation.org/publication/closing-the-emissions-gap-between-the-ira-and-ndc-policies-to-meet-the-moment/ , this reduction is consistent with the Kyoto agreement.

Mitigating the Growing Risks of Fenceline Monitoring Chemical Plants

Laura Blohm

Unlike the Refinery Sector Rule, with the development of Ethylene Production NESHAPs (EMACT) and Miscellaneous Organic NESHAPs (MON) Risk and Technology Reviews (RTR), industry was successful in negotiating fenceline monitoring requirements out of the final promulgated rules. However, the U.S.EPA clearly had other plans and bought more flexibility with these requirements by forcing many of the major chemical manufactures to implement fenceline monitoring programs via Consent Decrees (CDs). Most recently, nine (9) chemical manufacturers subject to HON received an Information Collection Request (ICR) requiring a pilot fenceline study specifically for benzene, 1,3 butadiene, ethylene oxide, ethylene dichloride, and vinyl chloride.

In addition to adding new constituents to fenceline monitoring requirements, with the U.S.EPA's Environmental Justice (EJ) initiatives growing at the federal, state and local levels, it is expected that facilities will continue to be forced to adapt fenceline programs and troubleshoot the growing burden of implementing new enhanced monitoring capabilities into routine compliance efforts and incorporate fenceline monitoring results into the management of community relations.

Based on years of experience helping refiners design, implement, and manage fenceline programs and recent assistance with the implementation of CD required programs at chemical plants and HON ICR fenceline pilot studies, this presentation will discuss observations and lessons learned in implementing a fenceline program, beyond only benzene. In addition, this presentation will cover strategies to best negotiate testing of specific chemicals and monitoring locations, how to manage difficulties related to collocated facilities, suggestions on adapting to new technologies and unspecified expectations from EPA, and how to best prepare for inspections that may be driven by EJ initiative efforts.

EPC PAPER NUMBER 123b

OLEFIN AND OLEFIN DERIVATIVE CARBON FOOTPRINT OF PRODUCT CONSIDERATIONS

Jennifer Port

Chief Process Sustainability

ExxonMobil Technology and Engineering Company, Spring TX

Luis Sotomayor

Sustainability Engineer

ExxonMobil Technology and Engineering Company, Spring TX

Prepared for Presentation at the 2023 Spring National Meeting

Houston, TX

March 13 – 16, 2023

EPC PAPER NUMBER 123

OLEFIN AND OLEFIN DERIVATIVE CARBON FOOTPRINT OF PRODUCT CONSIDERATIONS

Jennifer Port

Chief Process Sustainability

ExxonMobil Technology and Engineering Company, Spring TX

Luis Sotomayor

Sustainability Engineer

ExxonMobil Technology and Engineering Company, Spring TX

Abstract

As society works toward the goals of the Paris Agreement and mitigating the effects of climate change, carbon footprint of product (CFP) estimates have become a frequently requested aspect of product representation. At this time, however, globally aligned and accepted norms and conventions for calculating olefins and olefins derivatives CFP estimates are still evolving, and the producer is left with several choices under the various methods and standards available. Given the complex nature of the olefins supply chain and manufacturing process, different choices can lead to considerably different values for CFP estimates. This makes meaningful analysis for comparisons of different assets, process configurations and company baselines and decarbonization approaches challenging. This paper will present current options for methods and standards used to calculate olefin and olefin derivative CFP and provide ideas for managing the complexity of choices on the path to having clarity and transparency.

Introduction - What is Carbon Footprint of Product?

ISO 14067 defines Carbon Footprint of a Product (CFP) as the sum of GHG emissions and GHG removals in a product system expressed as CO_2 equivalents and based on a life cycle assessment using the single impact category of climate change expressed in mass of CO_2eq per functional unit.[1]

ISO 14067 is based on principles, requirements, and guidelines identified in existing ISO standard on life cycle assessment (LCA), ISO 14040[2] and 14044[3].

ISO 14067 addresses only a single impact category: climate change.

Why is CFP important to olefins producers?

As society works toward the goals of the Paris Agreement and mitigating the effects of climate change, CFP estimates have become frequently requested aspects of product representation. Common questions related to CFP estimates can include:

- My customer is asking for my CFP estimate. How will they use it?
- Are there standards to estimate CFP?
- How do CFP calculations work with co-produced products?
- My customer has a CFP reduction ambition for its product. How does my supply affect their CFP?
- How does my CFP estimate compare against other materials?
- How could my CFP estimate change as I continue to reduce GHG emissions in my operations?

Olefins CFP estimates are of particular interest due to the multitude of downstream derivatives and value chains that use olefins and products from olefins production as their feedstocks.

Illustrative Examples of Key Olefins Downstream Value Chains

Value chains shown are not all inclusive.

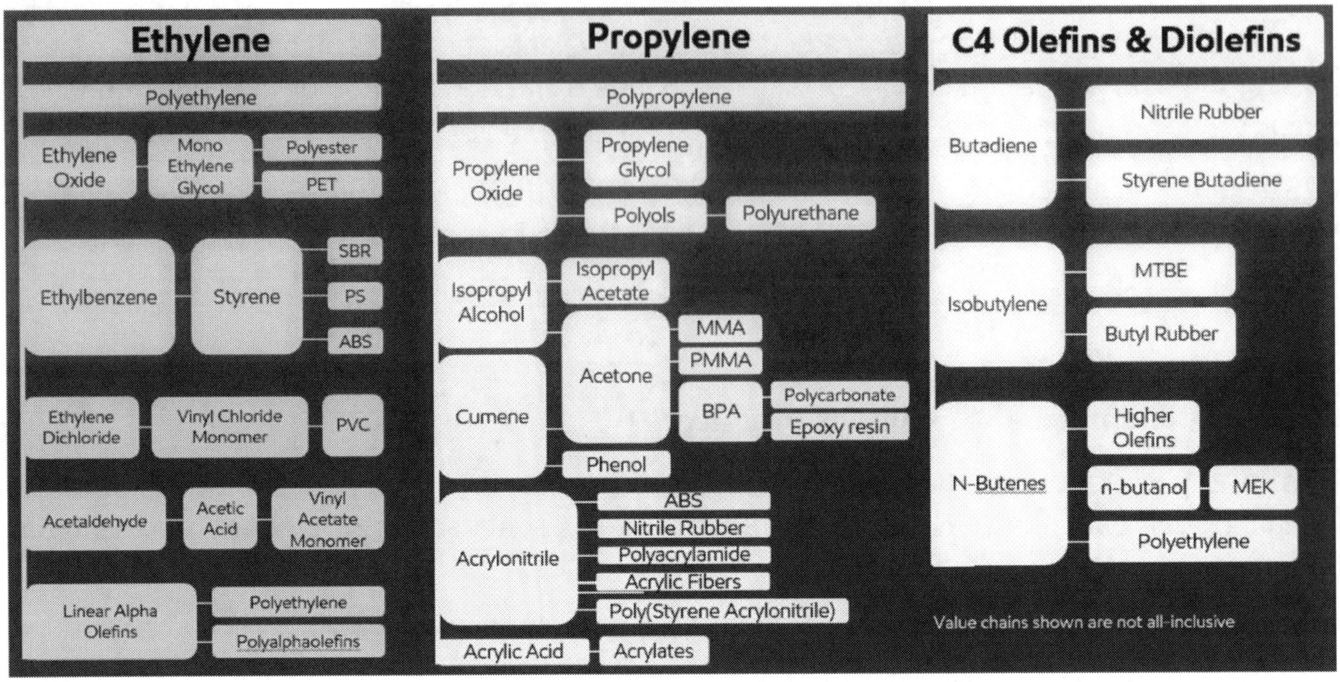

Estimating olefins' CFPs requires considering the following aspects:

- **Olefins production is energy intensive**. In 2023 most olefins are produced from cracking and dehydrogenating fossil feedstocks. Hydrocarbons in the feedstock are cracked into smaller molecules, and saturated hydrocarbons are dehydrogenated into olefins species. These processes are extremely endothermic and require large energy inputs.
- **Olefins can be produced from many different feedstocks, and the process makes many different products**. Olefins can be produced from very light feedstocks (such as ethane or propane) and heavier feedstocks derived from crude oil (such as naphtha, gas oils, or even crude oil itself). Some olefins plants are focused on a single feedstock type, and some are feed flexible and able to crack many different types of feedstocks. Feed flexible crackers frequently tailor their feed slates according to feedstock and product pricing changes.
- **The olefins industry has a complex supply chain**. The olefins industry is part of a complex global energy and industrial system interconnected with gas and liquid hydrocarbon production. Natural gas liquids, crude oil, crude oil products, and olefins derivatives are shipped and traded globally. Feedstock origin and how it arrives at the

steam cracker gate are important factors to consider for olefins production and CFP estimation.

Given these complexities plus the magnitude and diversity of the olefins derivatives value chains, it is critical to understand how these aspects can affect CFP estimates for olefins production.

For example, this table shows two publicly available references for ethylene CFP.

	PlasticsEurope[4]	ACC[5]
GWP, kg CO_2eq / kg ethylene (Note 1)	**1.44**	**1.239**
System Boundary	Cradle-to-Gate	Cradle-to-Gate
Life Cycle Inventory Source	Multiple, mostly EU references	Multiple, mostly US references
Feedstock	Mostly naphtha	Mostly NGL
Allocation	High Value Chemicals (Note 2)	Mass with avoided fuel product credits (Note 3)
Report Year	2012	2020
Data Collection Years	2007-2010	2015
Standard	ISO 14040-44	ISO 14040-44
H2 Disposition	Not stated	Not stated

Note 1 – GWP stands for Global Warming Potential.

Note 2 - High Value Chemicals for this example is defined as ethylene, propylene, butadiene, acetylene, benzene, toluene, xylenes, ethyl benzene, styrene, purified hydrogen

Note 3 - Page 102 of ACC publication[5]. Propylene, pyrolysis gasoline, and butadiene are among the coproducts of ethylene production and a mass basis was used to allocate the credit for the coproducts. For coproducts sold for fuel use in other processes, these were treated as an avoided fuel product and were given credits based on the fuel they would replace.

The ethylene CFPs in these examples differ by about 15%. The cause of the difference is unclear. It could result from inherent technical differences in the feeds and processes being modeled, or it could result from differences in the choices used for feedstock data sources and steam cracker process CO_2eq allocation, or some combination of both.

For life cycle inventories (LCI), the PlasticsEurope example uses Association of Petrochemicals Producers in Europe (APPE), ecoinvent 2.2, and Ullmann's Encyclopedia of Industrial Chemistry while the ACC example uses US LCI, GREET 2017, NETL, and Ecoinvent 3. Depending on the methodologies and background data used to generate these LCI databases, this could lead to differences in the CFP estimates.

The second difference is the allocation procedure. PlasticsEurope allocated steam cracker process CO_2eq only to High Value Chemicals. The ACC procedure used a mass allocation for materials products and incorporated a fuel coproduct credit. These allocation procedures are different and inconsistent, making comparison between the two examples challenging.

To explore these topics in more detail, this paper will first outline the key inputs to estimate olefins CFP. The paper will also examine how different data sources and different allocation procedures can affect CFP estimates.

To accomplish this, this paper will:

- Step through example calculations to provide illustrative ethylene CFP estimates
- Use only publicly available references for key inputs
- Discuss key concepts relevant to CFP estimates for olefins producers and illustrate the impacts of the available choices when estimating a CFP

This paper will not:

- Represent these CFP estimates as specific to ExxonMobil or any other producer
- Provide CFP estimates that can be used for anything outside the purpose of discussion within this paper

Olefins CFP Estimate Steps

ISO 14040[6] defines 4 stages of conducting an LCA study:

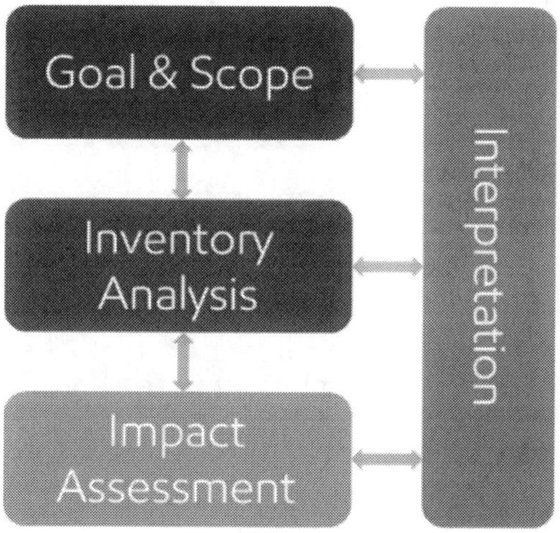

For this paper, the goal and scope can be defined as estimating the cradle-to-gate CFP of ethylene produced from steam cracking expressed as kg CO_2eq / kg ethylene. We will consider three separate ethylene CFP estimates:

1. Ethane feedstock cracked on an ethane cracker where hydrogen is left in the tail gas and burned in the cracker.
2. Ethane feedstock cracked on an ethane cracker where hydrogen is recovered at chemical grade purity and exported.
3. Naphtha feedstock cracked on a naphtha cracker where hydrogen is recovered at chemical grade purity and exported.

Consistent with the ACC and PlasticsEurope examples above, we will use a system boundary of cradle-to-gate. For fossil-based feedstocks, the intent is to account for the production of olefins starting with extracting oil and gas from the earth, purifying those molecules to become suitable steam cracker feedstocks, and steam cracking those feedstocks to produce olefins and other products. The gate is defined where the products leave the steam cracker boundary.

Example Cradle-to-Gate Scope for Ethylene CFP Estimates

The system boundary defines which unit processes are a part of the system under study. The boundary includes fossil feedstock extraction, transportation, and purification to steam cracker feedstock. The steam cracker boundary is the equipment required to produce polymer grade ethylene and propylene. C4+ coproducts will be produced in their raw state, further processing specific to coproduct purification like butadiene extraction or pyrolysis gasoline hydrotreatment will not be included for simplicity.

Cracker Boundary Includes	Cracker Boundary Excludes
Feed through DeC4	BD Extraction
Cold Box + PSA	Other C4 separation
C2 and C3 splitters	Gasoline Hydrogenation
Utilities (electricity, steam, CW)	Aromatics Recovery

Once goal, scope, and boundary are defined, the next stages include:

- Life cycle inventory (LCI) analysis which quantifies the relevant inputs and outputs of a product system and selects data sources to represent inputs to the system.
- Decisions around how to allocate raw material inputs and process emissions to products.

According to ISO 14067, a sensitivity analysis should be conducted to assess the effects of the choices made regarding methods and data on the CFP estimate. Two examples of choices made in the sensitivity analysis include allocation procedures and data sources, and these choices must be clearly disclosed under the transparency guidelines in ISO 14067.

Olefins LCI Analysis

Key steps specific to olefins LCI analysis include the following:

- **Material Balance** – a full material balance around the steam cracker boundary is required including all incoming feedstocks (rate and type) and all products that cross the cracker boundary. Tail gas produced from the cold ends section of the plant and burned onsite does not cross the cracker boundary and is excluded from the product list. For the examples in this paper where hydrogen is recovered as a product and exported, we assume 70% recovery of hydrogen from furnace cracked gas exits the gate with a sensitivity for the ethane case of leaving the hydrogen in the tail gas to be burned.

- **Energy Balance** – a quantification of the energy (amount and type) consumed by all units within the cracker boundary is required.
- **Data sources for raw material inputs** – the carbon footprint of feedstocks and raw materials that are inputs to the steam cracker boundary need to be quantified.

A typical temporal representation for CFP can be on a year's worth or more of production. Gathering all these data required for CFP estimates for a complex process like olefins production can be challenging. Olefins plants are frequently integrated with other facilities such as refineries, downstream derivatives units, and utility plants. Separating the olefins plant material and energy balances according to the defined system boundary can be difficult but is required for CFP estimates with consistent system boundaries.

Material Balances

The paper uses yield references from Ullman's Encyclopedia of Industrial Chemistry for the material balances.[7]

The ethane case assumed ethane furnace yields at 65% conversion and 0.4607 seconds residence time. The naphtha case assumed naphtha furnace yields at 0.45 P/E ratio and 0.1170 seconds residence time. Ultimate yields are calculated assuming recycle cracking ethane to extinction using the ethane yields referenced above.

The table below summarizes the ultimate feedstock and product rates for a steam cracker producing around 1 MTA of ethylene. The rates are reported on a contained molecule basis meaning these are molecular flowrates for these products, not actual stream flow product rates. For example, the benzene rate shown represents the flowrate of all benzene molecules leaving the steam cracker boundary. In reality those benzene molecules typically leave the steam cracker boundary as part of the mixed pyrolysis gasoline C5 – C10 product stream according to site specific separation factors and configuration. Nevertheless, for purpose of the illustrative CFP estimates in this paper, simple contained molecular balances will suffice.

Feed and Product Rates for in metric tons/hour, 1 MTA ethylene steam cracker

Component	Disposition	Ethane	Naphtha
Feedstock		155	367
Products			
Hydrogen (Note 1)	Note 1	10	3
CO	Burned onsite	0.1	0.2
CH4	Burned onsite	9	60
Ethylene	Battery limit product	125	125
Propylene	Battery limit product	3	56
Propane	Battery limit product	0.3	7
Butadiene	Battery limit product	4	21
Butenes	Battery limit product	0.5	13
Butanes	Battery limit product	1	2
Benzene	Battery limit product	1	27
Toluene	Battery limit product	0.2	10
Xylenes	Battery limit product	0	4
Other C5+	Battery limit product	1	40

Note 1 – Full contained hydrogen yield is shown, final yield and disposition depends on the case. For the naphtha case and the ethane case where hydrogen is recovered, 70% of the rate shown crosses the steam cracker boundary and the reminder is burned onsite. For the ethane case where hydrogen is burned, 100% of the rate shown is burned onsite.

Energy Balances

We used several references[8,9,10,11] to provide perspective for steam cracker energy consumption. For this paper we selected data from the Griffen reference because as it was a consistent mid-range value for both the ethane and naphtha cases. The specific energies used for this paper were 20,400 kJ/kg ethylene and 32,900 kJ/kg ethylene for the ethane and naphtha cases respectively. The paper assumed the specific energies were consistent with the steam cracker boundary defined above. Selecting a mid-range value and assuming a consistent boundary were deemed sufficient to provide the energy consumption to estimate an illustrative CFP.

Representative Steam Cracker Specific Energy Consumption, kJ/kg Ethylene

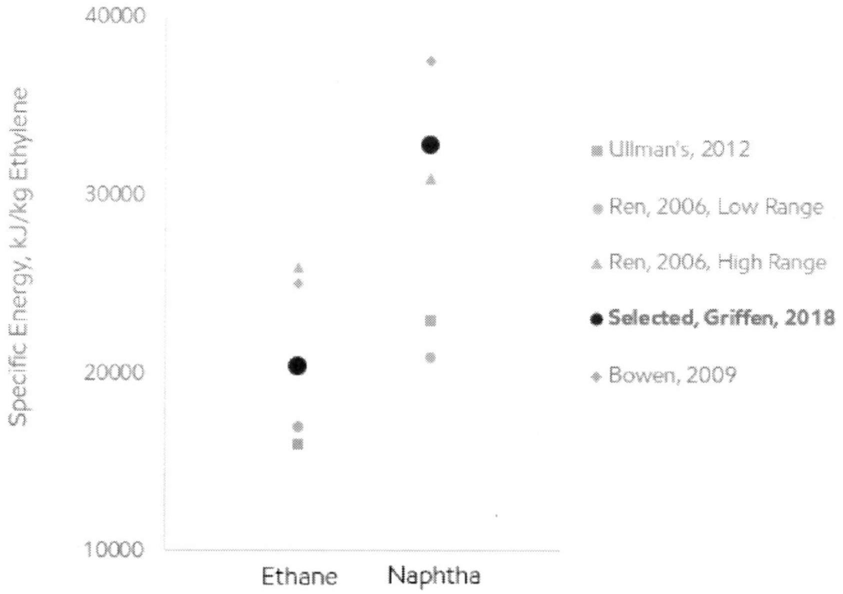

The energy demand of the steam cracker was assumed to be provided first by the available tail gas generated in the steam cracking process, and second by import natural gas to fulfill the remaining demand. For the ethane case with hydrogen recovery and the naphtha case where hydrogen recovery was assumed, we assumed 70% of the hydrogen molecules present in the furnace effluent were recovered as a high purity hydrogen product and exit the cracker boundary. These recovered hydrogen molecules were therefore removed from the tail gas stream, and natural gas was used as make up to satisfy the remaining combustion heat demand.

Data Sources for Raw Material Inputs

CFP estimates for raw material inputs like feedstocks and natural gas entering the olefins plant are key inputs to the ethylene CFP estimates.

CFP estimates for feedstocks can be found in sources such as industry association publications, models such as GREET[12], and 3rd party LCI databases. Note that some 3rd party LCI databases may require a license fee to use. This paper uses publicly available information to provide a perspective of the challenges and ranges found when selecting feedstock CFP estimates.

The basis for input natural gas to be burned inside the cracker boundary comes from GREET 2022.

We could not find any CFP references specific to ethane feedstock in the public domain. Most ethane feedstock is derived from natural gas production where natural gas liquids (NGL) containing ethane is first separated from methane / natural gas (NG). The NGL stream is subsequently fractionated to produce ethane. As such we used both NG and NGLs as proxies for ethane for our CFP estimates. The ACC reference is a calculated reference extracting information from Tables 18 and 54 from the ACC publication and represents this NGL mix: 50% ethane / 32% propane / 6% butane / 12% E/P mix.

Source	Stream	kg CO2e / kg feed
GREET 2022[12]	NGL	0.30
PlasticsEurope[13]	NG	0.52
ACC 2020 Report[6]	Processed NG 50/32/6/12 E / P / B / E/P mix	0.35
Contreras, et al[14]	NGL	0.52

We found limited CFP references for naphtha feedstock in the public domain. For this reason we provided gasoline CFP examples for reference plus a calculated reference extracting information from Tables 18 and 54 from the ACC publication for a mix of naphtha and refinery gases characterizing cradle-to-gate crude to refinery and then refinery process emissions to produce naphtha and refinery gases.

Source	Stream	kg CO2e / kg feed
GREET 2022[13]	Crude Naphtha	0.68
PlasticsEurope[15]	Naphtha	0.34
ACC 2020 Report[2]	Refined Petroleum Products (42% naphtha, 58% refinery gas)	0.76
EC-JRC[16]	Gasoline	0.73
US EPA[17]	Gasoline	0.76

In both the ethane and naphtha cases, there are vast differences between various references, up to around 70% for ethane and around 100% for naphtha. It is unclear what is driving these differences.

The reader should also note we were not able to find publicly available sources for ethane or naphtha feedstock CFPs for regions outside the US and EU. Given more than half of the world's ethylene is consumed outside the US and EU[18] this provides a challenging situation for producers outside these regions to both find this information and ensure consistent comparisons between all regions.

It should be noted that selecting different feedstock CFP estimates will cause different ethylene CFP estimates, but different selections will have no impact on the actual cradle-to-gate absolute CO_2eq emissions.

Steam Cracker GHG Allocation

The material balances above show the olefins process makes many different products, most apparent in the naphtha case. CO_2eq emissions from both the feedstock CFPs and the emissions from the cracker process itself are then allocated to the products.

Similar to the concept highlighted above regarding raw material input data source differences, for a given steam cracker heat and material balance, the actual absolute CO_2eq emissions are the same regardless of allocation procedure. The allocation procedure simply determines how the emissions are assigned to the various products for product CFP estimate purposes.

<u>Cracker Feedstock Emissions Allocation</u>

Since all the products leaving the steam cracker boundary are derived from the feedstock entering the boundary, we allocated feedstock raw material inputs to each product leaving the cracker boundary on a mass basis. For olefins production the CFP component resulting from feedstock will typically be somewhat greater than the feedstock CFP itself because the product rate leaving the steam cracker boundary is less than the feed rate (tail gas derived from feedstock is burned).

<u>Cracker Process Emissions Allocation</u>

The ISO standards themselves do not prescribe specific allocation procedures that should be used for a particular product or process, but they do require the producer be transparent about which procedure is used.

Some examples of allocation procedures are listed below.

Example allocation procedures

Choice	Description
Mass	All products leaving cracker boundary on a mass basis
Energy	All products leaving cracker boundary on an energy content basis
Economic	All products leaving cracker boundary on an economic basis
PlasticsEurope 2012 Ecoprofile[5]	"High Value Chemicals" Ethylene, propylene, butadiene, acetylene, benzene, toluene, xylenes, ethyl benzene, styrene, purified hydrogen
PlasticsEurope 2017/8 recommendation[19]	"Main Products" Ethylene, propylene, benzene*, butadiene*, hydrogen, toluene*, xylenes*, butenes * if separated from mix
ACC 2020[6]	Mass with avoided fuel product credits

For mass, energy, and economic allocation procedures, allocation is applied consistently to all products according to the selected allocation metric. For instance, in the case of mass allocation, all the cracker CO_2eq emissions are applied consistently across all the products leaving the steam cracker boundary on a mass basis. For a given olefins material balance, the properties of mass and energy do not change with time and are fixed according to the intrinsic properties of the products. Economic allocation on the other hand can vary greatly with time because feed and product prices are constantly changing according to market dynamics.

The PlasticsEurope 2012 Ecoprofile example and 2017/8 recommendation both use a different approach and allocate cracker process CO_2eq emissions only to selected products (although please note the products are slightly different between these two examples). The 2012 report allocates to a set of "high value chemicals" while the 2017/8 recommendation delineates between "main products" and "additional products." In the 2012 reference, it is not clear whether the molecules listed in the "high value chemicals" list are on a contained basis or whether they must be separated within the steam cracker boundary to be counted as "high value chemicals." The PlasticsEurope 2017/8 reference seeks to add clarity on this point by stating a producer can include a product in "main products" for allocation if they are "separated from the mix." However, the reference provides no clarity on the boundary under which that separation must occur.

For example, the naphtha case material balance above shows considerable amounts of benzene molecules produced by steam cracking. These benzene molecules are typically contained within the pyrolysis gasoline stream ("pygas") with other molecules in the C5 – C10 region. This stream can be further processed to extract the and purify the benzene into a finished benzene product, but note this process may or may not be located within the steam cracker boundary. If fact, steam cracker pygas is openly traded, and it is common that the units that further purify benzene into a finished product are part of an aromatics recovery unit, not a steam cracker. In that case it is unclear whether the producer can allocate to contained benzene if the producer sells their pygas stream and the benzene is eventually "separated from the mix" outside their boundary.

The ACC publication takes a different approach regarding allocation. Coproducts are separated into Material coproducts and Energy coproducts. System expansion credits were given for avoiding the energy-equivalent quantity of fuel production and combustion displaced by the exported coproduct energy. The ACC report does not provide sufficient granularity and data around these procedures, and hence we are unable to replicate the allocation procedure in this paper.

Impact Assessment and Interpretation

Based on the above material and energy balances, raw material input ranges, and two different allocation procedures, we then estimated indicative illustrative CFP estimates for ethylene.

Illustrative Ethylene CFP Estimates

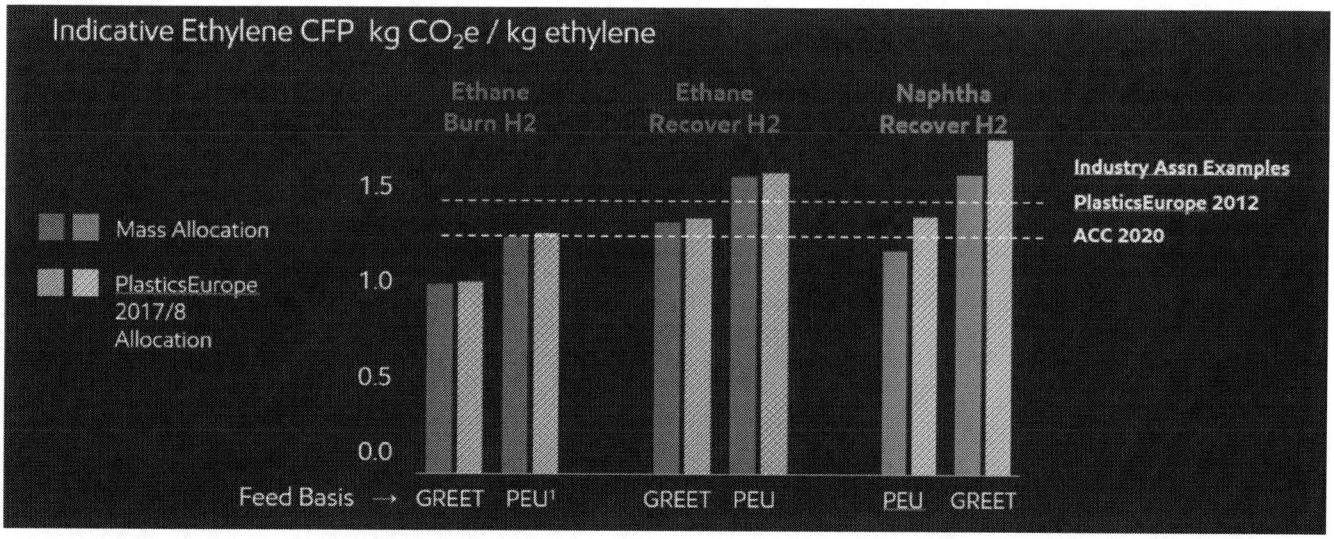

PEU = PlasticsEurope feedstock reference

Detailed Results

Feed	**Ethane**			
Hydrogen Disposition	**Burned**			
Feedstock CFP Source	GREET	GREET	PlasticsEurope	PlasticsEurope
Allocation Method	Mass	PlasticsEurope	Mass	PlasticsEurope
Feedstock Contribution to CFP (kg CO2eq/kg C2=)	0.34	0.34	0.59	0.59
Steam Cracker Contribution to CFP (kg CO2eq/kg C2=)	0.65	0.66	0.65	0.66
Total Ethylene CFP Estimate (kg CO2eq/kg C2=)	**0.99**	**1.00**	**1.24**	**1.26**

Feed	**Ethane**			
Hydrogen Disposition	**70% Recovery**			
Feedstock CFP Source	GREET	GREET	PlasticsEurope	PlasticsEurope
Allocation Method	Mass	PlasticsEurope	Mass	PlasticsEurope
Feedstock Contribution to CFP (kg CO2eq/kg C2=)	0.32	0.32	0.56	0.56
Steam Cracker Contribution to CFP (kg CO2eq/kg C2=)	1.00	1.02	1.00	1.02
Total Ethylene CFP Estimate (kg CO2eq/kg C2=)	**1.32**	**1.34**	**1.56**	**1.58**

Feed	**Naphtha**			
Hydrogen Disposition	**70% Recovery**			
Feedstock CFP Source	PlasticsEurope	PlasticsEurope	GREET	GREET
Allocation Method	Mass	PlasticsEurope	Mass	PlasticsEurope
Feedstock Contribution to CFP (kg CO2eq/kg C2=)	0.41	0.41	0.81	0.81
Steam Cracker Contribution to CFP (kg CO2eq/kg C2=)	0.80	0.99	0.80	0.99
Total Ethylene CFP Estimate (kg CO2eq/kg C2=)	**1.21**	**1.40**	**1.61**	**1.80**

For these cases we applied the PlasticsEurope 2017/8 allocation procedure assuming these "main products:"

- Recovered Hydrogen, Ethylene, , Propylene, Butadiene, Benzene, Toluene, Xylene

Feedstock and Configuration Impact

Three cases are compared, and each of these three cases have different material and energy balances and therefore different absolute CO_2eq emissions.

Hydrogen disposition for ethane cracking has a major effect on the ethylene CFP estimate. Burning hydrogen byproduct produced in the cracker versus recovering hydrogen and supplementing with natural gas for combustion heat demand results in around a 25-30% lower CFP estimate as compared to both the ethane and naphtha cases where hydrogen is recovered as a product.

The CFP estimates from the ethane case where hydrogen is recovered fall between the extremes of the naphtha case. For a given rate of ethylene, ethane cracking has lower absolute CO_2eq emissions than naphtha (numerator) but produces less products (denominator) and therefore results in similar CFP estimates when compared to naphtha.

LCI Feed Sources Impact

The selection of the LCI data source for a given steam cracker feedstock has no impact on absolute CO_2eq emissions, but it does impact the CFP estimate. In these examples the differences between GREET and PlasticsEurope references result in around a 25% difference for the ethane CFP estimate and around 35% difference for the naphtha CFP estimate.

Allocation Procedure Impact

Differences between mass and energy allocation were found to be negligible.

Feed Hydrogen Disposition	Naphtha 70% Recovery	
Feedstock CFP Source	GREET	GREET
Allocation Method	Mass	Energy
Feedstock Contribution to CFP (kg CO2eq/kg C2=)	0.81	0.86
Steam Cracker Contribution to CFP (kg CO2eq/kg C2=)	0.80	0.76
Total Ethylene CFP Estimate (kg CO2eq/kg C2=)	**1.61**	**1.63**

Due to the lower "additional" product make for steam cracking ethane, using mass allocation versus PlasticsEurope 2017/8 allocation procedures did not result in major CFP estimate differences for ethylene, but note the CFP estimates for "additional products" would be considerably different between the two cases. "Additional products" using the PlasticsEurope 2017/8 allocation procedure will have much lower CFP estimates than those same products calculated with the mass allocation procedure. The different allocation procedures resulted in around 10-15% difference in ethylene CFP estimates in the naphtha cases, and similarly there would be a considerable difference in the CFPs for "additional products." Again we note the selection of allocation procedure has no bearing on actual absolute CO_2eq emissions.

Feed	**Naphtha**			
Hydrogen Disposition	**70% Recovery**			
Feedstock CFP Source	GREET			
Allocation Method	**Mass**		**PlasticsEurope**	
Product Type	**Main**	**Additional**	**Main**	**Additional**
Feedstock Contribution to CFP (kg CO2eq/kg C2=)	0.81	0.81	0.81	0.81
Steam Cracker Contribution to CFP (kg CO2eq/kg C2=)	0.80	0.80	0.99	-
Total Ethylene CFP Estimate (kg CO2eq/kg C2=)	**1.61**	**1.61**	**1.80**	**0.81**

Aggregation of Choices Impact

For a given material and energy balance, the CO_2eq emissions do not change, but ethane CFP estimates can vary by up to 25% and naphtha CFP estimates can vary by up to 50% depending on the combination of the LCI data source for feedstock and the allocation procedure used.

Summary

Customers are increasingly requesting carbon footprint of products from olefins suppliers. Olefins are an important first step along a multitude of derivatives and value chains highlighting the need for accurate and consistent CFP estimates. This paper stepped through the basic stages required to estimate a CFP for ethylene and tested CFP impacts for different feedstock cases, configurations for hydrogen disposition, data sources for feedstocks, and allocation procedures.

Three configurations were considered which have different CO_2eq emissions and material balances. Burning the hydrogen produced in steam cracking ethane results in the lowest CFP estimates. Ethane cracking where hydrogen is recovered as a product and naphtha cracking have similar CFP estimates despite having very different CO_2eq emissions and production profiles for a given ethylene product rate.

Feedstock data source and allocation procedure choices do not impact CO_2eq emissions directly, but can have a dramatic impact on CFP estimates, up to 25% and 50% for ethane and naphtha cases respectively depending on the combination of choices applied.

We encourage all olefins producers to:

- Read and understand the CFP and LCA standards
- Carefully consider data sources used in life cycle inventories
- Understand and ISO sensitivity analysis requirements
- Aim to be transparent about your methodology

References

[1] ISO 14067:2018, (2018), Greenhouse Gases- Carbon Footprint for products- Requirements & Guidelines for Quantification

[2] ISO 14040:2006+Amd 1: 2021, (2020), Environment Management-Lifecyle Assessment- Principles and Framework

[3] ISO 14044:2006+Amd 2:2020, (2020), Environment Management-Lifecycle Assessment- Principles and Framework

[4] https://legacy.plasticseurope.org/application/files/8315/1783/7824/20130114101708-plasticseurope_eco-profile_ethylene__others_2012-11.zip

[5] https://www.americanchemistry.com/better-policy-regulation/plastics/resources/cradle-to-gate-life-cycle-analysis-of-olefins

[6] Figure 1 from ISO 14040:2006+Amd 1: 2021, (2020), Environment Management-Lifecyle Assessment- Principles and Framework

[7] Zimmermann, Walzl, Ethylene, *Ullmann's Encyclopedia of Industrial Chemistry*, Wiley-VCH, Germany, 2012.

[8] Table 2 from Griffen PW, Hammond GP, Norman JB. Industrial energy use and carbon emissions reduction in the chemicals sector: A UK perspective. Applied Energy 227 (2018) 587-602.

[9] Table 11 from Zimmermann, Walzl, Ethylene, *Ullmann's Encyclopedia of Industrial Chemistry*, Wiley-VCH, Germany, 2012.

[10] Table 2 from Ren T, Patel M, Blok K. Olefins from conventional and heavy feedstocks: Energy use in steam cracking and alternative processes. Energy 31 (2006) 425-451.

[11] Bowen, CP. Energy Incentives and Emissions Reductions. Ethylene Producers' Committee Conference Proceedings Volume 21, 2009, Session 110.

[12] https://greet.es.anl.gov/greet_excel_model.models

[13] https://legacy.plasticseurope.org/application/files/8015/1783/7825/20101226013326-plasticseurope_natural_gas_lci_data.zip

[14] Contreas W, et. al. Life cycle greenhouse gas emissions of crude oil and natural gas from the Delaware Basin. Journal of Cleaner Production 328 (15 Dec 2021) 129530

[15] https://legacy.plasticseurope.org/application/files/3215/1783/7827/20101226005412-plasticseurope_naphtha_eco-profile.zip

[16] European Commission's Joint Research Centre https://publications.jrc.ec.europa.eu/repository/handle/JRC119036

[17] United States Environmental Protection Agency https://www.epa.gov/sites/default/files/2016-07/documents/select-ghg-results-table-v1.pdf

[18] https://www.spglobal.com/commodityinsights/en/ci/products/ethylene-chemical-economics-handbook.html

[19] https://plasticseurope.org/wp-content/uploads/2021/12/PlasticsEurope_recommendation_on_Steam_Cracker_allocation-_Juillet_2018.pdf

How Is the U.S.Epa's Pollution Accountability Team (PAT) Using "Next Gen Enhanced Monitoring" in Its Investigations?

Joe Ibanez

In addition to its ongoing Nation Compliance Initiative, U.S. EPA recently announced broad policy actions to protect communities including a new multi-scale monitoring project called the Pollution Accountability Team (PAT). This initiative expands EPA's air monitoring capacity by utilizing assets such as a GMAP (Geospatial Measurement of Air Pollutants) mobile air monitoring vehicle and additional air pollution inspectors to enhance compliance inspections and enforcement. Numerous petroleum refining and petrochemical facilities have been visited by the U.S. EPA during this effort.

This presentation will discuss:

1. Real case examples of how the U.S. EPA recently used a GMAP unit, OGI monitors and other enhanced technologies when conducting inspections as part of its multi-scale monitoring project.
2. Areas of the facilities targeted during the inspections, some of the findings, and how facilities can better prepare for these types of audits.
3. Overlapping risks of air compliance programs such as fenceline monitoring, LDAR, BWON and air permits relative to these types of investigations.

123c Ethylene Risk and Technology Review Reconsideration, What NOW?

Steve Smith

EPA promulgated the Ethylene Risk and Technology Review (RTR) National Emission Standard for Hazardous Air Pollutants (NESHAP, 40 CFR, Part 63, Subpart XX and Subpart YY) on July 6, 2020. It is applicable to new and existing ethylene production units. The final rule included requirements for furnace and decoke operations, flares, process vents, heat exchange systems and pressure relief devices. Environmental non-governmental organizations filed a petition requesting EPA to reconsider certain sections of the rule. EPA granted the request and is working through the reconsideration process. This paper will review the potential impact of the reconsideration request.

123d Recent Ethylene Producer's Consent Decree Fenceline Monitoring and Experience Using Next Gen Enhanced Monitoring

Troy Boley, Laura Blohm, Joe Ibanez

Over the past several years, the USEPA has issued flare related consent decrees to both the refinery and ethylene producer groups that incorporated fenceline monitoring as an additional mitigation measure. Flare improvements are required as part of the Ethylene Production MACT. However, not all ethylene producers have entered into a flare consent decree. A review of these consent decrees and their varying fenceline monitoring requirements will be presented. The recent EMACT regulation did not include any fenceline monitoring provision. Some non-industry groups have petitioned USEPA to incorporate additional fenceline monitoring into all new revised MACT regulations.

Several industry consenters are now providing their Method 325A/B passive tube fenceline monitoring results on various public websites. This data will be summarized and compared to the Petroleum Refinery dashboard now available from EPA wherein monthly fenceline data from all refinery locations is share publicly.

Additionally, USEPA recently announced broad policy actions aimed to protect communities including a new multi-scale monitoring project called the Pollution Accountability Team (PAT). This initiative expands USEPA's air monitoring capacity by utilizing assets such as a GMAP (Geospatial Measurement of Air Pollutants) mobile air monitoring vehicle and additional air pollution inspectors to enhance compliance inspections and enforcement. Numerous petrochemical facilities have been visited by the USEPA during these recent efforts. Items to be presented will include: real case examples of how the USEPA recently used a GMAP unit, OGI monitors and other enhanced technologies when conducting inspections as part of its multi-scale monitoring project; areas of the facilities targeted during these inspections; and overlapping risks of air compliance programs such as fenceline monitoring, LDAR, BWON and air permits relative to these types of investigations.

Troy Boley from Spectrum Environmental Solutions LLC will provide a summary of the current publicly-available fenceline monitoring results at Ethylene producer facilities resulting from recent EPA enforcement actions.

Laura Blohm and Joe Ibanez with Tricord Consulting LLC will share recent experiences involving these newer, non-passive tube technologies.

AIChE Paper Number: 123e

Carbon Dioxide Mitigation Using Renewable Power: Life Is All about Priorities!

David J Brown
Senior R&D Manager
AVGI BV, Ghent, Belgium

Oleksii Mynko
PhD Researcher
Laboratory for Chemical Technology (LCT), Ghent University, Ghent, Belgium

Mike Bonheure
Doctoral researcher – FWO Fellowship
Laboratory for Chemical Technology (LCT), Ghent University, Ghent, Belgium

Ismaël Amghizar
Modeling Researcher
AVGI BV, Ghent, Belgium

Tamás Buzogány
PhD Researcher
Laboratory for Chemical Technology (LCT), Ghent University, Ghent, Belgium

Guy B Marin
Emeritus Professor
Laboratory for Chemical Technology (LCT), Ghent University, Ghent, Belgium

Kevin M Van Geem
Professor
Laboratory for Chemical Technology (LCT), Ghent University, Ghent, Belgium

Prepared for Presentation at the 2023 Spring National Meeting
Houston, TX, March 12-16 , 2023

Carbon Dioxide Mitigation Using Renewable Power: Life Is All about Priorities!

David J Brown
Senior R&D Manager
AVGI BV, Ghent, Belgium
Oleksii Mynko
PhD Researcher
Laboratory for Chemical Technology (LCT), Ghent University, Ghent, Belgium
Mike Bonheure
Doctoral researcher – FWO Fellowship
Laboratory for Chemical Technology (LCT), Ghent University, Ghent, Belgium
Ismaël Amghizar
Modeling Researcher
AVGI BV, Ghent, Belgium
Tamás Buzogány
PhD Researcher
Laboratory for Chemical Technology (LCT), Ghent University, Ghent, Belgium
Guy B Marin
Emeritus Professor
Laboratory for Chemical Technology (LCT), Ghent University, Ghent, Belgium
Kevin M Van Geem
Professor
Laboratory for Chemical Technology (LCT), Ghent University, Ghent, Belgium

Abstract:

Steam cracking is the leading technology to produce light olefins, and the most energy-consuming process in the chemical process industry, using approximately 10% of the sector's total primary energy globally. We examine three technological improvements that should reduce greenhouse gas emissions: a low-emission cracking furnace combined with an electrified separation train; the RotoDynamic Reactor; and a blue hydrogen fired furnace. Life cycle assessment is used to compare the efficacy of the technologies. The carbon footprint of the electricity grid defines the applicability of the electrified solutions. The RotoDynamic Reactor, coupled with an electrified separation train, reduces the carbon footprint of high value chemicals by almost 30%, the best result in this study, provided the electricity is generated from largely renewable sources (like the Norwegian grid). If the power generation is mixed, such as the Belgian grid, a low-emissions furnace with air preheat does best. Our results also highlight the potential of blue hydrogen fired furnaces equipped with integrated methane reforming units and carbon capture and storage, as a promising alternative to electrification.

Introduction

Since the Paris Agreement aims to reduce human-caused greenhouse gas (GHG) emissions to net zero in the second half of this century [1], the rapid transformation of various sectors including the chemical process industry (CPI) is required. [2]. Steam cracking is the most widely used production process for chemical building blocks, accounting for 8% of the primary energy consumption [3] and 7% of GHG emissions [4] in the CPI. Electrification has emerged as a promising path towards a more sustainable industry, substituting fossil fuels with renewable low-carbon electric power [5].

Only a few relevant life cycle assessment(s) (LCA) studies have been published on electrified steam cracking [5] or using hydrogen as fuel [6]. We evaluate three promising solutions: a low-emission (L-E) cracking furnace with an electrified separation train, a RotoDynamic Reactor (RDR), and a blue hydrogen fired furnace. The study uses a typical propane steam cracker studied in the previous work [7] as the base case scenario (BASE).

The electrification of olefin production is a focus of both industry and academia, with innovations aimed at electrifying furnaces [8, 9] or developing alternative electrified processes like plasma-based methods [10]. Light olefin plants may use low-efficiency, condensing steam turbines (overall efficiency 'below or equal to 45%') to drive the cracked gas (a.k.a. process gas) and refrigerant compressors [11]: this is an opportunity for electrification. The challenge is to replace the main steam turbine with an electric motor and use smaller extraction steam turbines for optimal steam application, while preserving heat for downstream equipment. A novel firebox design, TLE (generic term used for all types of quench exchangers), and heat recovery arrangement· proposed by Technip Energies (T.EN) can potentially reduce CO_2 emissions by 30%. The study evaluates the impact of two L-E furnace embodiments and the effect of the electricity mix on the results.

Several electrified cracking reactors have been proposed [12-14], with Coolbrook's patented RDR cracker [15] being the most advanced. The advantage of the RDR is its short residence time, resulting in higher light olefin yields making it potentially the most selective cracking technology. In contrast to the L-E furnace, the RDR-based cracking plant relies on electricity as the energy source for both the reactor *and* the separation section. Several RDR scenarios have been considered to determine the effect of the electricity mix on their environmental impact.

According to preliminary calculations and literature [9], electrifying steam cracking would be ineffective with a fossil-based electricity grid. It is better to use fossil fuel with a carbon capture and storage (CCS) system until low-carbon electricity becomes prevalent. Previously, post-combustion CCS has been evaluated [7], but a pre-combustion option known as blue hydrogen firing [16] is also worth examining. Blue hydrogen, produced via steam methane reforming (SMR), is selected for its backward compatibility and high industrial interest. The study evaluates two options, SMR with CCS and the advanced gas heated reformer (GHR), the latter offering better thermal efficiency and carbon capture, but at cost of pure oxygen consumption. Two scenarios, 'H2-SMR' and 'H2-GHR', were generated to perform a LCA of the integrated solution.

This work explores the most promising options to reduce environmental impact of olefin industry and considers factors beyond carbon footprint, such as resource consumption and health effects. The results will provide valuable insights for policymakers, researchers, and industry.

Modelling approach and technologies evaluated

L-E furnace

Due to the process gas temperatures of 850°C (1550°F) or higher that are required in the reactor coil, the temperature of the flue gases at the radiant arch of a steam cracking furnace (referred to as the bridgewall temperature) is typically around 1150°C (2100°F). The latter implies that the firebox fuel efficiency (a.k.a. 'radiant efficiency') of a steam cracker is rather low and varies from 38% to 47% [17] depending on the furnace and burner configuration. The main idea behind the L-E furnace concept is to reduce the overall firing duty by increasing the share of heat absorbed by serpentine reactor coils that are positioned in the radiant box to provide enough heat to the cracking process with less fuel. To increase firebox fuel efficiency, combustion occurs in the lower part of the firebox, in a firebox designed with floor burners only or wall burners mounted close to the bottom of the furnace. However, an increased share of the radiant duty in the overall furnace heat balance implies that the convection section duty is reduced and therefore it is insufficient to preheat the feed and the dilution steam (DS) mixture. To solve this challenge, T.EN has proposed a novel heat recovery scheme [18]. This arrangement uses two TLEs in series. In the primary TLE, the cracked gas leaving the reactor coils is quenched by exchanging heat with the preheated feed and the DS mix flow which is later fed to the reactor coil. This TLE allows for a larger share of the recovered heat to be utilized for feed preheating, and thus avoids using additional boilers even when the firebox efficiency is increased by 30%. The secondary TLE further cools the cracked gas. The convection section is altered as well. Since the mass flow and temperature of the flue gases are reduced, preheating the boiler feedwater is no longer possible. Thus, it is fed directly into the steam drum. Therefore, steam output is reduced and is no longer sufficient to drive the turbines of the separation train, facilitating the need for electrified downstream. Such an arrangement allows for 20% firing rate reduction for a basic design (L-E furnace) and up to 30% if the combustion air preheater is added [18].

RDR

The RDR is a concept based on the original patent [19, 20] that has been developed over many years by Coolbrook Oy. The RDR will (a) increase the selectivity toward light olefins (e.g., ethylene and propylene); (b) reduce the yield of secondary products; and (c) reduce coke formation and deposition. This is achieved by minimizing residence time and reducing the partial pressure of the hydrocarbons while increasing the process temperature, leading to increased severity without the drawbacks that plague classical high-severity crackers. The scheme of such a reactor is shown in Figure 1. The apparatus consists of a spinning rotor and two series of stationary blades: stator and diffuser (Figure 1a). The transonic blades of the ultrahigh load rotor accelerate the preheated process gas to a supersonic velocity using mechanical energy (pure exergy) provided by an electric motor that drives the rotor.

Then the process gas is immediately decelerated in a diffuser using a shockwave system which converts the mechanical energy of the flow into heat. Blades are designed to maximize flow separation in the blade passage which leads to better mixing required to maintain constant static pressure. As the gas flow passes through the diffuser, it is turned back to the axial direction before entering the vaneless space. In other words, this apparatus can be seen as a deliberately

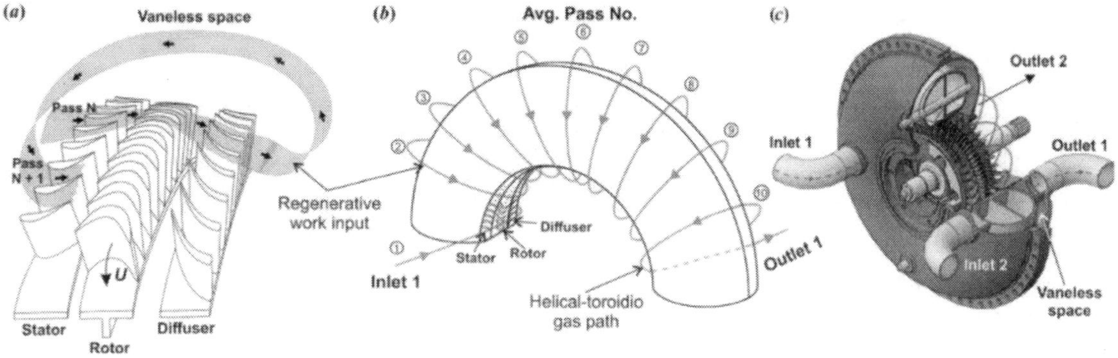

inefficient compressor designed to maximize the amount of dissipated heat instead of increasing gas pressure leading to virtually isobaric heating.

Figure 1. Scheme of the RDR reactor: a – regenerative pass, b – hemi-annulus view, c – full-annulus view [21]

If the process gas passes only once through the cycle that we described (elementary stage), an insufficient amount of energy is introduced in the system for the endothermic reactions to take place. Consequently, the RDR has a multistage configuration. Gas is guided through a toroid-shaped vaneless space into the following stator passages (Figure 1b). Such a design leads to a rapid stepwise increase in the process gas temperature, while keeping the static pressure constant and low. Since the heat is introduced to the process gas via mechanical work instead of indirect heat transfer, the solid surfaces containing the gas have roughly the same temperature as the process gas. Low gas–metal surface temperature and high viscous shear stress caused by supersonic gas flow help to inhibit coke deposition [21]. Moreover, this heating process avoids heat transfer limitations leading to the low inner volume of the multistage turbo reactor (500 times less than the volume of a tubular coil of the same capacity). This way the residence time is significantly reduced, preventing unwanted reactions.

Blue hydrogen combustion

Burning hydrogen in fired equipment results in zero CO_2 emissions from the stack (scope 1). However, a complete LCA or a comprehensive GHG accounting must also account for CO_2 emissions from the hydrogen production plant (scope 3 emissions). Therefore, blue hydrogen is considered. It is produced via steam methane reforming (SMR) with CCS technology. Two pathways for producing blue hydrogen have been considered: with traditional SMR and novel gas heated reformer (GHR).

Nearly half of the world's hydrogen production originates from SMR [22], which is often operated on-site to eliminate the need for transportation. The traditional SMR process typically emits 8.5 kg of CO_2 per kg of hydrogen produced, whereas blue hydrogen production results in only 1 kg of CO_2 emissions [16]. However, the classical methane reforming, even with CCS, has a major disadvantage: process heat is provided by a natural gas-fired furnace that emits unrecovered CO_2 into the atmosphere. Therefore, a novel arrangement is proposed to improve carbon dioxide recovery. It consists of two connected catalytic reactors: an ATR and a GHR. First, the feed is injected into the GHR, then it is fed into the ATR. Partial oxidation of methane in the ATR provides sufficient heat to complete the reforming reactions. The syngas leaving the reactor at ~950°C (1750°F) is then sent back to the shell side of GHR to recover its enthalpy by heat transfer to the tubes containing the catalyst. In this way, heat is provided for the endothermal reforming reaction.

Carbon footprint of the electricity mix

As a result of the high energy demand needed for the endothermic cracking reactions, and the fact that modern state-of-the-art ethylene plants already use well-established and optimized technical solutions for both the reactor and separation train, there is no viable opportunity to significantly reduce the overall specific energy consumption. Shifting the fraction of heat available in the radiant box to the process (feed preheating and cracking) itself can lead to a significant reduction of the firing rate; however, lack of steam output will require an alternative way to drive the compressors of the separation section. A typical light olefin plant requires around 621 MW of fired duty and 90 MW of total compressor power for a production capacity of 125 tons of ethylene per hour and starting from propane as feed [17]. The idea behind the L-E furnace concept is to replace compressor-turbines with electric motors. To minimize CO_2 emissions, these motors must be powered by renewable or low–carbon electricity; however, the utilization of fossil electricity produced with highly efficient power generators such as CCGT [23] will also lead to greater overall system efficiency [11]. The RDR concept, on the other hand, is aimed at replacing both process and compressor duty with electricity. Although offering a certain increase in thermal efficiency and selectivity, it is highly unlikely to provide a lower carbon footprint of cracking products than a conventional furnace if fossil electricity is used. Therefore, the carbon footprint of the electricity mix becomes a determining factor that defines the efficacy of the technologies that we have considered. An overview of the electricity mixes in our study is presented in Table 1.

Table 1. Carbon footprint of various national electricity grids for the year 2016 [24].

Region	**Carbon footprint [kg CO2 eq. per kWh]**
European Union (average)	0.5
Belgium	0.335
Netherlands	0.677
Norway	0.0345

Methodology

Scope

This study was carried out in accordance with the ISO 14040 standard [25]. The goal is to perform a comparative LCA of three electrified ethylene plant concepts and two hydrogen-fired steam cracking furnace concepts. The results are benchmarked against previous findings on furnace improvements [7]. Due to the absence of industrial data for the L-E furnace and RDR, which are still in the early stages of development, this study is based on first principles-based models. Moreover, the capital investment and construction-related environmental impacts of the proposed solutions cannot be accurately assessed, thus this study is a second order LCA limited to the operational phase only.

The functional unit has been defined according to PlasticsEurope recommendations [26] as one kilogram of high-value chemicals (HVCs) at the 'gate'. According to PlasticsEurope, HVCs are defined as follows: ethylene, propylene, benzene portion of pyrolysis gasoline, hydrogen fraction in the hydrogen-rich product gas, and 1.3-butadiene (within C4 product). Further treatment steps such as hydrogen valorization, C4 separation, aromatic extraction, and hydrotreatment of pyrolysis gasoline are a part of the downstream value chain which is outside of selected boundaries.

LCI

Due to the broad range of technical solutions, LCI had to be prepared via various tools. First, a simulation of a classically fuel-gas fired steam cracking furnace, used as a baseline scenario (or base case), has been carried out using COILSIM1D [27]. This fundamental model has been used to simulate an industrial furnace under its typical operating conditions. The results have been verified by benchmarking against plant operating data. The capacity of one furnace is 10 tons of propane feed per hour; the same capacity is assumed for all following cases for ease of comparison. The separation of cracking effluents (commonly referred to as 'cold train' or 'separation train') has been simulated with the Petro-SIM® process simulation package to estimate recovery rates as well as compression (kinetic), heat, and cooling duties. A generic demethanizer-first separation train [28] with front-end hydrogenation has been selected. More details on this scenario may be found in [7].

Table 2. Results of whole plant simulation used as LCI: 'L-E',– low–emission furnace, 'L-E APH' – L-E furnace with combustion air preheater, RDR, 'H2-SMR' – hydrogen firing furnace with steam methane reformer, 'H2-GHR' – hydrogen firing furnace with gas heated reformer.

Flow name	Flow type	Unit	BASE	'L-E'	'L-E APH'	RDR	'H2-SMR'	'H2-GHR'
Fossil C_3H_8		kg	8599.41	8599.41	8599.41	8587.71	8599.41	8599.41
Grid electricity	Inlet	kWh	-	3387.93	3935.22	11962.06	313.66	637.24
Boilers duty (natural gas)[a]	Inlet	MJ	46106.06	-	3750.11	6368.00	42928.24	50096.92
Oxygen (gaseous, from integrated ASU)	Inlet	kg	-	-	-	-	-	1048.85
Hydrogen	HVC product	kg	93.73	93.73	93.73	107.37	93.73	93.73
Ethylene	HVC product	kg	3404.23	3404.23	3404.23	3713.11	3404.23	3404.23
Propylene	HVC product	kg	1514.75	1514.75	1514.75	1515.83	1514.75	1514.75
Butadiene	HVC product	kg	234.89	234.89	234.89	266.6	234.89	234.89
Benzene	HVC product	kg	149.43	149.43	149.43	96.04	149.43	149.43
H_2 product, other	By-product	kg	38.69	38.69	38.69	35.41	38.69	38.69
Ethylene product, other	By-product	kg	1.03	1.03	1.03	0.91	1.03	1.03
Ethane	By-product	kg	341.85	341.85	341.85	301.67	341.85	341.85
Propylene product, other	By-product	kg	110.19	110.19	110.19	111.09	110.19	110.19
Crude BBB product, other	By-product	kg	174.65	174.65	174.65	183.46	174.65	174.65
Untreated pygas product, other	By-product	kg	362.83	362.83	362.83	280.34	362.83	362.83
Cracked gasoil product	By-product	kg	32.52	32.52	32.52	3.04	32.52	32.52
Captured CO_2	By-product	kg	-	-	-	-	1737.49	2734.94
HVC yield	Parameter	%	62.76	62.76	62.76	66.24	62.76	62.76
Fuel gas export	Energy export	kg	957.68	1194.27	1312.56	1472.75	605.10	980.71
Fuel to furnace	Inner flow	kg	1182.93	946.34	828.05	497.18	1535.52	1159.91
On-site CO_2 emissions	Emissions	kg	3178.32	2542.66	2224.83	1326.84	2419.56	451.54
On-site CO emissions	Emissions	kg	0.20	0.16	0.14	0.08	0.15	0.03
On-site NO_x emissions	Emissions	kg	1.25	1.00	0.87	0.48	1.62	0.99

Heat export (steam)	Energy export	MJ	36967.09	-	-	-	41984.93	37446.56

[a] – does not account for decoking steam generation

A hybrid approach has also been selected for the L-E furnace. Due to intrinsic limitations, only the process side has been simulated using COILSIM1D, while the mass and heat balance of the radiant box and the heat recovery system simulated with Petro-SIM®. Feed rate, conversion rate, STOR (steam-to-oil-ratio), coil inlet, and outlet temperature as well as pressure drop has been assumed to be equal to the base case. As a result of the negligible difference in product yields, the kinetic and thermal duties of a cold train are assumed to be equal to those of the base case as well. A conservative estimate of the efficiency of the electric drivetrain is taken as 92% [29]. Overall, such an approach represents a realistic scenario for separation train electrification.

The RDR has been simulated with COILSIM1D by imposing a temperature versus residence time profile [30]. According to the article published by the concept authors, such a reactor can realize a constant static pressure of the reactant gas mixture [21]. The reactor is assumed to be isobaric (2 bar a/29 psia). The shaft power needed for RDR operation is assumed to be equal to the enthalpy change of the reactant (8933 kWh per 10 tons of hydrocarbon feed). Heat losses to the environment are not accounted for. Due to the absence of a furnace and therefore, the convection section, a conventional feed – DS mixture preheating arrangement is not feasible. So, the authors have assumed the use of a typical natural gas-fired heater.

The furnace side of the blue hydrogen scenarios has been simulated identically to the 'baseline' scenario. However, it is the reformer itself that is novel. Two arrangements, SMR and GHR, have been simulated using Petro-SIM®, assuming full integration with the steam cracker. The methane-rich fraction recovered from the reactor effluent is used as both feedstock and fuel for the reformer furnace (in the SMR case), and the hydrogen-rich gas is sent to the burners of the cracking furnace.

Life cycle impact assessment (LCIA) method

This study has been carried out in accordance to [31] using the Environmental Footprint v2.0 method to analyze the environmental impact of the researched concepts. The method aggregates various midpoint LCIA models and covers 16 impact categories [32] ranging for climate change impact to freshwater acidification. Endpoint indicators were not considered as they involve subjective weighting factors and thus, are not recommended by industrial guidelines [33].

Results and Discussion

As can be seen in Figure 2, the carbon footprint of the electricity grid defines the applicability of the electrified solutions and, in the RDR case, becomes a major contributor if fossil energy is used to generate electricity. Such an outcome is expected since the amount of enthalpy that must be provided to the propane feed to convert it into high-value chemicals is roughly the same for all scenarios but the efficiency of fossil fuel powerplants is much lower than the thermal efficiency of a typical steam cracking furnace. The thermal efficiency of a modern natural gas-fired combined cycle powerplant reaches 55.5% [34] while according to our BASE case furnace simulations, the overall (radiant box and convection section) efficiency of a cracking furnace exceeds 90%. This unavoidable destruction of exergy suggests a preliminary conclusion: electrification of steam cracking makes sense (from an environmental perspective) only if low–carbon electricity is available. It can be noticed that in the case of mostly fossil grids, such as the grid in the Netherlands, all electrified options are worse than the baseline. If the electricity mix is mostly renewable, as the Norwegian grid, electrified scenarios show a tremendous advantage over traditional combustion options. The RDR reduces the carbon footprint of cracking products at the gate by 26.8%; the best result in this study. The L-E furnace has shown moderate results; the impact of electricity is higher for the solution with an APH due to the higher fraction of duty that is replaced by grid electricity. If mostly renewable electricity is used, the 'L-E' embodiment can lead to an overall reduction of 14.5% and by adding an APH of up to 16.5%.

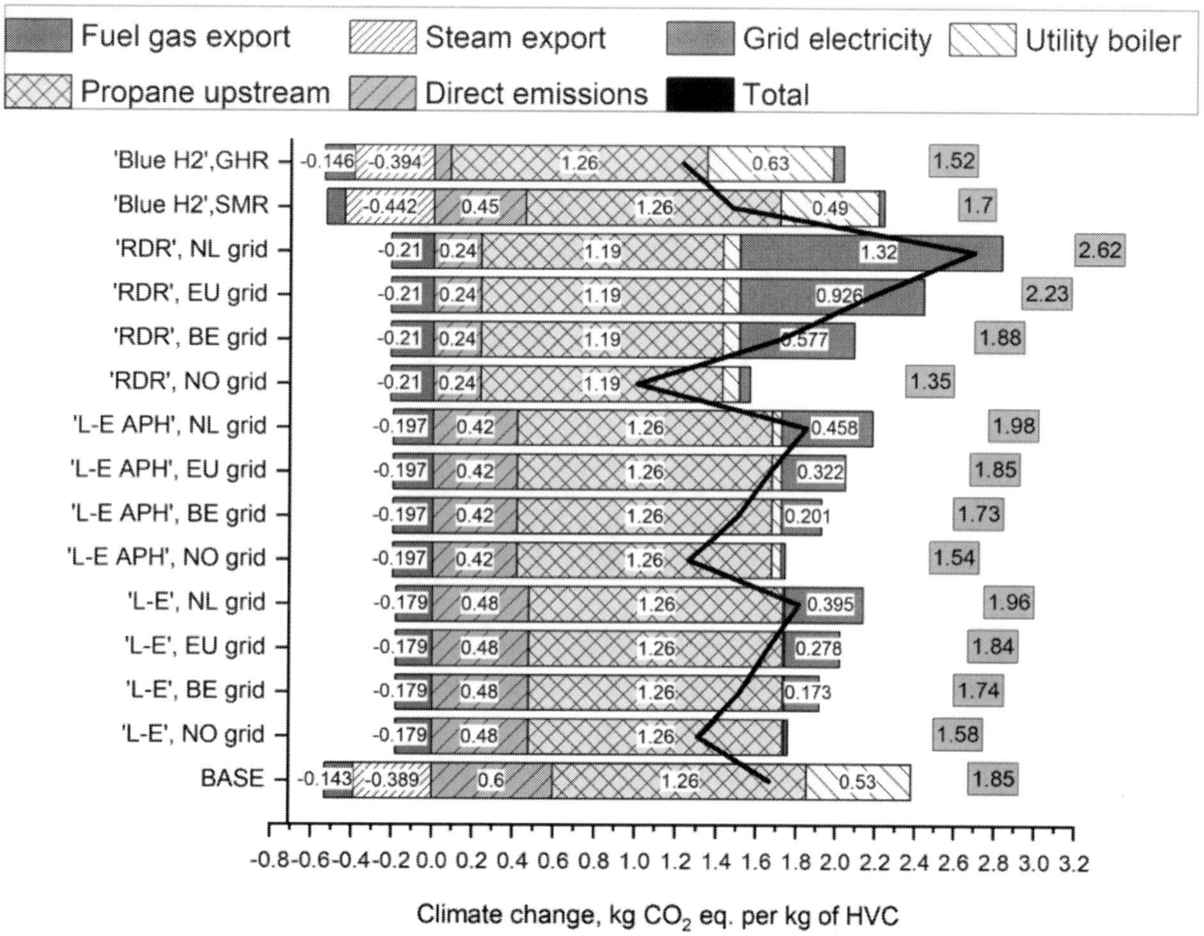

Figure 2. Contribution of each production stage to the carbon footprint of a functional unit at gate (GWP 100a).

It can be noticed that Figure 3A, B, and C are rather similar. The variation is more pronounced for the 'L-E APH' option and the RDR since more of the overall energy demand is replaced by grid electricity. With the highly renewable Norwegian electricity mix, both electrified crackers outperform a basic furnace in all categories except for land use and water scarcity. In the case of the Belgian national grid, ionizing radiation stands out mostly due to a significant share of nuclear power. The indicators of eutrophication, land use, and water scarcity are partially driven by biofuels [35], which are fairly widespread in both the electricity grid in Belgium and the Netherlands, as well as on average in the EU. In the case of the modern-day, power grid in the Netherlands, all electrified crackers perform worse than the base case due to a significant share of fossil fuels that cause emissions of CO_2, NO_x, and PM 2.5 during combustion, leading to terrestrial, marine, and freshwater eutrophication [36]. It is important to note that despite the significant variation, absolute values of the impact on eutrophication are low. All issues are amplified for the EU average grid due to its larger share of solid fuels – mostly brown coal and anthracite – that produce higher flows of NO_x, SO_x, and particulate matter when burnt.

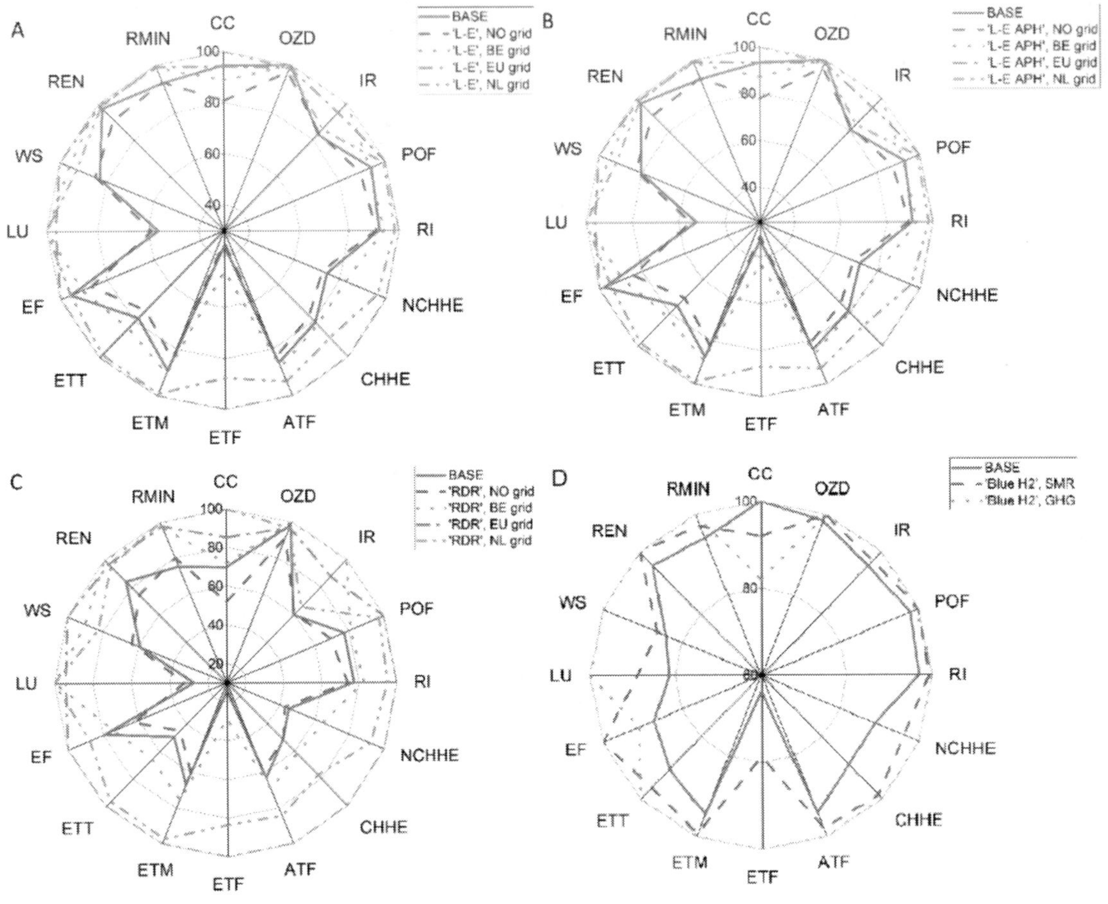

Figure 3. Normalized environmental footprints of one functional unit at gate for electrified steam crackers. Figure A represents a basic L-E furnace; Figure B – a L-E furnace with air preheater; Figure C – RDR; Figure D – blue hydrogen fired furnace.

Letters on each axis stand for impact category: CC – Climate change [kg CO2 eq.], OZD – Ozone depletion [kg CFC11 eq],IR – Ionizing radiation, HH [kBq U-235 eq.], POF – Photochemical ozone formation (human health impact) [kg NMVOC eq.], RI – Respiratory inorganics [disease inc.], NCHHE – Non-cancer human health effects [CTUh], CHHE – Cancer human health effects [CTUh], ATF – Acidification terrestrial and freshwater [mol H+ eq.], ETF – Eutrophication freshwater [kg P eq.], ETM – Eutrophication marine [kg N eq], ETT – Eutrophication terrestrial [mol N eq], EF – Ecotoxicity freshwater [CTUe], LU – Land use [Pt], WS – Water scarcity [m3 depriv.], REN – Resource use, energy carriers [MJ], RMIN – Resource use, mineral and metals [kg Sb eq].

Human health impacts are primarily caused by propane extraction and grid electricity generation. The RDR reduces non-cancer health effects by 6% when connected to a renewable grid, while L-E and L-E APH result in a 3% and 4% reduction, respectively. The impact of electricity production offsets the benefits if grey grid is used, with 'RDR, EU grid' having a 120% higher impact factor value. For cancerogenic emissions, RDR offers a minimal (1.3%) advantage over a basic steam cracker with the Norwegian power grid, while 'L-E' and 'L-E APH' reduce the impact by 4% and 6% due to reduced flue gas flow and low electricity consumption.

CAPEX-related impacts on resource depletion can only be accounted for in the background processes since this is a second order LCA study. Electrified scenarios do not improve the "resource use, mineral and metals" impact factor, and 'RDR, NO grid' actually increases it by 5%. However, the RDR is expected to be at least one order of magnitude smaller and lighter than a conventional steam cracker of the same capacity [21]. All three electrified concepts reduce energy resource depletion by 6.5%, 7.5%, and 13% when connected to a renewable grid, but electrification becomes unfavorable when using a non-renewable grid, leading to an increase in impact factor value by up to 17% for the RDR, BE grid scenario. This is due to the low thermal efficiency of nuclear power plants, which rely on steam turbines alone.

Blue hydrogen combustion is an alternative pathway to reduce GHG emissions from steam cracking which becomes especially relevant if low-carbon electricity is not readily available. An integrated methane reformer is effectively a pre-combustion CCS with the potential to reduce all scope 1 emissions (GHG Protocol) to zero. However, due to techno-economic constraints and high purity requirements for the captured CO_2 stream, a perfect carbon capture rate is not achievable, and hence, hydrogen-rich fuel flow still contains a significant fraction of CO_2 and CH_4. This and the emissions from the reforming furnace itself are the reasons why the 'Blue H2, SMR' scenario can only reduce the on-site emissions by 24% and the overall carbon footprint of a functional unit by 8%. Furthermore, the low thermal efficiency of SMR leads to substantial additional firing duties. For the entire plant, it is 30% higher than the plant based on a basic furnace and causes a 4.5% increase in the impact indicator of 'resource use, energy carriers'. The novel 'gas heated reformer' solves most of the problems. Since the design integrates the heat supply and reforming process into a single ATR, all the CO_2 that is generated during the process is fed to the CCS, resulting in the absence of stack emissions and a better carbon capture rate. Moreover, due to a better heat integration scheme that uses process heat for reforming reactions that occur in the GHG case, the overall heat duty is reduced. When integrated with a cracking furnace, this reformer leads to a reduction in Scope 1 GHG emissions by 86% and 18%, respectively, for an overall carbon footprint of 1 kg of HVC. This is the second-best result after the 'RDR, NO grid' scenario. The captured CO_2 can be further stored in underground caverns such as depleted oil or gas wells, used in the chemical industry, or for EOR. A detailed comparison between blue hydrogen-fired furnaces and the baseline is presented in Figure 3D. Impact factors such as ozone depletion, photochemical ozone formation, respiratory inorganics, both cancer and non–cancer human health effects, acidification terrestrial and freshwater, eutrophication terrestrial, and marine as well as resource use factors are dictated by the avoided burden from fuel gas exports and electricity consumption. Specifically in the case of 'H2, SMR' the reformer furnace is a significant contributor to NO_x emissions. On the other hand, the 'H2, GHR' technology does not have an additional fired heater and is more fuel efficient than a basic furnace, but requires more electricity to produce and preheat the oxygen needed to operate the ATR. In this case, these two effects cancel each other out so similar results as for the SMR and GHG furnaces are obtained. In the case of freshwater eutrophication, the outcome is strongly influenced by the footprint of the electricity which is used. This can lead to a 67% increase in the GHG scenario. The same trend is observed for land use and water scarcity. The situation is reversed for the freshwater ecotoxicity indicator which is mostly driven by the footprint of the hydrocarbon extraction so the plant with SMR has a 15% higher impact than the baseline due to the higher fuel consumption.

Conclusion

The comparative study of three major low-carbon steam cracking technologies reveals some insightful results. The LCA carried out using the Environmental Footprint method reveals that the RDR coupled with an electrified separation train reduces the carbon footprint of high value chemicals by 27% and the consumption of non-renewable energy resources by 13%, without any significant side effects. The only exceptions are the slight increases in water scarcity and inorganic resource consumption, which have risen by 6% and 5% respectively, due to the widespread use of hydropower in the reference renewable grid scenario, the Norwegian grid, and the corresponding infrastructure and heavy machinery required for hydroelectricity production.

However, if the available grid is grey, the RDR-based cracker becomes less sustainable due to unavoidable exergy destruction in the heat engine of any power plant. In contrast, if mixed networks are used, such as the Belgian grid, a L-E furnace proves to be a viable option for reducing the carbon footprint by 6% and 6.6% for basic and APH embodiments respectively, by replacing inefficient steam turbines with electric motors and providing process duty with a hydrocarbon gas furnace that exceeds 92% efficiency.

Our results also highlight the potential of hydrogen-fired furnaces equipped with integrated methane reforming units and CCS, as a promising alternative to electrification. This solution provides the added advantage of allowing steam crackers to run on indigenous fuel from the methane-rich fraction of the cracked gas, while reducing the global warming potential (carbon footprint) by 18%. However, it is important to note that these solutions also have their disadvantages, such as an increased impact on most NO_x, SO_x, and PM-related impact categories, as well as a 4.5% higher consumption of energy resources. Additionally, the relatively pure CO_2 stream produced can be used in the chemical or petroleum industry, and there is also the option of shifting towards green hydrogen produced via water electrolysis. We believe that this technology will be implemented in the near future.

Acknowledgement

This study is supported by the IMPROOF (Integrated Model Guided Process Optimization Of Steam Cracking Furnaces) project as a part of the Horizon 2020 framework program (H2020 Grant Agreement N° 723706) of the European Union. Mike Bonheure also gratefully acknowledges financial support from the Fund for Scientific Research Flanders (FWO) through project 1SD7121N. The research leading to these results has received funding from the European Research Council under the European Union's Horizon 2020 research and innovation program/ERC grant agreement n° 818607.

Abbreviations:

APH – combustion air preheat (a.k.a. 'CAP')
ATR – autothermal reformer
CCGT – combined cycle gas turbine
CCS – carbon capture and storage
CPI – chemical process industry
DS – dilution or process steam
EOR – enhanced oil recovery
GHR – gas heated reformer
GHG – greenhouse gas
LCA – life cycle assessment
LCI – life cycle inventory
LCIA – life cycle impact assessment
L-E – Low-Emission (as in Low-Emission furnace)
RDR – RotoDynamic Reactor
SMR – steam methane reforming
STOR – steam-to-oil-ratio (wt/wt)
TLE – generic term used for all types of quench exchangers
T.EN – Technip Energies

References:

1. Masson-Delmotte, V., et al. [Eds] (2018), "Global warming of 1.5°C An IPCC Special Report on the impacts of global warming of 1.5°C above pre-industrial levels and related global greenhouse gas emission pathways, in the context of strengthening the global response to the threat of climate change, sustainable development, and efforts to eradicate poverty," *© 2019 Intergovernmental Panel on Climate Change.*
2. Anonymous (2022), "Petrochemistry in Europe - Petrochemicals Europe - Petrochemicals Europe". [cited 2022 1.12]; Available from: *https://www.petrochemistry.eu/*.
3. Amghizar, I., et al. (2020), "Sustainable innovations in steam cracking: CO_2 neutral olefin production." *Reaction Chemistry and Engineering*, 5(2): pp. 239-257.
4. Isella, A. and Manca, D. (2022), "GHG Emissions by (Petro)Chemical Processes and Decarbonization Priorities - A Review," *Energies,* 15(20), pp. 7560-7560.
5. Layritz, L.S., et al. (2021), "The potential of direct steam cracker electrification and carbon capture & utilization via oxidative coupling of methane as decarbonization strategies for ethylene production," *Applied Energy*, 296(7), p. 117049.
6. Weydahl, T., et al. (2013), "Pursuing the pre-combustion CCS route in oil refineries – The impact on fired heaters." *Applied Energy*, 102, pp. 833-839.
7. Mynko, O., et al. (2022), "Reducing CO_2 emissions of existing ethylene plants: Evaluation of different revamp strategies to reduce global CO_2 emission by 100 million tonnes," *Journal of Cleaner Production*, 362, pp. 132127-132127.

8. Bonheure, M., et al. (2021), "Dream or reality? Electrification of the chemical process industries," *Chemical Engineering Progress (CEP)*, 117(3), pp. 37-42.
9. Eryazici, I., Ramesh, N. , and Villa, C. (2021), "Electrification of the chemical industry - materials innovations for a lower carbon future," *MRS Bulletin*, 46, 1197-1204.
10. Delikonstantis, E., Scapinello, M., and Stefanidis, G.D. (2019), "Process Modeling and Evaluation of Plasma-Assisted Ethylene Production from Methane," *Processes*, 7 (2), p. 68.
11. Durantay, L., et al. (2021), "A success story of steam turbine replacement by high speed electric system driven compressor." *PCIC Europe*.
12. Tijani, M.E.H., Zondag, H., and Van Delft Y. (2022), "Review of Electric Cracking of Hydrocarbons," *ACS Sustainable Chemistry & Engineering*, 10(49), pp. 16070-16089.
13. Tullo, Alexander H. (2021), "Sabic, BASF, Linde join for electric cracking," *Chemical & Engineering News (C&EN)*, 99 (11).
14. Gross, Samantha (2021), "The Challenge of Decarbonizing Heavy Industry," *© Brookings India. http://hdl.handle.net/11540/13929.*
15. Seppala, J., J. Hiltunen, and V.-M. Purola (2016), "Process and rotary machine type reactor," *US Patent 9,234,140.*
16. Yu, M., Wang, K., and Vredenburg, H. (2021), "Insights into low-carbon hydrogen production methods: Green, blue and aqua hydrogen." *International Journal of Hydrogen Energy*, 46(41), pp. 21261-21273.
17. Zimmermann, H. and Walzl, R. (2009), "Ethylene", *Ullmann's Encyclopedia of Industrial Chemistry*, pp. 465-526.
18. Oud, P., (2020), "Cracking Furnace System And Method For Cracking Hydrocarbon Feedstock Therein", *European Patent EP 3 415 587 B1*.
19. Bushuev, V.A. (2000), "Method for producing lower olefins, reactor for the pyrolysis of hydrocarbons and device for quenching pyrolysis gases," *European Patent Application EP 1 063 273 A1.*
20. Bushuev, V.A. (2007), "Process for producing low-molecular olefins by pyrolysis of hydrocarbons," *US Patent 7,232,937 B2*.
21. Rubini, D., et al. (2022), "A New Turbomachine for Clean and Sustainable Hydrocarbon Cracking," *Journal of Engineering for Gas Turbines and Power*, 144(2), 021024 (11 pages).
22. Suleman, F., Dincer, I., and Agelin-Chaab, M. (2015), "Environmental impact assessment and comparison of some hydrogen production options," *International Journal of Hydrogen Energy*, 40(21), pp. 6976-6987.
23. Godoy, E., Scenna, N.J., and Benz, S.J. (2010), "Families of optimal thermodynamic solutions for combined cycle gas turbine (CCGT) power plants," *Applied Thermal Engineering*, 30(6-7), pp. 569-576.
24. Wernet, G., et al. (2016), "The ecoinvent database version 3 (part I): overview and methodology," *International Journal of Life Cycle Assessment,* 21(9), pp. 1218-1230.
25. Anonymous (2006), " Environmental management — Life cycle assessment — Principles and framework," *ISO 14040:2006.*
26. Anonymous (20017), " PlasticsEurope recommendation on Steam Cracker allocation, Life Cycle and Sustainability working group of PlasticsEurope, 14 September 2017," *https://plasticseurope.org/wp-content/uploads/2021/12/PlasticsEurope_recommendation_on_Steam_Cracker_allocation-_Juillet_2018.pdf.*

27. Plehiers, P.P., et al. (2019), "Artificial Intelligence in Steam Cracking Modeling: A Deep Learning Algorithm for Detailed Effluent Prediction," *Engineering*, 5(6), pp. 1027-1040.
28. Van Geem, K.M. et al. (2008), "PETR 156-energy efficient design of the cold train of a steam cracker," *Presented at the 235th ACS National Meeting, New Orleans, LA, USA.*
29. Gómez, J.R., et al. (2022), "Assessment criteria of the feasibility of replacement standard efficiency electric motors with high-efficiency motors," *Energy*, 239 (A), p. 121877.
30. Johannesdahl, H. (2018), "RDR - Revolution in ethylene yield," *Paper 129a presented at the 30th Ethylene Producers' Conference, Orlando, FL, USA.*
31. Anonymous (2006), " Environmental management — Life cycle assessment — Principles and framework," *ISO 14040:2006*, pp. 652-668.
32. Fazio, S., et al. (2019), "Supporting information to the characterisation factors of recommended EF Life Cycle Impact Assessment methods," *EUR 29600 EN, Publications Office of the European Union, Luxembourg, https://publications.jrc.ec.europa.eu/repository/handle/JRC114822*
33. Anonymous (2014), "Life Cycle Metrics for Chemical Products. A guideline by the chemical sector to assess and report on the environmental footprint of products, based on life cycle assessment," *https://www.wbcsd.org/Projects/Chemicals/Resources/Life-Cycle-Metrics-for-Chemical-Products.*
34. Babaei Jamnani, M., and Kardgar, A. (2020), "Energy-exergy performance assessment with optimization guidance for the components of the 396-MW combined-cycle power plant," *Energy Science & Engineering*, 8, pp. 3561-3574.
35. Van Wijnen, J., et al. (2015), "Coastal eutrophication in Europe caused by production of energy crops," *Science of the Total Environment*, 511: pp. 101-111.
36. Jonson, J.E., et al. (2017), "Impact of excess NO_x emissions from diesel cars on air quality, public health and eutrophication in Europe," *Environmental Research Letters*, 12(9), 094017.

Reliability Improvement of a Dilution Steam System for an Ethylene Unit

Deepak Sharma, Mohammed Imran Ali Khan

Introduction

Dilution steam system (DSS) is designed to separate and recover quench water from hydrocarbons, recover heat, and generate steam for pyrolysis furnaces. Dilution steam is essential to reduce hydrocarbon partial pressure, promote the formation of ethylene, reduce the formation of undesirable heavier compounds, and reduce coke formation in the furnace tubes. The DSS incorporates several individual functions including process water recovery, hydrocarbon stripping, and dilution steam generation. The process water from the Quench Water Stripper (QWS) contains dissolved, suspended, and emulsified hydrocarbons, together with small amounts of free (entrained) oil, and for a gas cracker, suspended solids (mainly coke fines), polymer oils, tar, and heavier hydrocarbon components with densities close to that of water. Naphtha Cracker - Ethylene Plants usually face frequent cleaning of DSG reboilers. This leads to fouling issues in the downstream - in the reboilers / steam generators due to the carryover of tar fines, coke particles, hydrocarbon with water. The run length of the reboilers may be reduced to as low as one month due to fouling issues. There is a penalty on energy, cleaning cost and excess utilities usage and effluent treatment due to these issues. Usage of effective antifoulant and antipolymerant helps in reducing the fouling control in DSG reboilers by extending reboiler run length up to 1-2 years. This will eliminate the downstream energy penalty. Fouling in the quench water and dilution steam systems (DSS) of an ethylene plant are an ongoing problem within the petrochemical industry.

To minimize organic fouling in the Dilution Steam Generation (DSG) system and thereby increase plant reliability innovative antifoulant/ antipolymerant treatment program was developed, evaluated through lab experiments, and applied at JCP-DSS with a target to improve the run lengths at the DSG reboiler & overall system reliability over the currently applied chemical technology. DSG generates dilution steam, which is added with hydrocarbon feed to furnaces. Dilution steam is added to reduce partial pressure of hydrocarbon, consequently increasing olefins yield. Dilution steam is recovered in quench water column and is fed to the LP (Low Pressure) water stripper through filters & coalescer to remove solid particles, oil etc. Feed to the DSG mainly comes from LP water stripper and blowdown from the steam drums from the furnace area. Medium Pressure (MP) steam passes through the Dilution Steam Generators reboilers to generate dilution steam from the saturated liquid which comes from Low pressure water stripper bottom, with a certain degree of superheat in DSG drum. To minimize organic fouling in the Dilution Steam Generation (DSG) system and Process Water Stripper (PWS) to increase plant reliability, it is desired to remove these suspended solids, free oil, and any suspended or emulsified hydrocarbon upstream of the DSG System. It is also important to monitor the same characteristics to take corrective actions, otherwise at the operating temperatures of the PWS and DSG these precursors such as dienes, styrene, and indenes, can lead to organic fouling in the quench loop, process water stripper, or the DSG reboiler. Consequences of this fouling include reduced throughput, increased differential pressure, decreased heat transfer, and occasionally blockages. Fouling can also create an environment for under deposit corrosion, even when the bulk pH is properly controlled.

Problem Background:

This ethylene plant had experienced fouling in the Dilution Steam Generator system which resulted in foaming issues in DSG drum, major plugging of blow down lines and lower run length of its dilution steam reboilers, resulting into costly consequences accompanied the fouling of Dilution Steam Generator System (DSG) and negatively impacting the downstream operations. The Blowdown coolers lines & the heat exchanger were plugged resulting in the bypassing the entire cooler. Considering the criticality of the situation, a joint RCA (root cause analysis) report was submitted, after carrying out the foulant speciation. A thorough investigation & analysis were carried out by both JCP & Nalco Technical team to provide a comprehensive technical recommendation for the DSG issue resolution. JCP, DSG reboiler was swapped and a new reboiler was taken in line on 05th Sep'21. The cleaned online reboiler performance deteriorated very rapidly & it became limiting in the short span of just 25 days. MP steam to reboiler control valve opening was increased from ~ 58 to 100% in a weeks' time. Also, frequent plugging was observed in DSG BD cooler, Level transmitter (LTs) & its associated BD line. These parameter & condition resulted in less steam generation, increased MP steam make-up pushing up the additional steam demand to the furnaces from the utility boilers. With a detailed foulant speciation clearly indicating that the foulant was a crosslinking of styrene copolymers & pytar a revised application & monitoring was supposed to be introduced making sure that reboilers/steam generators are giving the desired run without increased of energy costs by importing MP steam from the utility boilers as makeup. Also frequent reboiler cleanings had its safety concerns with high Benzene exposures while cleaning the generators.

Operational & Chemical Action

On 23rd March 2022 after all the analysis for the foulant, a new treatment program was initiated as a trial to tackle & achieve the following KPIs:

1. Increase the run length of the Dilution Steam Generator System (DSG) reboilers
2. Reduce the DSG level transmitters plugging
3. Reduce utility make up steam flow and valve opening
4. Improve the B/D cooler runlengths without bypassing the system loading up the ETP

Conclusion

1. Significant improvement on the Dilution Steam Generator System with an increase in total steam production indicating not only sustained steam generation but also cleaning of the fouled reboiler as the trial was conducted on an old reboiler running for more than 7 months.
2. MP steam makeup to the DS header from the utility section is reduced to zero as make-up steam valve was completely closed.
3. The frequency of alarms deviations is significantly reduced and reduction in plugging and cleaning frequency of DSG drum level transmitters.
4. An improvement in the DSG blow down cooler performance with lower plugging extent along with the main BD line was achieved.
5. Reduction of overall maintenance job on DSG system resulted in less maintenance cost with an improvement on safety rating at site with lower human exposure.

AIChE PAPER NUMBER: 26a

REMAINING LIFE ASSESSMENT OF TUBES IN AN AGING CONVECTION SECTION

Linda Robinson, PE
Assistant Chief Engineer
KnightHawk Engineering, Inc.

Cliff Knight, PE
Chief Engineer
KnightHawk Engineering, Inc.

Prepared for Presentation at the 2023 Spring Meeting
Houston, TX, March 13 - March 16, 2023

REMAINING LIFE ASSESSMENT OF TUBES IN AN AGING CONVECTION SECTION

Linda Robinson, PE
Assistant Chief Engineer
KnightHawk Engineering, Inc.

Cliff Knight, PE
Chief Engineer
KnightHawk Engineering, Inc.

Abstract:

KnightHawk Engineering, Inc. (KHE) performed a remaining life assessment for the heat transfer tubes in a convection section of an ethylene steam-cracking furnace. KHE undertook this assessment for a client with a steam-cracking furnace, originally commissioned in the early 1990s, as it was nearing the end of a prior remaining life assessment that had been performed by others. The case study of this assessment is presented here.

KHE conducted this tube life assessment using methods and acceptance criteria from the 2016 Edition of API-579-1 Fitness-for-Service, with oxidation corrosion rates taken from API RP 581. The convection section tubes were evaluated for (i) wall thinning by oxidation corrosion and plastic collapse, (ii) creep damage accumulation and creep rupture, (iii) fatigue and creep-fatigue interaction due to cyclical operation, and (iv) deflection by static loads and accumulated creep strain. Tube fins were also evaluated for oxidation corrosion.

Based on the results of the analyses performed, KHE calculated the remaining life of the convection section and identified components that would potentially require replacement to ensure reliable operation for the next 100,000 hours of service. KHE further provided recommendations for future inspection and monitoring plans based on the results of the remaining life assessment.

Introduction

Many ethylene production facilities are faced with determining how to handle equipment that is near the end of its original design life. This paper presents a case study of a remaining life assessment KnightHawk Engineering, Inc. (KHE) performed on the tubes of an ethylene steam-cracking furnace convection section. For the purposes of confidentiality and protection of proprietary technologies, this paper will not include details of the furnace technology or the facility at which it is located. Rather, this paper is intended to provide a general overview of a life assessment methodology, which can be applied to the convection sections of other ethylene steam-cracking furnaces.

KHE performed this assessment for an ethylene steam-cracking furnace (the "Furnace") that was originally commissioned in the early 1990s. At the time of assessment, the Furnace was nearing the end of a 20-25-year design life that had been determined in a prior remaining life assessment performed by others.

This tube remaining life assessment was conducted using methods and acceptance criteria from the 2016 edition of API-579-1 Fitness-for-Service.[1] API RP 581 'Risk-Based Inspection Methodology'; was also referenced for oxidation rates for the tube materials. The tubes were evaluated for (i) wall thinning by oxidation corrosion and plastic collapse, (ii) creep damage accumulation and creep rupture, (iii) fatigue and creep-fatigue interaction due to cyclical operation, and (iv) deflection by static loads and accumulated creep strain. Tube fins, where they existed, were evaluated for oxidation corrosion. The assessment was performed to determine the likelihood of reliable operation of the convection section tubes for the next 100,000 hours of service.

Equipment and Process Description

The Furnace consists of a radiant and a convection section. In the convection section, the hydrocarbon process feed and several other streams are heated inside of a series of heat transfer tube coils by flue gas flowing out from the radiant section of the Furnace. Figure 1 below displays a schematic diagram of the general configuration of the convection section assessed. The various streams heated in the convection section include boiler feed water (BFW), hydrocarbon feed (HC), dilution steam (DS), and high-pressure steam (HPS). The mixed feed (MF) coil heats a mixture of the HC and DS streams. The flue gas flows vertically upwards from the radiant section over the successive rows of tubes of the heat transfer coils. The tubes in the coils are supported by intermediate and end tubesheet supports. The design code for the process coils (HC, DS, and MF) is ASME B31.3, and the design code for the other coils (BFW and HPS) is ASME BPVC VIII Div.1.

Typical flue gas temperatures are about 1,150 °C (2,100 °F) and the bottom inlet to the convection section to about 250 °C (480 °F) at the outlet. The materials of construction for the tubes and tube fins (where they exist) vary with the location (height) in the convection section

[1] Note: All references to API-579 in this paper will refer to the 2016 edition.

to provide adequate creep strength and resistance to oxidation for the coils operating at elevated temperatures. Table 1 below lists the materials that were evaluated as part of the convection section life assessment.

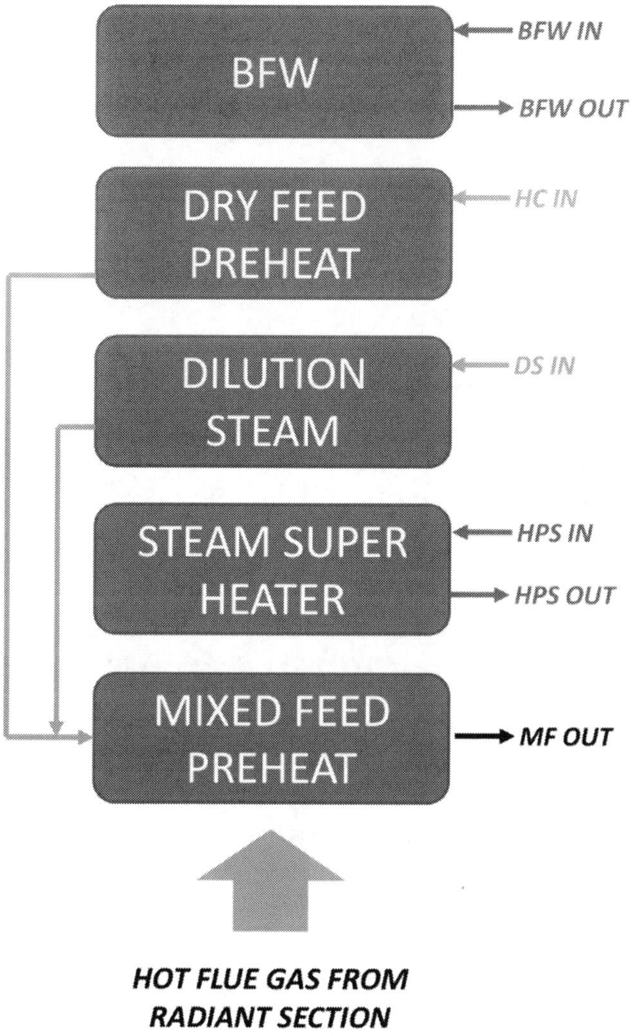

Figure 1: Diagram of General Convection Section Configuration

Table 1: Convection Section Materials Evaluated in the Life Assessment

Component	Form	Material/Composition
Tubes	Seamless Pipe	A312-304H (18Cr-8Ni Stainless) A335-P11 (1.25Cr-0.5Mo Alloy) A106-B (Carbon steel)
Fins	Plate	11-13Cr 304H Carbon Steel

Historical Operating Conditions

The convection section life assessment was executed as a coordinated effort between KHE and their client (the "Plant"). The Plant performed an analysis of its historical operating data which served as a key basis for the input parameters to the remaining life calculations. From the Plant's historical data analysis, which spanned 28 years of operation from 1991 to 2019, five historical operating periods were defined (T1 through T5). Within each these operating periods, operating conditions were relatively consistent, and a set of typical operating conditions was defined by the Plant. A sixth period (T6) was also defined for future operation through 2029.

The Plant provided KHE with tube, fin, and flue gas temperatures for the various tube rows for each of the six operating periods based on the Plant's operating data analysis and process simulation. The Plant also provided a documented history of inspection and repairs including when any tube rows were replaced.

Material Properties

KHE developed material property models for the convection section tubes' materials of construction which are summarized in Table 2 below. Material properties were sourced from the following applicable and relevant codes:

- ASME II Part D, Table 1-A for allowable stresses (Sa) for BFW and HPS tubes.
- ASME B31.3, for allowable stress (Sa) of HC, DS, and MF tubes.
- ASME II Part D, Tables TM-1, TE-1 and TC-1 for elastic modulus, thermal expansion and thermal conductivity.
- API 579-1 Table 10B.2, for the Larson Miller Parameter (LMP) for creep assessment.
- API 579-1 Table 10B.1, for Project Omega for creep assessment.
- API 579-1 Table 2E.2 and 2E.4, for temperature dependent yield and tensile strength values.
- API 579-1 Table 4.1, for the temperature limit used to define the Creep Range.
- API 579-1 Table 10B.3, for the Limiting Design Metal Temperature.

Third-order polynomial functions were developed and applied in the assessment model by KHE for temperature-dependent properties such as the yield stress and elastic modulus as well as the Larson-Miller Parameter and Project Omega constants.

Table 2: Material Properties

Material	Service	Composition	S_u MPa	S_y MPa	E GPa	TLCR °C	LDMT °C
A312-304H	Seamless Pipe, Tubes	18Cr-8Ni	517	207	195	510	816
A335-P11	Seamless Pipe, Tubes	1.25Cr-0.5Mo	414	207	204	427	649
A106-B	Seamless Pipe, Tubes	Carbon steel	414	241	203	343	538

Where:

- *S_u is The Minimum Ultimate Tensile Strength or UTS at ambient Temperature*
- *S_y is the Minimum Yield Strength at ambient Temperature*
- *S_a is the Allowable stress at ambient Temperature*
- *E is the Youngs Modulus at ambient Temperature*
- *TLCR is the Temperature Limit to define Creep Range*
- *LDMT is the Limiting Design Metal Temperature*

Oxidation Corrosion

Wall thinning by oxidation-corrosion was simulated in the assessment model using oxidation rates from API RP 581 'Risk Based Inspection Methodology' ('API RP 581'), Section 2.B 9.1 for High Temperature Oxidation. See Table 3 below for a summary of the applied oxidation corrosion rates. Oxidation rates were simulated using the API 581 values for the BFW and HPS tubes and two times the API RP 581 values for all process tubes. Functions were derived and applied in the assessment model for the temperature-dependent corrosion rate of a given material using 3rd-order polynomial curve fits.

Tube Remaining Life Assessment

The following assessments were made for each of the tube rows for the operating conditions associated with each of the historical and projected future operating time periods:

- Protection against plastic collapse considering wall thinning by oxidation corrosion,
- Protection against creep rupture considering creep damage accumulation,
- Protection against Creep-Fatigue interaction due to cyclic operation (decoke cycles),
- Ability to maintain form and function considering deflection from static loads and accumulated creep strain damage,
- Oxidation corrosion of the tube fins (where present).

Further details on each of the assessments are provided in the following sections.

Table 3: Oxidation Corrosion in mm/yr

$T_{service}$ °C	CS Material Carbon Steel CS	P11 Material 1.25Cr-0.5Mo	12Cr Material 12Cr	304H Material 18Cr-8Ni
0	0.0	0.0	0.0	0.0
496	0.05	0.05	-	-
524	0.1	0.08	-	-
552	0.15	0.1	-	-
579	0.23	0.18	-	-
607	0.36	0.3	-	-
635	0.56	0.46	-	-
663	0.84	0.76	-	-
691	1.22	1.17	0.03	-
718	-	-	0.08	0.03
746	-	-	0.2	0.05
774	-	-	0.38	0.08
802	-	-	0.76	0.1
829	-	-	1.27	0.15
857	-	-	-	0.23
885	-	-	-	0.33
913	-	-	-	0.46
941	-	-	-	0.64
968	-	-	-	0.89
996	-	-	-	1.22
1024	-	-	-	-
1052	-	-	-	-
1079	-	-	-	-
1107	-	-	-	-
1135	-	-	-	-
1260				

Where:

- *TService is the Middle Of Run (MOR) Service temperature for the relevant time period.*
- *CS is for Carbon Steel A106-B tubes and fins.*
- *P11 is for 1.25Cr-0.5Mo materials A335-P11 tubes.*
- *12Cr is for fin material. Note: Fin material was specified as 11-13Cr.*
- *304H is for austenitic stainless steel A312-304H tubes and fins*

Tube Plastic Collapse and Creep Rupture Assessment

A creep assessment for the tubes was performed for each row using the methodology from API 579-1 10.5.2.3 The operating time periods, T1 to T6, were broken into 12-month increments (or parts thereof) for adequate resolution of operating cycles and wall thinning. Each time increment was assessed individually per the following steps.

1. Service temperature for the time increment was applied as per the corresponding value from the Plant's process analysis.
2. The original code (ASME B31.3 or ASME BPVC VIII Div.1) minimum required tube wall thickness was calculated and verified against the nominal design thickness.
3. A 12.5% mill tolerance was subtracted from the nominal design thickness to define a minimum, uncorroded, Start-of-Life (SOL) wall thickness for each tube row.
4. Taking into consideration the service temperature and material of construction, tube wall thinning was calculated for the given time increment using the rates described in the Oxidation Corrosion section above. This wall thinning was added to wall thinning from previous time increments for the given tube row to calculate a total corroded wall thickness for the time increment under consideration.[2]
5. Taking into consideration the service temperature and material of construction the physical properties of Allowable stress (S_a), Yield strength (S_y), and Elastic Modulus (E) were derived as per the Material Properties section above.
6. Principal stresses were calculated depending on the design pressure, calculated corroded wall thickness (Step 4), and tube diameter. Stresses were calculated for the tubes per API-579 para. 10.5.2.5.a.1 using the following equations from API-579 Table 10.2

$$\sigma_1 = \frac{P_d \cdot D}{2 \cdot t_c} \cdot L_f \qquad \sigma_2 = \frac{P_d \cdot D}{4 \cdot t_c} \cdot L_f \qquad \sigma_3 = 0$$

Where:

σ_1, σ_2, σ_3 are the principal stresses
P_d is the design pressure
D is the pipe diameter[3]
t_c is the corroded thickness
L_f is the Lorentz Factor (taken as 1 for a straight section of tube)

---

[2] Tube row replacements per the Plant's maintenance records were applied at the appropriate corresponding time increments.

[3] API-579 specifies the mean diameter for use in these stress equations; however, KHE conservatively applied the outside diameter for this assessment based on agreement with the Plant.

Primary bending stress (P_b) was calculated from standard beam deflection with simply supported ends (spanning the tubesheet support locations) with tube mass, fin mass (if applicable), and fluid mass applied as a uniformly distributed load.

7. The primary load reference stress was calculated per API-579 equation 10.9 (shown below) and evaluated against elastic limits for protection against plastic collapse.[4]

$$\sigma_{ref} = \frac{P_b + \left(P_b{}^2 + 9 \cdot P_L{}^2\right)^{0.5}}{3}$$

Where:

P_L is the primary local membrane stress, equal to σ_1 calculated in Step 6 above.
P_b is the primary bending stress calculated in Step 6 above

8. The effective stress, S_{eff}, for the Larson Miller Parameter (LMP) was calculated using API-579-1 Equations 10.12, 10.23, 10.24, and 10.25 all shown below. KHE applied a 1.25 factor to the calculated effective stress, S_{eff}, as a safety margin to account for stress concentrations and other variables.

$$\sigma_e = \frac{1}{\sqrt{2}} \cdot [(\sigma_1 - \sigma_2)^2 + (\sigma_1 - \sigma_3)^2 + (\sigma_2 - \sigma_3)^2]^{0.5}$$

$$S_{eff} = \sigma_e \cdot e^{0.24 \cdot \left(\frac{J_1}{S_s} - 1\right)}$$

$$J_1 = \sigma_1 + \sigma_2 + \sigma_3 \qquad S_s = (\sigma_1{}^2 + \sigma_2{}^2 + \sigma_3{}^2)^{0.5}$$

Where:

σ_1, σ_2, σ_3 are the principal stresses calculated in Step 6 above
σ_e is the effective stress
S_{eff} is the effective stress used to compute the LMP

9. The Larson Miller Parameter (LMP) was calculated with the effective stress (S_{eff}) calculated in Step 8 and the parameters from API-579 Table 10B.2 appropriate for the material of construction.

10. The time to rupture (L) was calculated based on LMP and temperature with API-579-1 Equation 10B.16 as shown below.

$$L = 10^{\left(\frac{1000 \cdot LMP}{T_f + 460} - C_{LMP}\right)}$$

11. The Creep Damage Fraction (D_c) was calculated as the ratio of the duration of the time increment to the calculated Time to Rupture (L) from Step 10 above.

---

[4] API-579 Equation 10.10 specifies the yield stress at temperature, Sy, as the allowable limit for the primary load reference stress for austenitic stainless and nickel-based alloy materials; however, KHE conservatively applied a 90% factor to Sy for this assessment based on agreement with the Plant. Similarly, API-579 Equation 10.11 specifies 75% of Sy as the allowable limit for the primary load reference stress for all other materials (including carbon steels); however, KHE conservatively applied 60% of Sy.

Repeating Steps 1 to 11, the calculated Creep Damage Fraction is accumulated over the time increments from the historical operating periods T1 to T5 to determine the historic creep damage accumulation. This accumulated value was evaluated against an allowable Dc of 0.8[5] for assessment of protection against creep damage. The accumulated Creep Damage Fraction was calculated for the future operating period, T6, to evaluate its fitness for continued service. The above assessment method was carried out for all of the tube rows of the convection section.

Tube Fatigue Assessment Method

A tube creep-fatigue assessment was applied for each row using the methodology from API 579-1 10.5.3 for creep-fatigue interaction. The creep damage was evaluated in the Tube Plastic Collapse and Creep Rupture Assessment section above. Fatigue damage was calculated for each tube row and time increment as per ASME BPVC VIII Div. 2, Par. 5.5.5, Fatigue Assessment of Welds per the following steps:

1. The maximum membrane and maximum bending stresses were applied as calculated in the Tube Plastic Collapse and Creep Rupture Assessment section above (Step 6).
2. The elastic and non-linear structural stress and strain ranges were calculated conservatively assuming that each decoke cycle results in reducing the tube stress to zero (i.e., maximum possible stress/strain range considered).
3. The equivalent structural stress range ΔS_{ess} was calculated.
4. The number of permissible cycles (N) was calculated including appropriate materials and temperature correction factors, environmental factors, and fatigue improvement factors as per ASME BPVC VIII Div. 2, Par. 5.5.5.
5. The fatigue damage fraction (nDf) for the time increment under consideration was calculated as the ratio of the cycles applied to the number of permissible cycles (N).

Repeating steps 1 to 5, the accumulated fatigue damage fraction Df was calculated over the historical and projected future time periods and evaluated against an allowable limit of 1.0.

The accumulated fatigue damage fraction (Df) and the accumulated creep damage fraction (Dc) that was calculated in the preceding section were plotted on a Creep-Fatigue Envelope as per API 579 Fig. 10.29, see Figure 2 below for an example. If the point of intersection between Df and Dc falls on or below the curve corresponding to the material of construction, the tube is considered to have adequate protection against damage from creep-fatigue interaction. The above assessment method was carried out for all of the tube rows of the convection section.

---

[5] Note: This 80% Creep Damage threshold was applied in this assessment as a conservative measure and may be adjusted upwards or downwards in the future based on inspection results (e.g., actual wall thinning) and/or defects observed during service.

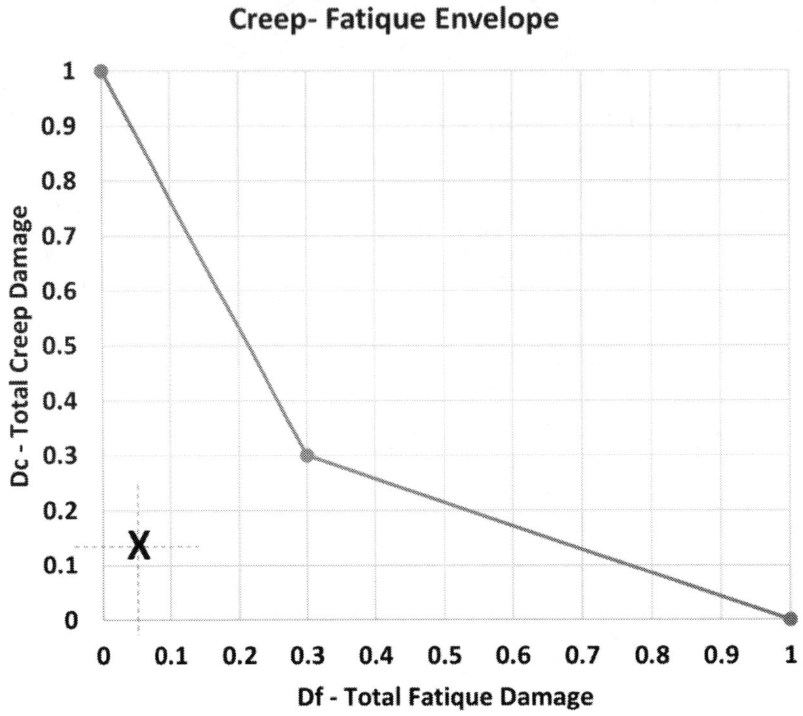

Figure 2: Creep-Fatigue Envelope per API-579 Fig. 10.29 (shown here for Type 304H SS material)

Tube Strain Assessment

A tube strain assessment was applied for each row using the MPC Project Omega strain rate methodology from API 579-1 Annex 10B para. 10B.2.1 as follows:

1. The effective stress, σ_e, applied was the primary bending stress, P_b, calculated in the Tube Plastic Collapse and Creep Rupture Assessment section above (Step 6).
2. The effective stress, S_1, was calculated as the logarithm ($\log_{10}$) of the effective stress, σ_e, per Equation 10B.21 in ksi.
3. The MPC parameters (A_0 through A_4 and B_0 through B_4) were taken as specified for each material of construction in API 579-1 Table 10B.1.
4. A Prager factor, β_Ω, of 0.33 was applied as well as a state-of-stress parameter, α_Ω, of 2 for pressurized cylinders as defined in API-579-1 10B.8.
5. The adjustment parameter for creep strain rate, Δ_Ω^{sr}, which accounts for the material scatter band in the Project Omega Model, was applied as -0.475.[6]

[6] A range of -0.5 for the bottom of the scatter band to +0.5 for the top of the scatter band can be used for this value. KHE's applied value of -0.475 corresponds to the lower 95 percentile of the scatter range.

6. The adjustment parameter for creep ductility, Δ_{Ω}^{cd}, in the Project Omega Model was applied as -0.3.[7]
7. The Omega uni-axial omega damage parameter, Ω, Bailey-Norton coefficient, n_{BN}, MPC parameter, δ_{Ω}, minimum Omega uniaxial damage parameter, Ω_n, and Omega multi-axial Omega damage parameter, Ω_m, were all calculated using the following equations from API-579-1 10B.3 through 10B.7:

$$\Omega = 10^{\left[(B_0+\Delta_{\Omega}^{cd})+\left(\frac{B_1+B_2\cdot S_1+B_3\cdot S_1^2+B_4\cdot S_1^3}{T+460}\right)\right]}$$

$$n_{BN} = -\left(\frac{A_2+2\cdot A_3\cdot S_1+3\cdot A_4\cdot S_1^2}{T+460}\right)$$

$$\delta_{\Omega} = \beta_{\Omega}\cdot\left(\frac{\sigma_1+\sigma_2+\sigma_3}{\sigma_e}-1\right)$$

$$\Omega_n = max(\Omega - n_{BN}, 3) \qquad \Omega_m = \Omega_n^{(\delta_{\Omega}+1)} + \alpha_{\Omega}\cdot n_{BN}$$

Where:

T is the service temperature (in °F) during the time increment considered
σ_1, σ_2, σ_3 are the principal stresses with $\sigma_1=\sigma_e$, and $\sigma_2=\sigma_3=0$
All other parameters are as defined in the preceding steps above

8. The creep strain rate, ε_{co}, was calculated from API-579-1 Equation 10B.2 shown below

$$\varepsilon_{co} = 10^{-\left[(A_0+\Delta_{\Omega}^{sr})+\left(\frac{A_1+A_2\cdot S_1+A_3\cdot S_1^2+A_4\cdot S_1^3}{T+460}\right)\right]}$$

Where:

All parameters are as defined in the preceding steps above

9. The vertical deflection of the tube due to the creep strain rate over the time increment under consideration was calculated.

Repeating steps 1 to 9, the tube's accumulated vertical deflection was calculated over the historical and projected future time periods. This creep strain deflection was added to a static deflection for the tube under its own weight that was calculated with standard beam deflection methods to obtain a total vertical deflection of the tube. The total deflection was evaluated against a maximum allowable limit of 1 inch or 25.4mm. The above assessment method was carried out for all of the tube rows of the convection section.

[7] A range of -0.3 for ductile behavior to +0.3 for brittle behavior can be used for this value.

Tube Fin Assessment

The fins on all the finned tube rows were assessed for protection against oxidation corrosion for the historical and projected operating periods. As with the previous tube assessments, the time periods T1 through T6 were applied in 12-month time increments. Fin temperatures were taken from the process analysis results provided by the Plant. Wall thinning by oxidation corrosion was applied at the API RP 581 rates described in the Oxidation Corrosion section of this paper. The corroded fin thickness was compared against a minimum allowable fin thickness of 0.5mm.

Results

Tube Life Assessment

The tube life assessment for the aging convection section was carried out as described in the preceding sections of this paper. The results of the assessment were as follows:

- Service temperatures for the Feed and Mixed Feed Preheat coils and some rows of the Steam Superheater and Dilution Steam coils exceeded the temperature limit for the creep range (TLCR) but were below limiting design metal temperature (LDMT) for the materials of construction and were be assessed for creep damage. Service Temperatures for all other tube rows were below the TLCR and long-term creep damage is not anticipated.
- Corroded wall thicknesses based on the estimated oxidation corrosion rates resulted in induced stresses that were adequately below the elastic limit for protection against plastic collapse. The calculated peak stress was observed in one of the BFW tube rows with a peak stress of 97 MPa (or 86% of the allowable limit).
- The calculated remaining life (time to rupture) for all of the rows of the convection section exceeded the 100,000 hours criteria for the assessment. The minimum remaining life was calculated to be 107,600 hours in the top rows of the Mixed Feed Preheater.[8]
- The accumulated creep damage for all rows of the convection section through the historical operating periods (T1 to T5) was below 0.8. The maximum projected accumulated creep damage at the end of the projected future operating period (T6) occurred in the top rows of the Mixed Feed Preheater and was 0.89. This was rated to PASS the assessment for adequate protection against creep damage.
- For all rows of the convection section, the accumulated fatigue damage was found to be negligible, and the Creep-Fatigue interaction was within the acceptable range.

[8] KHE's assessment predicted a limited remaining life for the bottom two rows of the Mixed Feed Preheater at the end of 2004; however, these tubes were replaced in 2005. Cracks were observed in the midwelds of some of the replaced tubes.

- Deformation by creep and static deflection was deemed acceptable with a maximum calculated deflection of 9.2 mm at the bottom rows of the Mixed Feed Preheat (i.e., shield rows).
- The fins for the BFW pass the tube fin assessment with at least 0.5mm of fin thicknesses remaining after the estimated loss due to oxidation corrosion. However, the fins for all the other finned coils failed the assessment.

Recommendations

Based on the results of the remaining life assessment for the convection section tubes KHE recommended inspection, thickness measurement, and in-service monitoring of the fins on all the finned tube rows except the BFW. The oxidation rates applied in the remaining life assessment of the fins can be updated based on the actual observed oxidation rates from these inspections.

KHE also recommended random inspection of other locations as per the inspector's discretion and reasonable access to confirm the oxidation rates applied in the assessment. Any thickness readings that are less than the minima indicated in KHE's assessment should be investigated and properly assessed.

Additional Assessments

KHE also assessed the remaining life of the intermediate and end tubesheet supports of this convection section using similar methods as those presented in this paper for the remaining life of the tubes. Wall thinning was applied at two times API RP 581 rates for the corresponding materials of construction, with the exception that oxidation rates for high-temperature casting materials not addressed in API RP 581 were sourced from supplier datasheets.

Principal stresses were calculated for each tubesheet dependent on deadweight and frictional loads. Dead loads included the tube mass, fin mass (where appropriate), and fluid mass. Primary bending stresses were calculated from standard wide beam deflection with simply supported ends. A casting Margin of 0.8 was applied to the reference and equivalent stress to account for inherent casting defects and a stress concentration factor was applied to account for the wide beam design. An equivalent elastic strength was applied to account for holes in the tubesheet as a ligament efficiency.

Several of the lower end tubesheets failed the remaining life assessment, and the lowest intermediate tubesheet was indicated to have a marginally acceptable remaining life. It is anticipated that the methods applied in the assessment were conservative, possibly overly conservative, specifically the application of two times oxidation rates to the tubesheets, the service temperatures applied to the end tubesheets, and the application of creep rupture acceptance criteria to structural components. KHE recommended targeted inspections of these tubesheets at the earliest inspection opportunity and during all shutdown opportunities. The assumptions applied in the remaining life assessment of the fins can be updated based on data from these inspections.

Defect Free Commissioning: Challenges and Opportunities amidst COVID'19

Sundaravel Singaravelu

Background: OQ commissioned its first steam cracking unit (SCU) in Oman amidst COVID pandemic. Commissioning of new plant requires support from process licensors, critical equipment or machinery subject matter experts and engineering consultants on site. With partial commissioning in progress, COVID Pandemic stroke, impacting the mobility of critical and specialist resources across globally and posed a challenge to the Project team in moving forward with commissioning.

Abstract: Despite all challenges amidst COVID'19, OQ adopted a risk-based approach and successfully managed to start up the SCU with remote support from across. In this paper, the challenges that Asset Integrity and Materials Corrosion Team faced during the pre/commissioning and the opportunities that were exploited to overcome the challenges are briefly outlined.

This article highlights the key issues associated with materials and corrosion standpoint in project context. Pitfalls in the commissioning processes with lessons learnt from the project phase to post commissioning, Operational phase are reflected as case studies from a Petrochemical facility. **Case:1** Acoustic induced vibration (AIV) failure of small-bore connections (SBC) in a dilution steam header-design review. **Case: 2** Microbial Induced Corrosion (MIC) leaks in run down line - SS 316L - in cryogenic service. **Case 3**: Fatigue failure / cracking of a spent caustic piping connected to PD Pumps discharge - Dampener failures.

AIChE Paper Number: 26b

Ethylene Cracker Radiant Tube Life Optimization.

Abdul Shukoor Vemmarath
Staff Engineer-Fixed Equipment, Reliability
Saudi Polymers Company

Mohammed Al-Qarni
Superintendent - Fixed Equipment, Reliability
Saudi Polymers Company

Prepared for Presentation at the 2023 Spring National Meeting
March 12 – 16, 2023

Ethylene Cracker Radiant Tube Life Optimization

Abdul Shukoor Vemmarath
Staff Engineer-Fixed Equipment, Reliability
Saudi Polymers Company

Mohammed Al-Qarni
Superintendent - Fixed Equipment, Reliability
Saudi Polymers Company

Abstract

Ethylene cracker radiant tubes are exposed to some of the most severe operation conditions in the chemical process industries. These include elevated temperatures, oxidizing reducing atmospheres, high carbon potentials, etc. In addition to that, during steam/air decoking; burning-off the accumulated carbon on tube internal walls after each run, initiate and accelerates the active damage mechanisms until eventually to the tube fails. Early identification of the primary active damage mechanisms and timely replacement of affected tubes are important for reliable and safe furnace operation.

The Saudi Polymer Company (SPCo) ethylene plant started operation in 2011 with industry typical alloy (25/35 Cr/Ni) radiant tubes. An initial two-year PM and four-year retubing cycle was established based on the furnace design as per industry best practice. Since startup, furnace run length (decoking frequency) had been significantly lower than design. A root cause analysis and extensive analytical study was performed on collected coke and tube samples during preventive maintenance (PM) outages. Tube analysis showed oxide spalling with chrome oxide layer depletion. Based on these learnings, the radiant tubes were upgraded with an alumina forming (oxide layer) metallurgy which improved coking resistance and resulted in significantly longer run lengths. The improved decoke frequencies, minimized the effects from the high temperature exposure of the tube ID wall. This resulted in no tube failures during the four-year expect life of the tubes with the alumina forming alloy. An initial remaining life assessment of the four-year aged tubes provided sufficient confidence to increase tube life by 2 more years. However, a number of tube failures were reported after 5th year, therefore, another study was performed to further understand tube life expectations. The presence of ID pitting was identified during further metallurgical analysis in the failures. All the failures were found with a range of pitting from just starting in the alumina protective layer to through the tube wall. While the traditional chromium oxide tubes easily regenerate their oxide layer, the regeneration of alumina oxide layer may require special procedures, so early identification during the PM, allows for addressing this issue as way to address life limiting damage mechanism that can end up in unplanned furnace outages.

This paper discusses the Ethylene Cracker Radiant Tube Life Optimization and inspection methods for early identification of life limiting pitting corrosion of the alumina protective layer.

Introduction and Background

The Saudi Polymer Company, SPCo started operation in 2011. The olefin products are produced via steam cracking using nine Lummus SRT-VI furnaces. The original radiant tubes were conventional alloy (Cr-25/Ni-35). After startup, the run length for ethane feed was around 20 days and propane were 25-30 days, which was significantly lower than the design run length of 60 days. In most cases, the furnaces were operated at low conversions (58-61% for ethane, 85-88% for propane) and higher than design steam to hydrocarbon (S/HC) ratios (0.4-0.5). Furnaces reached end of run within 20-30 days based on a critical pressure ratio (CPR) limit at 0.90.

To address short run times, potential root causes and avenues were explored. Extensive analytical study has been performed on collected coke and radiant tube samples during a two-year PM outage in 2013. The tube indicated depleted chrome oxide surface and spalled oxide layer on tube inner surface. Coke samples showed a characteristic "green" color – conclusively identified as chromium (**Figure-1**).

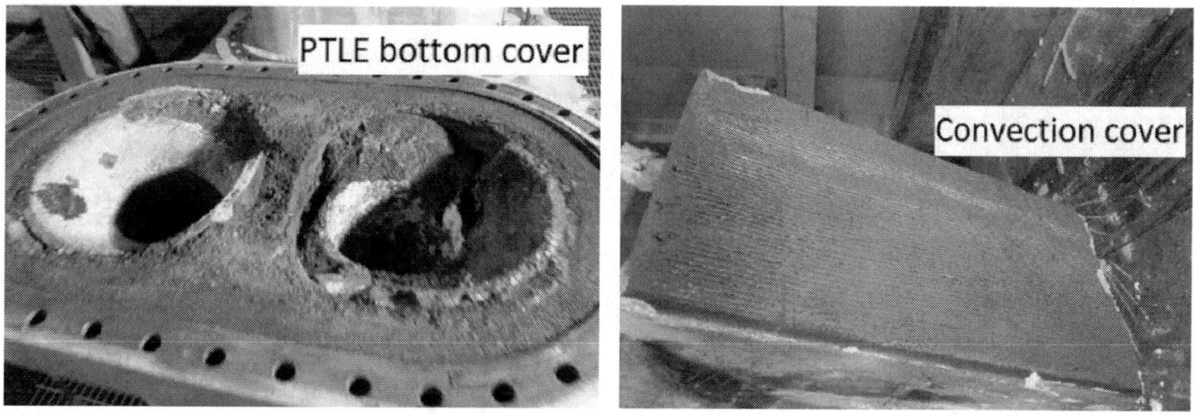

Figure 1 – Greenish deposit-presence of chromium oxide.

Based on these findings, a coil section was replaced in a single furnace. The first run after the coil replacement shows excellent performance. A subsequent run showed the replaced coil's performance degraded immediately and by the third run the replaced coil showed no difference compared to coils that have been in service for two years. All these findings were the result of low performance of the conventional radiant tube against oxidizing reducing atmosphere and promoted high coke formation and eventually low run length. Based on these studies and findings, a strategical decision was made to trial and upgrade with alumina forming metallurgy (Cr-30/Ni-45/Al-04 alloy), based on the proven mechanical and thermal stability at higher temperatures, such as higher structural stability, better high temperature stress rupture strength and enhanced carburization /oxidation resistance. With alumina forming metallurgy, the run length significantly improved even beyond the initial design run length of 60 days. As a result, the decoking (a detrimental operation sequence for tube life) frequency decreased significantly.

Radiant tube life optimization

SPCo established an initial re-tubing frequency of four years based on the furnace design per industry practices after commissioning. A furnace PM every two years was also established to be in line with the retubing cycle. The philosophy worked fine with conventional radiant tubes. After upgrading with alumina forming metallurgy no tubes were reported with either a plug or failure in the first four years. A couple of samples were collected after four years of service outage with the alumina forming radiant tubes to study the mechanical and metallurgical performance to assess the remaining life.

Visual inspection of the four-year tube sample did not indicate any significant damage. A dimensional (creep) inspection revealed no relevant outer diameter expansion, thickness reduction, bulging, or swelling. The samples were found to be within the tolerances as per the drawing. The chemical analysis by Optical Emission Spectroscopy confirmed the alloy composition was within the tolerance. Accelerated creep rupture tests were performed on three specimens as per the parameters taken from production test plan (ITP), two specimens ruptured as expected and one ruptured below the 80% expected life.

4Y tube Specimen	Temp (^{o} C)	Stress (MPa)	Duration (Hr)	Result (Hr)	% min required time
Sample-1	1150	7.0	1000	763.5	**76.35**
Sample-2	1150	7.0	1000	985.7	**98.57**
Sample-3	1150	7.0	1000	875.2	**87.52**

Table- 1 Mechanical Analysis (Creep Test)

Macro-etching shows a 100% columnar structure and a sound wall free of visual defects. Micro structural inspection was conducted with inverted optical microscope and Scanning Electron Microscope (SEM). The microstructure consists of rounded and coarsened primary carbides which are surrounded by Beta phase (Ni-Al) and a huge number of secondary carbides and acicular precipitations (Al, Cr) in austenitic matrix. Isolated creep voids were detected in the tube wall and weld heat affected zones. Most of the tube inner surface was found to be covered by a protective Al oxide scale. Some areas (dark spots on visual inspection) revealed internal oxidation with adjacent minor carburization. Aluminum nitrides were detected close to outer surface.

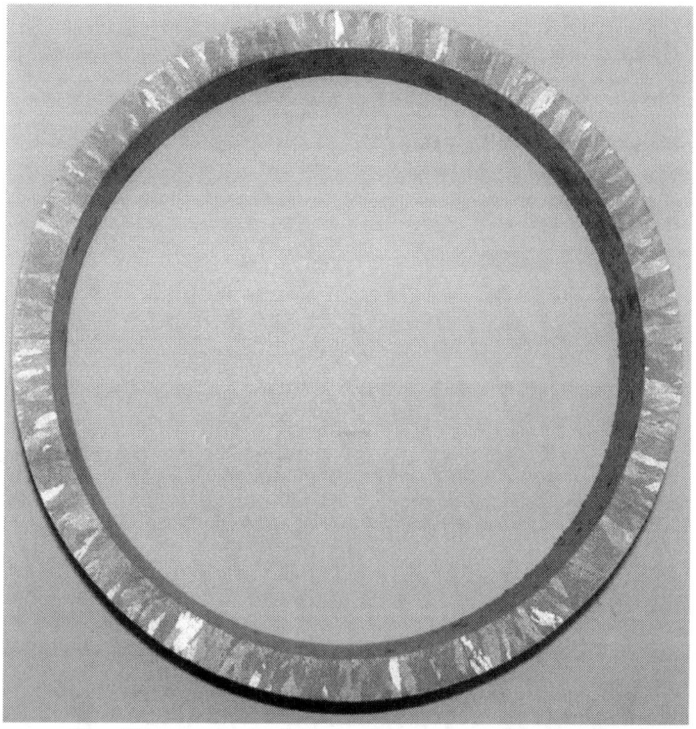

Figure 2– Macro-etching shows columnar structure.

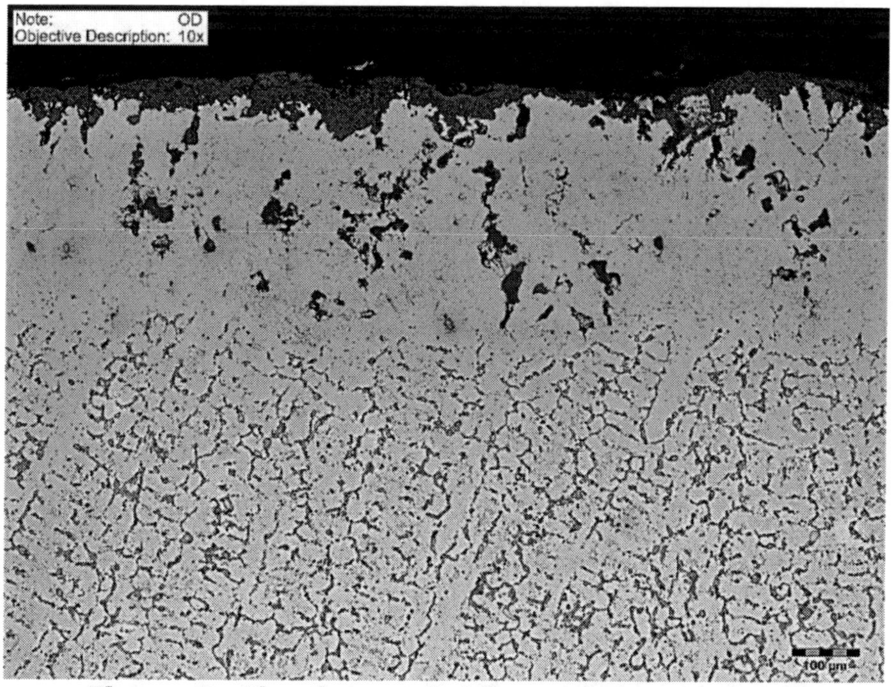

Figure 3 -Aluminum nitride precipitates close to OD

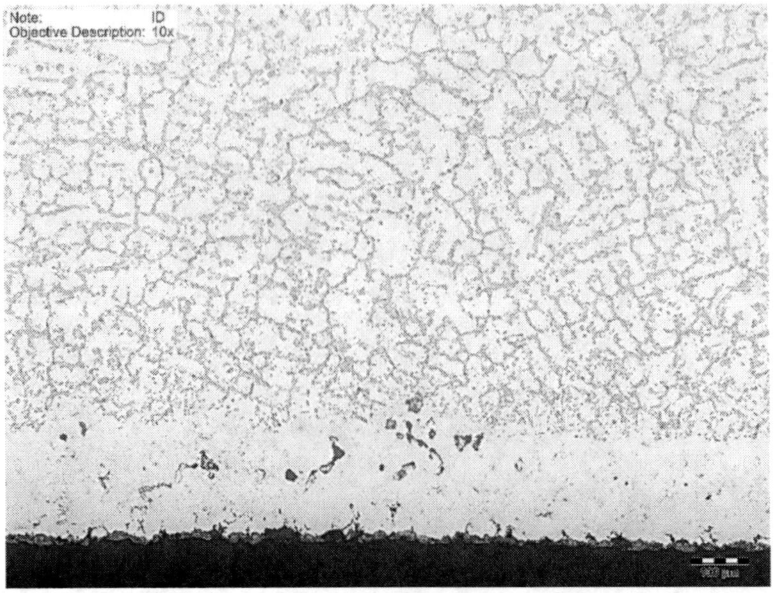

Figure 4-Carburized layer indications detected close to the ID

From the collected mechanical and metallurgical experiment data for life limiting active damage mechanism, it could be inferred that the radiant tube life has been affected, but tube should have considerable remaining life, if the future operating conditions are the same as the historic conditions. It should be noted that numerous isolated cavities in the heat affected zone (HAZ) of each tube weld could be potential concerns and would eventually form to circumferential creep cracking at the HAZ.

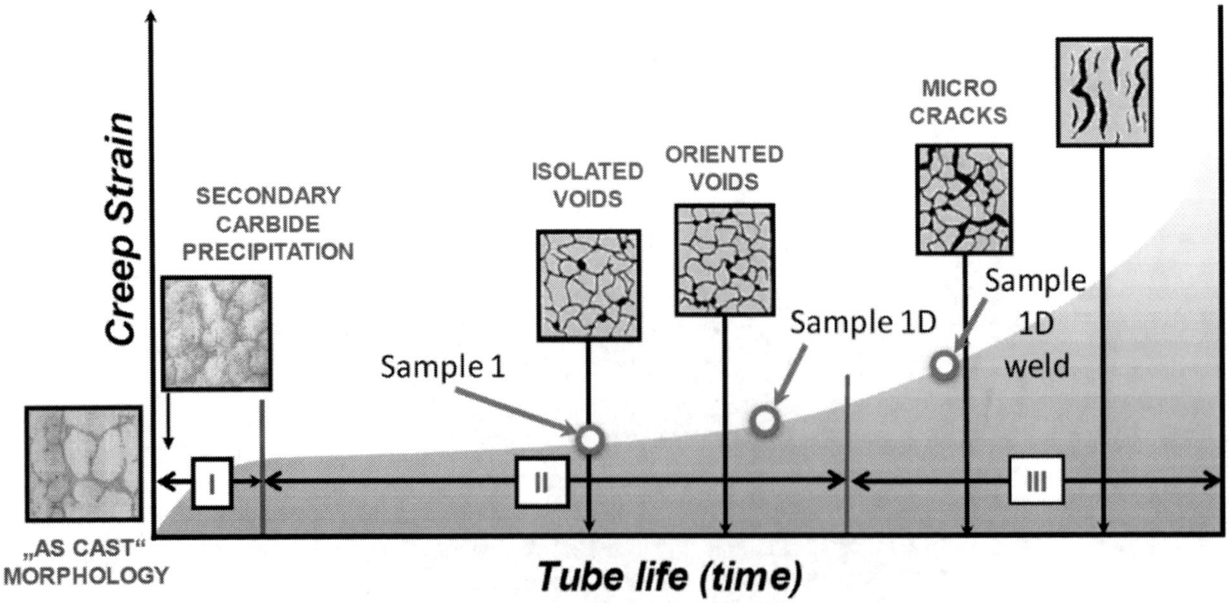

Figure 5 – Tube Creep vs. Life

The conclusion was that the tube has significant remaining life, and the alumina protective oxide layer was intact on tube ID. As a result, the decision was to extend the retube cycle to five years and then do a further assessment during another outage. During the 5-year outage findings lead to extending the expected life to six years. This brought the retubing schedule in line with the furnace PM frequency, which was also optimized from two-year to three-year intervals after assessing steam drum pressure relief devices (PRDs) and burner performance.

Tube Failure during Extended Life

However, number of tube failures occurred during the extended duration. A failure analysis was performed on samples from the failed tubes. Each tube failure happened during last quarter of the expected tube life and initiation of ID pitting was noticed on the 2" radiant tube (**Figure 6**). The pitting appeared to progress into a bulge and then rupture over time. Pitted areas were found to be magnetic and metallurgically degraded. Every failure (ID pitting) zone was located in the high heat zone near the floor burners, typically within 5-meters from the floor burners. The failure was found on both the base tube and weld zones. These impacts were only found in the lower 5-meter zone of the 14-meter overall length tubes.

In the Lummus SRT-6 configuration (**Figure 8**), each furnace contains seven coils with a combined total of 196-2" inlet tubes (14-meter length). Each tube is fabricated from typically 4 - 5 centrifugally cast segments. The failure of any individual tube segment will result in a furnace stoppage to attend the failure. These failures reduced the benefits of the extended tube life to six years. Maintaining tube integrity is particularly important to reliable and safe furnace operations.

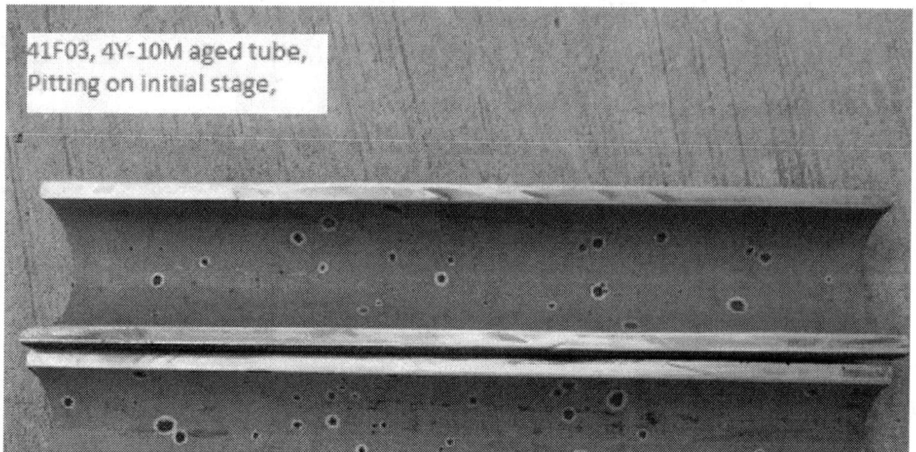

Figure 6 –Visual Appearance of ID Pitting on Initial Stage

Metallurgical Studies

For identification of the root cause, tube and coke samples were collected on each failure and studied. In each case of failure, the tube ID found to be affected with pitting corrosion in various stages which resulted in tube bulging or tube rupture over time. In the initial stage of the alumina pitting, it had a morphology of bright and dark contrast on tube ID surface (**Figure 6)** and moderate attraction to the magnet. As per the metallurgical assessment, the alumina protective layer was found to be depleted and pitting would be initiated due to internal high temperature exposure during decoking. Elemental Mapping revealed a closed and protective aluminum oxide scale at the tube inner surface areas without dark spots. However, at dark spotted areas, reveals internal oxidation and absents of closed protective aluminum oxide scale (**Figure 7)**. After the removal of the alumina protective layer, the tube behaves like a conventional tube and further regeneration of alumina or chromium oxide layer was found not to be possible. ID pitting damage mechanism found to be irreversible and susceptible to the accelerated pitting corrosion once it being initiated. It considered a life limiting damage mechanism for tube sections.

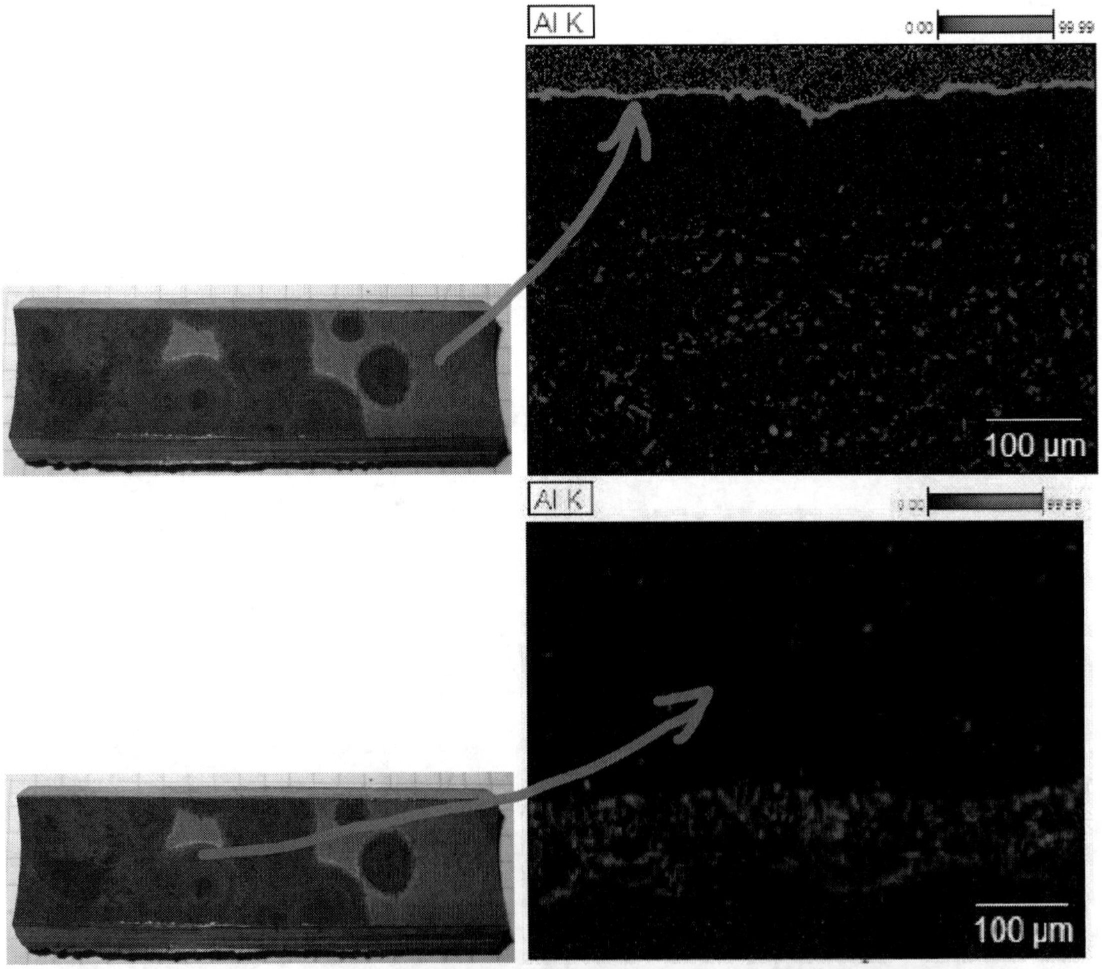

Figure 7 – Elemental Mapping of Bright and Dark Spots

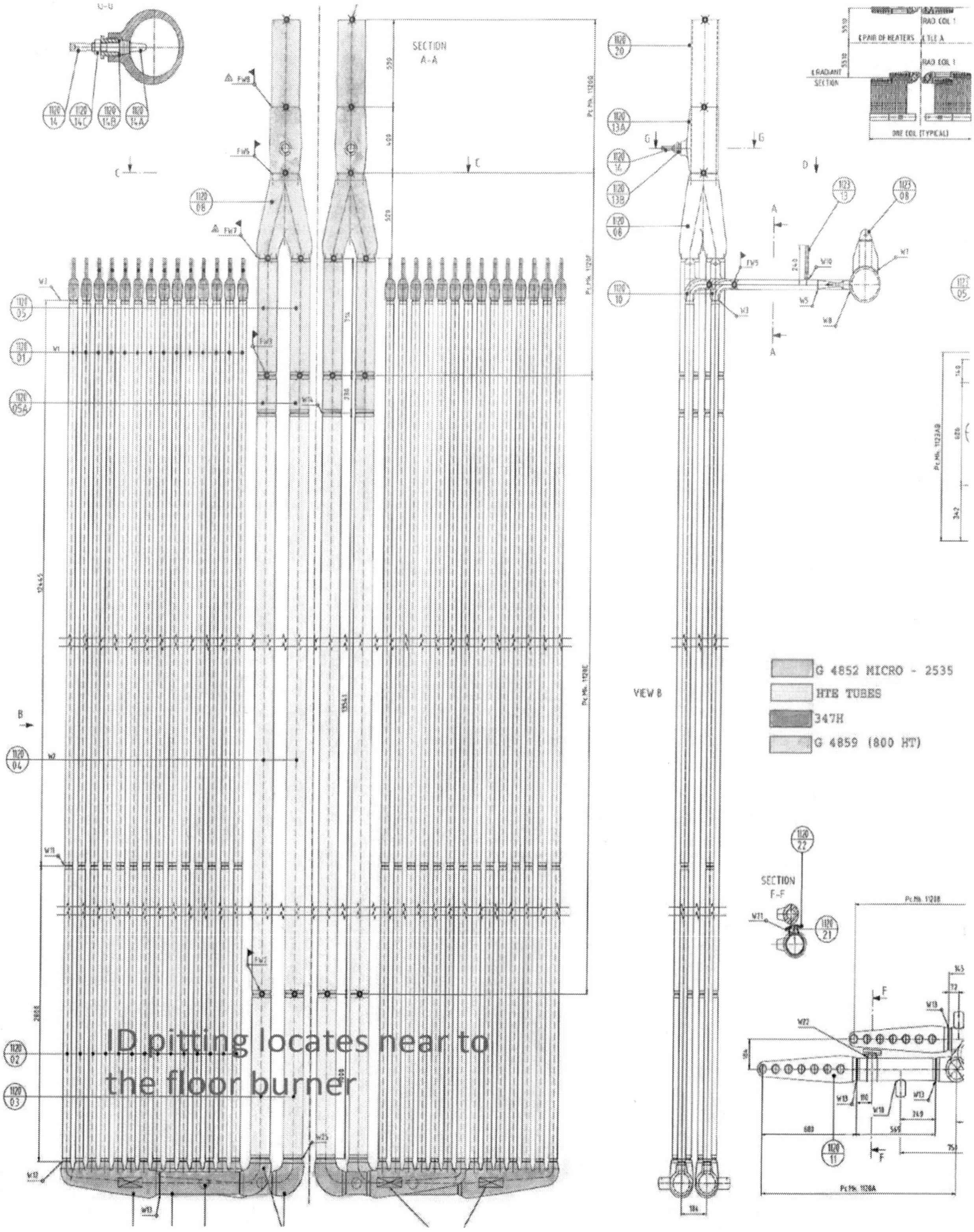

Figure 8 – Typical Radiant Coil Arrangement

Inspection Challenges

Each furnace radiant section has numerous tubes segment which consists of thousands of weld joints. The pitting observation found to be on localized locations closer to the floor burners where coke deposits expected to be developed and high temperature exposure of internal burning during the decoking. Other tube sections away from the high temperature exposure, the alumina protective layer found to be intact, and metal found not affected mechanically or metallurgically (**Figure 15**). Failure of any of such tube sections or weld joints required unplanned shutdown of furnace for tube section replacement. As per the furnace tube failure history, most of the tube failures reported occurred during the fifth and sixth years of service. If ID pitting can be identified and replaced at an early stage, unplanned furnace failures can be avoided. Three-year PM window was found to be optimal for identification and replacement of affected tubes in the same window.

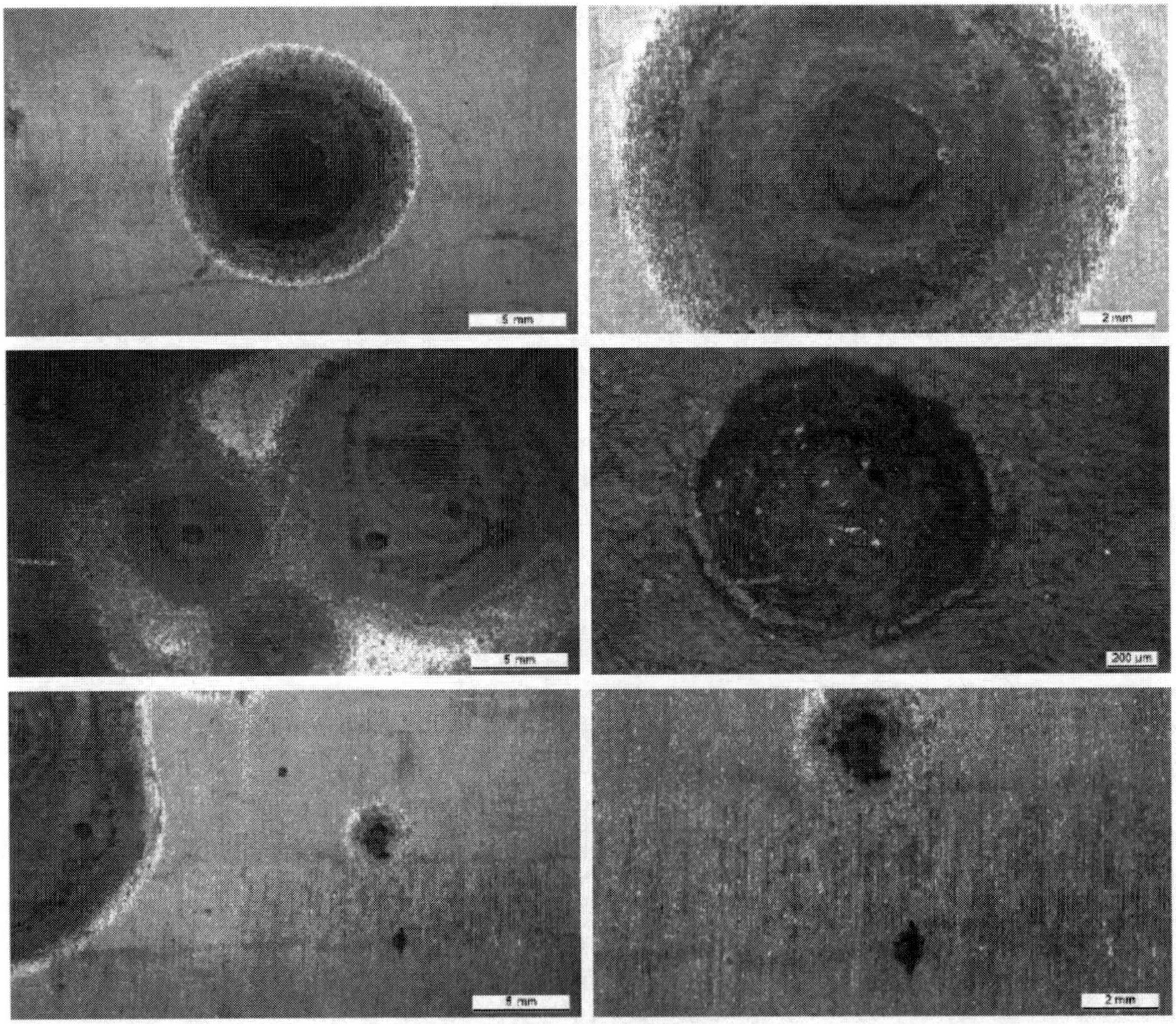

Figure 9 – ID Pitting of a Moderately Affected Tube (Magnified View)

Since the ID pitting has no practically measurable thickness dip during the initial stage, any external inspection methods found to be ineffective during a three-year PM. The complex routing of Lummus SRT-VI furnaces and critical flow venturi throat diameter of 15mm in the inlet of radiant tube makes most of the conventional tube inspection methods such as IRIS, RFT impossible due to access. However, borescope inspection was found to be an effective method due to the bright and dark morphology of early-stage pitting on alumina layer and iron oxide stains from the moderate pitting (**Figures 6 and 9**). The borescope tube of 10mm diameter can be inserted through feed strainer flange and pass through the critical flow venturi orifice to fetch tube internal videos and images of radiant tubes. The images found to be clear, distinctive, and traceable on machine with length indicator and articulate facility and found to be dependable for locating affected tube section. Since the ID pitting found to be localized, the data collection of 100% tube is important.

With the help of borescope inspection, the affected tubes portions replaced and the prepared the furnace for next half of the life with confidence. To see the condition of six-year aged radiant tube and room for further life optimization, samples were collected again. However, the tube found to be affected and alumina protective layer depleted in larger area. That's make further increase in the tube life has no benefit. and found some tube OD locations affected (**Figure 19**) with nitridation and carburization on tube ID due to depletion of alumina protective layer.

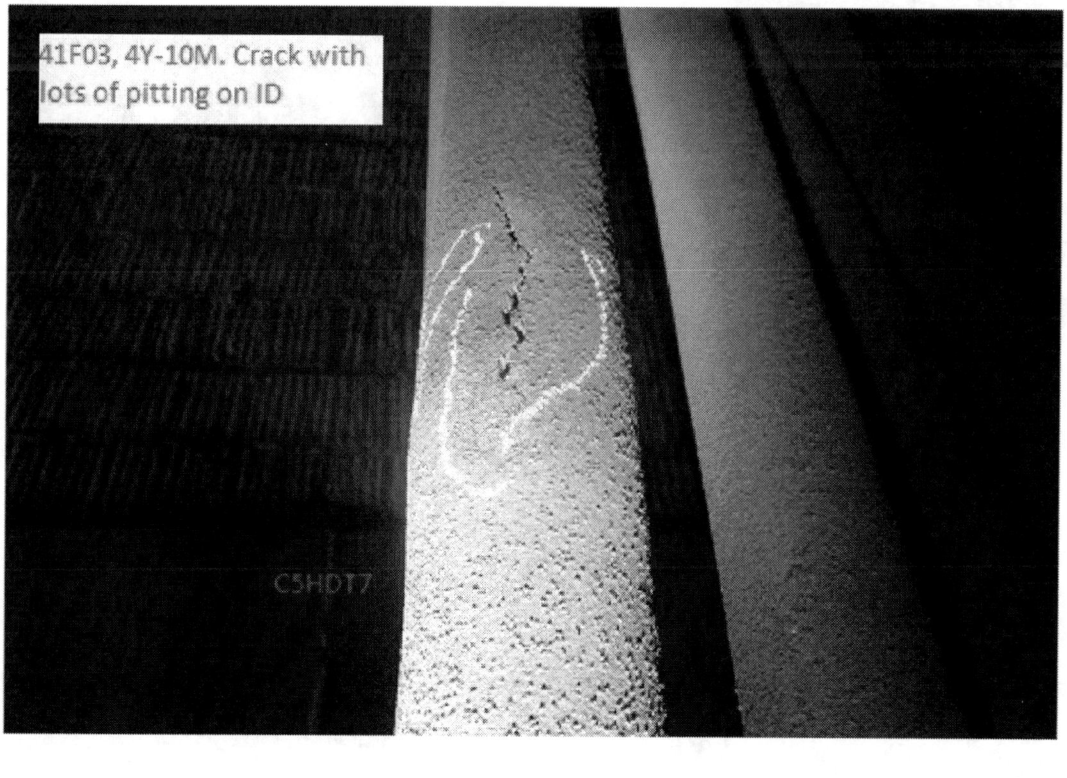

Figure 10 – ID Pitting Developed into Longitudinal Crack

Figure 11 – ID Pitting Developed into Bulge

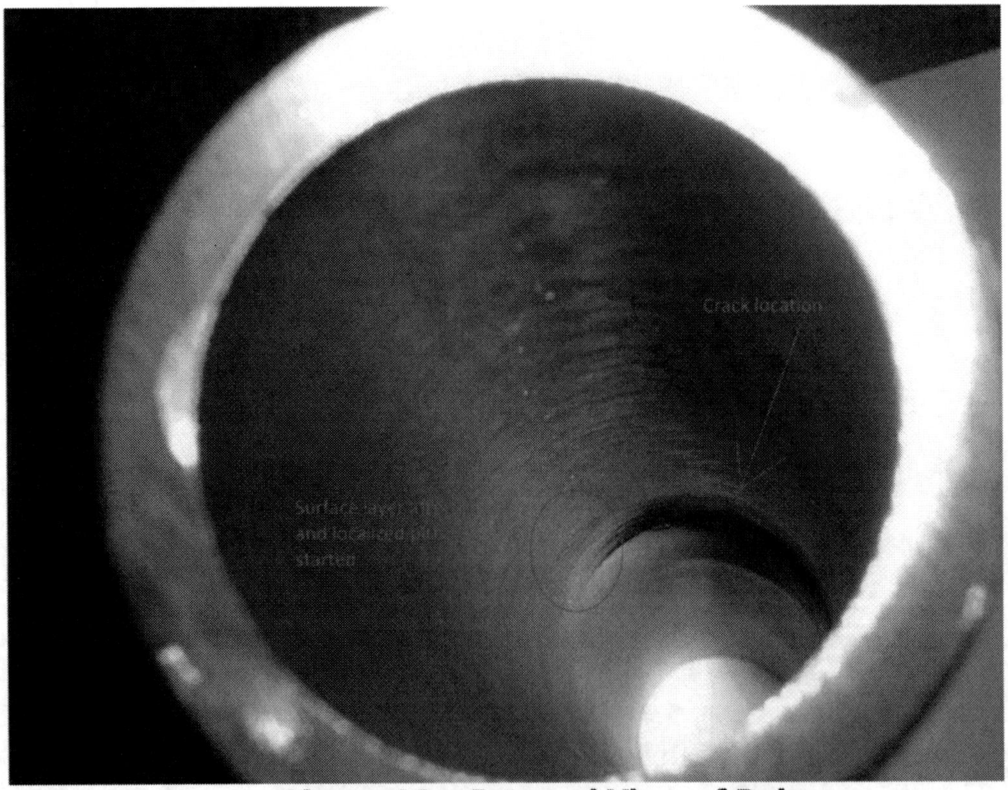

Figure 12 – Internal View of Bulge

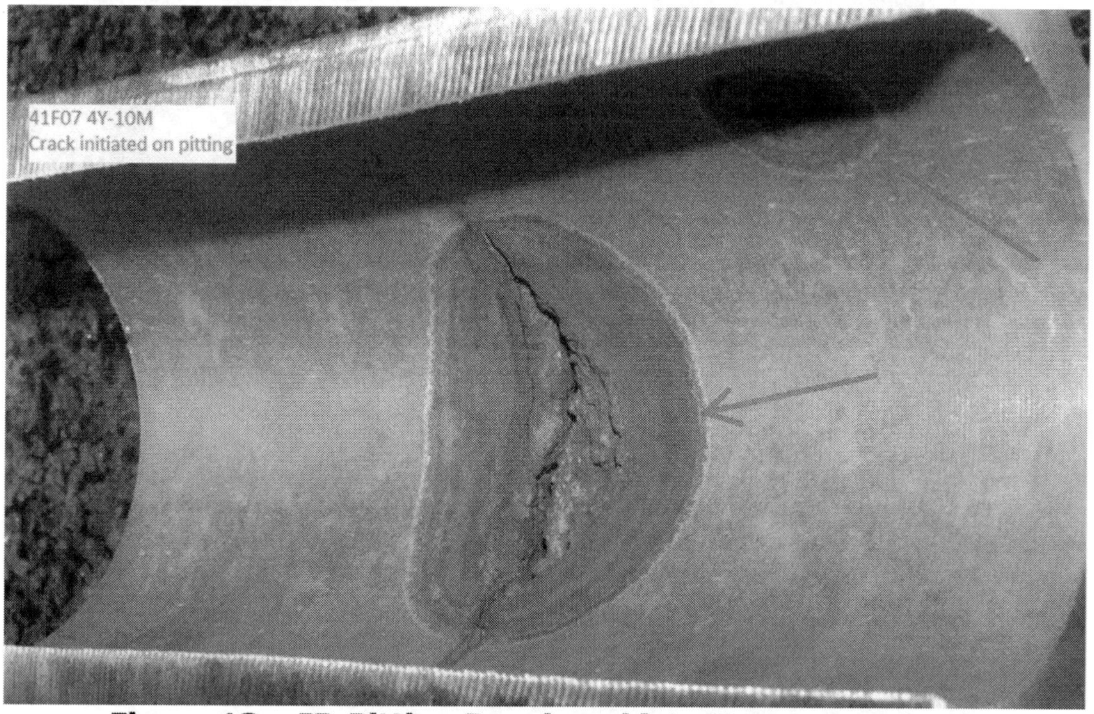

Figure 13 – ID Pitting Developed into Transverse Crack

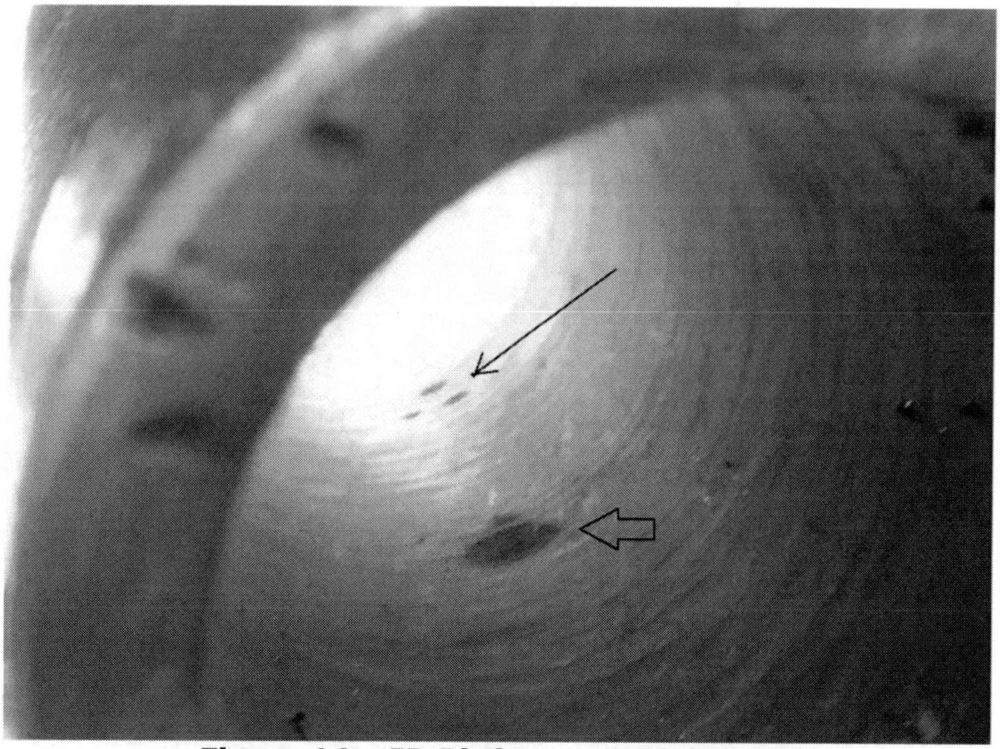

Figure 14 – ID Pitting on Initial Stage

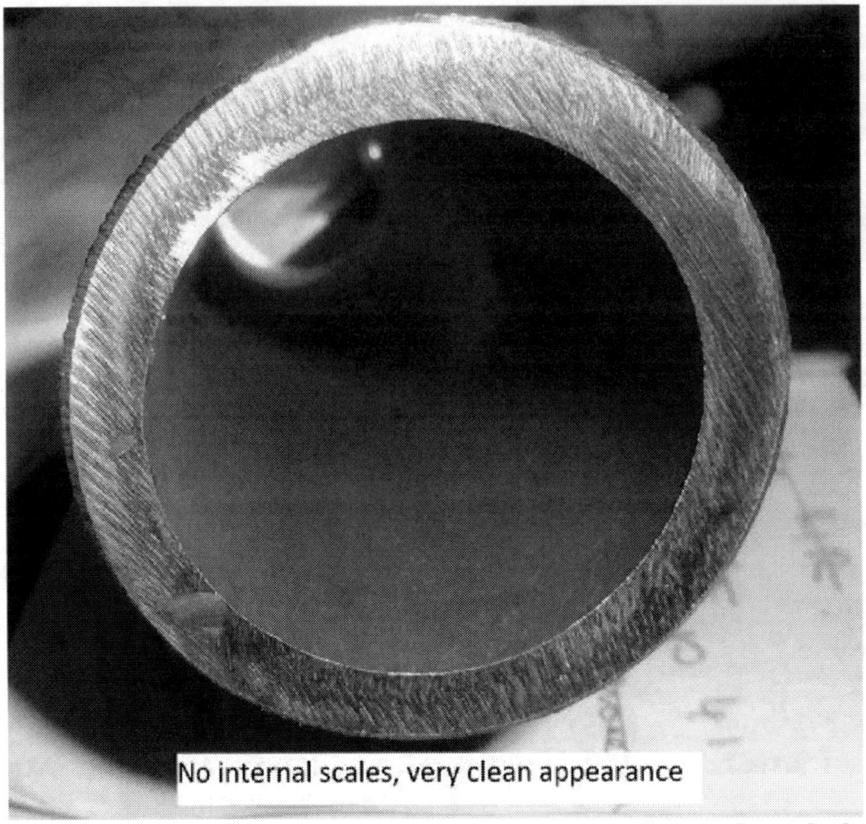

Figure 15 – Very Clean Appearance Away from the Pitting

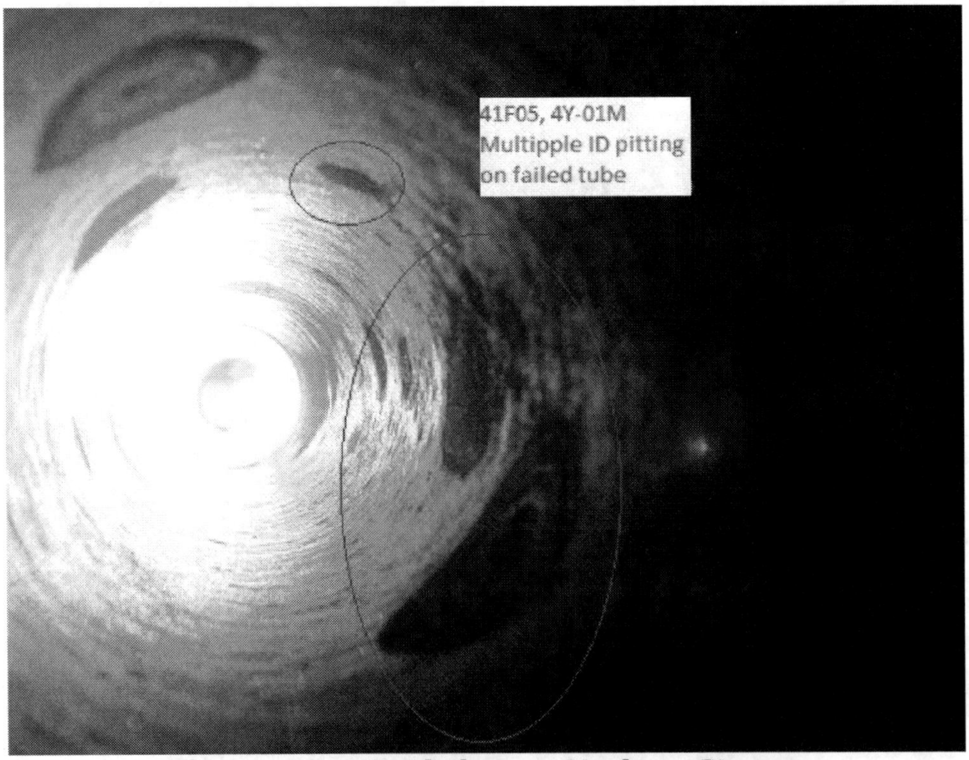

Figure 16 – ID Pitting on Various Stages

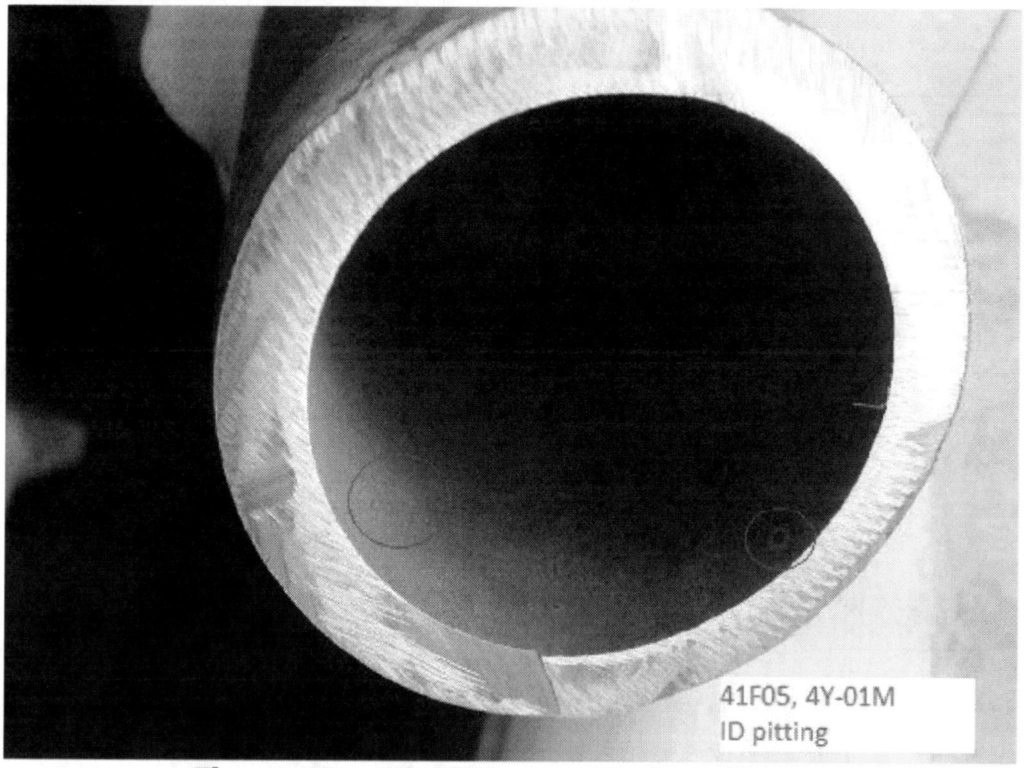

Figure 17 – ID Pitting on Various Stages

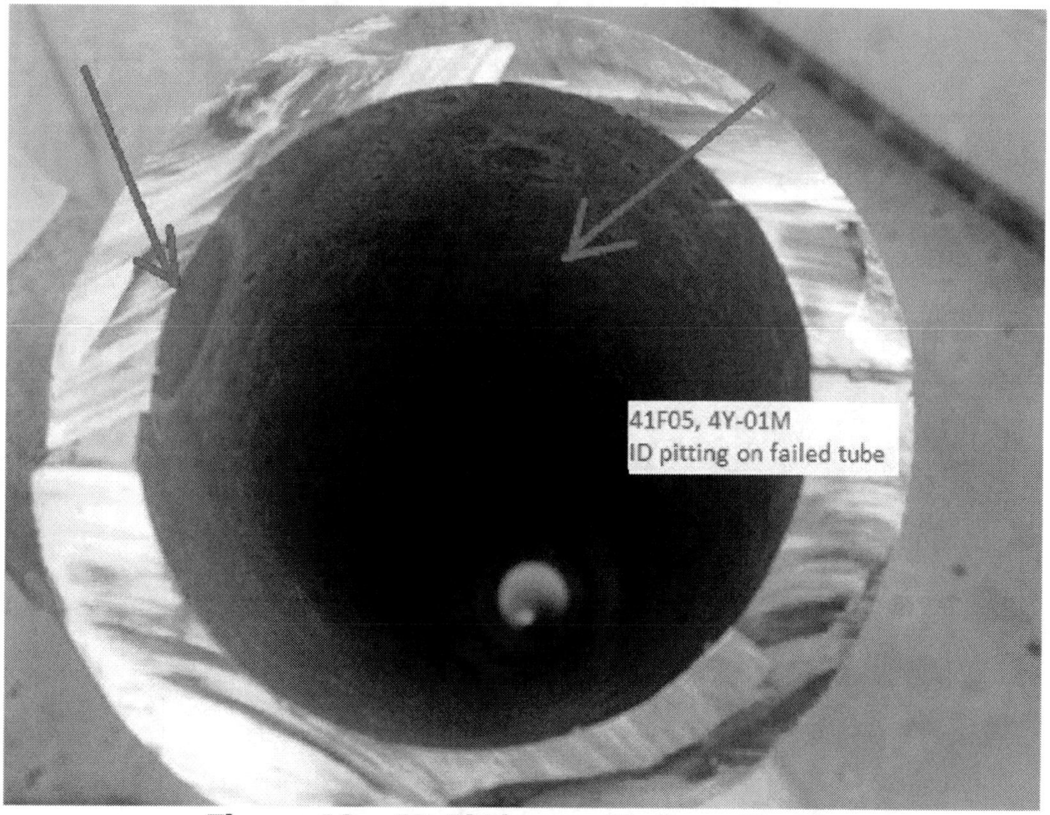

Figure 18 – ID Pitting on Various Stages

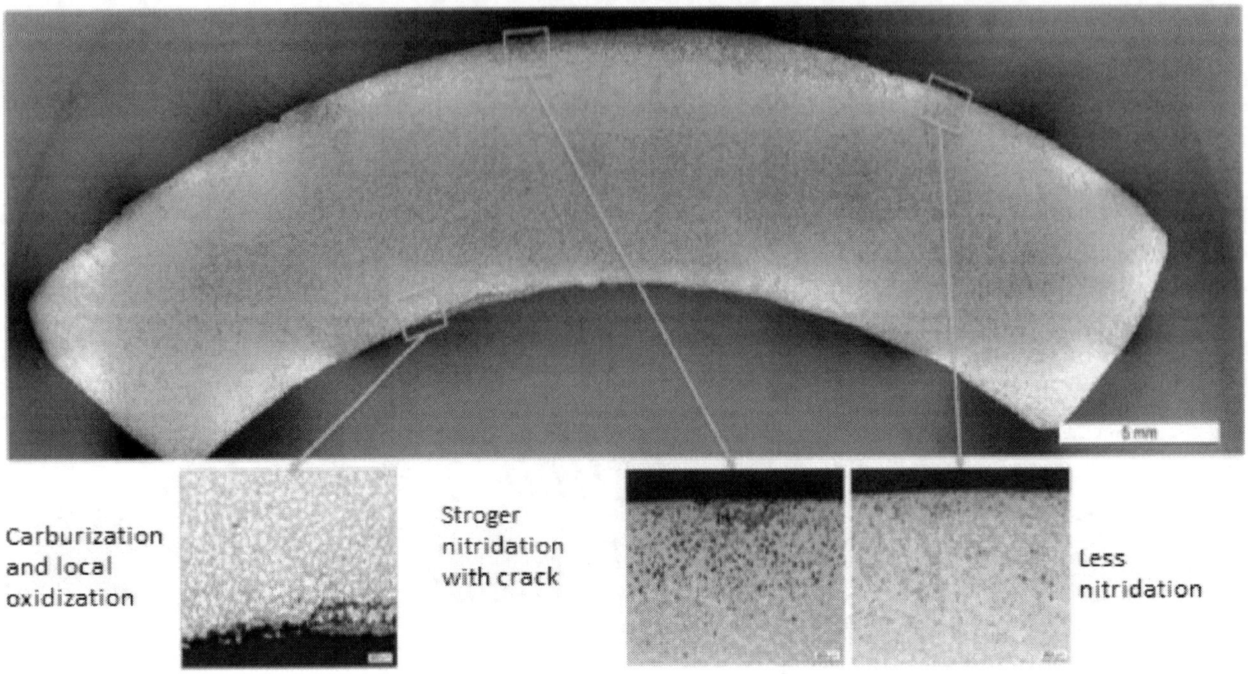

Figure 19 – Macro-etching Shows Nitridation and Carburization

Figure 20 – Borescope Accessing through Feed Strainer.

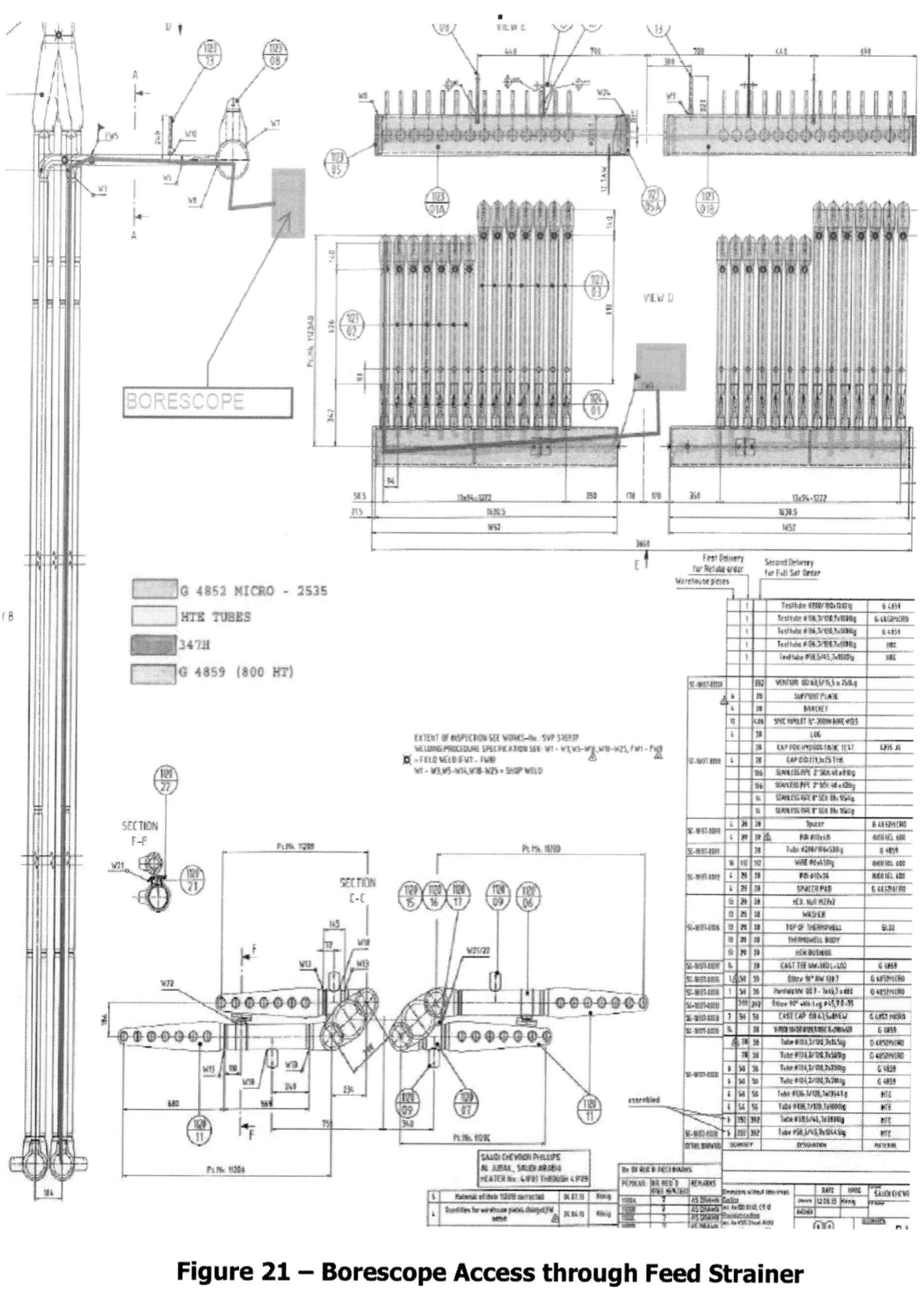

Figure 21 – Borescope Access through Feed Strainer

The collected borescope images during the three-year PM window should be assessed with reference images and sorted as low, medium, and major pitting. Medium and major pitting observations should be arranged for cutting and replacement in the same PM window. As per the experience the affected tube section for replacement is only a few. By replacing those tubes during PM window, the furnace can be prepared for reliable and safe operation for next three plus years. SPCo operated some of its nine furnaces up to 7 years with confidence due to tube supply schedule impact during recent geological issue in Europe.

Conclusions

Following are the main conclusions that can be drawn from this work:

1. Depletion and spalling of chromium oxide layer on conventional radiant tube has serious impact on furnace run length.
2. High temperature exposure during decoking promotes radiant tube ID protective layer depletion and pitting.
3. Pitting on alumina protective layer is localized, and no regeneration is possible once it formed.
4. Unplanned shutdown due to tube failure can be avoided by early identification of ID pitting and replacement during PM window.
5. Although the optimum life of a radiant tube is established at six years due to the integrity of other furnace components, tubes verified free of ID pitting can confidently exceed the established life of six years in any unforeseen event.

AIChE Paper Number: 26c

Tube-to-Tubesheet Weld Inspection for Secondary Transfer Line Exchangers (STLEs) – Innovative PAUT Inspection Technique

David Bajula - CEng
NDT Subject Matter Expert
ASNT/ACCP Level III
Acuren Inspection, Inc.

Thomas Garza, PE
Fixed Equipment Engineering Manager
Chevron Phillips Chemical

David Plunkett
Senior Inspection Specialist
Chevron Phillips Chemical

Prepared for Presentation at the 2023 Spring National Meeting
March 12 – 16, 2023

Tube-to-Tubesheet Weld Inspection for Secondary Transfer Line Exchangers (STLEs) – Innovative PAUT Inspection Technique

David Bajula - CEng
NDT Subject Matter Expert
ASNT/ACCP Level III
Acuren Inspection, Inc.

Thomas Garza, PE
Fixed Equipment Engineering Manager
Chevron Phillips Chemical

David Plunkett
Senior Inspection Specialist
Chevron Phillips Chemical

Abstract:

Chevron Phillips Chemical (CPChem) operates numerous cracking furnaces at its various ethylene plants in the Gulf Coast and the Middle East which utilize proprietary design Secondary Transfer Line Exchangers (STLE) provided both by BORSIG Process Heat Exchanger GmbH (BORSIG) and SCHMIDTSCHE SCHACK. The purpose of the STLEs (and Tertiary Transfer Line Exchangers, if present) is to further cool the heater effluent from the Primary Transfer Line Exchangers to the desired quench system inlet temperature and to recover this heat in the generation of very high-pressure steam, boiler feed water preheat, or increase feed preheat to the furnace convection section.

Due to the uniqueness of the tube-to-tubesheet weld design, standard inspection techniques, such as visual inspection (VT) of the root of the tube-to-tubesheet weld may only be performed up to 60% of the welds and a 100% liquid penetrant (PT) on the completed welds does not assess the critical root zone, only the external weld surface and tubesheet area. As a result, a Phased Array Ultrasonic Testing (PAUT) technique was developed to provide a near full volumetric inspection of the weld, with emphasis on high resolution coverage of the critical root zone.

This paper discusses the development, testing, proof-of-concept, and implementation of this PAUT technique to ensure the integrity of the welds during fabrication of the STLEs and field repairs, when required, so that the equipment could be returned to service with the assurance of safe, reliable operations.

Introduction and Background

Chevron Phillips Chemical operates a number of cracking furnace technologies (e.g., Technip Energies, ABB Lummus, etc.) at its various ethylene plants located in the Gulf Coast and in the Middles East, which includes Primary, Secondary, and at times Tertiary Transfer Line Exchangers to cool the heater effluent to the desired quench system inlet temperature and to recover this heat in the generation of very high-pressure steam, boiler feed water preheat, or increased feed preheat to the furnace convection section. **Figure 1** shows a simplified overview of the ethylene process which shows the respective transfer line exchangers (e.g., Primary and Secondary).

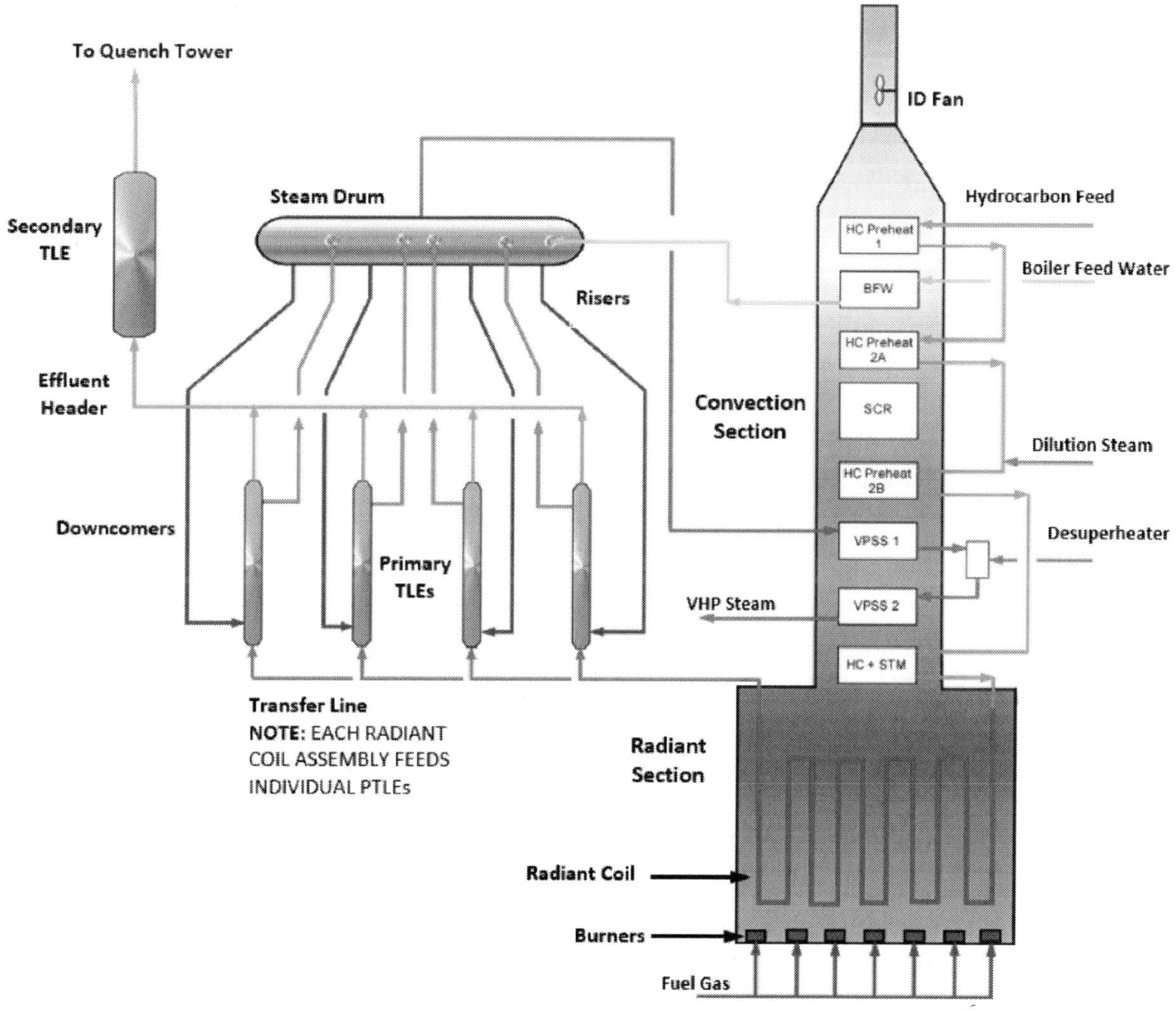

Figure 1 – Simplified Overview of an Ethylene Process

The tube-to-tubesheet weld configuration of these proprietary transfer line exchangers has a unique configuration, where the tubes are welded on the waterside of the tubesheet by a full penetration weld via an automated orbital welding process. Limited non-destructive examination of the root of the tube-to-tubesheet welds is feasible. For example, visual inspection (VT) only allows up to 60% of these welds to be inspected during the fabrication stage. Radiography of these particular welds is limited due to the weld configuration. **Figure 2** shows a typical tube-to-tubesheet weld of a BORSIG Transfer Line Exchanger.

Courtesy of BORSIG Process Heat Exchanger GmbH

Figure 2 – Tube-to-Tubesheet Weld of BORSIG Transfer Line Exchanger

For clarification, the initial pass, or the root pass of the tube-to-tubesheet weld is fused with a filler wire with shielded gas to avoid any crevices on the waterside of the exchangers. This is a multi-weld process which may include two or three intermediate weld passes and one cap weld. Once the welding process has been completed, 100% liquid penetrant (PT) on the completed welds (face of tubesheet) can be performed. **Figure 3** shows penetrant testing of tube-to-tubesheet welds.

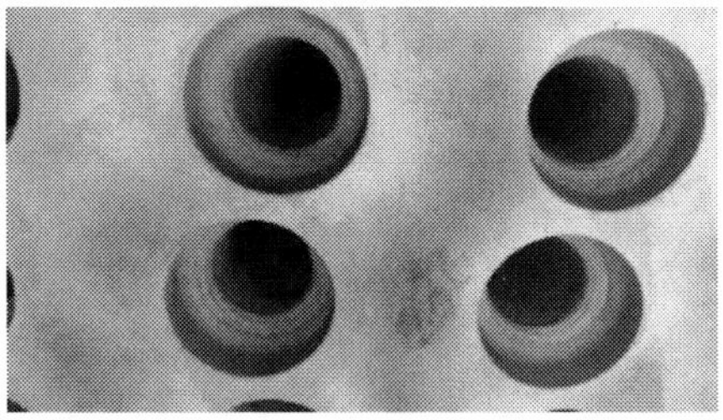

Courtesy of BORSIG Process Heat Exchanger GmbH

Figure 3 – Penetrant Testing of Tube-to-Tube Sheet Welds

There are numerous benefits of this particular full penetration weld design:

- Automatic orbital welding
- Minimum weld volume and heat-affected zone
- Notch-free
- Tube-to-tubesheet weld is located on the waterside of the tubesheet which reduces the stress levels compared to the traditional tube-to-tubesheet welds

As a result of the limited conventional non-destructive inspection techniques to ensure the integrity (e.g., flaws, indications, lack of penetration, etc.) of the root pass of the tube-to-tubesheet weld, CPChem in conjunction with Acuren Inspection developed a Phased Array Ultrasonic Testing (PAUT) technique to identify root type flaws typical of micro non-fusion.

Heat Exchanger Data

From a process design standpoint, STLEs are typically designed for an approximate gas inlet (nozzle) temperature of 690°F and 360°F on the outlet side. The typical mechanical design conditions of the STLEs are approximately 850°F for the inlet side with a 500°F on the outlet side. STLEs are designed to ASME Section VIII, Division 1.

Common Materials of Construction of key components of the transfer line exchangers are noted in **Table 1** below.

Table 1 – Typical STLE Materials of Construction

Component	Material of Construction
Gas Inlet Cone (Refractory Lined)	SA 204 Gr. A
Shell	SA 516 Gr.70
Tubesheet	SA 336 T1
Tubes	SA 209 T1

Development of PAUT Technique

Initial phased array ultrasonic inspection had been undertaken by CPChem Cedar Bayou facility utilizing various NDT service providers. These efforts were limited with handheld probes being utilized to position the phased array probe on the tube inside diameter surface with manual manipulation while observing the ultrasonic phased array instrument and resultant live images. An Acuren Subject Matter Expert (SME) was consulted to advise on improving the technique performance and probability of detection, as well as to enhance the overall productivity and quality of the examination. The result was the development of an inspection tool that included a special semi-automated encoded scanner for manipulation of the phased array probe on the I.D. surface of the tube with a specially prepared wedge that could be used for specific diameters of tubes.

Testing of Protocol

BORSIG provided welded coupon samples that were used to finalize the inspection technique development and to fabricate a calibration standard. Calibration standards are a critical part of any ultrasonic inspection method and are used to validate technique performance. They are also frequently used to establish appropriate detection thresholds and acceptance/rejection criteria.

Proof-of-Concept

The prototype scanner and technique were developed including the drafting of a procedure, scan plans, and initial flaw acceptance/rejection criteria, per ASME Section VIII, Div. 1, Mandatory Appendix 12. **Figure 4** provides an example of the scanner that was developed, **Figure 5** provides an example scan plan that was implemented, and **Figure 6** provides an example of the calibration standard that was developed.

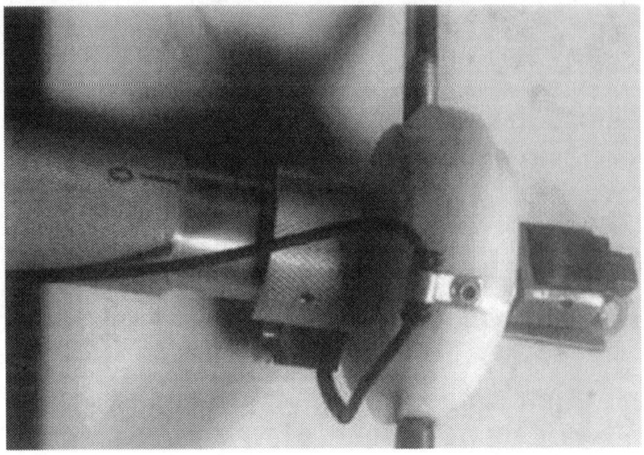

Figure 4 – Tube-to-Tubesheet Scanner

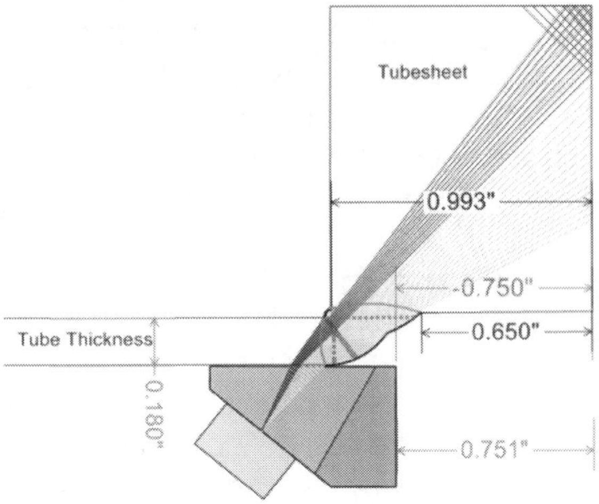

Figure 5 – PAUT Scan Plan

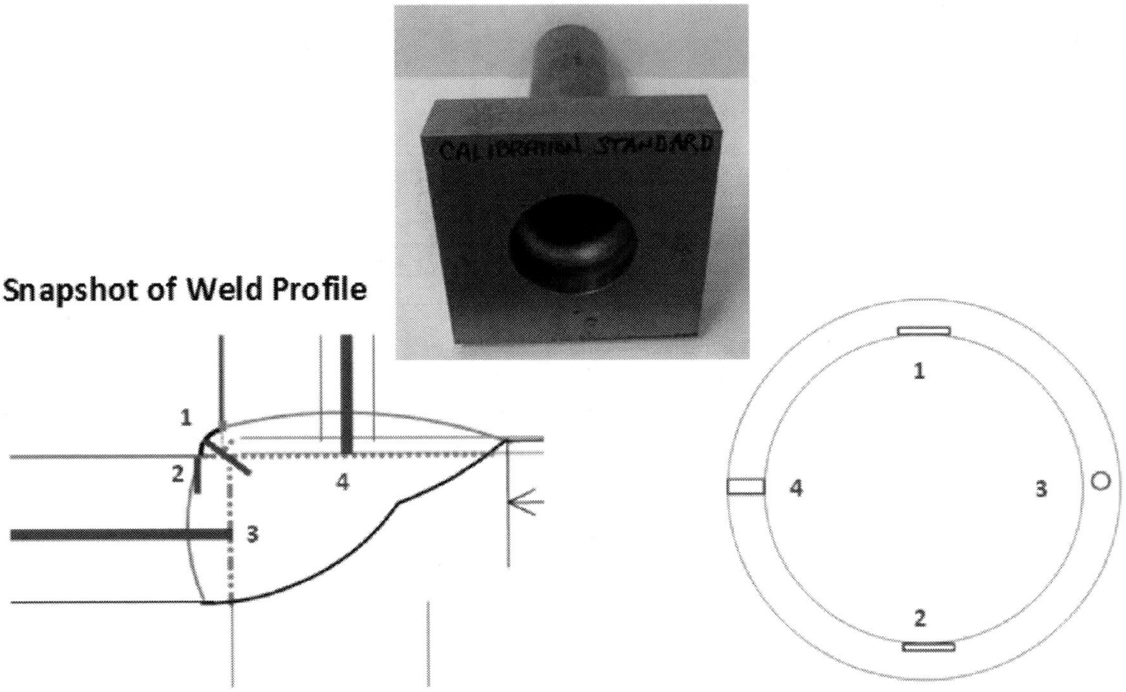

	#	Inch	mm	Length/Dia.	
ID Root Notch	1	0.020	0.5	0.25	excluding weld root reinforcement
ID Tube Notch	2	0.020	0.5	0.25	at tight edge to penetration
Tube side NF Reflector	3	0.079	2	0.079	may consider with 1.5mm
Tubesheet Side NF Reflector	4	0.079	2	0.079	may consider with 1.5mm

Figure 6 – Tube-to-Tubesheet Welds Calibration Standard

Implementation of PAUT Technique

Following successful proof-of-concept efforts, examinations were implemented at the BORSIG facility in Germany. The field tasks included performing a full examination on a single exchanger having 612 tube welds on the inlet side and an equal number on the outlet side of the exchanger; 1224 total tube welds were inspected.

Initially, examinations performed revealed numerous potential flaws, categorized as indications subject to disposition. Subsequent follow up included diligent removal of weld metal (grinding favored over arc gouging) to reveal the indication(s), followed by rewelding and reexamination to confirm indication removal and weld integrity. **Figure 7** depicts typical weld root flaws that were detected; screen presentation, radial distance or length of the flaws, and corresponding cross-sectional view utilizing a beam tool generation program to depict the location of the flaw(s). Additionally, 2nd pass flaws (embedded) were detected and initially subjected to the same acceptance/rejection criteria. Over time, these criteria were modified to put priority on the root flaws (1st pass) only.

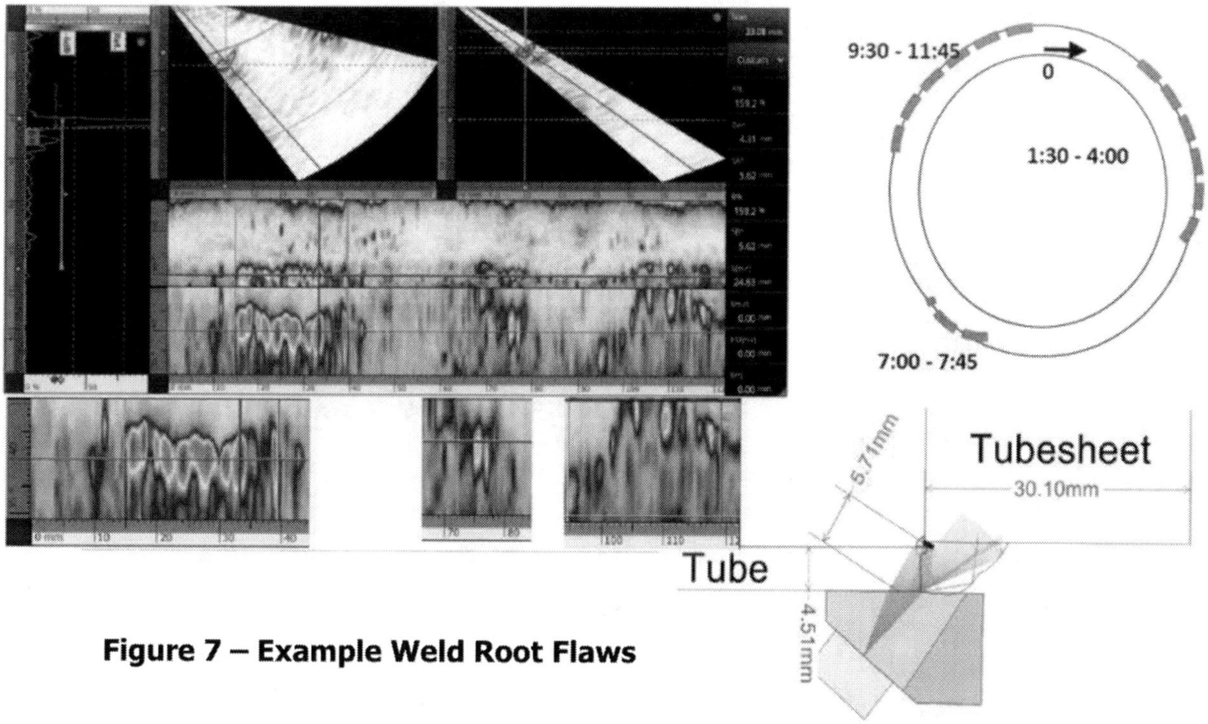

Figure 7 – Example Weld Root Flaws

To better understand the sensitivity of this specialized volumetric examination, inspection was also performed on welder prepared coupons where detection of indications/flaws could be explored destructively with greater precision. **Figure 8** shows the results of an indication that was detected and subjected to a controlled liquid penetrant examination, incremental weld material removal by grinding, and intermediate phased array inspections. This resulted in the determination that inside diameter surface root related flaws could be detected as small as 1.5mm in length and 0.20mm in depth, as shown in **Figure 9**.

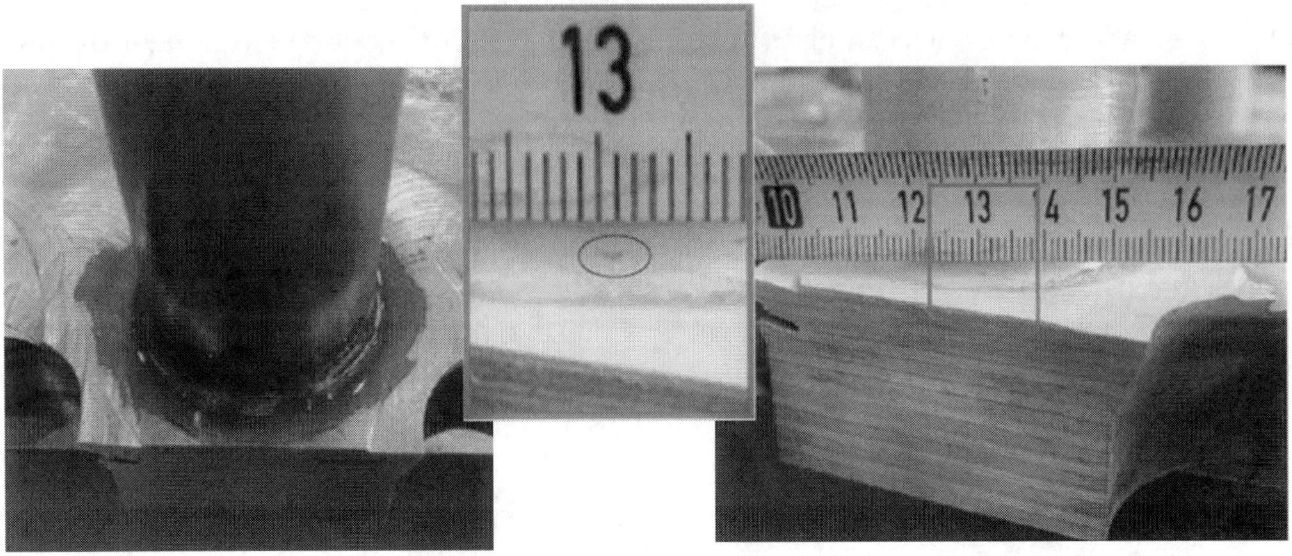

Figure 8 – Welded Coupon / PT Exam / Small Root Flaw

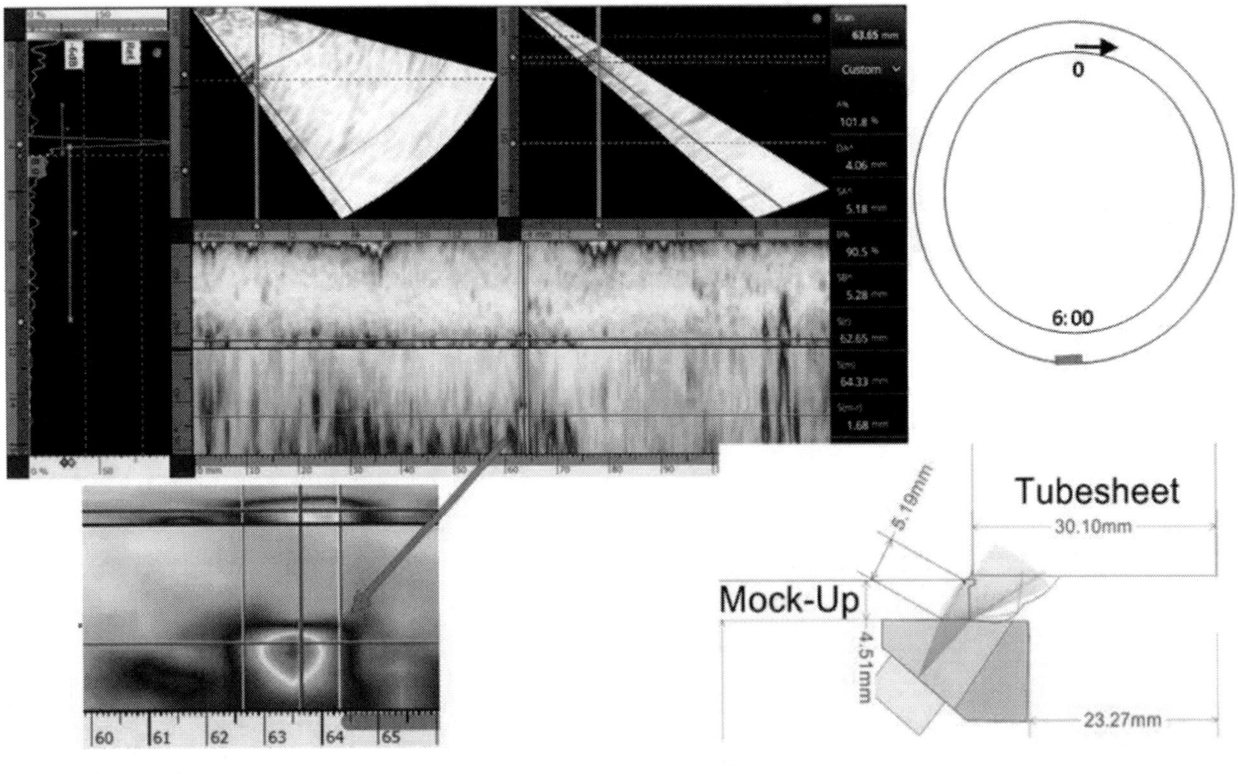

Figure 9 – Data Output & Results Corresponding to Figure 8

Conclusions

The success of the field inspections resulted in modification of the CPChem NDT requirements for all future TLEs. This included revision of the procedure and acceptance/reject criteria to align with the expected quality of welding that could be expected for these exchangers and the tube-to-tubesheet weld designs. The specialized phased array ultrasonic examination was proven to enhance the quality of welding and better assure the serviceability of the equipment in the plant.

26d Inspection Methods for Schmidt'Sche® Transfer Line Exchangers (TLEs) during Shut-Downs

Martin Stößel, Wei Sun

ARVOS SCHMIDTSCHE SCHACK (formerly part of ALSTOM) has supplied over 9000 TLEs to the Ethylene industry since 1959. Schmidt'sche® TLEs are installed directly downstream of the furnace radiant coils, and rapidly cool the cracked gas by producing high pressure steam. Deterioration of TLEs occurs over time, due to gas-side erosion, or water-side corrosion.

To understand the status quo and ensure the lifetime of the heat exchangers, periodic inspection and maintenance is required. The following paper describes SCHMIDTSCHE SCHACK's recommended best practices for inspection of TLEs during shutdowns. The paper will provide inspection options to assess the condition of TLEs depending on the plant's operating parameters and feasible inspection points depending on shut-down duration.

The guidelines discussed will include inspection methods, for example, water-side endoscopy, wall-thickness measurement testing, and tube straightness evaluation. Based on the inspection results the suggested follow-up actions will be defined, such as water side flushing, inner tube replacement and plugging, and partial oval-header replacement.

26e Aluminum Core Coldbox Failure Case Study

Kevin Sprague

This paper addresses the failures of the aluminum coldbox cores due to thermal stresses coupled with internal pressure stresses from the process and determination of likely future failure locations.

Ineos Olefins and Polymers operates numerous aluminum core coldbox units globally. The coldbox units operate at temperatures ranging from +90 °F to -150 °F, with adjacent streams experiencing differential temperatures greater than the manufacturers recommended limit of 50 °F.

Failure of a coldbox core unit occurred with subsequent investigation determining the leak was from a crack between passes of the coldbox core upper section. Investigation into the process data, core construction factory records and similar failures globally, indicated that a combination of the factory repairs and thermal cycling led to a crack forming in a weld bead near the end of a repaired section. Open areas in the core illustrate areas of inadequate or non-fusion between the plates and side bars, resulting in a reduced cross section weld repair being the main pressure boundary. Similar failures have occurred in units globally with new acceptance criteria and inspection hold points determined to understand the effect of factory repairs on core life.

Revisions to the initial inspections and acceptance criteria from the manufacturer have been determined along with new inspection hold points and documentation criteria.

AIChE Paper Number: 26f

UPDATE ON PROPOSED MODIFICATIONS TO API 579 PART 3 BRITTLE FRACTURE SCREEENING PROCEDURES

Brian R. Macejko, P.E.

Consulting Engineer II

The Equity Engineering Group, Inc.

Prepared for Presentation at the 2023 Spring National Meeting

Houston, TX, March 12 – March 16, 2023

UPDATE ON PROPOSED MODIFICATIONS TO API 579 PART 3 BRITTLE FRACTURE SCREEENING PROCEDURES

Brian R. Macejko

Consulting Engineer II

The Equity Engineering Group, Inc.

Abstract:

Significant inconsistencies and inadequacies exist in the current brittle fracture screening procedures in the 2021 edition of API 579-1/ASME FFS-1 (API 579), Fitness-For-Service as well as the various pressure equipment and piping design codes. Results from an evaluation completed using the current published guidance could prove unsafe in some instances. In 2016, the ASME/API Joint Committee on Fitness-For-Service (FFS) initiated a project to rewrite API 579 Part 3, Assessment of Existing Equipment for Brittle Fracture. The intent of the project is to establish new Level 1 and Level 2 evaluation procedures and acceptance criteria developed using state-of-the-art fracture mechanics consistent with the Failure Assessment Diagram (FAD) approach in API 579 Part 9 Assessment of Crack-like Flaws, which is currently employed in Level 3 Fitness-For-Service assessments.

An AIChE paper published in 2019 (182d) summarized the deficiencies and concerns with current published methods for evaluating susceptibility of pressure equipment and piping to potential brittle fracture failures. The paper also offered insight on proposed modifications to the assessment procedures to mitigate those concerns through employment of modern fracture mechanics to establish new screening criterion. Subsequently, several technology enhancements were recently developed and incorporated into the 2021 edition of API 579 Part 9 and its associated annexes to reduce conservatism in previous fracture mechanics-based approaches to establish low temperature limits for equipment and piping.

This paper will provide an update on recently proposed modifications to API 579 Part 3 brittle fracture screening assessment methods. The paper will outline alternate Level 1 and Level 2 procedures and acceptance criteria developed using the latest state-of-the-art fracture mechanics methodologies. At the time this paper was published, these alternate procedures were being reviewed by the ASME/API Joint Committee on FFS. Lastly, the paper will include case study examples to demonstrate the difference in assessment results obtained utilizing the proposed modified procedures compared to results obtained employing current brittle fracture screening methodologies.

Nomenclature:

a	Crack depth (inches).
C_{T_o}	Constant for Master Curve toughness index temperature based on ASME exemption curve designation (°F).
ΔT_o^C	Master Curve reference temperature shift due to constraint effects (°F).
f_{Kavg}	Factor for stress intensity factor solutions.
$K_{mat-req}$	Material toughness required (ksi√in).
L	Crack length parameter (inches).
L_r	Load ratio.
MAT	Minimum allowable temperature (°F).
$MAT_o^{Level\ 1}$	Parameter used to establish Level 1 MAT (°F).
$MAT_o^{Level\ 2}$	Parameter used to establish Level 2 MAT (°F).
P_f	Probability of failure.
σ_{msys}	Code minimum specified yield strength at temperature (ksi).
σ_P	Primary membrane stress (ksi).
σ_{SR}	Secondary plus residual membrane stress (ksi).
σ_{ysA}	Actual material yield strength at temperature (ksi). If unknown, it is estimated as the code minimum specified yield strength plus 10 ksi.
T	Assessment temperature (°F).
T_σ^P	T-stress credit for primary stress (°F).
T_σ^{SR}	T-stress credit for secondary plus residual stress (°F).
t	Thickness (inches).
T_{LS}	Temperature corresponding to the material lower shelf toughness limit (°F).
T_0	Master Curve toughness index temperature (°F).

τ_m Factor for t-stress credit.

Φ Plasticity interaction factor.

Introduction

Brittle fracture is the sudden rapid propagation of a crack-like flaw under stress (residual or applied) where the material exhibits little or no evidence of ductility or plastic deformation. Although brittle fractures are rare, the consequences are typically catastrophic. The risk for such a failure to occur in an ethylene plant can be significantly higher due to the potential for auto-refrigeration from depressurization of liquefied compressed gases (such as ethylene, ethane, propylene, etc.). Brittle fracture failures experienced within industry have resulted in costly unplanned repairs, extensive production downtime, and most unfortunately loss of life [1][2].

There are three (3) key components that drive susceptibility to brittle fracture failure in carbon and low alloy steels:

- Stress (residual and/or applied): Stress provides the energy necessary to drive a defect to fracture. Typical sources for stress include pressure, weight, and thermal loads, in addition to residual stress from the welding processes. A post weld heat treatment (PWHT) operation will significantly reduce weld residual stress.
- Material fracture toughness: Fracture toughness is the ability of a material to absorb energy and plastically deform without fracturing. It is a function of material strength, ductility, chemistry, grain size, and heat treatment. The majority of pressure equipment utilized in the oil & gas and petrochemical industries is constructed of carbon or low alloy steel. For these materials, material toughness is a function of metal temperature. At low temperatures, the material tends to behave in a brittle manner and has a high susceptibility to fracture. At high temperatures, the material tends to behave in a ductile fashion. The material fracture toughness of some materials can be altered (i.e., degraded) by the operating environment (e.g., carburization, hydrogen embrittlement, etc.).
- Crack-like defect (size and location): Defects can result from environmental damage (such as wet H_2S or caustic exposure), mechanical damage (such as gouges, dents, or fatigue), or from original fabrication (such as laminations, lack of fusion, lack of penetration, slag inclusions, porosity, etc.).

Scenarios or circumstances that tend to have an increased risk of brittle fracture failures in the oil & gas and petrochemical industries include:

- Start-ups and shutdowns
- Weld repairs, particularly repairs without PWHT
- Depressurization with Liquid Petroleum Gas (LPG) present
    - Lowering pressure below liquid hydrocarbon boiling point
    - De-pressuring before draining liquids
    - Equipment malfunction (relief valve failing, etc.)
- Flashing across flow obstruction when the downstream pressure is below the LPG boiling point, such as during the following:
    - Leaking isolation block valve
    - Manually letting down LPG into a flare header

    - Overfilling a vessel and having the gas outlet control valve flash LPG to downstream equipment
- Operational upsets
    - Overfilling vessels or drums
    - Back-flowing material via a common header (flare system, liquid drain header, or recycle header)
    - Recycling material back from a higher-pressure source
    - Loss of reboiler heat input on a distillation column
- Hydrostatic or pneumatic pressure or proof tests

API 579-1/ASME FFS-1

API 579-1/ASME FFS-1 (API 579) [3] is the industry accepted code used to make run, repair, replace decisions for in-service equipment. Part 3 of API 579 is dedicated to the evaluation of existing equipment for brittle fracture. The API 579 Part 3 Level 1 and Level 2 Method A procedures are based upon the ASME Section VIII, Division 1 Code (ASME VIII-1) [4] Part UCS-66 design philosophy. The Level 3 method (which references Part 9 of API 579) is a detailed fracture mechanics approach. For all assessment Levels, a critical exposure temperature (CET) is compared to a minimum allowable temperature (MAT) to determine acceptability. The CET represents the driving force for fracture and consists of the lowest potential metal temperature from all operating, upset, or atmospheric conditions. The CET can consist of a single temperature and coincident pressure or an envelope of temperature and pressure combinations. The MAT represents the material resistance to fracture. The MAT for a pressure vessel or piping system is the highest (i.e., warmest) value of the MAT for any of its components. An API 579 Part 3 brittle fracture assessment may become necessary for any of the following scenarios:

- Brittle fracture was not formally addressed in the original equipment design (e.g., equipment built prior to 1987)
- Process hazards review identifies scenarios where process temperatures can fall below original design considerations (e.g., auto-refrigeration, upsets, pressure relief, etc.)
- Change in process operating conditions (e.g., re-purposed equipment, etc.)
- Re-rating or FFS for other damage
- Start-up/shutdown optimization
- Hydrostatic pressure test
- Repairs or alterations (with or without hydrostatic pressure tests)

Significant inconsistencies and inadequacies exist in the current Level 1 and Level 2 brittle fracture screening procedures in API 579 as well as the various pressure equipment design Codes. The issues are thoroughly documented in [1] and [5]. Results from an evaluation completed using the current published guidance could prove unsafe in some instances and may be overly conservative for other applications. For critical applications, it is advisable to utilize an alternative approach to assess susceptibility to brittle fracture.

PROPOSED MODIFICATIONS TO FFS PROCEDURES

To mitigate the numerous discrepancies and deficiencies with the current design Codes and the Level 1 and Level 2 FFS procedures in API 579 Part 3, it is proposed that new brittle fracture screening procedures and acceptance criteria are employed based on state-of-the-art fracture mechanics consistent with API 579 Part 9 and WRC Bulletin 599 [1]. The newly proposed procedures were developed similarly to the procedures included in ASME Section VIII, Division 2 (ASME VIII-2) [6] which were fully documented in WRC Bulletin 528 [7]. However, the new procedures include several significant technological upgrades that have gained industry acceptance within the past several years.

A fracture mechanics-based approach is the most accurate and most reliable means to determine low temperature operating limits for pressure equipment and piping. API 579 Part 9 utilizes the failure assessment diagram (FAD) as shown in Figure 1 to assess the 3-parameter failure criterion: brittle or unstable fracture; elastic-plastic fracture; and plastic collapse.

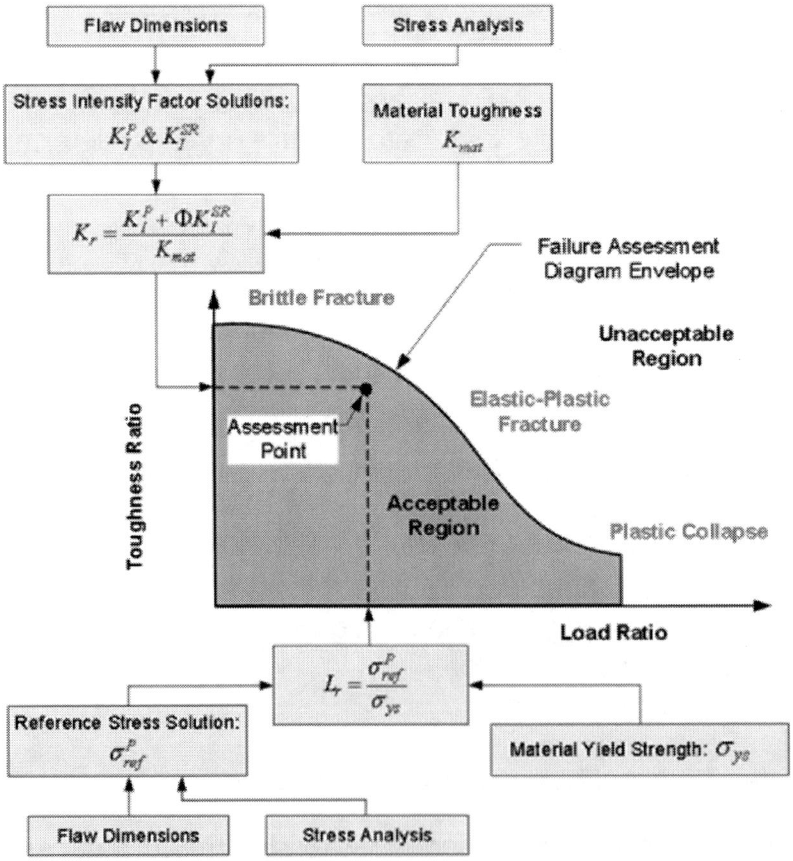

FIGURE 1: Failure Assessment Diagram (FAD)

Since the temperature is implicit in this criterion, iteration is required to establish the MAT for a component with a crack-like flaw. The general procedure is documented in WRC Bulletin 562 [8] and summarized as follows:

- Define the applicable loads and reference flaw size
- Postulate a temperature for the MAT
- Determine the appropriate material properties (e.g., fracture toughness, yield strength, etc.) at the assessment temperature
- Perform the FAD assessment
  - If the limiting assessment point lands on the FAD envelope, the postulated assessment temperature is the MAT
  - If the limiting assessment point falls within the FAD envelope, the postulated assessment temperature should be decreased and the FAD assessment should be repeated
  - If the limiting assessment point falls outside the FAD envelope, the postulated assessment temperature should be increased and the FAD assessment should be repeated

The 2021 edition of API 579 introduced a new Annex 9H (Constraint Effects for Surface Flaws in Carbon and Low-Alloy Steel Components in the Ductile-Brittle Transition Region) and a new Annex 9I (Alternative Estimate of Mode I Stress Intensity Factors for FFS Assessments in the Brittle-Ductile Transition Region) to reduce conservatism in Level 3 brittle fracture assessments. The background and technical basis for these Annexes are documented in WRC Bulletin 577 [9]. API 579 Annex 9J (Determination of the Minimum Allowable Temperature Using Level 2 Fracture Mechanics Approach Consistent with Parts 3 and 9) and WRC Bulletin 599 provide simplified procedures to establish the fracture mechanics-based MAT for a component, including the modifications from API 579 Annex 9H and Annex 9I.

General Overview of Proposed Level 1 MAT Procedure

The following is an abbreviated overview of the proposed Level 1 MAT procedure. The detailed background, technical basis, assumptions, and comprehensive step-by-step procedures are thoroughly documented in WRC Bulletin 599 [1].

- Based on the extent of inspection and the type and sensitivity of NDE performed on the component, determine an appropriate reference flaw size. Historically, a reference flaw depth equal to 25% of the governing thickness of the component (i.e., t/4 flaw) has been used when no information was available regarding previous inspections for crack-like defects. However, a smaller reference flaw can be justified with inspection.
- Determine the Master Curve reference transition temperature, T_0, for the material using API 579 Annex 9F paragraph 9F.4.3.2 and the available toughness data.
  - Option 1 – Determine the Master Curve reference temperature utilizing direct fracture toughness data per API 579 paragraph 9F.4.3.2.d.
  - Option 2 – Estimate the Master Curve reference temperature utilizing Charpy V-notch (CVN) impact test transition curve data per API 579 paragraph 9F.4.3.2.c.

- Option 3 – Estimate the Master Curve reference temperature utilizing CVN impact test data at one or more temperatures per API 579 paragraph 9F.4.3.2.b.
- Option 4 – If no toughness data or CVN impact test data is available, estimate the Master Curve reference temperature based on the material impact test exemption curve designation (i.e., A, B, C, D) and assuming high sulfur carbon steel per API 579 paragraph 9F.4.3.2.a as shown below.

$$T_0 = C_{T_0} + 1.2629\sigma_{ysA} \tag{1}$$

$$C_{T_0} = \begin{pmatrix} 5.4^\circ F & \textit{for ASME Exemption Curve A} \\ -32.8^\circ F & \textit{for ASME Exemption Curve B} \\ -70.8^\circ F & \textit{for ASME Exemption Curve C} \\ -96.5^\circ F & \textit{for ASME Exemption Curve D} \end{pmatrix} \tag{2}$$

- Using the tables and figures presented in [1], determine a constant, $MAT_o^{Level\ 1}$, as a function of the component governing thickness, stress relief condition (as-welded or PWHT), and the postulated reference flaw depth. The following Figure 2 is provided as an example.

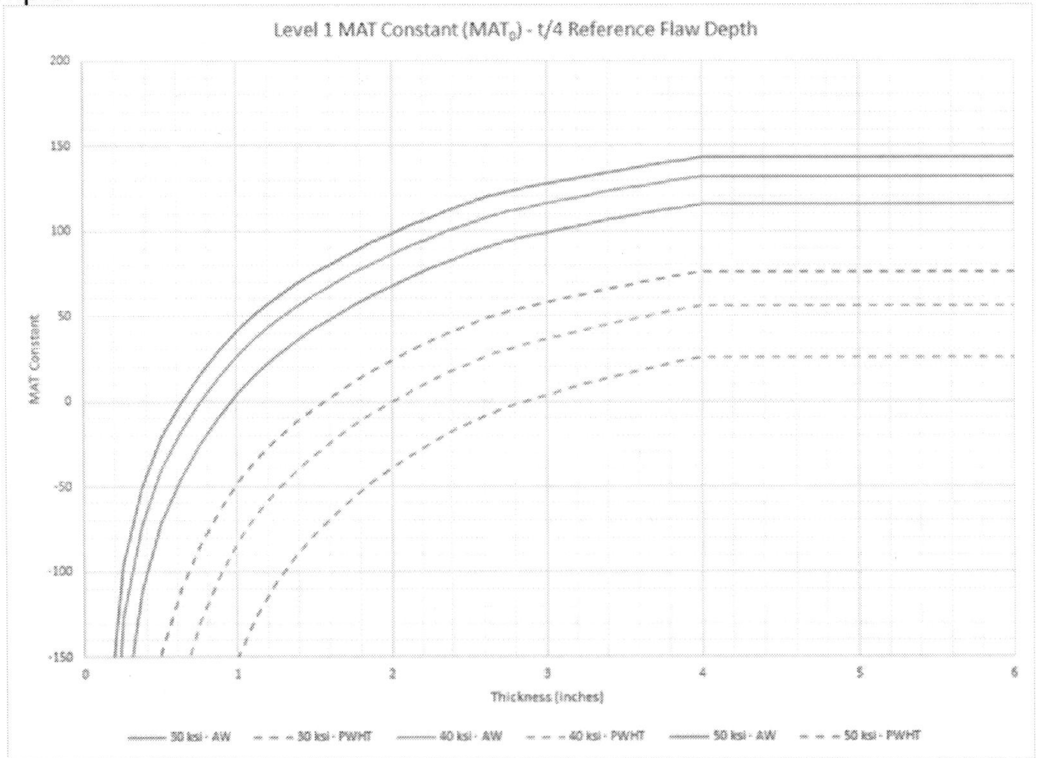

FIGURE 2: Example Level 1 $MAT_o^{Level\ 1}$ for t/4 reference flaw

- Determine the temperature corresponding to the material toughness lower shelf limit, T_{LS}, as shown below:

$$T_{LS} = T_o + \frac{\ln\left(\left[\dfrac{20.008}{70.1\left(\dfrac{0.0513}{L}\right)^{0.25}}\right] - 0.1412\right)}{0.0106} \quad (3)$$

- Determine the Level 1 MAT, using the following equation.

$$MAT = \max\left[MAT_0^{Level\,1} + T_0, T_{LS}, -155^\circ F\right] \quad (4)$$

General Overview of Proposed Level 2 MAT Procedure

The following is an abbreviated overview of the proposed Level 2 MAT procedure. The detailed background, technical basis, assumptions, and comprehensive step-by-step procedures are thoroughly documented in [1].

- Similar to the Level 1 procedures, an appropriate reference flaw size must be chosen based on the extent of inspection and the type and sensitivity of NDE performed on the component. Additionally, the Master Curve reference transition temperature, T_0, for the material must be determined using API 579 Annex 9F paragraph 9F.4.3.2 and the available toughness data.
- Calculate the primary, secondary, and residual stress in the component for the applicable loading condition(s) being assessed using the guidance in [1].
- Determine the required fracture toughness for the conditions being assessed using the following equation.

$$K_{mat-req} = \frac{f_{Kavg}\left(\sigma_P + \Phi\sigma_{SR}\right)\sqrt{\pi a}}{\left(1 - 0.14L_r^2\right)\left(0.3 + 0.7\exp\left[-0.65L_r^6\right]\right)} \quad (5)$$

- Calculate the constant, $MAT_o^{Level\,2}$, as a function of the component governing thickness, stress relief condition (as-welded or PWHT), and the postulated reference flaw depth.

$$MAT_0^{Level\,2} = \begin{bmatrix} \dfrac{1}{0.0106}\ln\left[\dfrac{\left(K_{mat-req} - 18.2\right)\cdot\left(L^{0.25}\right) - 4.711404}{33.360547}\right] & if\ \left(K_{mat-req} - 18.2\right)\left(L^{0.25}\right) > 4.711404 \\ -500 & if\ \left(K_{mat-req} - 18.2\right)\left(L^{0.25}\right) \le 4.711404 \end{bmatrix} \quad (6)$$

- Calculate the Master Curve reference temperature shift, ΔT_o^C, due to constraint effects.

$$T_{\sigma}^{P} = \tau_m \sigma_P \quad (7)$$

$$T_{\sigma}^{SR} = \max\left[\tau_m\left(\sigma_S + \sigma_R\right), -0.4\sigma_{msys}\right] \quad (8)$$

$$\Delta T_0^C = \frac{T_{\sigma}^{P} + T_{\sigma}^{SR}}{0.806} \quad (9)$$

- Determine the temperature corresponding to the material toughness lower shelf limit, T_{LS}, as noted in the Level 1 MAT procedure.
- Determine the Level 2 MAT, using the following equation.

$$MAT = \max\left[MAT_0^{Level\ 2} + T_0 + \Delta T_0^C, T_{LS}, -155^{\circ} F\right] \quad (10)$$

- For components subject to environmental cracking or toughness degradation, a modified procedure is provided in [1].

EXAMPLES

The following examples are provided to illustrate results obtained using the proposed modifications to the Level 1 and Level 2 procedures and acceptance criteria of API 579.

Example 1: Non-Impact Tested Pressure Vessel Shell

Calculate the MAT for a non-impact tested pressure vessel cylindrical shell section.

- Construction Code: ASME VIII-1
- Component: Cylindrical Shell
- Material: SA-516 Grade 70 (MSYS = 38 ksi, Curve B)
- Inside Diameter: 60 inches
- Nominal Thickness: 1.0 inches
- PWHT: No
- Design Pressure: 600 psig
- Impact Tested: No
- Radiography: 100%

Results:

- Proposed Level 1 MAT
  - t/4 Reference Flaw 58°F
  - t/8 Reference Flaw -17°F

2021 Edition API 579 Part 3 Level 1 MAT 31°F

- Proposed Level 2 MAT at 600 psig
  - t/4 Reference Flaw 29°F
  - t/8 Reference Flaw -58°F

2021 Edition API 579 Part 3 Level 2 MAT 23°F

The plot of the Level 2 MAT results as a function of pressure are shown in Figure 3.

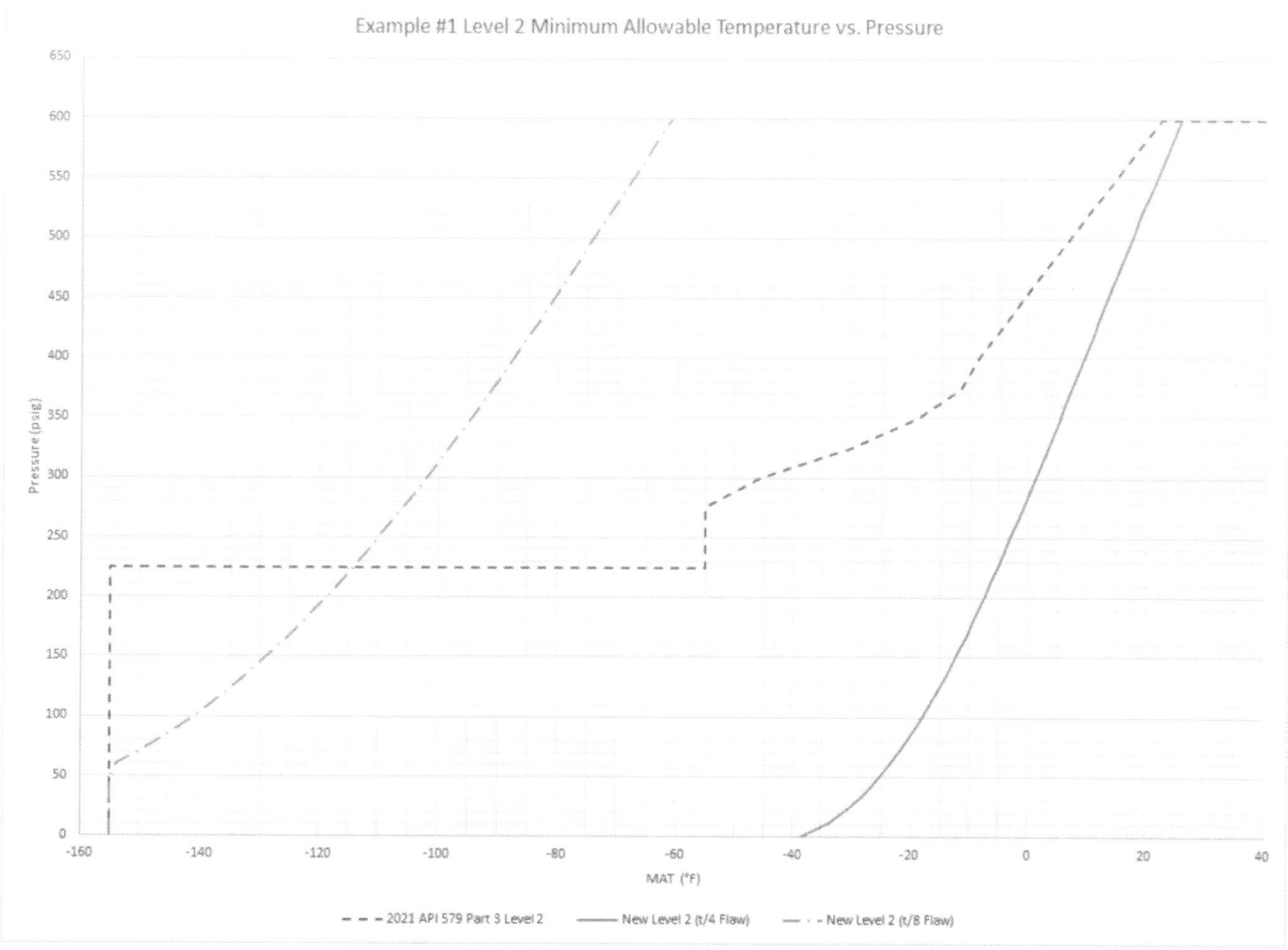

FIGURE 3: Example 1 Level 2 MAT Results

In this example, the results from the newly proposed Level 2 brittle fracture screening criteria are more limiting than the existing API 579 Part 3 screening procedures when employing a t/4 reference flaw size. For this example, the t/4 reference flaw has a depth equal to 0.25 inch and a length equal to 1.5 inches. As shown in the plot, results obtained using a t/8 reference flaw (i.e., flaw depth of 0.125 inch and flaw length of 0.75 inch) are much more favorable than the t/4 reference flaw results. With inspection and NDE, it would be expected that the t/8 reference flaw may be adequately justified for use in establishing the MAT for the vessel. It is worth noting that the current API 579 Part 3 Level 2 screening results are potentially unsafe for some operating pressures in comparison to the fracture mechanics-based results obtained using the t/8 reference flaw. Again, the deficiencies, discrepancies, and concerns with the current API 579 Part 3 brittle fracture screening criterion are fully documented in WRC Bulletin 599 [1].

Example 2: Impact Tested Pipe

Calculate the MAT for an impact tested section of pipe.

• Construction Code:	ASME B31.3
• Component:	Pipe
• Material:	SA-333 Grade 6 (MSYS = 35 ksi)
• Inside Diameter:	NPS 18
• Nominal Thickness:	Sch. 120 (1.375 inches)
• PWHT:	No
• Design Pressure:	2000 psig
• Impact Tested:	13 ft-lbs (avg) at -50°F per Material Specification
• Radiography:	100%

Results:

• Proposed Level 1 MAT	
o t/4 Reference Flaw	9°F
o t/8 Reference Flaw	-62°F
2021 Edition API 579 Part 3 Level 1 MAT	-50°F
• Proposed Level 2 MAT at 600 psig	
o t/4 Reference Flaw	-52°F
o t/8 Reference Flaw	-141°F
2021 Edition API 579 Part 3 Level 2 MAT	-87°F

The plot of the Level 2 MAT results as a function of pressure are shown in Figure 4.

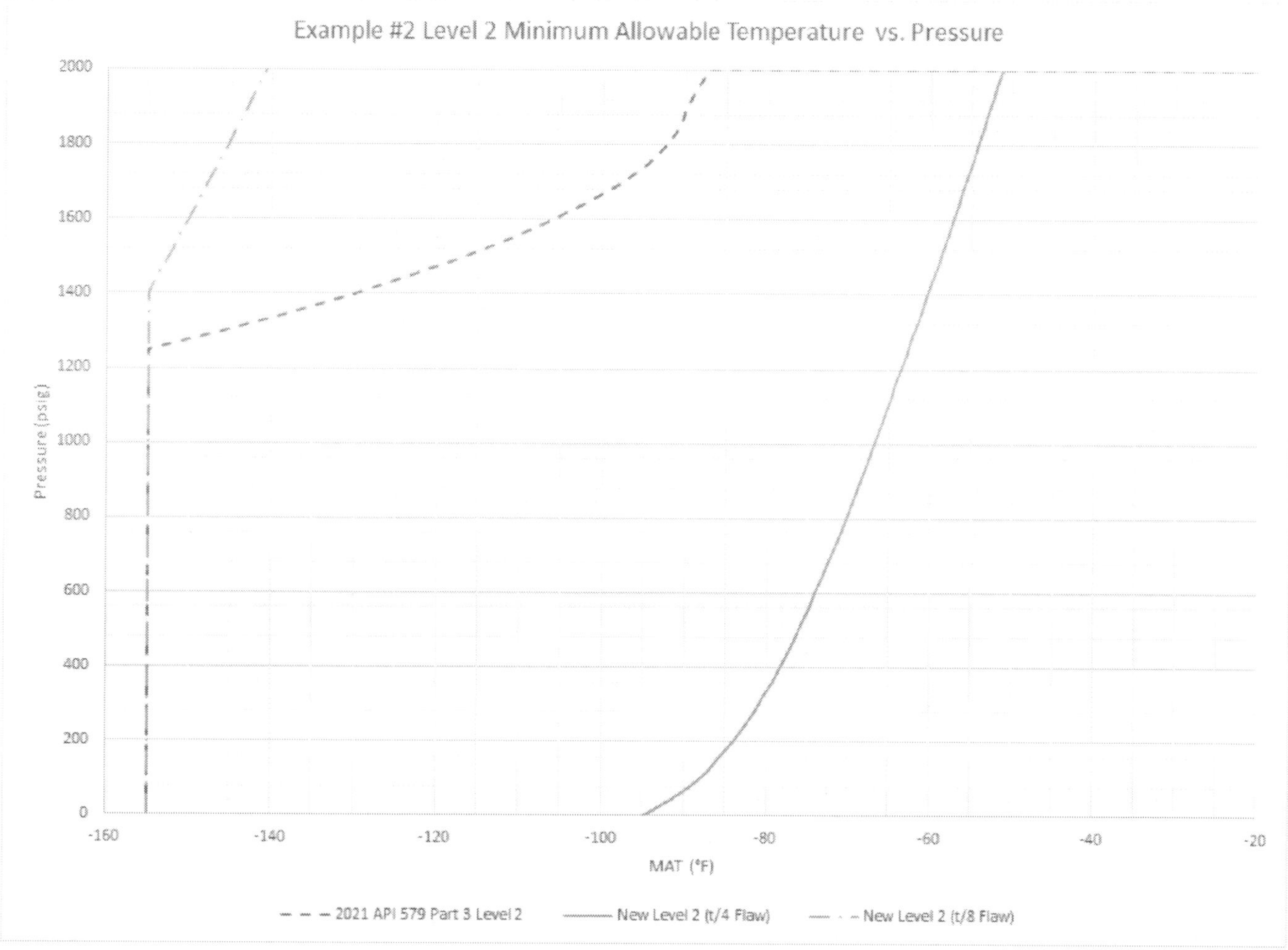

FIGURE 4: Example 2 Level 2 MAT Results

As shown in Figure 4, results from the fracture mechanics-based screening indicate that the MAT for the component may be higher or lower than the impact test temperature for the material depending on the reference flaw size employed in the assessment. For this example, a t/8 reference flaw (i.e., flaw depth equal to 0.344 inch and flaw length equal to 2.063 inches) would be justifiable with minimal inspection and NDE. The results obtained using the t/8 reference flaw are significantly less limiting than the results obtained using the current 2021 edition of API 579 Part 3 Level 2 procedures.

SUMMARY

This paper provides a general overview of the newly proposed modifications to the brittle fracture screening assessment procedures used to establish low temperature operating limits for pressure equipment and piping. Employing a modern fracture mechanics-based approach will significantly mitigate issues with current industry design and FFS screening methodologies. Examples were provided in this paper to demonstrate the differences in results comparing the newly proposed procedures to current industry guidance. At the time this paper was published, these alternate procedures were being reviewed by the ASME/API Joint Committee on FFS for incorporation into a future release of API 579.

REFERENCES

1. Macejko, B.R., Kummari, S.R., Osage, D.A., 2023. Welding Research Council Bulletin 599: Recommendations for Evaluating Resistance to Brittle Fracture for Carbon and Low Alloy Steel Equipment and Piping. Welding Research Council, New York.

2. T. Munsterman, J. Aller, M. Gonzalez, Safety Experience of Light Liquid Hydrocarbon Containment, AICHE Ethylene Producers Symposium, February 1996.

3. API/ASME, 2021. API 579-1/ASME FFS-1 Fitness-For-Service. The American Petroleum Institute and The American Society of Mechanical Engineers, Washington D.C./New York.

4. ASME, 2021. Boiler and Pressure Vessel Code, Section VIII Rules for Construction of Pressure Vessels, Division 1. American Society of Mechanical Engineers, New York.

5. Macejko, B., Kummari, S., and Prueter, P.E., "Current Deficiencies and Proposed Modifications to API 579 Part 3 Brittle Fracture Screening Procedures." Proceedings of the AIChE 2019 Spring National Meeting. AIChE Paper Number 182d, March 31 - April 4, 2019, New Orleans, LA.

6. ASME, 2021. Boiler and Pressure Vessel Code, Section VIII Rules for Construction of Pressure Vessels, Division 2 Alternative Rules. The American Society of Mechanical Engineers, New York.

7. M. Prager, D.A. Osage, J. Staats, B. Macejko, 2010. WRC Bulletin 528, Development of Material Fracture Toughness Rules for the ASME B&PV Code, Section VIII Division 2. The Welding Research Council, New York.

8. Osage, D., Spring, D., Anderson, T., Kummari, S., Prueter, P., and Wallin, K., 2017. Welding Research Council Bulletin 562: Recommendations for Establishing the Minimum Pressurization Temperature (MPT) for Equipment. Welding Research Council, New York.

9. Anderson, T., Constraint Effects on Fracture Toughness in Ductile-Brittle Transition Region, WRC Bulletin 577, 2020, Welding Research Council, Inc., New York, NY.

124a: SAFEpark IMPLEMENTATION AT GREENFIELD STEAM CRACKER

Dan Swatske, ExxonMobil Technology and Engineering Company Spring TX

Prepared for Presentation at the Year 2023 Spring National Meeting, March 13-16, 2023

124a: SAFEpark IMPLEMENTATION AT GREENFIELD STEAM CRACKER

Dan Swatske, ExxonMobil Technology and Engineering Company Spring TX

The ExxonMobil SAFEpark concept was developed to mitigate the consequences of major reliability incidents potentially originating from utilities, furnace or recovery areas. This paper will introduce the SAFEpark concept and further elaborate on how it was implemented at a greenfield steam cracker in three parts: development, application, and sustainment.

INTRODUCTION

In 2000, ExxonMobil set forth to address the subject of reliability and operability driven by a period of operating challenges and events. ExxonMobil addressed the concern via development of a best practice (BP) known as SAFEpark.

Despite the name, SAFEpark is not strictly a process safety practice although reduction of transient and prolonged abnormal operations are the target of this practice. It is akin to a shed list, commonly used in conjunction with monitoring blend gas or steam cushion. The output from SAFEpark development is a matrix and associated procedures. The matrix defines a variety of scenarios or operability events. It also lays-out the states for the steam cracker and triggers for when to move the steam cracker to a given state; while associated SAFEpark procedures define how to enact the change. States are commonly defined by various levels of turndown, however the critical quality of each state is the inherent stability provided by achieving that state.

The SAFEpark concept is general enough to allow application at any steam cracker regardless of process sequence. This is important because each steam cracker is unique – not only in process sequence, but also in utility equipment availability and the way in which the primary unit operation process is (or is not) integrated into a large chemical plant with co-product, derivative, refinery, and/or parallel steam cracking train interconnects; each steam cracker may have different potential major reliability events, triggers, SAFEpark states, and strategies for attaining each state.

SAFEpark is primarily utilized to assist operations with decision making in response to operability events which need to be acted upon within the shift. ExxonMobil recently commissioned a greenfield steam cracker on the gulf coast using the SAFEpark concept. In a unique application, SAFEpark was premeditated from the early phases of the project but there was no experience based data to draw from when developing the system. For purposes of this paper, the plant will be referred to as EMSC. EMSC is an ethane steam cracker and several derivative units on the same plot where utilities are shared between the units.

This paper will discuss the SAFEpark development process employed at EMSC. It will focus on examples of how SAFEpark was applied to the steam cracking operation post-startup. Finally, we will discuss how the system can be sustained throughout the life of the plant and continue to bring value.

SAFEpark DEVELOPMENT

SAFEpark can be a positive influence on project design from an early stage. The ability to influence any project design is highest at the start. During front-end engineering, the process design team evaluated the initial design against typical utility events to gauge the possible impact on the plant from a high level. For instance, what is the impact of a total power failure or loss of natural gas? Or, given the size of the polished water tank, how much time would the plant have to safely shutdown if polished water supply to the tank was interrupted?

Steam crackers will experience utility failures which impact not only the reliability of the unit but could also escalate into safety issues. Using our knowledge of SAFEpark scenarios, the team evaluated multiple steam cracker events against the EMSC initial design and identified areas of possible changes/adjustment.

As the project approached start-up, the organization grew and, more personnel with manufacturing experience were added. With the overarching design fixed, SAFEpark development then shifted focus to ensuring operations had the information required for sound decision making during operation – less about design and more about practicable operations and start-up preparation.

PLANNING

It was desired to have the SAFEpark process ready well ahead of start-up. The start-up team expected there would be little time to dedicate to SAFEpark development in the time period following start-up while the months leading to start-up would be dedicated to making the final tweaks on start-up and commissioning plans. The project milestone to complete the development of operating procedures was nearly one year ahead of the planned start-up. SAFEpark development fit neatly into this category.

Gathering a knowledgeable team is critical to the success of SAFEpark implementation. This should be a mixture of experienced process operations personnel and manufacturing/technical engineers familiar with the key areas of the plant. The team at EMSC found value in keeping an open line of communication with those involved in front-end engineering design during the process.

During the development of SAFEpark, other operating procedures were in development. By the time the effort was completed, many of these procedures were final drafts. EMSC found that some emergency procedures needed to change as a result of the SAFEpark work. A process for formally adding-to or modifying procedures should be in place ahead of the effort. An existing site might take this for granted, but work processes need to be built from the ground-up along with the actual work product at a greenfield site.

DEFINING SAFEpark SCOPE

Since SAFEpark has been practiced at existing ExxonMobil steam crackers and the process design team had the process in-mind during front-end engineering, EMSC had a head-start on developing a seed list of potential SAFEpark scenarios or events. EMSC reviewed the matrices of other steam crackers in the circuit and added-on scenarios specific to the EMSC flow plan. At this point, EMSC developed a list of scenarios to be evaluated later. Some examples can be found in the Table 1.

Table 1. Example SAFEpark Scenario Seed List

Loss of liquid feed	Single steam producer trip
Loss of gas feed	Partial loss of boiler feed water (BFW)
Single furnace trip	Total loss of BFW
All furnace trip	High liquid level in drum or tower
Partial loss of quench water / oil cooling	Partial cooling water loss
Partial loss of quench water / oil flow	Total cooling water loss
Total loss of quench water / oil flow	Partial power failure
Loss of chemical additive	Total power failure
Loss of a single pump (anywhere)	Loss of raw water supply
Loss of dilution steam generator	Loss of feed to demin water unit
Drier load exceeding regen capability	Loss of deaerator feed
Loss of tower reboiler	Loss of ethylene consumer
Major compressor trip	Loss of co-product consumer
Minor compressor trip	Loss of natural gas supplier
Reactor trip	Loss of nitrogen supplier
Product off-spec	Loss of instrument air

This is not a comprehensive list, but it is intended to give the reader a flavor of the scenarios which might show up in a seed list. Ultimately, EMSC ended up with a seed list of about 80 scenarios. From ExxonMobil experience, it is typical to have 50-100 scenarios generated for the seed list.

NARROWING SAFEpark SCOPE

Many of the seed-list scenarios will not make it into the final SAFEpark matrix. This section of the paper will discuss how the seed list was pared down to a pragmatic size. The team kept two questions in mind while paring the list:

- Is there a simple response to the scenario such that it would not result in a major environmental, safety, reputation or reliability impact?
- Does this scenario need to be resolved within the shift?

Each of the scenarios which had a simple response; did not have a major reliability, environmental, safety or reputation impact; or were emergency response (e.g. loss of containment, fire) were stricken from the list of potential SAFEpark scenarios.

The team considered the following examples to have a simple response. A single pump is lost, but a spare pump is available. A single cooler is lost, but a spare is available. A piece of equipment which is not normally online is found to be lost (e.g. reactor regeneration heater). If there was a clear disposition to manage material imbalances, it was also considered a simple response: a stripper tower's heat or stripping medium is lost, but it can be bypassed and effluent maintained in the normal disposition for at least one shift (or an alternative disposition exists which can contain the new influx for longer than a shift). Similarly, if product is off-spec but a disposition exists which would result in minimal or no flaring, the scenario was removed from the SAFEpark list.

The answer to the first question will remove the majority of scenarios from the list. The team started with 80 scenarios and, using the simple response criteria, the list was reduced to about 30. The next step to whittling down the list was to determine if the initiating event has a major reliability, environmental, safety or reputation implication ***which must be resolved within the shift.*** Typically, those scenarios which do not require action within the shift would be removed from the list.

An example of a scenario which would need to be resolved within the shift was the total loss of process condensate outlet scenario. When the process condensate pumps are lost, the quench tower will begin to fill. If out-flow from the system cannot be reestablished, and the quench tower overfills, there is the potential to damage the trays or packing or downstream equipment The furnaces are the main water input to the tower, so furnaces must be removed in this scenario to maintain material balance. During furnace transitions, additional water is added to the quench tower for a short period. Because of this, the water accumulation gets worse before getting better. Operations must know ***when to start*** pulling furnaces and ***how quickly*** to pull them. Depending on the capacity of the quench water separator and the size of the furnaces, this could feasibly need to start immediately after the process condensate pumps are lost. If operations is unsuccessful at pulling the furnaces at the required rate and liquid continues to build, it is essential to understand when the quench tower is critically high and the remaining furnaces (and potentially the process gas compressor) need to be tripped. This is a good example of how a simple event (pump loss) can cascade into a larger event or extended outage (quench tower packing replacement required due to damage on overfill). This scenario also clearly needs to be resolved within the shift.

On the other side of the time spectrum, EMSC decided to keep the following scenario on the list even though the material balance analysis showed the problem did not necessarily need to be resolved within the shift: loss of feed to polished water tank. Raw water is brought into the industrial water tank. The raw water is treated in the demin unit and is pumped into a polished water tank. The polished water is pumped to the deaerator, and finally, the boiler feed water is pumped to the furnace and boiler steam drums. Reference Figure 3.

Figure 3: Generalized Water System Sketch

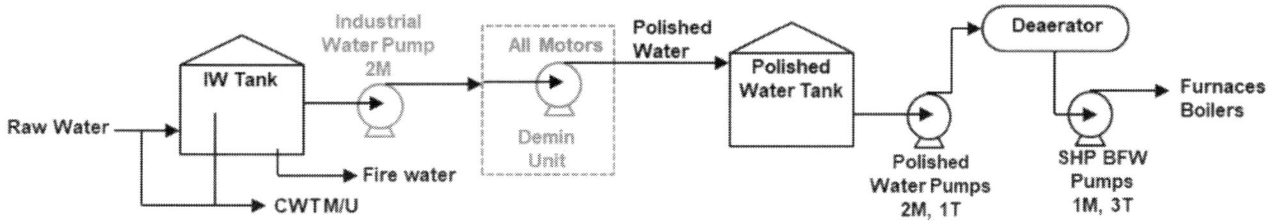

Given the size of the polished water tank; available bypasses (and willingness of the site to bypass the demin unit); and availability of temporary demin trailers the polished water tank was expected to take longer than one shift to reach critically low levels if the appropriate shed steps were taken. The team choose to keep this scenario in the list due to the impact of water loss to the site (i.e. total loss of steam). This scenario is also particularly interesting because zero inventory in the polished water tank is not the time to start shutting the plant down – rather it is the time to complete the shutdown. The team determined what level in the tank should be considered the low operating envelope based on how much water was required to safely remove all of the furnaces from operation. This is known as a trigger point. Trigger points will be discussed in the next section.

Ultimately, when scenarios are removed from the list, it is at the discretion of the team developing the SAFEpark process. Each site will have a different tolerance for what is simple and what is not; what needs to be dealt with immediately and what can wait.

DEFINING SAFEpark STATES AND TRIGGER POINTS

Once the seed list of scenarios has been pared down to a critical number of scenarios which would result in a major reliability, environmental, safety or reputation loss and need to be dealt with within the shift, the team turned to setting trigger points for each scenario and defining the EMSC's SAFEpark states.

Each plant will have a different definition of the SAFEpark states according to the variability of the design, and each state defined should be a stable operating point. Typically, the states are defined as various levels of turndown and specify which major pieces of equipment (e.g. critical compressors, towers, reactors, and furnaces) are online. A naming convention for each state can be useful to ensure everyone is communicating effectively (e.g. State 1, State 2, etc).

Trigger points tell operations when to enact a specific SAFEpark state. For each remaining scenario, the team performed detailed heat or material balances to determine when and what SAFEpark state needs to be enacted.

One unique quality of a greenfield start-up is that there is no history or previous event-based data on the scenarios at hand. However, the operations team used a 3D model of the plant for training and gained an intuition for the time required to perform many tasks.

For the total loss of process condensate outlet scenario discussed in the previous section, the team had to use a combination of operational "experience" along with material balances to determine where the trigger points should be placed. The first step in this process was to understand how the quantity of water added to the system contributes to the water level relative to the location of the transfer and kickback line. Figure 4 shows how the water level rapidly approaches the transfer and kickback piping after the quench water separator drum is completely full due to the shape of each vessel.

Figure 4. Quench Tower and Quench Water Separator Drum

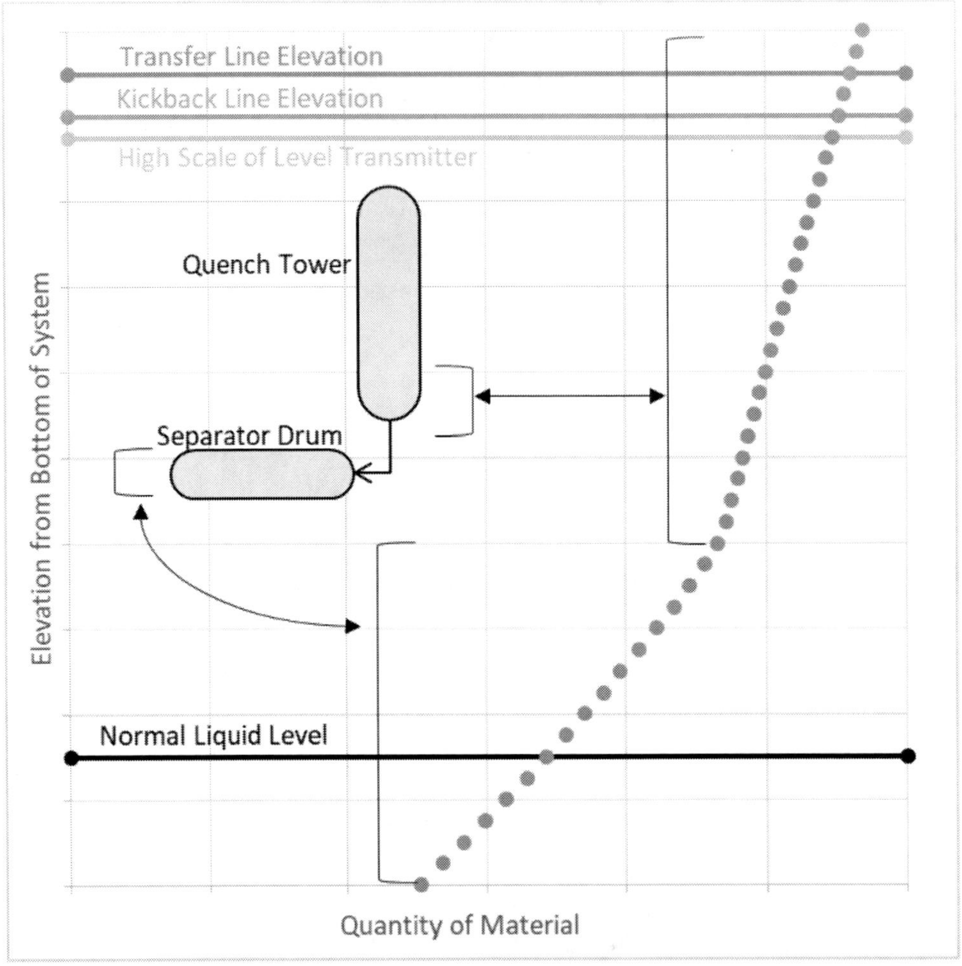

Understanding the water input to the system during a loss of condensate outlet allowed the team to run a series of "simulations" to determine how best to handle this situation. Figure 5 demonstrates how quickly the level builds if the furnaces are moved to a state of minimum dilution steam input.

Figure 5: Loss of Condensate Outlet

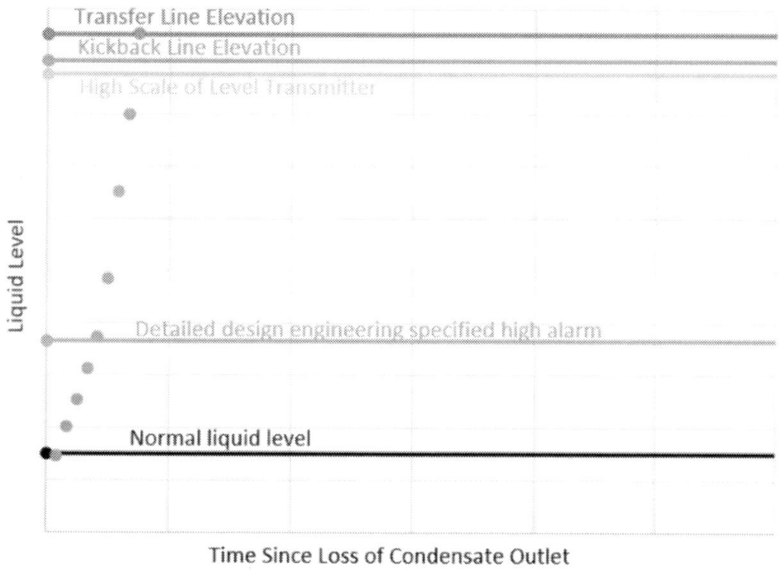

Figure 6 shows that the system will still overfill if an alternate disposition is available but the online furnaces are only taken to minimum dilution steam input conditions. Figure 7 builds on the previous one, removing a single furnace. The system would still overfill in this case if the normal condensate outlet could not be established. The impact of the system geometry referenced in Figure 4 can be clearly seen as an inflection point, and faster rising liquid level. Figure 8 builds on both Figure 6 and 7 – a second furnace is removed from the system. The decreasing liquid level shows that the material imbalance was resolved.

Figure 6: Establishing An Alternate Disposition

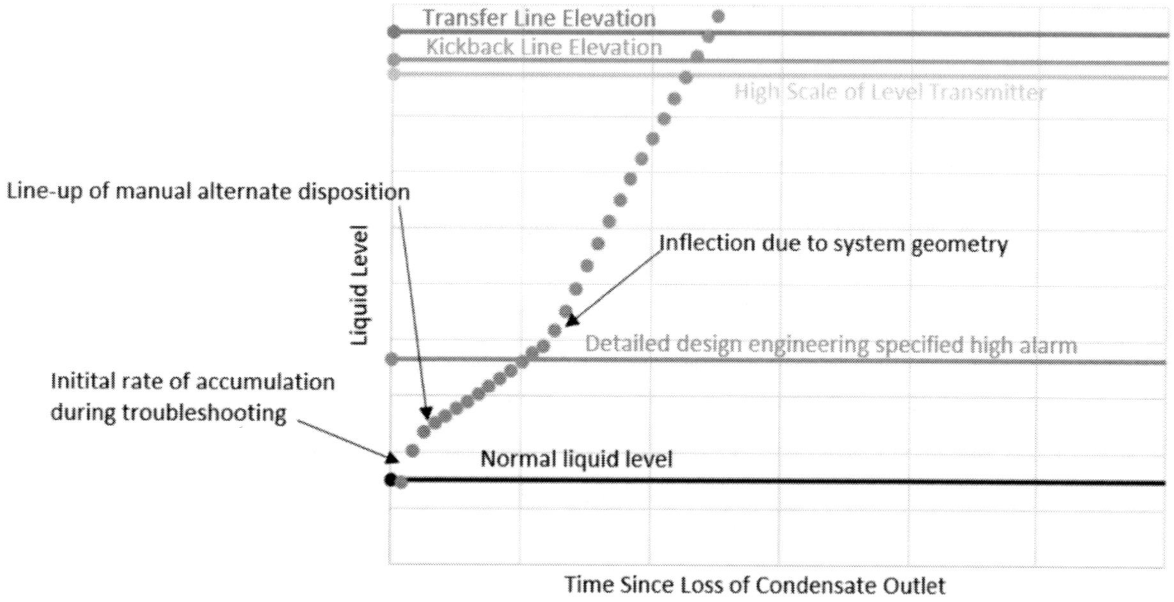

Figure 7: Building on Figure 6, Pulling a Single Furnace

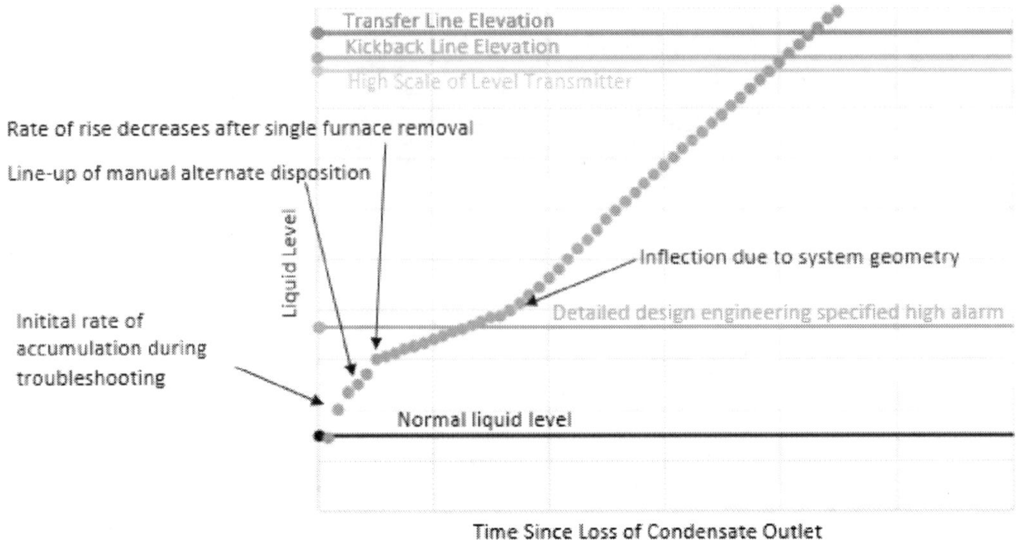

Figure 8: Building on Figure 7, Pulling a Second Furnace

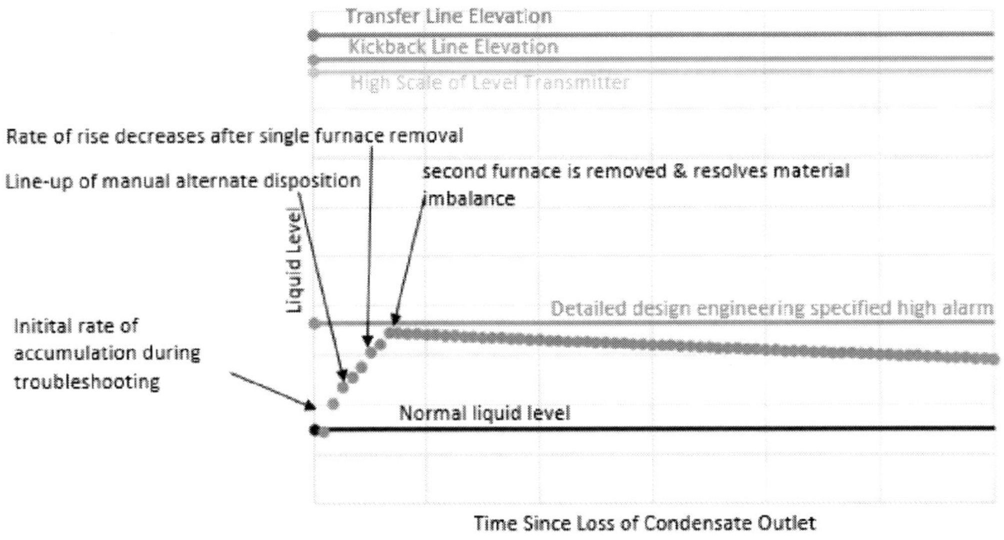

Each of the above simulations are idealistic. They show what will happen if operations immediately takes action on pump loss rather than waiting until the high level alarm or spending time troubleshooting. Similar to Figure 8, Figure 9 also shows what would happen if two furnaces were pulled, however the first furnace is not pulled until the high alarm point. The inflection point due to system geometry is obscured in Figure 9 because a furnace is being pulled at the transition point. Note however, the impact of pulling a furnace on the liquid level becomes more severe due to the liquid building in the quench tower rather than the drum. This highlights the need to take action quickly, rather than spending additional time troubleshooting.

Figure 9: Waiting Until High Alarm Point to Pull the Furnaces

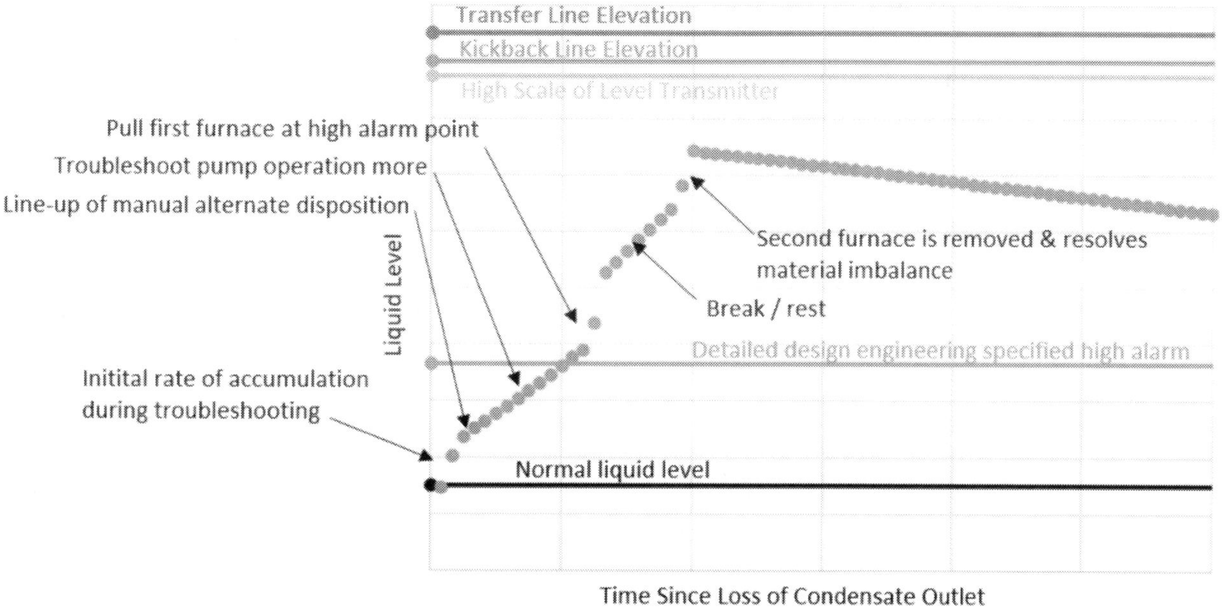

Figure 10 shows the same actions as Figure 9, however it assumes the alternate line up is only achieving 90% of the design capacity. In this case, pulling two furnaces out of the system is not sufficient to reestablish material balance. Both Figures 9 and 10 have a break or rest period included. The length and sequence of the rest period is arbitrary in these examples. It is included only to highlight that the team should consider human factors in scenarios which require a significant amount of physical activity.

Figure 10: Alternate to Figure 9 with 90% Capacity of Alternate Line-up

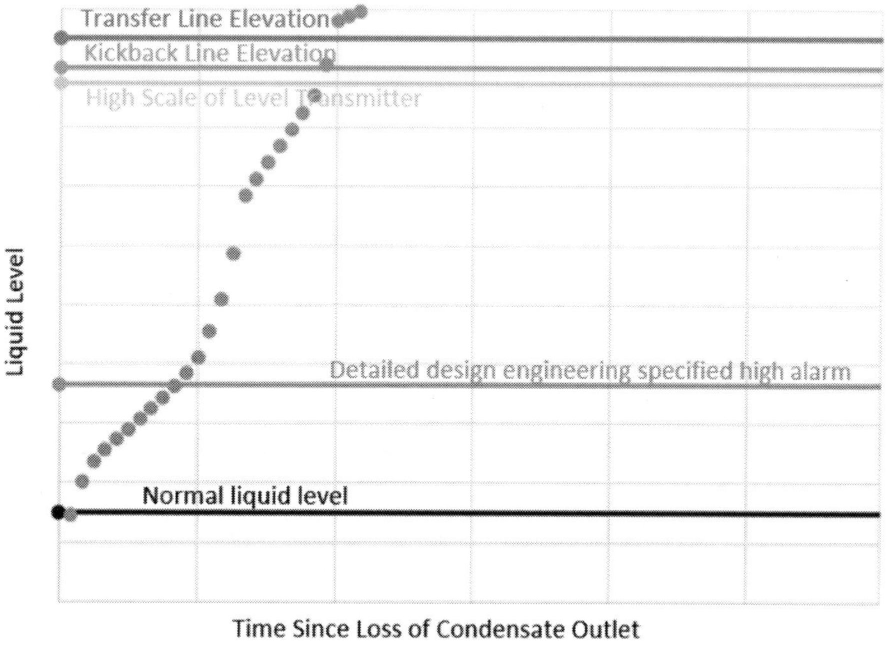

Each remaining scenario on the list should have trigger points defined along with an associated SAFEpark state at this point. While setting trigger points, the EMSC team documented the assumptions and rationale employed. There are many reasons a trigger points may need to change over time: equipment degradation resulting from higher utility consumption, differences between design and actual operation, or projects added onto the plant – changing the original heat and material balance. Understanding the basis for each trigger point will make future updates smoother.

SAFEpark PROCEDURE DEVELOPMENT

Once the scenario paring, trigger points, and state definitions are completed, the team transitioned the remaining seed-list scenarios into the final SAFEpark matrix format. The final formatting allows for quick reference by the operations team (see Table 3).

For each scenario, the team determined if a procedure should be required for the scenario or if SAFEpark matrix comments provided sufficient guidance. For those which required a detailed procedure, the name of the procedure was documented on the matrix. If a procedure already existed, it was edited to use agree with the SAFEpark matrix, otherwise the team assisted process operations with procedure development.

Table 3. Abbreviated Example of a Finalized SAFEpark Matrix

Scenario	Procedure Name	Notes	Trigger point	SAFEpark State	Comments
1. Loss process condensate pump	OP-1001	Loss of condensate pump resulting in accumulation of water in quench tower	Flow Meter <90% of Target	3	Min DS firing Est. Alt Line-up Normal transition
			Level Indicator >55%	3 → 2	Abbreviated furnace transition
			Level Indicator >85%	2 → 1	Cold Trip Remaining furnaces, PGC
2. Caustic System Upset	See Matrix Comments	Loss of caustic feed / export / CO2 breakthrough	Loss of import or export pump	3	Minimize furnace feed additive
			1hr after pump loss	3 → 2	
		Prev. demonstrated loss caustic 12h	Analyzer shows CO2 at Ovhd.	2 → 1	See off spec C2=, secure reactor in standby

In Table 3 (loss of process condensate pump scenario) trigger points are based on the process condensate pump flow meter and the quench system liquid level instruments. Table 3 also indicates which SAFEpark state to move to when the process variable reaches the trigger point. SAFEpark states 2 and 3 are specific levels of turndown where the plant is running normally, but with lower throughput than normal operation. The distinction between turndown states such as 2 or 3 can be defined, for example, by different kickback valve positions on the process gas compressor where state 2 is lower throughput than state 3. State 1 can be defined by the removal of all fresh feed from the plant with some or all major machines running on total recycle, reactors shutdown in standby position, and towers on total reflux.

In addition to ensuring adequate procedures for scenario-specific responses were in place, general procedures were written describing how to move the plant from normal operation to each SAFEpark state. Since there are many actions for each scenario which are the same to achieve a given SAFEpark state, the general procedures were developed to drive a consistent response and lower the potential for error when updating procedures in the future (i.e. an update may only need to occur in the general SAFEpark procedure rather than all of the emergency procedures found in the SAFEpark matrix).

Many scenarios require a reduction in furnace firing and often removal of furnaces from the quench tower or primary fractionator. While the matrix and procedures inform on the target furnace rate, it is imperative that operations understand which order the furnaces should be removed so as to preserve the health of the furnaces. This reduces the potential for breaking a tube (escalation of an event). In accordance with this philosophy, EMSC ensured a method of communicating the desired order of furnace removal was in place ahead of start-up.

SAFEpark APPLICATION

In plant operation, the SAFEpark matrix was used in response to a loss of caustic scenario. Several consecutive days of sub-freezing temperatures resulted a loss of fresh caustic. While this specific scenario was not on the EMSC SAFEpark list (loss of caustic import pump), it clearly fell into the “Caustic System Upset” scenario. Upon realization that fresh caustic import was lost, the operations team transitioned to a pre-determined SAFEpark state per the plan. The plant was able to demonstrate stable operation without fresh caustic input, and avoided further escalation.

A loss of the demin water plant pumps has also occurred. This scenario will be discussed further in the next section of this paper.

SAFEpark SUSTAINMENT

SAFEpark sustainment describes the methods used to ensure SAFEpark remains a living document which changes when the plant changes. There are three basic parts to sustainment: procedure management and accessibility; training; and follow-up review of events.

The SAFEpark general procedures and scenario response procedures are stored in accordance with the site-adopted procedure management system. This ensures they will not become lost or abandoned. In addition, the procedures will receive ongoing validation through established processes

Similarly, the SAFEpark matrix is treated as a procedure and is stored in the same system as a site's operating procedure. A mirror of the SAFEpark matrix was added as a schematic to the DCS for all relevant consoles with a live read-out of each trigger point variable shown next to the trigger point set point requiring action by the operator. This was done to ensure everyone associated with the plant operation has quick access to the SAFEpark information. Posting physical copies of the matrix is an adequate alternative to adding a schematic to the DCS – the main criteria is accessibility.

Sites which regularly train or review SAFEpark scenario responses receive the greatest benefit. While quick access to the information is critical, ideally the initial reaction from a SAFEpark scenario is achieved by memory, similar to emergency response. This can be achieved through site established protocols for emergency response or through a separately established process.

After each event where the SAFEpark matrix is called upon, it is good practice to reflect on the prescribed steps in the procedure: what went well –and what did not. Through this process, the site discovered that the material balance provided in the design documents did not accurately capture one of the derivative unit's boiler feed water consumption, resulting in an inaccurate SAFEpark response to a loss of demin water to the polished water tank. The SAFEpark scenario and trigger point rationale had been clearly documented allowing the error to be quickly identified. The existing trigger points would not have given operations adequate time to shut down since the actual polished water tank draw-down rate was 25% higher than originally anticipated. The SAFEpark matrix and trigger points for this scenario were re-tooled along with troubleshooting to determine what could be done about the off-design water consumption. The site designated a steward of the SAFEpark process to facilitate this type of review, ensuring its success during the early stages of implementation. Mature sites using SAFEpark have transitioned this to operations whereby process engineering provides input when required.

CONCLUSION

The ExxonMobil SAFEpark process has been used for several decades in the ExxonMobil steam cracking circuit – the benefits of installing a SAFEpark process at steam cracking sites has been demonstrated. EMSC, a greenfield steam cracker had SAFEpark in mind starting in the project development phase, allowing additional facilities for handling SAFEpark scenarios. The manufacturing team at EMSC started development early enough to ensure the process was ready well ahead of start-up. This allowed the team to focus on the start-up itself. They leaned into the SAFEpark experience of the ExxonMobil steam cracking circuit to develop the seed list, and used creative thinking to simulate the potential outcomes and required actions of each scenario when defining trigger points.

AIChE PAPER NUMBER: 124b

PUSH TESTING – IDENTIFICATION OF MODIFICATIONS TO INCREMENTALLY BOOST PLANT CAPACITY

Brian K Sullivan
Director of Operations & Services
Lummus Technology

Prepared for Presentation at the 2023 Spring National Meeting Houston, TX, March 12 – 16, 2023

PUSH TESTING – IDENTIFICATION OF MODIFICATIONS TO INCREMENTALLY BOOST PLANT CAPACITY

Brian K Sullivan
Director of Operations & Services
Lummus Technology

Abstract

Typically, maximum ethylene plant profit equates to maximum production. Every plant has constraints that limit production. Some of them require significant investments to eliminate, while others do not. A Push Test is a study procedure to identify minimal risk, low investment opportunities to stretch plant production capacity. While this concept can be applied to larger expansions, typically the objective of a Push Test is to find a set of potential incremental breakpoint modifications that can be easily implemented to boost plant capacity by five to fifteen percent, without a major investment.

The Push Test procedure combines the use of design capacity factors, operational reviews, and localized plant testing to "push" individual unit operations to their constraining limits to confirm their available remaining capacity with a high degree of confidence. Incremental breakpoints are identified, along with potential modifications needed to mitigate the associated constraints.

The paper will elaborate on the Push Test procedure for an ethylene plant, including descriptions of how to test individual unit operation capacity limits after reaching the initial plant constraint when increasing feed rate. Examples of potential Push Test applications will also be provided.

Introduction

When a new ethylene plant (or similar petrochemical plant) is designed, a design basis sets the intended annual production rate of the major products (i.e., ethylene, propylene, and key by-products) for a given on-stream factor (i.e., hours of operation per year). The licensor generally includes a capacity safety margin for all equipment to ensure that production guarantees can be met. When detailed engineering and procurement are performed, certain equipment is purchased "off the shelf" in the next available size. Other equipment may be procured with more than the specified capacity on the licensor datasheet to ensure that the designed performance is met. Consequently, plants are often capable of achieving more than their nameplate capacity. However, due to the variety of safety margins that may be applied during the plant design, the additional capacity available in the various sections of the plant may vary. Moreover, in some cases, the current plant feed mix or quality is quite different from that used to develop the original plant design. The change in feed properties affects different sections of the plant differently.

When a plant undergoes a major expansion revamp (e.g., capacity increase of 30% or more) where a significant capital investment is expected, a similar engineering process occurs. It is assumed that the original design capacity of the major equipment will be exceeded, and the expanded capacity is typically designed using engineering principles and safety margins similar to those used during the original design.

However, if the objective is to maximize the potential production of the plant without major investment, then an alternate approach, called a Push Test, can be used to identify smaller incremental capacity increase breakpoints, and ultimately the modifications needed to achieve them, so that an optimum production increase vs investment tradeoff can be evaluated to achieve minimum capital expenditure per ton of capacity increase.

The following sections of this paper will discuss important aspects of this Push Test approach to studying limited breakpoint capacity increase possibilities:

- Definition of a Push Test
- Basis for a Push Test
- Step-by-Step Push Test Process
- Examples of Push Test Results

What Is a Push Test?

A Push Test is a study procedure to identify minimal risk, low investment opportunities to stretch plant production capacity. While this concept can be applied to larger expansions, typically the objective of a Push Test is to find a set of potential incremental breakpoint modifications that can be easily implemented to boost plant capacity by five to fifteen percent, without a major investment. The goal is to identify potential production capacity increase breakpoint steps, up to the maximum production target, while taking actual equipment capability into account.

It should also be noted that the motivation for a Push Test may be related to other factors, such as a shift in feed allocation or a change in product specification (i.e., from chemical grade propylene to polymer grade propylene), rather than a simple desire to maximize production under current conditions. For example, a plant that was originally designed to process a significant percentage of liquid feedstock, may want to transition to more gas feedstock. A Push Test can also be used to identify where and how the resulting change in combined cracking furnace effluent composition will impact the recovery section of the plant so that modifications needed to maximize the benefit of the shift in feed allocation can be identified.

The Push Test combines the use of design capacity factors, operational reviews, and localized plant testing to "push" individual unit operations to their constraining limits to confirm their available remaining capacity with a high degree of confidence. Incremental breakpoints are identified so that potential modifications needed to mitigate the associated constraints can be designed.

Maximizing re-use of existing equipment and limiting modification or replacement to only that equipment that is absolutely necessary, helps to minimize the capital investment related to the capacity increase. When re-using equipment, it is critical to confirm that it is capable of performing to specification under the expanded production capacity. For this level of capacity increase, the best way to confirm the plant's capability is to perform a "Push Test."

By virtue of its name, "Push Test" implies that it is a field test based upon increasing the load in the plant until limits are reached. However, that is just the culminating step in a long set of activities that lead up to the field test. While the process results in optimizing/minimizing the capital expenditure per capacity gain, a series of methodical and detailed engineering study procedures are still required to achieve the desired end result.

A capacity field test alone in which feed rate is increased until a limit is reached would only identify one or two plant constraints. With a true Push Test, once that first or second plant constraint is reached, individual systems must be artificially forced to higher capacities (to the extent possible) in order to test the other sections of the plant for capacity breakpoints. For example, the internal loading of a distillation tower can be increased by raising the reboiling and condensing rate at constant feed rate to evaluate its loading (i.e., approach to flooding), but more on that later.

The equipment examined during the Push Test can include reactors, distillation towers, pumps, compressors, and heat exchangers. The Push Test identifies the extent of the additional capacity available within the key equipment.

Push Test Basis

Like any debottleneck study, a basis is required to define the Push Test baseline. The basis defines the following information:

- The minimum desired increase in annual production target or maximum actual annual production target to be achieved, based upon a defined on-stream factor (i.e., number of hours of operation per year). Specifically, this must address the desired production rate, product specifications, and battery limit conditions for the ethylene, propylene, and critical byproducts, as well as any requirements for the relative production mix.
- The composition/qualities, maximum available flow rate, and battery limit conditions for each fresh plant feedstock, including external feeds to the recovery section. Note that this composition/quality and maximum rate must be available throughout the Push Test field activities.
- Minimum and maximum allowable cracking furnace conversion/severity or coil outlet temperature (COT), run length, and steam to oil ratio.
- Availability and battery limit supply conditions of all utilities, as well as any known limitations on the OSBL utility supply.
- Documentation of any changes made to plant equipment or operating basis that present significant deviation from the original plant design basis. Examples include:
    - Cracking feeds that are significantly lighter or heavier than considered in the original plant design basis, which can shift the loading of the cracking furnace effluent throughout the recovery section equipment in the plant.
    - Change in key product specifications or grades. For examples, older plants may have shifted from producing chemical grade propylene to polymer grade propylene, which will impact the loading in the C3s section of the plant recovery system.
    - Modifications to equipment, such as new radiant coil type, change of tower internals (i.e., from trays to packing or vice versa). This ensures that proper loading estimates are prepared for planning field testing.
- Identify any known existing plant constraints and conditions under which the constraint exists. This is helpful to confirm the capacity factor analysis (described later) and plan the field test activities more thoroughly.

Step By Step Push Test Procedure

The full set of Push Test steps, following definition of the basis, includes the following:

- Assemble key records
- Develop baseline material balance and expected combined cracking furnace effluent
- Perform capacity factor analysis
- Assess plant operations to identify potential operational and maintenance improvements

- Prepare data collection requirements and protocols
- Plan the Push Test field activities and prepare the plant
- Execute the Push Test field activity
- Evaluate the results

Assemble Key Records

Assembling accurate records of as-built design data for key equipment is essential. This information can be used to corroborate observed performance against expected performance to identify underperforming equipment. Some of the key records needed may include the following:

- As-built Piping & Instrument Diagrams (P&IDs).
- As-built datasheets and vendor drawings for major capacity limiting equipment, including information such as
  - General arrangement drawings for the cracking furnaces
  - Details (i.e., type, dimensions, etc.) of internals for distillation towers
  - Head/capacity curves for compressors and steam extraction-condensing maps for turbines or motor data sheets (if applicable)
  - Heat exchanger design values for area, duty, LMTD, etc.
  - Vessel and major drum drawings, including details of internals
  - Head/capacity curves and driver data sheet (rated horsepower) for major pumps.
  - Specifications of catalysts and desiccants in reactors and dryers.
- Up-to-date plot plan, indicating areas open for expansion.
- Current utility balances (i.e., steam, cooling water).

Develop Baseline Overall Material Balance and Combined Furnace Effluent

In order to determine the capacity factors (in the next step) used for identifying the most highly loaded units, an overall material balance must be developed for the desired target production using the baseline feedstock quality and quantity. An example of a simplified overall material balance for a Push Test target is shown in Table 1.

The purpose of developing the baseline overall material balance is to determine the combined cracking furnace effluent composition and flow rate required to meet the target production rate under the Push Test baseline conditions. The total combined cracking furnace effluent composition and flow rate is used in the next step to project the loading in the different recovery section areas of the plant.

In developing the baseline overall material balance and combined cracking furnace effluent, the impact of the higher production rate on the system pressure drop and therefore the coil outlet pressure (i.e., COP) on the yields must be considered, along with the expected operating conversion/severity of the cracking furnaces.

Table 1 Example Push Test Material Balance	
Feeds	**Rates, MT/h**
C2+ Feed	70.0
Naphtha	60.0
External Recycles to Recovery Section	3.2
Total	133.2
Products	
Export Hydrogen	0.1
Fuel Gas	25.9
Polymer Grade Ethylene	70.0
Polymer Grade Propylene	21.6
Pyrolysis Gasoline	12.8
Fuel Oil	2.8
Total	133.2
Recycles	
Ethane	35.0
Propane	5.2
C4s (and C5s)	15.8

Perform Capacity Factors Analysis

The capacity factor method used for Push Test preparation is a screening tool to evaluate the impact of the target capacity increase on different sections of the plant. Once the projected combined furnace effluent composition and flow rate is determined during the baseline overall material balance step, the process of developing capacity factors can begin. Using this approach, the ethylene plant is broken down into various areas and capacity factors are calculated for each of those plant sections downstream of the cracking furnaces. The capacity factors are derived by comparing the impact of the combined furnace effluent on the loading of various areas in the recovery section of the plant for the target production case relative to the expected loading of these recovery section areas at their full design capacity, including all as-built over-design. Licensors will have their own unique method for converting the combined cracking furnace effluent composition and flow rate to a loading factor for each recovery section area.

This method produces capacity factors for all key sections in the recovery systems of the plant. A capacity factor value of 1.00 represents loading that equals the design loading, including as-built over-design safety margins. Any capacity factor value that exceeds 1.00 indicates that the load on that system for the push test case may exceed the capacity of the currently installed hardware beyond their original design limits (including their original over-design factors).

Table 2 shows an example capacity factor analysis result, in which the bolded capacity factors highlight the plant areas that limit the plant the most. In this example, the capacity factor analysis indicates that the following equipment would potentially become bottlenecks as production capacity is increased toward the desired target:

- Charge gas compressor
- Chilling train and demethanizer
- Propylene fractionator
- Refrigeration system

The remaining equipment should not present a bottleneck and therefore would not need to be individually tested (although they should be monitored during the Push Test field activities, especially if their capacity factor is close to a value of 1.0).

Table 2
Example Capacity Factor Summary

PLANT AREA	Capacity Factor for Target Production
Transferline	0.80
Gasoline Fractionator	0.84
Quench Oil/DS Exchanger	0.92
Quench Tower	0.94
Quench Water System	0.71
PWS/Dilution Steam Drum	0.72
MPS/Dilution Steam Exchanger	0.92
Charge Gas Compressor	**1.16**
Caustic Tower and System	0.99
Charge Gas Drying	0.99
Chill Train & Demethanizer	**1.19**
Methanation System	1.00
Deethanizer	0.94
Ethylene Fractionation	0.95
Depropanizer	0.90
MAPD Converter System	0.97
Propylene Fractionation	**1.09**
Debutanizer	0.94
Refrigeration System	**1.15**

Assess Plant Operations

Up to this point in the process, the steps are theoretical in that they assume the following:

- The equipment is in a clean, as-built condition
- All maintenance and operating procedures are conducted as intended

- The control system and instrumentation are in good working order
- Actual equipment over-design is consistent with original basis

Therefore, the capacity factor analysis does not account for limitations due to fouling or equipment damage, improper operation and maintenance, differences in as-built design vs specification, etc. To compensate for this fact, it is recommended to perform a plant operations and maintenance review to assess known operating problems, unit by unit performance, and maintenance issues to expose potential capacity limitations that are not evident from the capacity factor analysis.

A plant operations assessment questionnaire is used to assess the operational performance and maintenance status of the cracking furnaces and selected recovery section areas. Any issues that are identified by discussing the questionnaire with operations personnel can be followed by site inspections of the cracking furnaces and key recovery section equipment for confirmation. An excerpt from an example operations assessment questionnaire is shown in Table 3.

Table 3 Example Excerpt from Operations Questionnaire

1.0 General Operations and Problems

1.1 Capacity

What is the maximum sustainable hourly ethylene production capacity that the plant has achieved? ____________

1.2 Plant Limitations and Issues

Please list the causes of any unplanned outages or 'off-specification' production during the past two years? ____________

List any bottlenecks (current limits to plant production) that prevent an increase in plant throughput.

Please make a list of your most significant operational problems or issues that affect the plant operation and provide a short explanation of each problem.

2.0 FURNACE OPERATION

Please respond to the following questions.

What typically limits the furnace capacity?

Run length	YES or NO
Crossover Temperature	YES or NO
Firing	YES or NO
ID Fan	YES or NO
Emissions limit	YES or NO
Other	YES or NO

Do your cracking furnaces experience any of the following problems?

Convection section process side fouling YES or NO

If so, please provide the location of the fouling and a description

Convection section flue gas side fouling YES or NO

If so, how much does the stack temperature increase in one year? _____

Burner fouling YES or NO

Is decoking done on a fixed schedule cycle? YES or NO

Operations Questionnaire page 3 of 16 LUMMUS TECHNOLOGY

Prepare Data Collection Requirements and Procedures

Once the previous steps of the process identify the areas of the plant that need to be tested for their capacity limitation, the requirements for monitoring the performance of the plant during the field testing must be specified prior to beginning the test.

A list of all data required for analysis should be prepared and configured in the historical data collection system for electronic retrieval following the test. This will include process and utility data. In addition to the typical flow, temperature, pressure, pressure drop, level and analyzer values, other atypical data types should be collected, such as control valve positions (or controller output value if valve position feedback is not available), controller modes, day of run for SOR/EOR equipment, etc.

In addition, a list of required local readings (i.e., field measurements) and laboratory analyses must be prepared and log sheets with required recording frequency created for use during the test.

- It is likely that it will be necessary to install some local pressure gauges on taps and thermocouples into thermowells to evaluate some individual equipment. These requirements should be identified and addressed prior to beginning the Push Test field activities.
- Where online analysis is not possible, samples should be taken around each of the potentially limited equipment, and laboratory analysis performed at regular intervals throughout the Push Test to ensure that capacity increase adjustments do not cause the process to go off-spec. Again, these requirements should be identified and addressed prior to beginning the Push Test field activities.

To derive maximum value from a Push Test, extensive review and rectification of the plant's instrumentation and process control system, similar to what is done in preparation for a Performance Guarantee Test Run (PGTR) is recommended. If a full rectification is not possible, then as a minimum, the accuracy of the key flow meter readings should be verified prior to beginning the Push Test field activities.

Plan the Push Test Field Activities and Prepare the Plant

Planning and preparation for the Push Test field activities is crucial for success. Once all plant areas/equipment (i.e., specific pumps, exchangers, towers) that must be tested have been identified, the following steps should be taken prior to starting the Push Test field activities:

- Prepare test log sheet for each piece of equipment to be tested.
- Prepare the data collection procedures for the Push Test field activities for each of the systems/plant areas targeted for individual evaluation per the previous steps.

- Confirm that all relevant instrumentation (measurements, analyzers, valves, etc.) are properly calibrated and in working order.
- Confirm that all relevant control strategies are properly tuned and operating in automatic mode.
- To minimize the possibility of operating upsets during the field testing, confirm that plant operation can be maintained steadily for the duration of the test such that no infrequent or batch operations will be performed during the test and that routine maintenance has been recently performed on the equipment (e.g., cracking furnaces have been decoked, TLEs and burners have been cleaned, etc.).
    - If the condition of any reactor or dryer equipment that requires periodic regeneration is close to EOR conditions, then the equipment should be swapped or regenerated prior to the start of the test.
    - Only frequent batch operations, such as the charge gas dryer regeneration, which are part of normal operation, should take place during the Push Test field activities.
- Identify which spares can be used for the test, if needed (i.e., spare pumps)
- Consider the impact of a potential global trip (i.e., power failure, cooling water failure) on the flare load, if it were to occur while the plant was operating beyond design capacity. It may be necessary to program or plan certain temporary modifications to avoid overloading the flare at the elevated throughput during the Push Test.

A detailed schedule of activities/testing steps must be prepared in advance of beginning the field testing, including the dates for which the plant will be available for the field activities. The test schedule must consider the following:

- Confirm that sufficient feed will be available and that the feed quality will be relatively constant and consistent with the baseline feeds throughout the duration of the Push Test field activities.
- Plant production must not be affected due to downstream customers or ISBL/OSBL product storage capacity during the Push Test. Coordination with suppliers and customers is needed.
    - Any production constraints related to product storage or downstream consumer's usage must be eliminated prior to the Push Test. Product storage tanks should be drawn down and downstream customers must be prepared to take as much product as possible so that the high capacity operation is not interrupted prior to its scheduled completion.
- Consideration must be given to the disposition of any spare cracking furnaces. How the spare cracking furnace is handled during the field Push Test depends upon the results of the previous steps and where the most likely capacity breakpoints may exist. For example, if the quench tower is identified as a capacity breakpoint, then operating a spare cracking furnace on high steam standby (HSSB) conditions during the test will allow evaluation of the quench tower loading without unduly overloading the rest of the plant. Similarly, if capacity is limited by lack of SHP steam to drive turbines, then again operating the spare cracking furnace on HSSB conditions would help to push the plant to a higher capacity. Alternatively, if the plant configuration is such that the recovery section is limiting and

cracking furnace capacity is in excess, then either the cracking furnaces should begin the Push Test at reduced capacity or the spare furnace should be shut down (if it is known that it will not be needed).

- No off-spec product or slops rerun shall be processed during the Push Test field activity.
- All cracking furnaces should have sufficient minimum number of run days remaining to meet the schedule (typically at least a week), to ensure that no decoking takes place during the Push Test.

The detailed schedule/execution plan for the field activities must be shared and discussed with operations. The detailed execution plan must address:

- Definition of initial conditions for starting the field activities and ultimate target production for the Push Test.
- List of all measurements (online and local) to be monitored throughout the field activities.
- Initial adjustments to be made to furnace conditions and feed rate.
- Definition of individual Push Tests on major equipment that were identified as a potential limit by either area factor analysis or plant assessment

Daily meetings should be held each morning to review progress to date and plan objectives and actions for that day. All changes made to increase plant production or individual equipment loading should be logged throughout the field testing.

Execute the Push Test Field Activity

To begin the field activities, the plant operation must be stabilized at the designated initial conditions prior to attempting any production increases. The first step of the field activities is to confirm the individual furnace capacity limitations for each type of cracking furnace. Depending upon whether the anticipated capacity limit for the plant is the cracking furnaces or the recovery section, the individual cracking furnace limitation test may require testing the achievable production capacity of each cracking furnace type/feedstock combination (one at a time) at the target conversion/severity (with consideration given to also being able to achieve the required run length) or it may only require confirming that the cracking furnaces are capable of achieving the production capacity required to test the limitations of the recovery section by testing one of each type, one at a time.

Depending upon whether the objective is maximum conversion or maximum throughput, which depends upon furnace limits and anticipated limits in the recovery section, either slowly increase the cracking furnace coil outlet temperature (COT)/conversion or feed rate and monitor the furnace constraints and recovery section constraints. Note that depending upon the situation, both actions may be required (see more below). Once at least one furnace of each type/feed combination is at its maximum capacity, if the recovery section has not yet reached a bottleneck, then the production rate and/or conversion/severity of the additional furnaces can be increased. These step increases should be performed slowly and cautiously by increasing the feed rate of the remaining cracking furnaces at the baseline cracking severity/conversion. The feed rate to the plant should be slowly increased while monitoring all systems. Each incremental increase in plant production must be followed by a period of stable operation to confirm that the increased production is sustainable. Note that all changes should be made to minimize disturbance to the operation and avoid approaching plant limits that could result in plant trips or damage.

Once the plant reaches a capacity limitation based upon increasing the load of the cracking furnaces, then efforts should shift to either changing the cracking furnace operating conditions (i.e., cracking conversion/severity) or the operating parameters of the limited equipment to improve its performance and alleviate the initial constraint. Note that this may also entail adjusting the conditions of utilities, where possible.

Once the plant reaches a capacity limitation that cannot be alleviated by adjustment of operating conditions, then the focus of the Push Test activities should shift to testing the individual equipment identified in the test plan, according to the following guidelines:

- Compressors can be tested by slowly opening the minimum flow valves to artificially increase compressor throughput. Adjusting the turbine extraction valve may also boost capacity, if possible. It is necessary to closely monitor the system pressures and temperatures relative to trip values during this type of test. The compressor and turbine-side measurements, including the surface condenser, should also be monitored for limitations.
    - In some instances, it may be difficult to open the minimum flow valves or adjust the turbine extraction steam valve in compressors with anti-surge or integrated turbo-machinery control packages. In this case, it may be necessary to consult with the compressor control system vendor in advance of the Push Test.
- Heat exchangers are best checked by calculating the heat transfer coefficient and temperature approaches from the operating data and comparing them to their design values. It will be necessary to measure flows and temperatures on both sides of the exchangers in order to calculate the necessary information, so this may not always be possible. Note also that local temperature gauges may need to be installed in existing thermowells to get the proper readings, depending upon how the heat exchanger instrumentation is designed.
- The distillation towers should be tested with the plant as close as possible to maximum production rates. Then slowly increase the reflux and reboil rates to over-reflux/over-reboil the tower and increase the internal loading to determine maximum loading without flooding. Tower pressure drop should be monitored and design simulations can be used to match the plant data at different loading breakpoints to estimate the internal loading against internals vendor recommended limitations. Condenser and reboiler loading should also be monitored for limitations.
- Pump capacity tests are performed by slowly increasing the suction supply levels and opening the control valves and minimum flow recirculation lines, where possible. In instances where there is no way to measure the flow from the pump discharge, the motor

amperage or turbine steam rate should be recorded. Alternatively, actual operating data can be plotted and compared to vendor pump head curves.

    - In cases in which the pump becomes the bottleneck, then the control valve bypass can be opened, or the spare pump placed in service, where possible, to continue to increase plant production.

- Note that data can be collected for drums, hydrogenation reactors and miscellaneous minor equipment such as mixers and ejectors, but specific tests are generally not done for this equipment.
- In cases where testing cannot be performed, historical data can be used to find maximum proven operating capacity. However, this may not represent maximum absolute capacity.

The Push Test field activities should be terminated once the following criteria are met:

- The target production rate for the Push Test is reached, or
- No additional plant production capacity can be achieved, and all individual equipment capacity tests have been completed to the extent possible.

Once the field activities are terminated, the relevant data should be collected for review.

Evaluate the Results

The results of a Push Test identify the production capacity breakpoints where certain equipment become a bottleneck for further production. The results will also identify equipment that could not be fully or properly tested (i.e., due to lack of instrumentation, potential damage, fouling, or other reason). In those cases, it is necessary to identify the path forward for evaluating that equipment. For example, the path forward could include:

- Compare historical operating data with the vendor design information and extrapolate maximum achievable capacity on that basis.
- Perform more detailed simulation and rating of as-built equipment design. In some instances, equipment manufacturers must be consulted.

The net result of the Push Test is a summary of:

- The equipment that would require modification at each production breakpoint.
- Recommended modifications to improve operability of the plant, including:
    - operating and maintenance procedures
    - process control and instrumentation upgrades

This summary is used to estimate the cost of the modifications required to achieve each production capacity breakpoint, so that an analysis can be performed to evaluate the economic cost/benefit tradeoff at each breakpoint.

Examples of Push Test Results

The results from a few Push Test case examples are presented below. In the first example, production capacity was "pushed" by increasing the feed rate to the cracking furnaces and by adjusting the cracking conversion/severity. In the second case example, individual equipment was "push tested," rather than the entire plant. Finally, some general examples of additional benefits derived from the push test procedure are highlighted.

Example Push Test Case 1

Chart 1 shows the results from a three day Push Test, in which the ethylene production was increased by approximately 17% above the normal ethylene production capacity. The Push Test was actually terminated due to a trip that was caused by a utility failure that was unrelated to the Push Test. Extrapolation of the Push Test data indicated that it would have achieved just under 20% above normal ethylene production capacity if the trip had not occurred.

While the ethylene production steadily increased during the Push Test, the propylene production remained steady. For this Push Test, both the feed flow rate and the cracking furnace conversion/severity were increased to maximize production against the initial constraints. Therefore, the combination of increasing the cracking severity and increasing the feed rate counterbalanced the effect on propylene production. Chart 2 shows the relationship between the increase in cracking furnace conversion/severity and the decrease in C2 and C3 recycle during the Push Test.

Although it was not part of the objective, due to the operational changes that were implemented as part of the push test field activities, the specific energy consumption of the plant (i.e., kcal/kg of ethylene produced) was reduced by 11% from the start of the Push Test to the end of the Push Test.

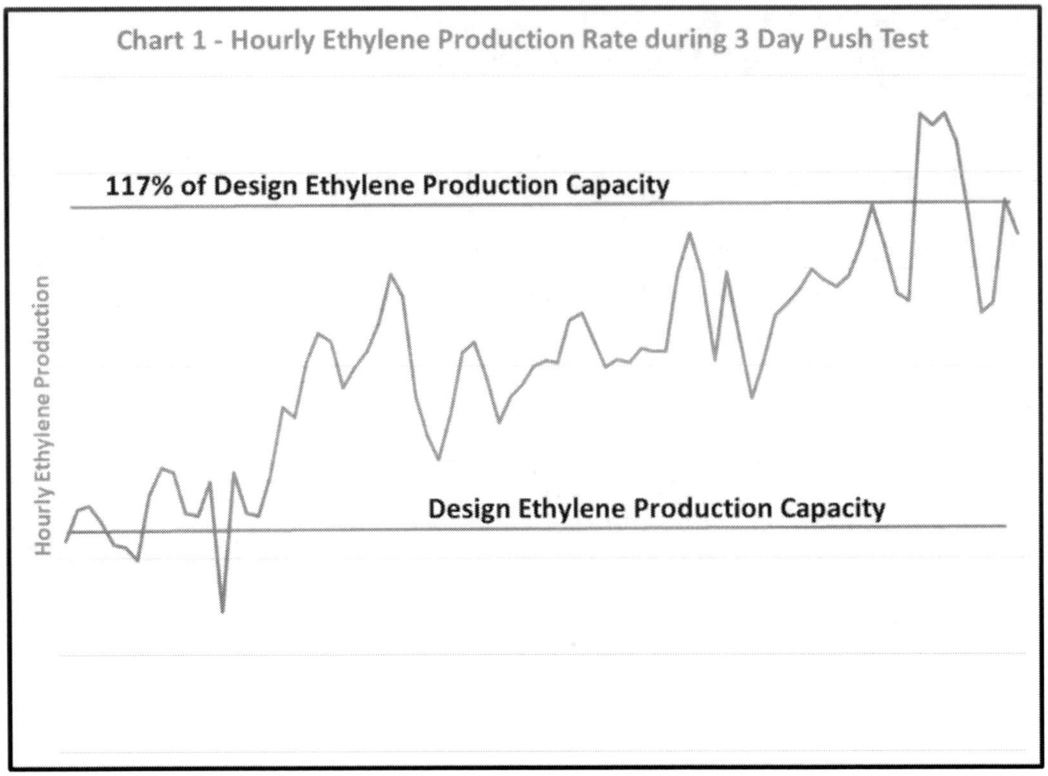

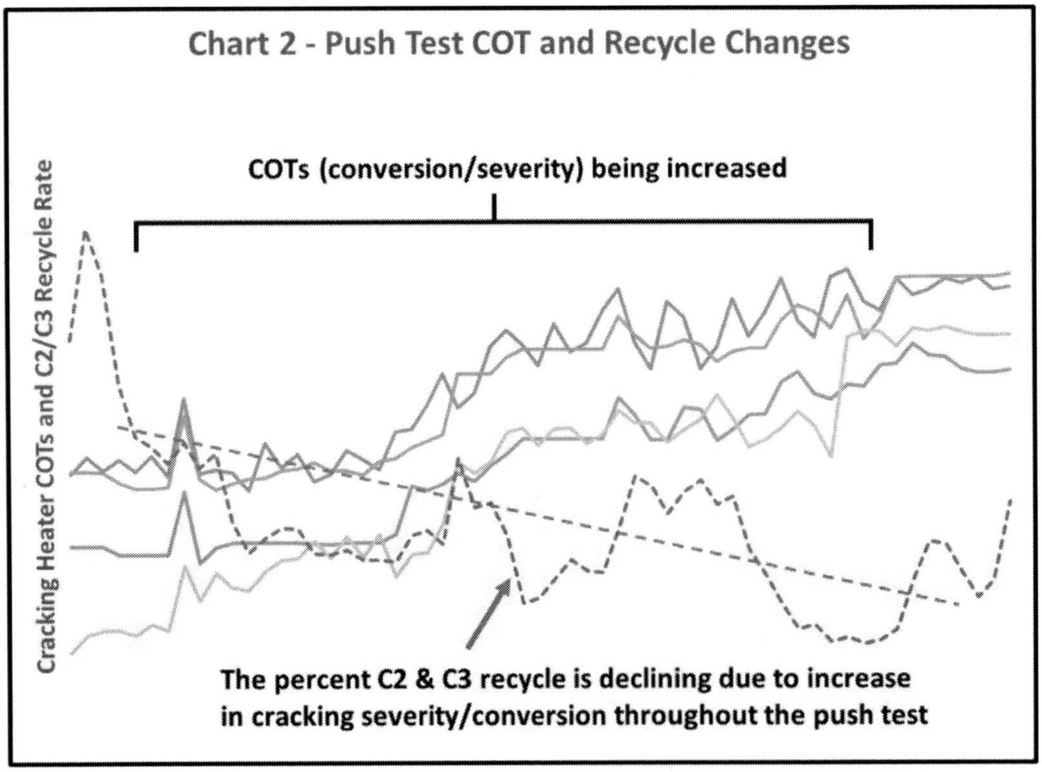

It is important to note that the maximum production rate and lower specific energy consumption values were achieved with the intention of identifying multiple plant constraints that were limiting production. It was not possible to sustain these rates/values on a continual basis without making the appropriate modifications to the plant that were found during the Push Test. Some of the activities performed to achieve the Push Test results included the following:

- Due to the predetermined plant limitations identified, the test plan included scheduling cracking furnace operation at higher cracking severities to shift production to lighter, higher value components and off-load the areas of the plant that processed the heavier components. This also reduced recycles, which had a significant impact on the specific energy of the plant.
- Several spare pumps were put into service during the Push Test to overcome limitations in pump capacities.
- Several control valves opened fully and had to be operated with their bypasses fully open during the Push Test.
- The upper quench water/cooling water exchanger on the quench tower was found to have insufficient surface area to meet the design temperature approach to the cooling water temperature at the target rates.
- The ethylene fractionator reached its fractionation limit during the push test. It was necessary to allow up to 6% ethylene to leak out into the recycle bottoms ethane to maintain the ethylene product on specification. Upgraded tower internals will be required to operate continuously at the new target production rate.
- The spare cracking furnace was placed in service to utilize all the cracking furnaces for the Push Test. Initially, this required the steam to oil (S/O) to be raised to keep the inlet tube venturis choked, but as the feed rate was increased, the S/O ratio was reduced again to reduce the load on the quench tower.
- The products all remained within specification throughout the test.
- As part of the preliminary operations review and subsequent Push Test data analysis, several measurement, instrumentation, and control problems were identified and recommended for rectification. Numerous operating and maintenance procedures were also upgraded. Potential equipment damage was found and recommended for investigation during the next turnaround.
- Due to the trip caused by the OSBL utility supply, the feed and furnace operating conditions portion of the Push Test was nearly completed, but it was not possible to perform the majority of the individual unit testing, beyond the pump limitations already noted.

Example Push Test Case 2

The objective for the Case 2 Push Test example was to identify potential plant bottlenecks that would prevent the plant from attaining a 30% increase in polymer grade ethylene production beyond its current nameplate production. Unfortunately, this goal proved to be much too ambitious. The Push Test procedure indicated that several intermediate capacity increase breakpoints (i.e., less equipment modification required) occurred at a capacity increase well below that target.

Moreover, prior to beginning the plant-wide capacity increase portion of the Push Test, the plant management decided not that they did not want to risk pushing the plant capacity to its limits and deferred to only performing individual equipment Push Tests. Examples are provided below for three individual equipment Push Tests:

- Ethylene Fractionator
- Propylene Fractionator
- Charge Gas Compressor

A push test was performed on the ethylene fractionator to test an increase in internal loading on the performance of the tower while the ethylene production rate was maintained constant at normal maximum production. Chart 3 shows the trends of the main variables during this push test. The reflux and reboiler flow rates were increased by about 7% to 8%. The pressure drop across the column increased by about 3% but remained within acceptable limits. As a result, the C2H6 content in the ethylene decreased from 700 ppm to 100 ppm, while the ethylene content in the bottoms remained relatively constant, after a disturbance due to an operational mistake. The results indicated that at least 7% additional capacity was available in the ethylene fractionator at the operating conditions present during the test. A limit was not reached, but the plant preferred to stop the test. A rating simulation of the tower was required to determine the additional available capacity.

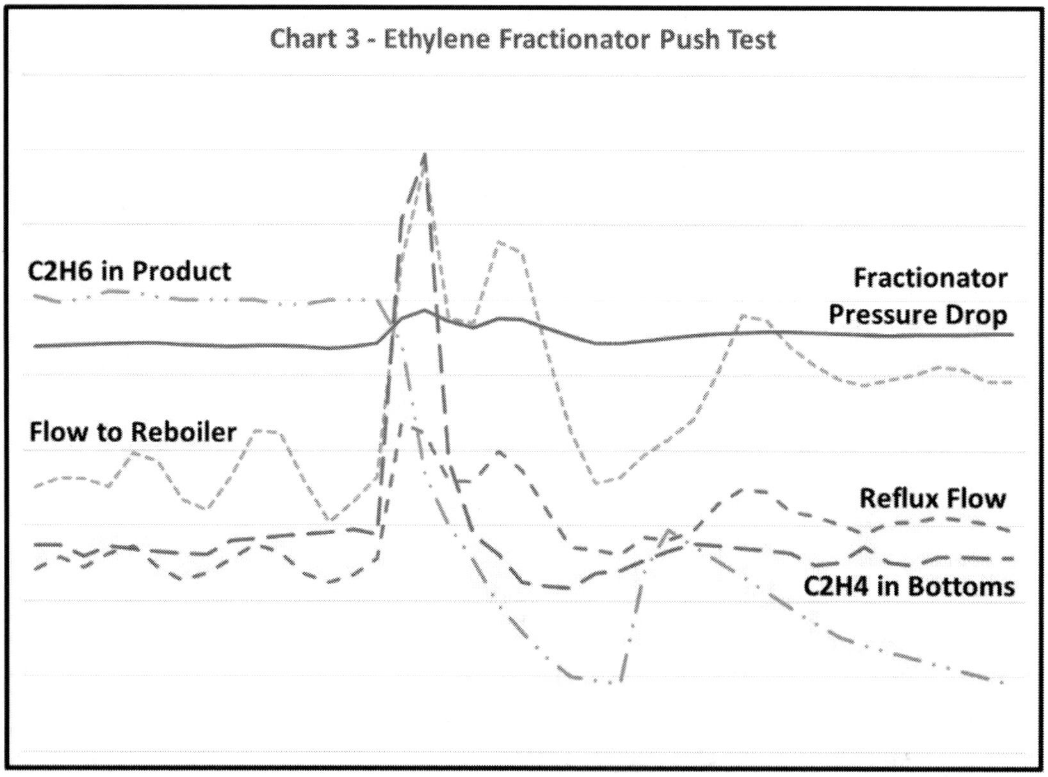

As shown in Chart 4, a push test was performed on the propylene fractionator to test an increase in internal loading on the performance of the tower while the propylene production rate was held constant. The reflux flow rate was increased in steps, which represented an increase of approximately 20% in liquid loading in the tower. The tower performed well, with corresponding improvements in fractionation and no significant increase in pressure drop across the tower. Therefore, it was concluded that there was at least 20% additional capacity available in the propylene fractionator at the operating conditions present during the test (i.e., available cooling water temperature). A further increase could have been estimated by rating the tower with detailed simulations to match plant data.

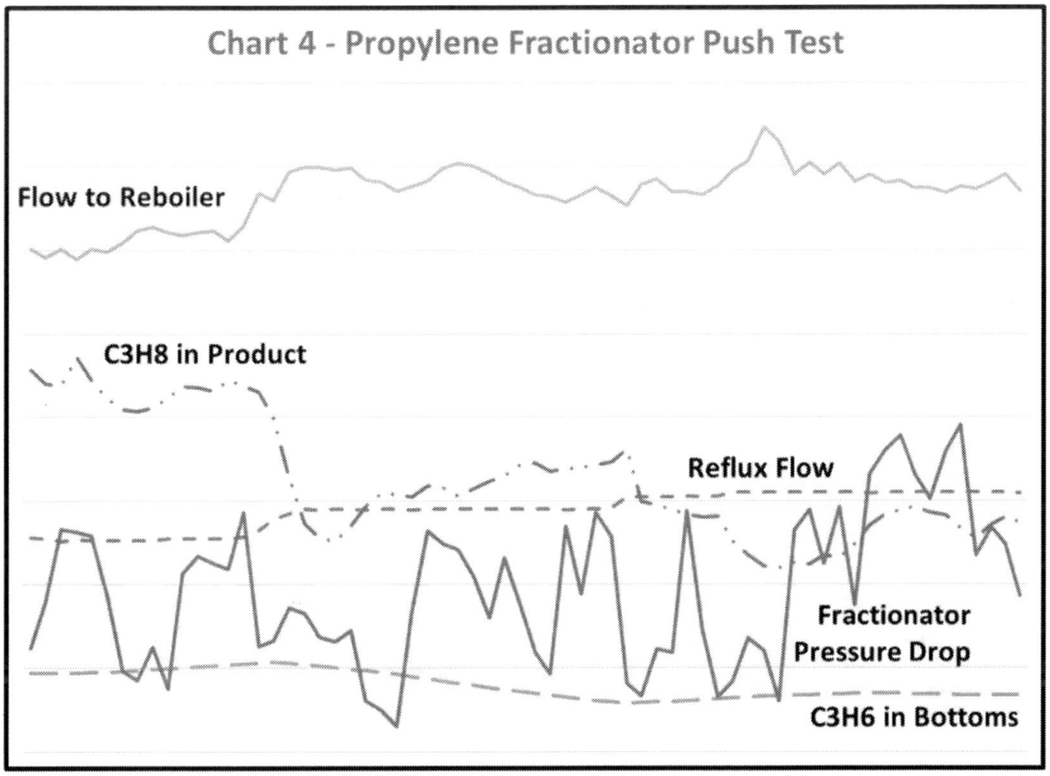

The Push Test operations review found that the SHP steam system normally operated about 8% below its intended design operating pressure, due to instability in the control system. Therefore, a Push Test was performed to gradually raise the pressure of the SHP steam system by about 3.6% to determine the impact on the major compressors with SHP steam drivers. During the test, the operators also reduced the extraction steam from the charge gas compressor turbine by about 40%. The net result was a reduction in SHP steam usage in the charge gas compressor turbine by approximately 15%, as shown in Chart 5.

In addition, during this test, the SHP steam flow to the propylene refrigerant compressor turbine decreased by about 8%. The propylene refrigerant compressor turbine had no extraction steam, so this demonstrated a direct correlation between raising the SHP steam pressure and increasing the power to the compressor turbines.

Note that in this case, the limitation was in the steam turbine, and not the compressor, so it was not necessary to open the minimum flow valves.

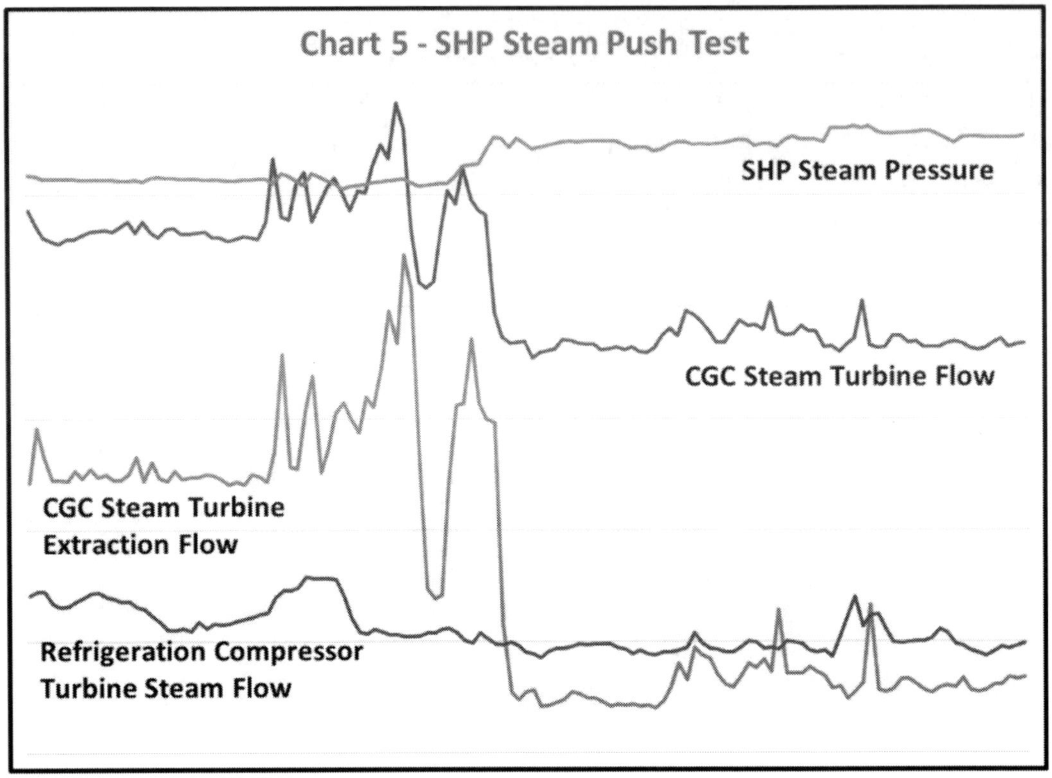

General Push Test Examples

Aside from recommendations to overcome the identified capacity breakpoints, the operational review that is conducted during the process of preparing for the Push Test provides an additional benefit of highlighting improvements in operating and maintenance procedures, as well as minor upgrades that improve the operability of the unit. Some examples of recommendations that were the direct result of the Push Test study procedure include:

- Cleaning of equipment and modifications to prevent/mitigate future fouling
- Confirm existence of damage via a gamma scan
- Better type of instrumentation for a given measurement application

Other interesting conclusions from Push Tests have included:

- Limitations due to return of storage tank boiloff gases (BOG) that create an internal recycle that adds load to some equipment.
- Use of ethylene as charge gas compressor seal gas that creates an internal recycle that adds load to some equipment.

Summary

A Push Test is a study procedure to identify low risk, low investment opportunities to stretch plant production capacity. The Push Test procedure combines the use of design capacity factors, operational reviews, and localized plant testing to "push" individual unit operations to their constraining limits to confirm their available remaining capacity with a high degree of confidence.

Each of the three main steps uniquely contribute to the identifying/confirming the key plant limitations.

- Capacity factor analysis identify most likely theoretical capacity bottlenecks for a given baseline
- Operational reviews identify production limitations due to equipment issues
- Localized plant testing confirms the available remaining capacity of individual equipment

Collectively, the various steps of the Push Test process provide the basis for determining the optimal equipment to modify to achieve an economical capacity increase.

Achieving 8-Year Cycle in Butadiene 1,3 Fractionation

Henrique Silva, Tatiana Barbosa, Débora Schonhorst, Sara Masiero, Carla Weiand

After 7 years of run, a butadiene 1,3 plant in southern Brazil had to face a new challenge due to maintenance restrictions by the times of the COVID-19 pandemic. So, a new mark was set, postponing the scheduled turnaround for one more year in oder to acquire workers availability to support the planned job. Although several fixes were faced in the extractive distillation section, a great result was built for the fulfillment of the campaing in 8 years, in the fractionation train, without any intervention of these equipment for popcorn cleaning. This paper aims to retrieve information about butadiene 1,3 popcorn formation in fractionation train, its severe consequences of this fouling process and how Veolia and Braskem managed the chemical program associated to operation practices, achieving an 8-year cycle

AIChE Paper Number: 124d

AUTO-REFRIGERATION DURING PRESSURING / RE-INVENTORYING NINTROGEN FILLED LINES WITH FLASHING SUPERCRITICAL ETHYLENE: THE MISCOBCEPTION OF THE ROLE OF PARTIAL PRESSURE

Afrooz Eshraghian
Research Engineer in Training
NOVA Chemicals Corporation

Kamal K. Botros
Fellow Research Engineer
NOVA Chemicals Corporation

Prepared for Presentation at American Institute of Chemical Engineers
2023 Spring National Meeting
Houston, TX
March 12 – March 16, 2023

AUTO-REFRIGERATION DURING PRESSURING / RE-INVENTORYING NINTROGEN FILLED LINES WITH FLASHING SUPERCRITICAL ETHYLENE: THE MISCOBCEPTION OF THE ROLE OF PARTIAL PRESSURE

Afrooz Eshraghian
Research Engineer in Training
NOVA Chemicals
Kamal Botros
Fellow Research Engineer
NOVA Chemicals

Abstract:

Upon pressurizing a nitrogen-filled pipeline with high-pressure supercritical ethylene, the ethylene flashes and mixes with nitrogen, resulting in auto-refrigeration cooling. These low temperatures of the mixture of ethylene and nitrogen can vary along the length of the pipe depending on the degree of mixing and local pressure. This is called the Critical Exposure Temperature (CET), which is always required to be above the pipe material minimum allowable temperature (MAT). The commonly held belief among engineers is that the minimum temperature of flashing ethylene and nitrogen mixture at any pressure is equal to the saturation temperature of ethylene at its partial pressure in the mixture. This approach was applied categorically to all mixture compositions and became a misconception. This misconception missed the fact that adiabatic mixing occurs, i.e. energy conservation during mixing for ethylene mole fractions up to a certain threshold when the mixture is in the vapor phase. The present paper, using simulations in Aspen HYSYS, elaborates on the thermodynamics of the process, including the effect of initial conditions of both nitrogen and ethylene, and analyzes the re-inventorying process under both isobaric (venting) and isochoric (shut-in) conditions. As a result of this work, clear insights into the misconception are provided, as well as a more representative approach to performing the pressuring/re-inventorying analysis.

Introduction

Brittle fractures are types of material failure that occur suddenly and without warning, resulting in potentially dangerous consequences for nearby personnel and equipment. Three key factors that contribute to brittle fracture are (a) high stress, (b) the presence of cracks or notches, and (c) susceptible material to brittle fracture, as shown as a triangle in Figure 1a [1]. Eliminating at least one of these elements reduces the risk of brittle fracture.

The presence of the first factor, high stresses, can occur due to welding, high internal pressure, or concentrated stresses at corners or notches, which are sometimes unavoidable. Furthermore, cracks can be formed during fabrication or as a result of service-induced operations. As no vessels or pipelines are fabricated perfectly, these cracks or notches are always present, making the elimination of the second element of the brittle fracture triangle nearly impossible. Therefore, brittle fracture triangles always have two sides, i.e. the stress and material susceptibility to brittle fracture. As a result, avoiding the material susceptibility to brittle fracture is the best strategy for preventing brittle fractures [2, 3].

Temperature directly affects carbon steel's brittle fracture susceptibility, with lower temperatures increasing fracture risk. Therefore, it is important to analyze the process temperatures and ensure that the temperature is above the temperature that causes brittle fracture. The concept of minimum allowable temperature (MAT) refers to the specific temperature at which brittle fracture of processing equipment and pipelines is likely to occur and is the lowest temperature boundary at different pressures such that the brittle fracture occurs below that. The MAT at maximum allowable working pressure (MAWP) is defined as the minimum design metal temperature (MDMT). Figure 1b illustrates an example of a MAT curve. A critical exposure temperature (CET) is the lowest temperature that can be expected in a process at different pressures. There is a risk of brittle fracture on equipment or pipes if the CET curve is at or below MAT curve [1, 4].

It is common to see low temperature exposure caused by auto-refrigeration, particularly during transitions between operating modes [5–8]. It is possible, for example, that this auto-refrigeration occurs during the startup of the ethylene plant after a shutdown, when supercritical ethylene is introduced into nitrogen-filled pipes. Since the pressure of ethylene is usually higher than that of nitrogen, the ethylene is throttled as it flows from the upstream high-pressure ethylene feed pipeline through the small size bypass valve into the nitrogen filled pipe. This throttling cools the ethylene, and auto refrigeration occurs.

Throttling the supercritical ethylene can result in a two-phase ethylene stream. Thermodynamically, the temperature of throttled ethylene from its supercritical state (e.g. at 8000 kPa-g, and 10 °C) can go to temperatures as low as -97 °C (at pressure of 50 kPa-g). However, during pressuring/re-inventorying the nitrogen filled lines with supercritical ethylene, the temperature of mixture of ethylene and nitrogen at specific conditions can go even lower than that of the throttled ethylene, because of the presence of nitrogen in the system that reduce the partial pressure of the ethylene and consequently result in evaporation of some of the liquid ethylene and reduction in temperature.

Unfortunately, there is a common misconception that the minimum temperature of flashing ethylene and nitrogen mixture at any pressure is equal to the saturation temperature of ethylene at its partial pressure. This approach is applied to all mixture compositions and is incorrect as it failed to take into account the fact that adiabatic mixing occurs, due to energy conservation for certain ranges of ethylene mole fractions when the mixture is in a vapor phase.

This project aimed to determine the temperature profile of the ethylene/nitrogen mixture during pipeline re-commissioning or re-inventorying and clarify misperceptions regarding minimum temperatures. By performing these evaluations, the risk of brittle fracture events caused by auto-refrigeration will be reduced and engineers will have a better understanding of best practices for different pressuring scenarios of nitrogen-filled lines with ethylene.

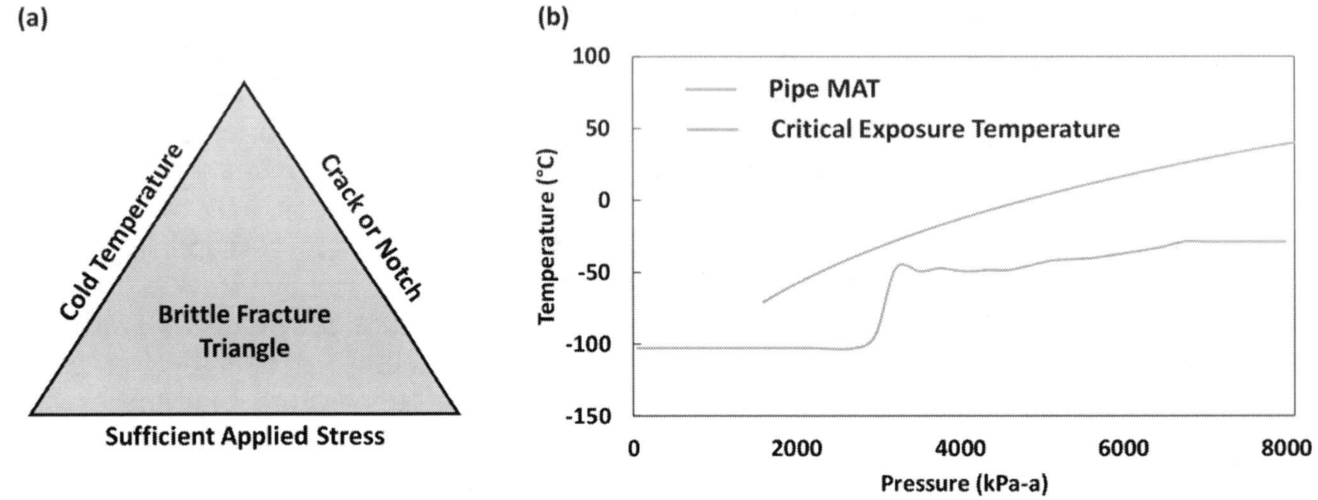

Figure 1. (a) Brittle fracture requirement triangle, and (b) MAT and CET comparison.

Analysis Approach

Identifying processes that can lead to brittle fracture as a result of auto-refrigeration requires evaluating the processes in terms of the occurrence of auto-refrigeration, generating the CET curves, mechanical evaluation of pipe materials to generate MAT curves, and comparing the generated CET with MAT curves. The pipeline is at risk of brittle fracture if the CET curve is at or below the MAT curve. For this purpose, Aspen HYSYS is used to calculate the temperature of the ethylene and nitrogen mixture during introducing flashing ethylene into the nitrogen filled line. The next section explains the calculation procedure and assumptions. As this study was intended to develop the analysis approach, we assumed a generic (hypothetical) MAT curve based on [9] and not the exact MAT pertaining to a specific pipe material.

Calculation Procedure in Aspen HYSYS

Consider pure ethylene at pressure of P_1 and temperature of T_1 in pipe 1 (see Figure 2), and a mixture of ethylene and nitrogen in pipe 2. The mixtures of ethylene and nitrogen in pipe 2 at any time represent sequential batches (dx), each containing specific concentrations of ethylene. As pure ethylene from pipe 1 passes through the feeding valve and flashes to a lower pressure, it blends with the first batch of the mixture of ethylene and nitrogen (batch dx1) in pipe 2.

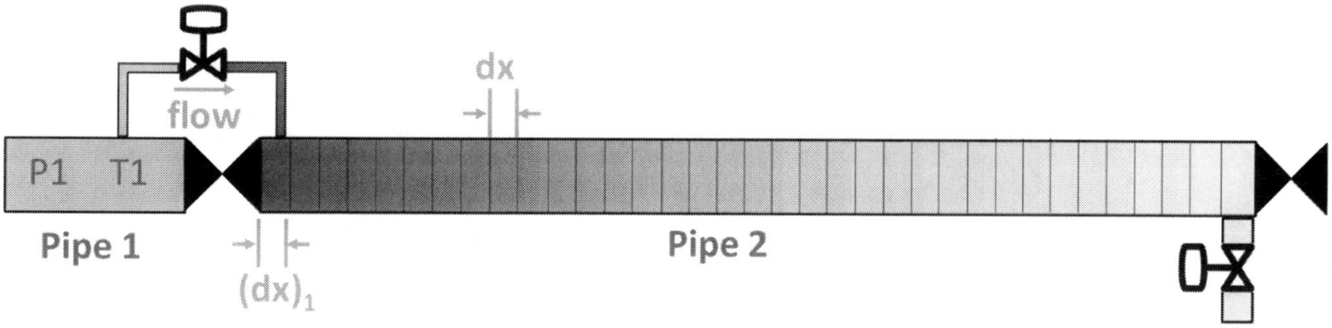

Figure 2. Pressurizing/re-inventorying process: the ethylene flashed from P1 and T1 to the line pressure of P2 and mixed with nitrogen to produce the sequential mixture of ethylene and nitrogen batches.

Figure 3 shows a block diagram illustrating the steps for calculating the temperature of the ethylene and nitrogen mixture in sequential batches of Pipe 2. First, pure ethylene from pressure of P_1 and temperature of T_1 is throttled to the initial condition of Pipe 2 (P_{2i}). Heat transfer to or from the pipe wall is not considered in any of the calculations, which yields a conservative result. Based on the isenthalpic pressure drop process, the temperature of the throttled ethylene ($T_{Ethylene}$) is then calculated.

Next, the throttled ethylene with known pressure P_{2i} and temperature $T_{Ethylene}$ is adiabatically mixed with the ethylene/nitrogen mixture present in the front batch of the pipe (dx1) at the pressure of P_2, which is equal to P_{2i} when venting is occurring in Pipe 2.

To simplify the process, it is possible to consider the adiabatic mixing of pure ethylene (at temperature of $T_{Ethylene}$ and pressure of P_{2i}) and pure nitrogen (at temperature of T_{2i} and a pressure of P_{2i}) at different mole ratios. To apply this simplified case, it is necessary to assume adiabatic mixing of ethylene and nitrogen, and account for adiabatic mixing between different sequential batches along the entire length of the line. Furthermore, it is necessary to assume that the local mixing of incoming flashing ethylene with nitrogen is immediate at any given time at the front side of the pipe (dx1). As a result, there is a mixture of ethylene and nitrogen with a mole fraction of ethylene ranging from 0 to 1 at any location (x) along the pipe and at any time. The temperature of the mixture at different ratios of ethylene/nitrogen (T_2) can be calculated based on these assumptions.

If venting occurs during re-inventorying, a constant pressure is assumed ($P_2 = P_{2i}$). In the case of a shut-in, the mixtures of ethylene and nitrogen at different ethylene/nitrogen ratios are pressurized to a known pressure of $P_{2\text{-}p}$ (where $P_{2i} < P_{2\text{-}p} \leq P_{2\text{-}final}$). The final temperature after pressurizing to a pressure of $P_{2\text{-}p}$ ($T_{2\text{-}p}$) can be calculated, assuming an isentropic pressurizing process.

For this study, three generic MAT curves were assumed for demonstration purposes only, corresponding to the three MDMT values of -29°C, -40°C, and -46°C.

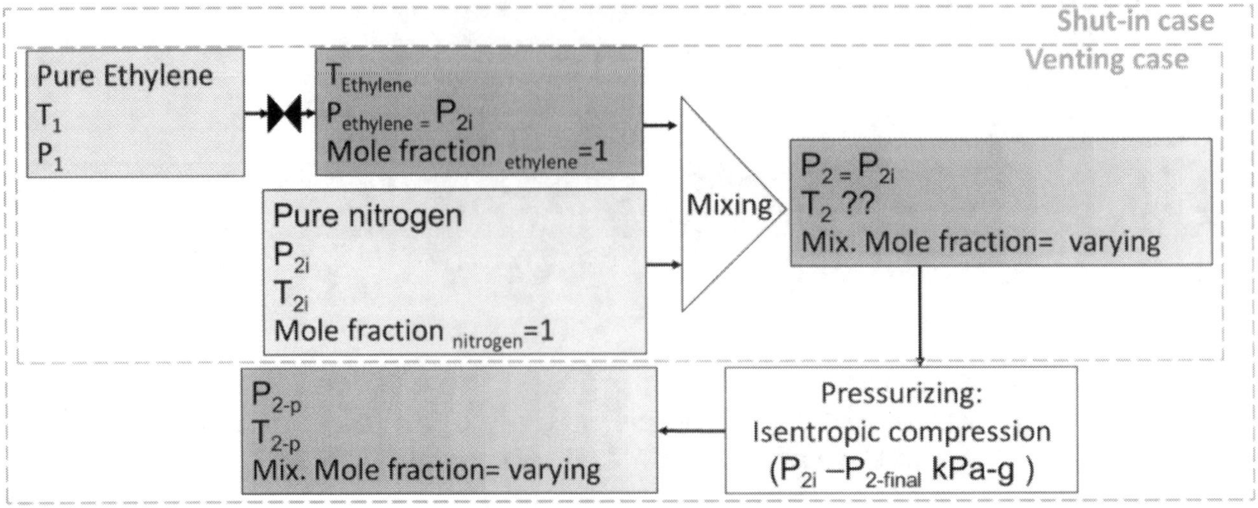

Figure 3. Calculation approach performed in Aspen HYSYS.

Study Scenarios

This study considers two main re-inventorying scenarios: (Scenario 1) the introduction of supercritical ethylene into a pipeline containing vapor nitrogen, and (Scenario 2) the introduction of vapor ethylene into a pipeline containing vapor nitrogen. In Scenario 1, sensitivity analysis of the resulting mixture temperature to the initial temperature of ethylene and nitrogen is also considered, as Scenario 1.b. The required conditions for each scenario are summarized in the Table 1.

Table 1. Scenarios investigated in this study.

	Scenario 1.a									Scenario 1.b	Scenario 2
P_1 (kPa-g)	8000	8000	8000	8000	8000	8000	8000	8000	8000	8000	1000
T_1 (°C)	10	10	10	10	10	10	10	10	10	0	10
$P_{Ethylene}$ (kPa-g)	50	500	800	1200	1500	2000	3000	4000	5000	1500	50
$T_{Ethylene}$ (°C)	TBC*	TBC	TBC	TBC	TBC	TBC	TBC	TBC	TBC	TBC	TBC
P_{2i} (kPa-g)	50	500	800	1200	1500	2000	3000	4000	5000	1500	50
T_{2i} (°C)	10	10	10	10	10	10	10	10	10	0	10
$P_{2\text{-}Venting}$ kPa-g)	50	500	800	1200	1500	2000	3000	4000	5000	1500	50

$P_{2\text{-Shut-in}}$ (kPa-g)	8000	8000	8000	8000	8000	8000	8000	8000	8000	8000	8000
T_2 (°C)	TBC	TBC	TBC	TBC	TBC	TBC	TBC	TBC	TBC	TBC	TBC

*TBC: to be calculated

Procedure to Study the Misconception of Partial Pressure

The purpose of this section is to demonstrate the approach for translating the misconception temperature, which is saturation temperature of ethylene at its partial pressure in the mixture of ethylene and nitrogen into a temperature profile and to compare it with the temperature profile of the ethylene and nitrogen mixture in Pipe 2, which was calculated based on the procedure in the previous section considering adiabatic mixing.

For an ideal gas and an ideal solution in equilibrium at a temperature of T and a pressure of P, the following thermodynamic equations (Eq 1-1, and Eq 1-2) are considered for a binary mixture [10].

$$y_1 P = x_1 P_1^{sat} \qquad \text{(Eq 1-1)}$$

$$y_2 P = x_2 P_2^{sat} \qquad \text{(Eq 1-2)}$$

where

y_1 and y_2 = molar concentration of component 1 and 2 in vapor phase

x_1 and x_2 = molar concentration of component 1 and 2 in liquid phase

P_1^{sat} *and* P_2^{sat} = partial pressure of component 1 and 2 in vapor phase

For such a binary mixture, $x_1 + x_2 = 1$. By substituting Eq 1-1 and Eq 1-2 into this equation, the formula for the mixture can be written as Eq 2.

$$x_1 + x_2 = 1 = \frac{y_1 P}{P_1^{sat}} + \frac{y_2 P}{P_2^{sat}} \qquad \text{(Eq 2)}$$

Using Antoine's equation, when component-specific constants of A, B, and C are known for components 1 and 2 [10], the below formula can be used to find P_1^{sat} and P_2^{sat}.

$$ln\, P_1^{sat} = A_1 - \frac{B_1}{T_{Dew} + C_1} \qquad \text{(Eq 3.1)}$$

$$ln\, P_2^{sat} = A_2 - \frac{B_2}{T_{Dew} + C_2} \qquad \text{(Eq 3.2)}$$

And if we assumed $x_1 = 1$, then from Eq 1.1, $P_1^{sat} = y_1 \times P$. Eq 3.1 can be rewritten to find T_{Dew} as:

$$T_{Dew} = \frac{B_1}{A_1 - ln\, P_1^{sat}} - C_1 \qquad \text{(Eq 4)}$$

For a binary mixture of ethylene (component 1) and nitrogen (component 2) at a constant pressure and varying ethylene composition, Eq 4 can be used to calculate T_{Dew}, which represents the saturation temperature of ethylene at its partial pressure. This value is mistakenly assumed to be the minimum temperature of the mixture of flashing ethylene and nitrogen.

Results and Discussion

Adiabatic Mixing: Scenario 1.a

In scenario 1.a, supercritical ethylene at a pressure of 8000 kPa-g and a temperature of 10°C is flashed and introduced into a nitrogen-filled line at different pressures ranging from 50-5000 kPa-g. When the supercritical ethylene is throttled to the initial condition of the nitrogen pipeline, which has a pressure ranging from 50-5000 kPa-g, the throttled ethylene becomes a two-phase system with a temperature range of -97 °C to 3 °C, depending on the throttled pressure. When this two-phase ethylene is mixed adiabatically with nitrogen, which is in the vapor phase at a pressure of 50-5000 kPa-g and a temperature of 10 °C, the liquid portion of the two-phase ethylene wants to evaporate to go to vapor phase because the partial pressure of ethylene in the vapor phase decreases as a result of adding vapor nitrogen. As the mixing is adiabatic, the required energy for evaporation of liquid ethylene is absorbed from the mixture, resulting in a reduction of the temperature of the mixture. Depending on the ratio of flashed ethylene and nitrogen being mixed and at a constant mixing pressure, the available enthalpy for evaporating the liquid phase of the ethylene is different, resulting in complete or partial evaporation of the liquid ethylene. Figure 4 shows the mixture temperature profile and vapor mass fraction results versus ethylene mole fraction for scenario 1.a at different initial pressures of the nitrogen line. The results show that for all initial pressures of nitrogen, increasing the mole fraction of ethylene in the mixture up to a certain mole fraction will provide enough available energy to evaporate all the liquid ethylene, resulting in the mixture being fully in the vapor phase, and the temperature decreasing as the mole fraction of ethylene increasing up to this certain point. However, beyond a certain point, there is not enough available enthalpy for evaporating all the liquid ethylene, and some liquid ethylene remains, resulting in a two-phase mixture. This defining point is where the minimum temperature of the mixture occurs. By further increasing the ethylene mole fraction beyond this certain point, part of the liquid ethylene evaporates, and the final mixture temperature is still less than the throttled ethylene and nitrogen at that pressure. However, the temperature profile versus the mole fraction of ethylene has an increasing behavior.

For all the mixture temperature profiles at different initial pressures of nitrogen, there is a minimum temperature at different mole fractions of ethylene that corresponds to where the mixture transitions from a single vapor phase to a two-phase mixture. The results show that this minimum temperature decreases as the initial pressure of the nitrogen decreases. The hypothetical MATs curve is plotted in Figure 4, which corresponds to -29 MDMT at pressures of 50-5000 kPa-g. Each of these mixture temperature profiles needs to be compared with the associated MAT at a specific pressure to analyze the chance of brittle fracture. As shown in Figure 4, a portion of the temperature profile for an initial pressure of 50 kPa-g is below the MAT at 50 kPa-g, introducing the risk of brittle fracture. By knowing the MAT, the minimum allowable pressure to avoid brittle fracture can be calculated. For example, for an MAT temperature of -102°C, the minimum allowable initial pressure of nitrogen is 160 kPa-g such that the minimum temperature of the process is just above the MAT. It should be emphasized that all the MATs reported in this study are assumed values and are for illustration purposes to show the analysis approach.

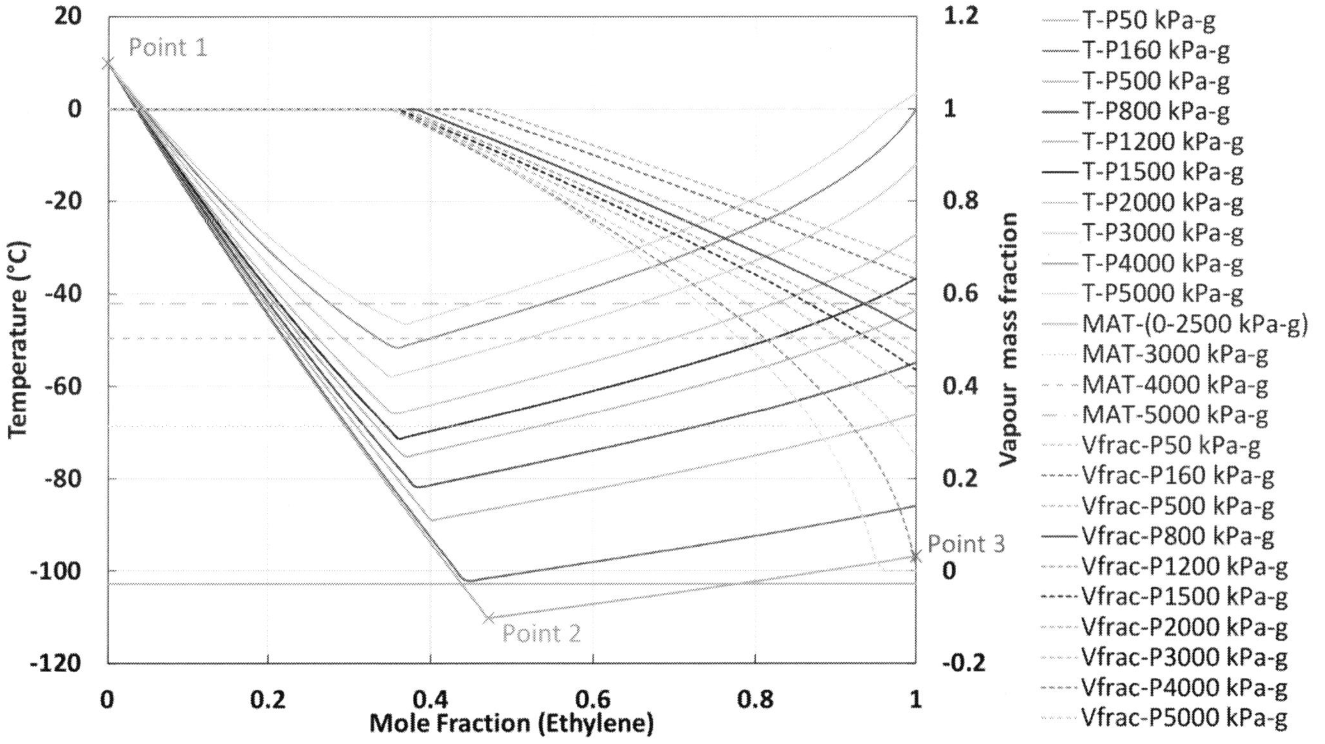

Figure 4. Temperature and vapor phase mass fraction versus mole fraction of ethylene in venting case. Point 1 (Pure nitrogen), Point 2 (minimum temperature), and pPint 3 (pure ethylene) is shown as an example for initial pressure of nitrogen at 50 kPa-g.

When there is a shut-in case, all mixtures in sequential batches with specific mole fraction of ethylene and previously calculated temperature (points in temperature profile of Figure 4) should be pressurized to a final pressure of 8000 kPa-g. To illustrate pressurizing result more clearly, three sets of points from Figure 4 were selected to show the pressurizing procedure: point 1 (pure nitrogen), point 2 (the point associated with lowest temperature) and point 3 (pure ethylene). The resulting temperature profile versus pressure for these different sets of points are plotted and compared with the MAT associated with MDMT of -29, -40, and -46 °C as can be shown in Figure 5. For pure nitrogen and ethylene, the CET curves are above the MAT curves, which shows there will be no concern with the brittle fraction. However, for point 2, the minimum temperature curves, there are some conditions where the CET curves cross or are below the MAT curve, and as a result there will be risk of brittle fraction.

The result of this Section suggests that there are two criteria to be observed when analyzing the best practice for re-inventorying/pressuring the nitrogen filled line with supercritical ethylene and analyzing the brittle fraction. The first is to make sure that the initial pressure of the nitrogen is above the minimum allowable pressure such that the temperature profile versus the mole fraction of ethylene is above the MAT at the venting pressure. The second is to watch for the temperature versus pressure (CET curve) to be above the MAT and not cross that during pressurizing in the shut-in case.

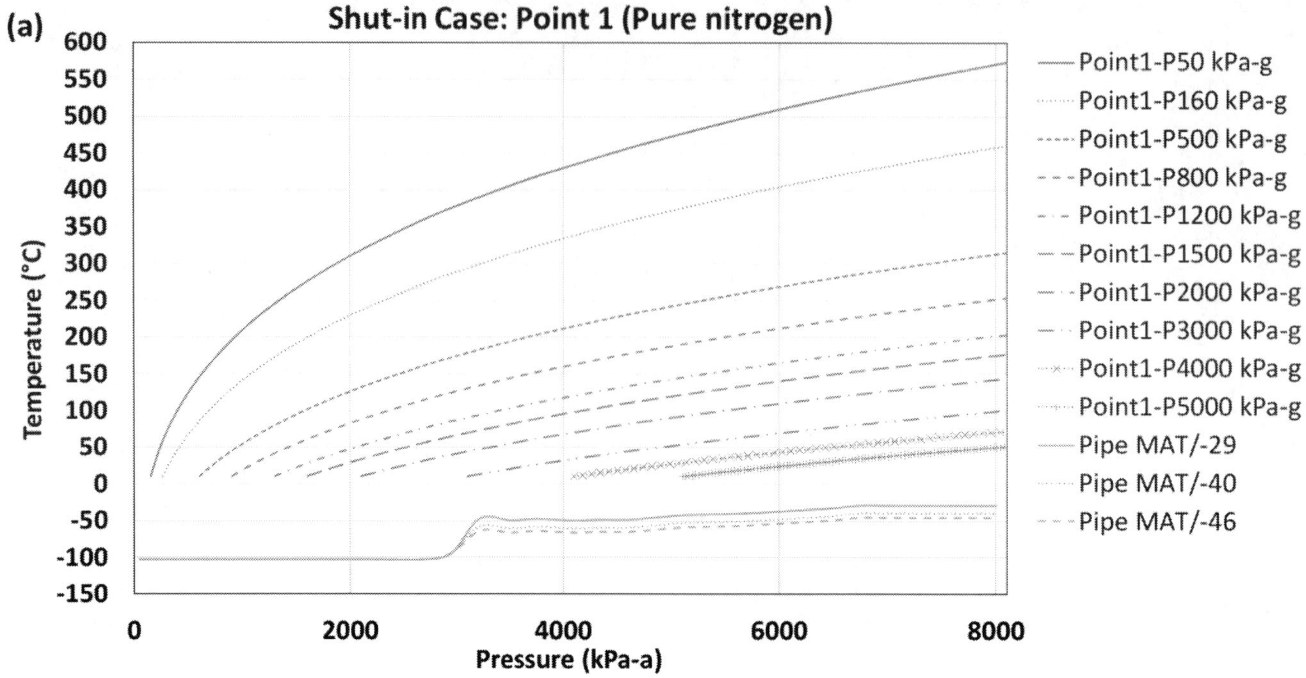
(a)
Shut-in Case: Point 1 (Pure nitrogen)
Temperature (°C)
Pressure (kPa-a)
Point1-P50 kPa-g
Point1-P160 kPa-g
Point1-P500 kPa-g
Point1-P800 kPa-g
Point1-P1200 kPa-g
Point1-P1500 kPa-g
Point1-P2000 kPa-g
Point1-P3000 kPa-g
Point1-P4000 kPa-g
Point1-P5000 kPa-g
Pipe MAT/-29
Pipe MAT/-40
Pipe MAT/-46

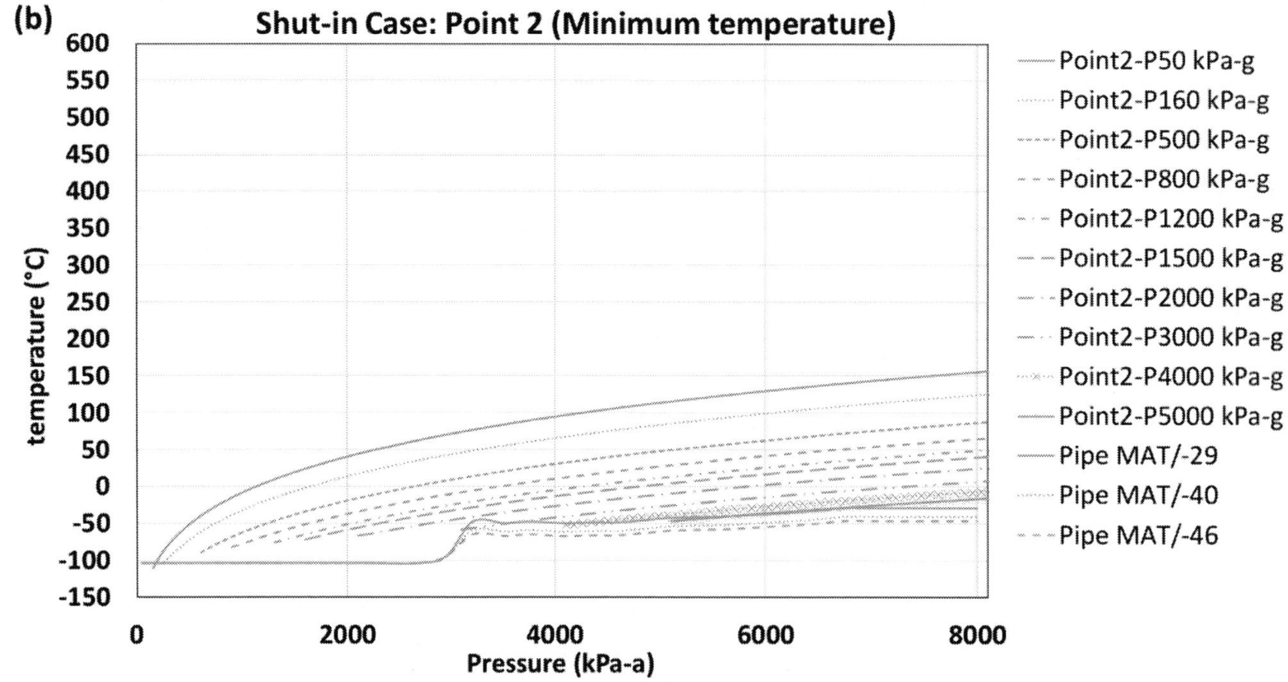
(b)
Shut-in Case: Point 2 (Minimum temperature)
temperature (°C)
Pressure (kPa-a)
Point2-P50 kPa-g
Point2-P160 kPa-g
Point2-P500 kPa-g
Point2-P800 kPa-g
Point2-P1200 kPa-g
Point2-P1500 kPa-g
Point2-P2000 kPa-g
Point2-P3000 kPa-g
Point2-P4000 kPa-g
Point2-P5000 kPa-g
Pipe MAT/-29
Pipe MAT/-40
Pipe MAT/-46

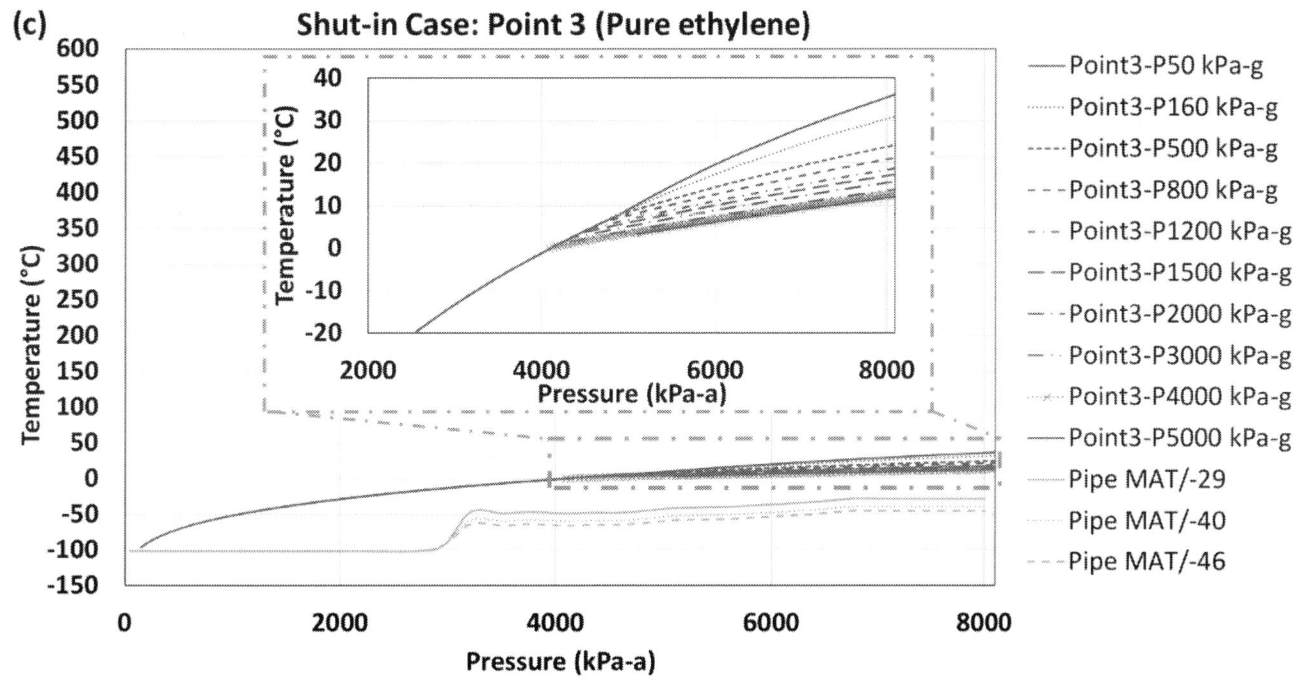

Figure 5. (a) Point 1 pressurizing profile, (b) Point 2 pressurizing profile, and (c) Point 3 pressurizing profiles from initial pressure of nitrogen to 8000 kPa-g.

Adiabatic Mixing: Scenario 1.b

To understand the sensitivity of the mixture temperature to the initial temperature of nitrogen and ethylene, we repeated the same analysis as in Scenario 1.a, but with an initial temperature of 0°C instead of 10°C, for an initial nitrogen pressure of 1500 kPa-g. Figure 6.a shows the temperature and vapor mass fraction versus mole fraction of ethylene in the mixture. By comparing the temperature profile in Figure 6.a with that in Figure 4 for an initial nitrogen pressure of 1500 kPa-g, it is observed that the minimum temperature decreased by 5°C, from -71°C for an initial temperature of 10°C to -76°C for an initial temperature of 0°C. In addition, Figure 6.b depicts the temperature versus pressure profile (CET) in the shut-in case. Comparing Figure 6.b and Figure 5.b, we can see that at a pressure of 1500 kPa-g, decreasing the initial temperature of ethylene and nitrogen from 10°C to 0°C leads to a lower Point 2 (minimum temperature) temperature profile that is closer to the MAT curve. For instance, the minimum temperature difference between the MAT curve and the Point 2 pressurizing in Figure 5.b at the initial pressure of 1500 kPa-g is approximately 20°C, while in Figure 6.b, it is around 16°C. These findings demonstrate the impact of the initial temperature on the temperature profile of the mixture, highlighting the need for careful consideration of initial conditions during pressurization and re-inventorying of nitrogen-filled lines with ethylene.

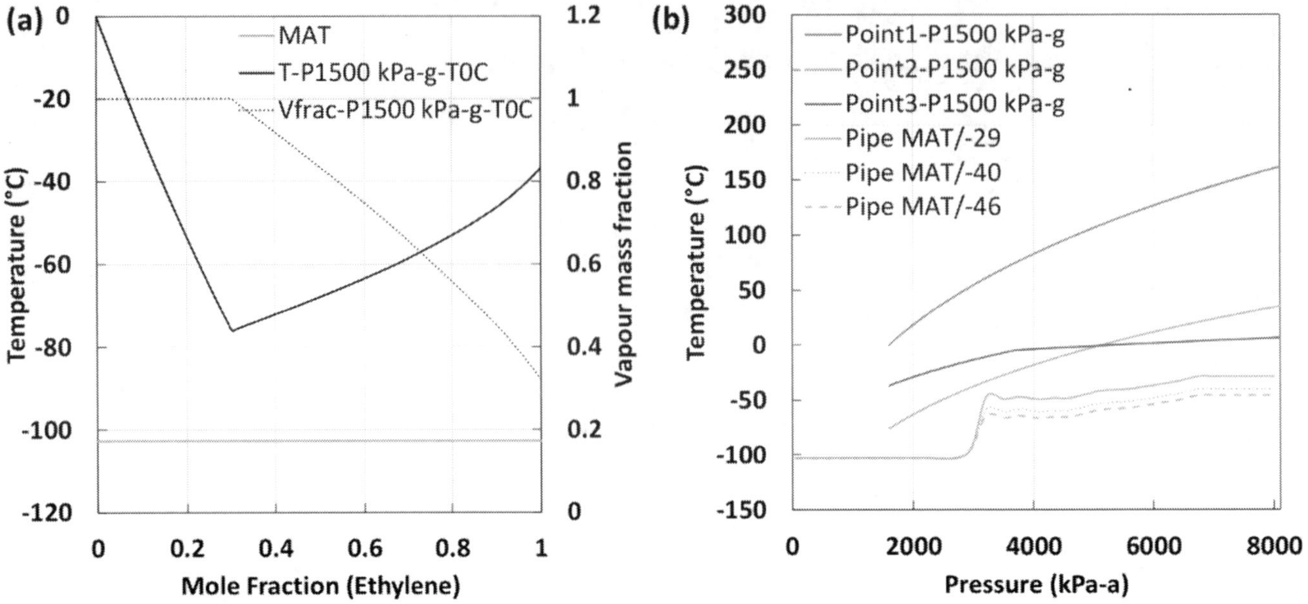

Figure 6. (a) Temperature and vapor mass fraction profile versus mole fraction of ethylene by adiabatic mixing of throttled ethylene and nitrogen at 1500 kPa-g in venting case, and (b) temperature versus pressure profile in shut-in case.

Adiabatic Mixing: Scenario 2

Scenario 2 involves throttling the vapor ethylene and introducing it into a pipe containing vapor nitrogen. Figures 7a and 7b show the results for the venting and shut-in cases, respectively. Unlike Scenario 1, there is no liquid ethylene to evaporate and cause a minimum temperature behavior. As a result, the temperature profile of the mixture in Figure 7a shows a decreasing trend with respect to the mole fraction of ethylene. The throttled ethylene, with a temperature of -12.8°C, is adiabatically mixed with nitrogen at 10°C at different ratios, resulting in a mixture temperature that falls between -12.8 °C and 10 °C, depending on the ratio of ethylene to nitrogen. The shut-in case results in Figure 7b show that all CETs are above the MAT curve, indicating no risk of brittle fracture. The results suggest that greater caution is necessary when re-inventorying nitrogen lines with ethylene in the supercritical region compared to the vapor region. This is because when ethylene is initially in the supercritical region and is throttled, it transitions to the two-phase region and its mixture temperature with nitrogen is further reduced due to the evaporation of its liquid phase.

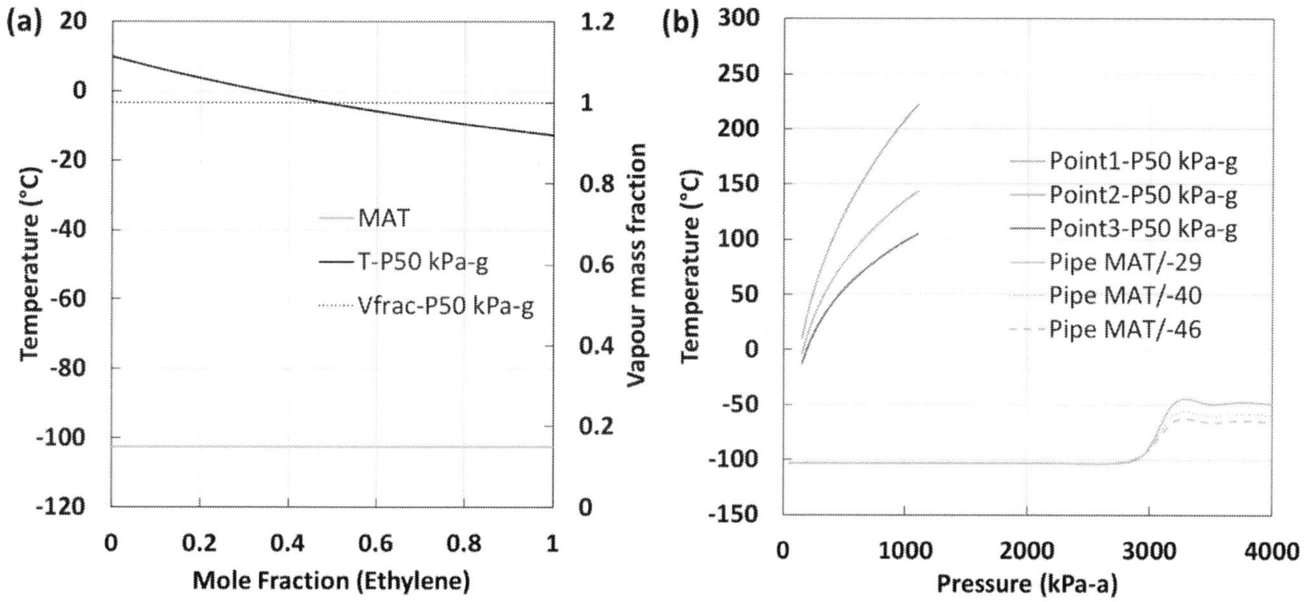

Figure 7. (a) Temperature and vapor mass fraction profile versus mole fraction of ethylene by adiabatic mixing of throttled ethylene and nitrogen at 50 kPa-g in venting case, and (b) temperature versus pressure profile in shut-in case.

Misconception of Partial Pressure:

To compare the misconception temperature with the actual mixture temperature obtained from the adiabatic mixing of ethylene and nitrogen, T_{Dew} can be plotted against ethylene composition. Table 2 summarizes the different conditions considered in this study for the comparison.

Table 2. Cases investigated in this study for misconception temperature.

	(a)	(b)	(c)	(d)
P_1 (kPa-g)	8000	8000	8000	8000
T_1 (°C)	10	10	10	10
$P_{Ethylene} = P_{2i}$ (kPa-g)	50	1500	2000	3000
$T_{Ethylene}$ (°C)	TBC*	TBC	TBC	TBC
P_{2i} (kPa-g)	50	1500	2000	3000
T_{2i} (°C)	10	10	10	10
P_2 (kPa-g) Venting	50	1500	2000	3000
T_2 (°C)	TBC	TBC	TBC	TBC

*TBC: to be calculated

The temperatures of the ethylene and nitrogen mixture, calculated based on adiabatic mixing at the conditions specified in Table 2, were compared with the dew and bubble curves calculated in Aspen HYSYS, as well as the misconception temperatures at different pressures calculated based on the procedure explained in the methodology section, as shown in Figure 8a-d. The results reveal a deviation between the misconception temperature and the calculated mixture temperature based on adiabatic mixing, especially at ethylene mole fractions where not all the liquid ethylene has evaporated. At mole fractions where there is a two-phase mixture, this deviation is smaller and closer to the dew point of the mixture. Moreover, as the pressure increases from 50 kPa-g (Figure 8a) to 3000 kPa-g (Figure 8d), the deviation between the misconception temperature, dew curve, and the mixture temperature based on adiabatic mixing at the two-phase region appears to be increasing.

These findings emphasize the need to distinguish between the misconception temperature and the actual mixture temperature when assessing the risk of brittle fracture as a result of auto-refrigeration during the re-inventorying of nitrogen-filled lines with flashing ethylene. For example, as shown in Figure 8.b, at the initial nitrogen pressure of 1500 kPa-g, the actual mixture temperature is consistently higher than the assumed MAT curve at that pressure, regardless of the ethylene mole fraction. Thus, relying on the misconception temperature profile below the MAT in Figure 8b, especially at ethylene mole fractions below around 0.1, could lead to unnecessary safety measures. Moreover, since the ethylene mole fraction in the pipe during re-inventorying varies, it is impractical to determine a single mole fraction for calculating the misconception temperature. This issue further underscores the limitations of using the misconception temperature as the minimum possible temperature during re-inventorying. However, the misconception temperature at the transition point between a single vapor phase and a two-phase mixture is closer to the actual mixture temperature compared to other mole fractions. Thus, if this mole fraction at transition is known, the calculated misconception temperature maybe used as the minimum possible temperature during the process.

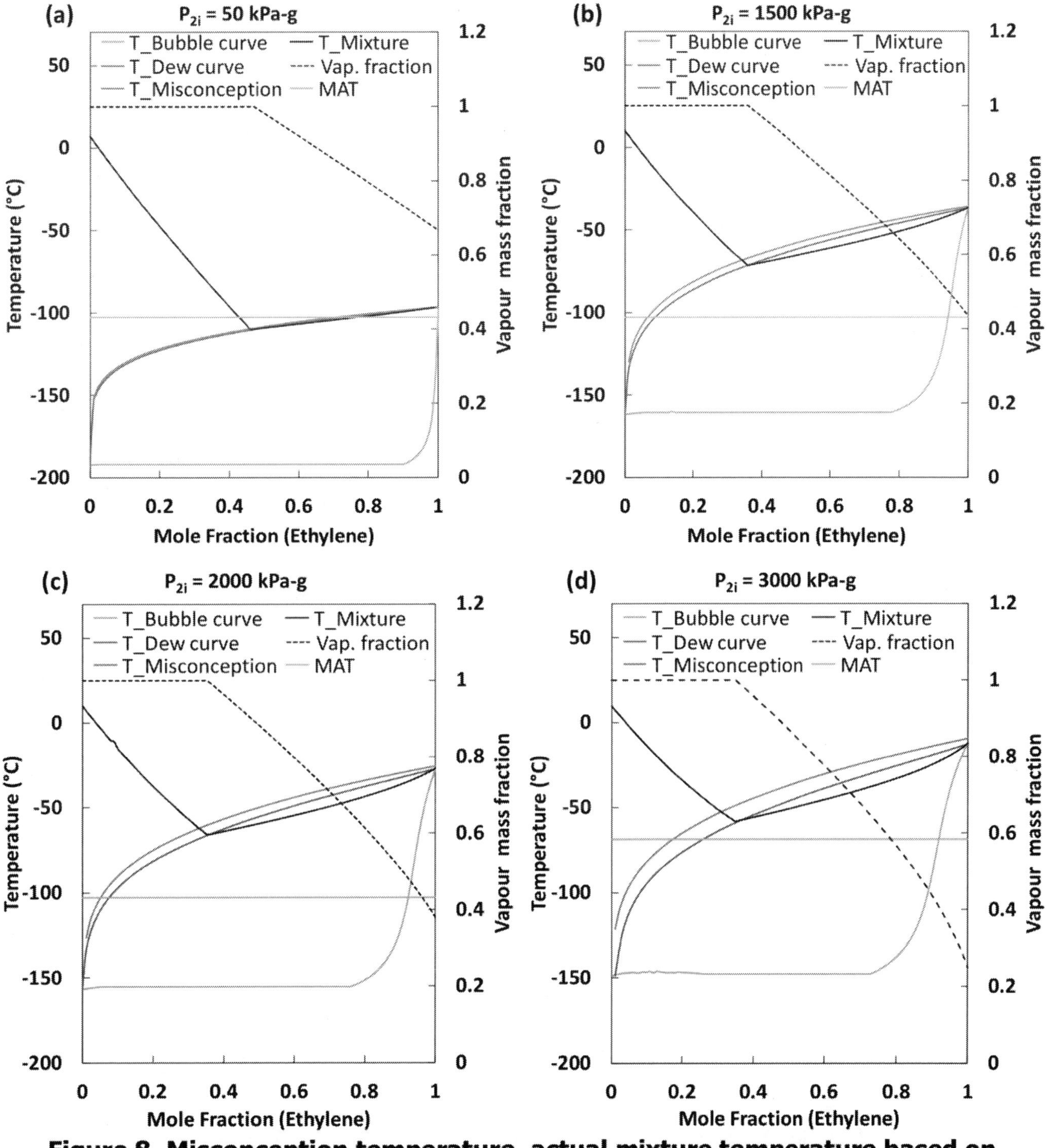

Figure 8. Misconception temperature, actual mixture temperature based on adiabatic mixing, dew temperature, and bubble temperature curve versus mole fraction of ethylene at constant pressure of (a) 50 kPa-g, (b) 1500 kPa-g, (c) 2000 kPa-g, and (d) 3000 kPa-g. The MAT is based on MDMT of -29 °C.

Conclusion

In this study, simulations were conducted in Aspen HYSYS to investigate the effect of initial temperatures and pressures of nitrogen and ethylene during the pressuring/re-inventorying of a nitrogen-filled line with ethylene under both isobaric (venting) and isochoric (shut-in) conditions on auto-refrigeration that can result in brittle fracture. Two main scenarios were studied: introduction of supercritical ethylene and introduction of vapor ethylene into the nitrogen-filled line. For the first scenario, a minimum temperature behavior was observed, dependent on the initial nitrogen pressure. A sensitivity analysis showed that decreasing the initial temperature of nitrogen and ethylene decreases the minimum temperature, increasing the chance of brittle fracture. For the second scenario, where the feed ethylene pressure is relatively lower (1000 kPa-g) no minimum temperature behavior was observed, and the temperature of the mixture decreased with increasing ethylene mole fraction. The study emphasizes the importance of two criteria: ensuring the initial nitrogen pressure is above the minimum allowable pressure for a specific MAT curve and observing the CET curve to be above the MAT during pressurizing in the shut-in case. Last but not least, results also showed a deviation between the misconception temperature and the calculated mixture temperature based on adiabatic mixing, with the deviation increasing as initial pressure of nitrogen increased. The study highlights the need to distinguish between the misconception temperature and the actual mixture temperature when assessing the risk of brittle fracture, as relying on the misconception temperature could lead to unnecessary safety measures.

References

1. Benac DJ, Cherolis N, Wood D (2016) Managing Cold Temperature and Brittle Fracture Hazards in Pressure Vessels. Journal of Failure Analysis and Prevention 16:55–66. https://doi.org/10.1007/S11668-015-0052-3/FIGURES/16

2. Bernard A (2015) Evaluating the potential for cold brittle fracture in ethylene plant through dynamic simulation. In: Ethylene Producer Conference, 2015 AIChE Spring Meeting. Austin, TX

3. The American Society of Mechanical Engineers (2020) ASME Code for Pressuring Piping, B31.8-2020

4. Koerts T (2012) Auto Refrigeration and Metal Embrittlement. In: AIChE Spring Meeting. Houston, TX

5. The American Society of Mechanical Engineers (2021) Fitness for Service (2021): API 579-1/ASME FFS-1

6. Thompson CR, Korst MW (2014) PHA Methodology and Training Practices Addressing Auto-Refrigeration Brittle Fracture Hazards - 25 Years Later. In: Ethylene Producer Conference, AIChE Spring Meeting and Global Congress on Process Safety. New Orleans, LA

7. Werner AV (2014) Prevention of Brittle Fracture Scenarios in Olefins Plants . In: Ethylene Producer Conference, 2014 AIChE Spring Meeting. New Orleans, LA

8. Kennedy MJ (2005) Auto-refrigeration Near-Miss Incident at Equistar Chemicals. In: Ethylene Producer Conference, 2005 AIChE Spring Meeting. Atlanta, GA

9. King RE (2007) Auto-Refrigeration/Brittle Fracture Analysis of Existing Olefins Plants—Translation of Lessons Learned to Other Processes. J Hazard Mater 142:608–617. https://doi.org/10.1016/J.JHAZMAT.2006.06.066

10. Sandler SI (1999) Chemical and engineering thermodynamics, Third edition. Wiley

Gap in pagination due to formatting issues.

Pages 312-335

AIChE Paper 124e

Planning an Olefins Unit Startup with an Operating Training Simulator as Part of a Plant Revamp Commissioning Strategy

André Bernard
Process Design Engineer
NOVA Chemicals Corporation

Graham Riddell
Training Specialist
NOVA Chemicals Corporation

Prepared for Presentation at the 2023 Spring National Meeting
Houston, TX, March 12th – March 16th, 2023

AIChE Paper 124e

Distillation in Ethylene Plants – An Operators' Overview

André Bernard
Process Design Engineer
NOVA Chemicals Corporation

Graham Riddell
Training Specialist
NOVA Chemicals Corporation

Abstract

Startup strategies, sequences and procedures were developed using an Operating Training Simulator (OTS) in preparation for starting up an olefins plant undergoing a significant revamp. The revamp involved the replacement of internals on numerous towers, re-wheel of compressors and a non-negligible amount of process integration. Most of those modifications being implemented during a major plant turnaround. This paper discussed the challenges of validating the OTS with pre-turnaround operating data no longer reflecting the revamped unit and process design simulations. A significant portion of this work was conducted during COVID-19 restrictions and added on the complexity. Execution of the actual startup is also discussed and compared to the expected sequence.

Introduction

Construction of a new polyethylene facility by NOVA Chemicals Corp. in the Sarnia, Canada area has driven the need for expanding the existing NOVA Corunna's ethylene cracker capacity by about 50%. The Corunna facility was originally design, build, and operated as a naphtha cracker in the late 70's. Through several revamp and expansion projects, feed flexibility was expanded over the years to natural gas liquid (NGL). In the last decade, the plant was converted to 100% ethane cracking with limited capability for propane and butane.

Beside additional cracking furnaces for the capacity increase, significant modifications were implemented in the recovery area. Multiple towers were re-trayed, 2 compressors were re-wheeled, and one replaced. A number of process integration were introduced and expected to change the operating and startup philosophy of the plant.

An Operating Training Simulator (OTS) was already in service for operator training at the site and had provided value since its introduction. The OTS system update requires to reflect the changes intended was included in the expansion project scope. Beside training operators in preparation for operating a new or revamp facility, OTS have been proven effective in validating and speeding up commissioning of control systems if properly scheduled. The resolution and realism of an OTS strongly depends on the data provided for its configuration. Development of an OTS for a new facility or updates of an existing one during the engineering phase of a plant revamp rely mostly on process design simulations. Although, existing operating data from similar or Pre-Revamp facilities can be use as reference, caution and judgment must be applied when extrapolating the behavior of the OTS to the actual plant.

A team of process engineers and operation coordinators started working on the OTS validation and development of the startup procedures 1 year ahead of the plant shut down needed for completing the construction activities required by the revamp project and plant turnaround maintenance work. The magnitude of the changes implemented by the revamp project far exceeds the scope of this paper. As such cases studies for the discussion will be limited to the propylene refrigeration system and parallel C2 Splitter startup. Those examples illustrate transient conditions in alignment and deviations from the OTS predicted behavior. It discusses how the startup team was able to recognize the symptoms and react to undesired transient that were threatening startup progress.

Expansion Project Scope Summary

The incremental capacity required the addition of 2 new cracking furnaces in the front end of the plant. Previous conversion of the plant to 100% ethane cracking left sufficient capacity in the quench system and as such required only minor changes.

The Crack Gas Compressor (CGC) 5th stage was re-wheeled. The Propylene Refrigeration Compressor (PRC) was re-wheeled, and the 3rd stage suction converted from a side load to an extracting stage. The Ethylene Refrigeration Compressor (ERC) was replaced by a new unit.

The Demethanizer and Deethanizer towers were re-trayed. The plant operates 2 parallel C2 Splitters, with the primary tower already equipped with high-capacity trays. The feed point was raised in the Primary tower internally, which required modifications of 10 trays around the feed location. A secondary side reboiler sub-cooling propylene refrigerant (C3R) was added to the tower and would operate in parallel with the existing primary side reboiler cross exchanging with crack gas. The parallel tower, equipped with conventional 2 pass trays has been re-trayed with high-capacity trays. A new side reboiler cross exchanging with ethane feed was also installed.

Numerous changes on heat exchangers through additional units and replacements along with significant piping changes were also implemented. As a result of those changes the plant behavior was expected to be different.

Operating Training Simulator (OTS) - Overview

Operation of a chemical process involves direct actions (inputs) which results in an outcome (outputs). Some of those inputs are directly initiated through human intervention or driven by a controller configured for meeting a desired outcome. Those outcomes manifest themselves in the form of flow, pressure, temperature, and compositions, which are measured by devices that correlates their physical value to a standard electric signal of 4 to 20 mA. The electric signal must then be converted to a digital signal that is compatible with a process computer. Similarly, inputs to the process initiated by a process operator or resulting from a PID controller calculated output, interface with the process through a digital signal that gets converted to an analog electrical signal of 4 to 20 mA. The electrical signal typically feeds the actuator of a control valve, which set the valve position at the value set by the process operator or output calculated by a PID controller. Figure 1 provide a high-level architecture of the Human/ Computer/ Plant interface for a plant and OTS environment. In an OTS, the process and related instrumentations are substituted by a computer model that emulate real plant behavior through first principle mathematical models.

Not shown on Figure 1 are control layers involving advanced process control (APC) and online optimizers that can always be integrated to the OTS system. Nevertheless, APC and optimizer minimizes the need for human intervention and would offer little value relative to operator training. Furthermore, OTS are frequently use for assessing startup and response to abnormal conditions, which are circumstances in which APC and optimizers are normally turned off. As such, those control layers are rarely included in an OTS system. PLC's which provides safety protection shall be physically included or emulated in the process model. Data storage, while nice to have for reviewing transient conditions and training performance are not always part of an OTS system.

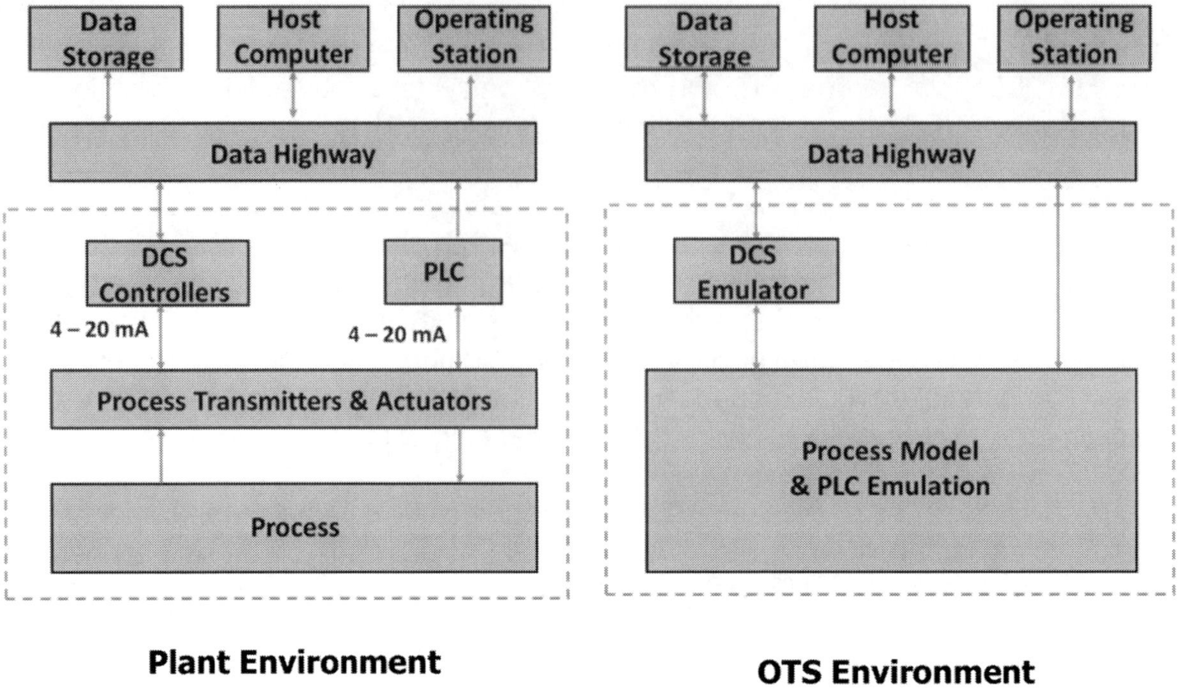

Figure 1: Process/ OTS – Control System Architecture

Operator training simulators (OTS) have proved successful in maintaining the skills required by board operators to fulfill their tasks. Doe[1] and Besuijen[2] have previously described a training structure with OTS system for qualifying/ re-qualifying process operators. They also described the benefits of OTS for evaluating startup strategies and response to abnormal conditions. With typical run length of more than 5 years in ethylene plants, exposure to startup and shutdown becomes limited. While abnormal situations could always arise, they are infrequent due to proactive and ongoing assessment of plants vulnerability through process hazard analyses (PHA) and maintenance/ reliability programs. As such, board operators have limited exposure to those special events. The OTS training structure described by Doe and Besuijen allows preservation of those skills. Their training structure recognize modeling limitations as the introductory segment described, touch on what the OTS can and cannot do.

OTS are typically developed from a steady state simulation model and process equipment drawings and datasheets. The resolution of the model is often improved through frequent utilization and observed behaviors deviating from plant experience. In the case of a significant expansion project associated with a revamp OTS, the resolution of the model depends solely on the design process simulations and equipment drawings and data sheets. Real plant behaviors are still to be experience. This, therefore, requires judgment in differentiating process comportments caused by model limitations and/ or inaccuracies from expected plant response.

Operating Training Simulator (OTS) Typical Modeling Limitations

OTS have been around for about 30 years. Early models relied extensively on simplifications due to computer limitations and the desire to build robust models. Improvements in computer capabilities over the years have enhance the resolution of OTS models significantly. However, as for most process simulator, there are still limitations that the user must be aware of, such as:

- Knock Out Drums often modeled by flash blocks assumed perfectly mixed environment.
- Distillation tower depending on software and configuration may
  - Not provide flow reversal.
  - Assume constant efficiency regardless of loads.
  - Flooding, weeping downcomer sealing is typically inconsequential in the model.
  - Thermosyphon reboiler are not limited by turndown.
- On saved conditions, shut down streams typically retain process value from the time at which the stream flow gets interrupted (pressure, temperature, composition).
- Movement of material from one process unit to the next can either be flow or pressure driven depending on model configuration. Capability for flow reversal may or may not be configure and could challenge model robustness if required.
- Piping holdup is often imbedded in process vessels, which may affect process lags and controller dynamics.

The modeler may often introduce simplifications intended to build robustness and avoid model crashes. In other cases, process model deviations from PFD's/ P&ID's might be require for effectively reproducing plant behaviors. For instance, as shown on Figure 2, compressor kick-back streams intended to provide anti-surge protection, while entering the knockout drums in the plant P&ID's may have to join the vapor outlet of the flash block in the model. As indicated above, flash blocks typically assumed a perfectly mixed environment and as such in the presence of liquid holdup would cause the vapor stream leaving the vessel to be at saturated conditions. In reality the hot vapor stream, while causing some liquid flash, exit the vessel at a warmer temperature than saturation. Creating that warmer vapor stream temperature is also critical for achieving realistic liquid quenching that is typically part of surge controls in refrigeration systems.

It is important to understand that distillation unit in an OTS model will typically achieve fractionation regardless of hydraulic loadings. Configuration of internals provides hold up and pressure drop calculations. However, fractionation efficiency, unless manipulated by some customize malfunctions, is insensitive to flooding, weeping and unseal downcomers. Failing to understand this model limitation could mislead process operators relative to effective tower operation, especially during startup transition.

The development and configuration of an OTS relies strongly on a steady state process simulation. In the context of an expansion project, the design simulation becomes the reference point and is develop using several assumptions and design practices. For instance, heat exchangers involve in condensing and vaporization services assume saturation conditions of the outlet stream. Thermosiphon reboilers are based on a desire vaporization and if condensing on the hot side assume saturated liquid in the outlet. Heat exchanger design will typically exceed the performance predicted by the design simulations as additional surface is normally provided. As a result, offset in degree of sub-cooling and superheating is expected when comparing the OTS to plant performance. Those offsets can be minimized once plant data becomes available for further model tuning but until then judgement must be applied relative to the potential impact of those offsets.

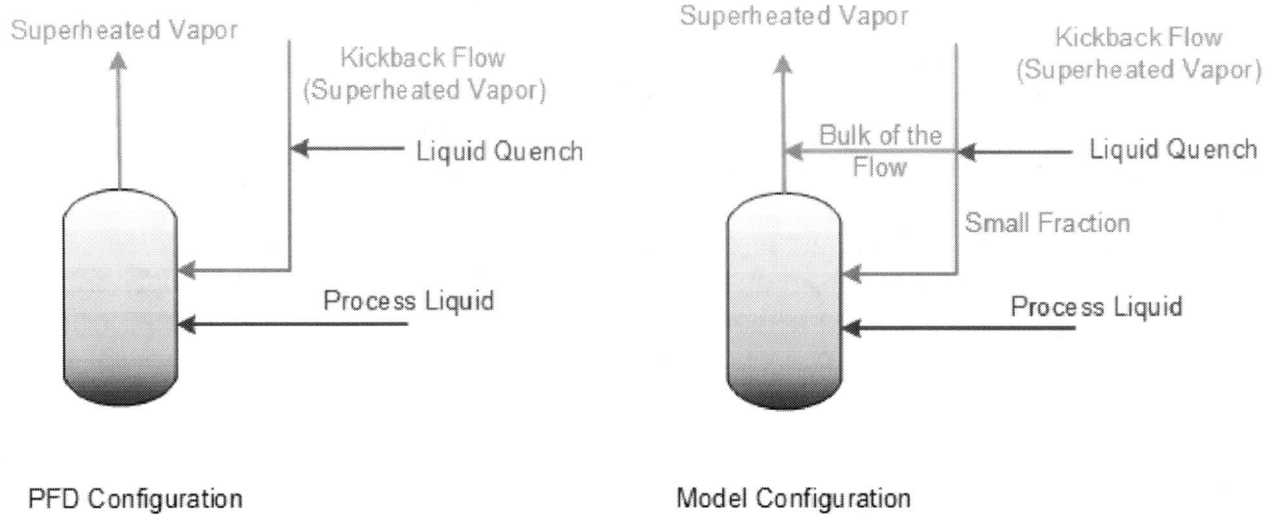

Figure 2: PFD vs Model Configuration Variation Example

Startup Procedure Development Strategy

Refrigeration Systems & C2 Splitter Total Reflux

The propylene refrigeration system (C3R) is the first unit being started and operated on circulation without any process loads. The only heat sink at this point is provided by fin fans and cooling water exchangers. As previously described, the main change on this system is the conversion of the 3rd stage suction to an extracting stage. Given that the plant was originally design as a naphtha cracker, the C3R system looks unconventional. As shown on Figure 3, rather than floating on compressor inter-stage pressure, the 3rd and 4th stage knockout drums are pressure controlled through split range controller that 1st regulate 3rd stage suction (1st extraction) and 4th stage suction (2nd extraction) control valves. Once at maximum extraction, additional propylene vapor is supplied by letting down 4th stage discharge to 3rd and 4th stage drums. The extraction lines are equipped with check valves preventing flow reversal to the compressor. As such, pressure dump valve between the 4th to 3rd, and 3rd to 2nd stage knockout drums are required to prevent over pressure caused by flashing liquids that cascades down from the accumulator to the 2nd stage knockout drums, when operating the C3R system on circulation without any process loads. The 4th to 3rd pressure dump valve was configured in a previous system revamp, in which the 4th stage suction was converted to extraction. The 3rd to 2nd stage pressure dump valve is new due to conversion of the 3rd stage suction to an extraction.

Prior to this revamp, startup of the C3R system was straight forward and required minimal technical support. For the sake of time, startup training on the OTS typically begins with the refrigeration systems operating on recirculation. Due to the modifications implemented by this revamp project, the startup team schedule OTS time dedicated to the C3R system startup.

As part of normal startup practice, the main C2 Splitter is normally started on total reflux. This practice, help the C3R system through the heat pump circuit created between the C2 Splitter reboiler and condenser. It also speeds up the time required for meeting ethylene specification once hydrocarbon feed is initiated to the cracking furnaces. The parallel C2 Splitter is put on total reflux later in the startup sequence, once plant rates reach level allowing effective parallel tower operation.

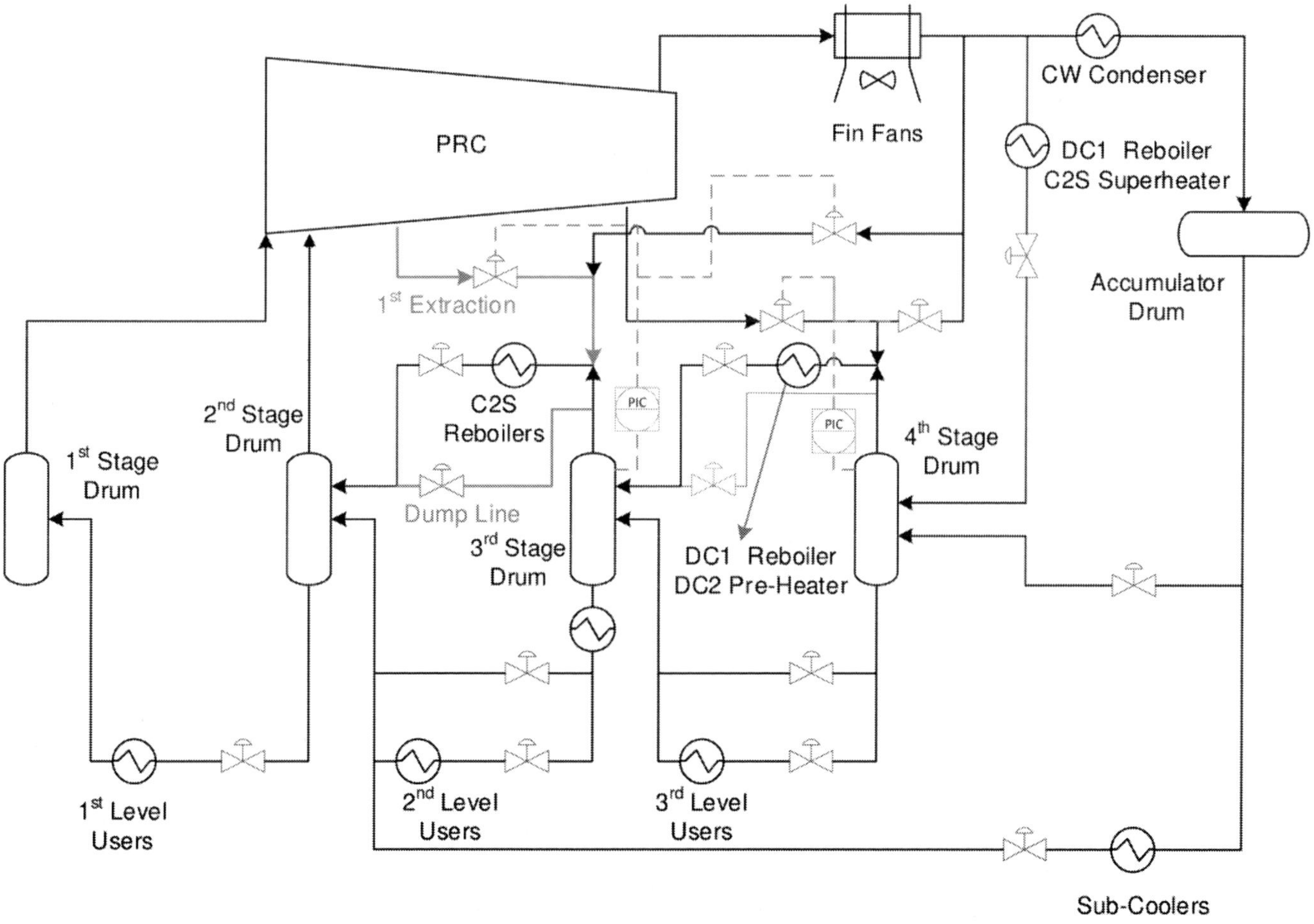

Figure 3: Propylene Refrigeration System (C3R)

The ethylene refrigeration system (C2R) is normally started shortly after stabilizing the C2 Splitter on total reflux. The startup sequence was expected to be similar to previous procedure as the system had been through only minor changes.

Chill Train & Towers

A first step in the development of the startup strategy was to identify pre-revamp startup issues and understand if they still applied. Difficulties in starting and sustaining operation of the Demethanizer Pre-Stripper reboiler had been experience in several pass startup and was expected to remain a challenge. Potential for cold temperature and over pressure was also a concerned. The approach promoted was based on the startup strategy described by Bernard[3], which as shown on Figure 4 leverage on the crack gas compressor (CGC) for purging nitrogen and pressurizing the chill train, Pre-Stripper, Demethanizer and Deethanizer with warm crack gas.

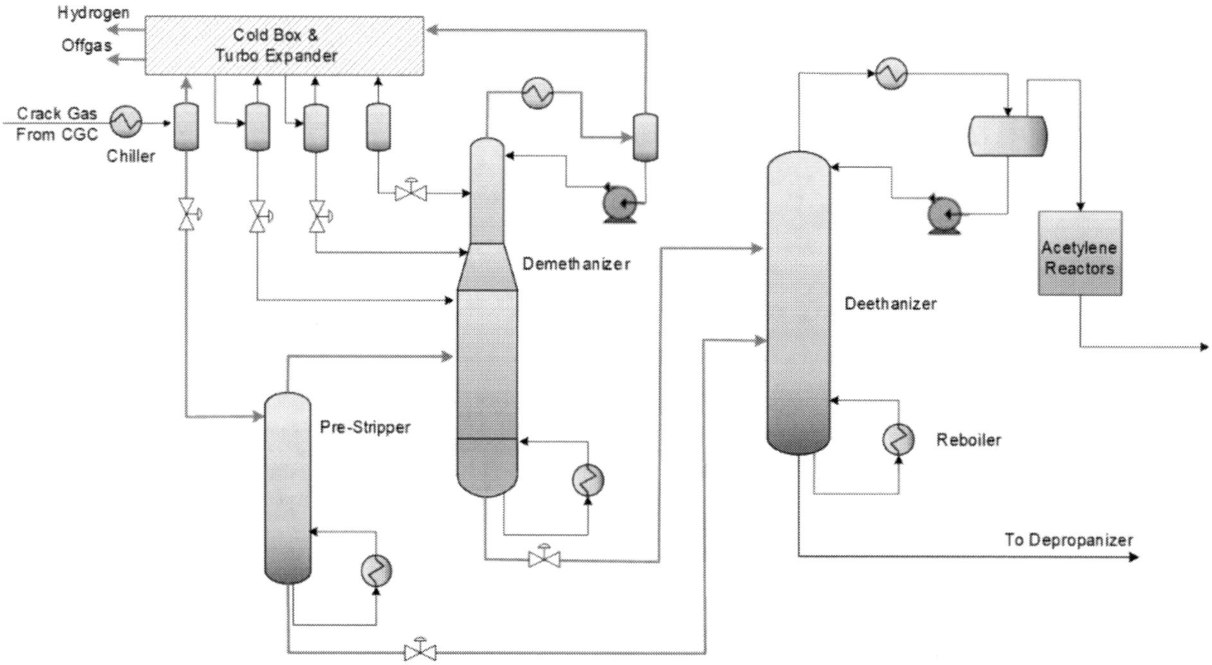

Figure 4: Chill Train, Pre-Stripper, Demethanizer, Deethanizer Pressurization

This sequence initially requires operating all the chillers and coolers upstream the 1st knockout drum without refrigerant and/ or cooling streams. The 1st knockout drum, Pre-Stripper and Demethanizer sump level control valves are in the fully open position at this stage. Doing so, allows displacement of nitrogen with warm crack gas flowing to the towers, hence avoiding cold temperature excursion cause by partial pressure imbalance when exposing liquid hydrocarbons to a nitrogen environment. A 30-minute purge is implemented prior to pressurizing the towers and cold box. The Deethanizer liquid drains are open to flare during the 30-minute purge to ensure displacement of nitrogen confined in the tower bottom. The liquid drains are close once the purge is complete and the pressurization step begins. As the Deethanizer pressure reaches its target, the level control valves between the Pre-Stripper, Demethanizer and Deethanizer are ramped in the close position. As the Demethanizer approaches its target pressure, refrigerant is introduced in the 1st level propylene refrigerant (C3R) chiller upstream the 1st knockout drum. At this point the crack gas flow path is transitioned to the next knockout drum by fully opening the level control valve of the 2nd knockout drum and pinching the level control valve of the 1st knockout drum to an output of about 5%. Once the crack gas temperature approached the refrigerant temperature, the crack gas flow path is transitioned to the 3rd stage knockout drum and the next refrigeration level commissioned (3rd level C2R). The refrigeration levels are sequence that way up to the 1st level C2R.

The above strategy beside preventing cold temperature excursion caused by liquid hydrocarbon/ nitrogen contact, speed up pressurization and provide some control on the cool down rate of the cold box by sequencing the refrigeration levels. With regards to Pre-Stripper reboiler issue, the composition of the first drops of liquid accumulating in the reboiler are more compatible from a boiling curve perspective with the hot stream supplying the reboiler then liquid that would accumulate at low pressure. Condensed liquid at low pressure would be heavier with the presence of nitrogen compounding this effect due to partial pressure imbalance causing further migration of light hydrocarbon to the vapor phase. As such the boiling curve of the liquid accumulating in the reboiler could get close to a temperature pinch with the hot stream causing the reboiler to stall. Note that at low pressure, once the chill train knock out drums fills with liquids, towers pressurization relies solely on the heat generated by the reboilers.

The hot streams to the Pre-Stripper, Demethanizer and Deethanizer reboilers are slowly commissioned as C3R is introduced in the chiller upstream the 1st knockout drum. The initial intent is to warmup the reboilers. Once cold liquid accumulates in the reboiler, the hot stream will start condensing as observed by liquid level rising in the reboiler condensate pot. The heat stream is then slowly increase up to its target value. The condensate pot level is used as reference relative to the heat stream ramp rate.

By looking at Figure 4, it is expected that by following this sequence, liquid will first accumulate in the Pre-Stripper and become available for feeding the Deethanizer. Ethane cracking yields generates less than 5% wt C3+ material. As such, if relying solely on forward feed, it may take a long time to build sufficient liquid inventory in the Deethanizer for stable reboiler startup. Furthermore, while the tower is already up to pressure and free of nitrogen, accumulation of a C2 rich mixture may result in cold temperature in the tower. Therefore, an external C3 stream is used for inventorying the Deethanizer sump prior to commissioning heat to the reboiler. The C3's are expected to be contained in the tower as reflux gets commissioned, until accumulation of C3+ material originating from forward feed cause the sump level to rise.

The above sequence occurs after startup of the Crack Gas Compressor (CGC). The startup procedure remains relatively similar to the pre-revamp era between the initial furnace feed and CGC startup steps. Flare points transitions from the Quench Water Tower overhead to the CGC discharge once the compressor goes to acceleration. Then the flare point transition from the CGC discharge to the Off Gas and Deethanizer overhead once hydrocarbons move forward to the chill train and towers. A key difference with the proposed sequence surrounding the chill train and towers lies around the magnitude of flaring and composition of crack gas travelling through the chill train and towers. The Off Gas flare valve is expected to handle the entire warm crack gas flow once the Deethanizer reaches operating pressure and the Pre-Stripper and Demethanizer level control valves gets close. The magnitude of the flow is expected to drop as refrigerant is introduce in the C3R chiller upstream the 1st knockout drum, but nonetheless the Off Gas flare valve must have sufficient capacity for this short period, or an over pressure excursion could be expected. Contingencies must therefore be developed to address this potential issue.

Acetylene Reactors

Crack gas use for purging nitrogen and pressurizing the tower as a high hydrogen content estimated at about 35% mole. As shown on Figure 4, the plant operates tail end acetylene reactors. The Deethanizer being initially inventoried with crack gas, has a high hydrogen content that must be flared upstream the acetylene reactors until it becomes hydrogen free. The flare point can then transition from upstream to downstream of the acetylene reactors. Premature transition can potentially result in runaway reaction in the acetylene reactor. Once the transition complete, the reactor is ready for hydrogen supply once the hydrogen stream from the chill train meets specification. At this point, the procedure become similar to the pre-revamp era and requires meeting the acetylene specification prior to feed the C2 Splitter.

Operating Training Simulator (OTS) Validation & Startup Procedure Testing

Factory Acceptance (FAT) & Site Acceptance (SAT) Testing

Due to COVID 19 restriction, the factory (FAT) and site acceptance testing (SAT) has been done remotely. The FAT was high level and focus on ensuring that the model met the intended scope. The startup team had less involvement during the FAT and did not physically operate the model. A maximum of 3 persons with face covering and social distancing restrictions were allowed in the simulator room during the SAT. The OTS supplier application engineer along with the rest of the startup team were attending remotely. Those physically present in the simulator room (referred as "test operators" in this discussion) had the ability to share one of the DCS screen to those attending remotely. The design heat and material balance was first compared against the model steady state conditions. The functionality of the model was then tested by performing the startup procedure. Problems related to model configuration were identified and resolve throughout the SAT. Problems believed related to the startup strategy were noted with resolution defer to a later time has the objective of the SAT was to validate and test the functionality of the OTS.

Startup Procedure Testing

Startup procedure testing began shortly after completion of the SAT, about 1 year prior to the scheduled plant startup that would follow the turnaround. Similar COVID 19 restrictions as for the FAT & SAT were in effect for the major portion of this work. The COVID 19 restrictions were relaxed just a few weeks before the plant turnaround. As such, except for those operating the OTS, the startup team view was limited to a single DCS page shared by the test operators. Good communication between the test operators and those participating remotely was key and effectiveness was dependent on the level of knowledge of the team members. Those involve in developing the startup sequence, conducting associated simulations, and establishing process targets had more facility in being actively involve and keeping track of progress. Those involve on a part time basis or with less insight in the revamp plant and/ or startup sequence felt more confused. However, they remained actively involve and their observations and comments provided a form of unbiased cold eye review.

Process targets associated with the startup sequence previously described were calculated through steady state simulations and equipment characteristics obtained from vendor. The sequence while understood and based on a number of steady state and dynamic simulations, still needed refinement and validation on the OTS. To effectively use OTS time, issues needed further investigation were capture on a log and worked outside the OTS schedule time.

OTS Startup Testing Key Learnings & Resolutions

Propylene Refrigeration Compressor (PRC) Startup

The PRC operation had been evaluated on Aspen Plus Dynamics™ standalone models prior to conduct the OTS startup rehearsal. The focus of that evaluation was to predict compressor power and operating points through the startup sequence and initial rate ramp up. Figures 5 and 6 summarizes the findings of that evaluation.

Those standalone models included similar surge protection kickback circuits as the plant. Surge protection of the PRC is ensured by 2 kickback circuits, 4th discharge to 1st stage suction and 4th discharge to 2nd stage suction. Those kickbacks circuits are regulated by a PLC that ensure operation above a minimum surge margin of 10%. As shown on the polytropic head curve illustrated on Figure 5, the standalone models suggests that by satisfying the 10% 1st stage surge margin, the 3rd and 4th stages surge margins gets well beyond 10%. In most cases the 2nd stage surge margins exceed 10% without requiring any flow through the 4th to 2nd stage kickback circuit. The operating points plotted on Figure 5 covers operation of the PRC without any process loads, C2 Splitter on total reflux and part of the incremental plant rates expected through the startup sequence. Figure 6 plots the predicted compressor power, 4th to 1st, 4th to 2nd kickback flows and compressor discharge temperatures for each data points plotted on Figure 5.

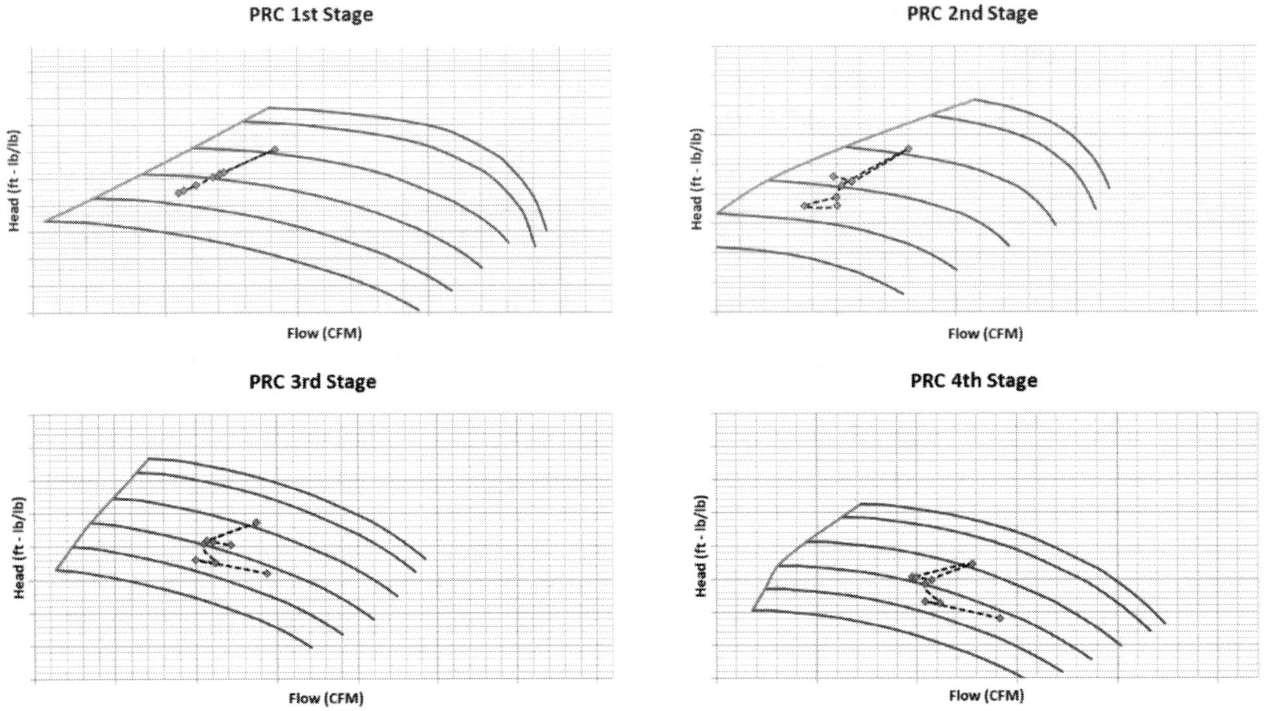

Figure 5: Propylene Refrigeration Compressor – Operating Points Predictions

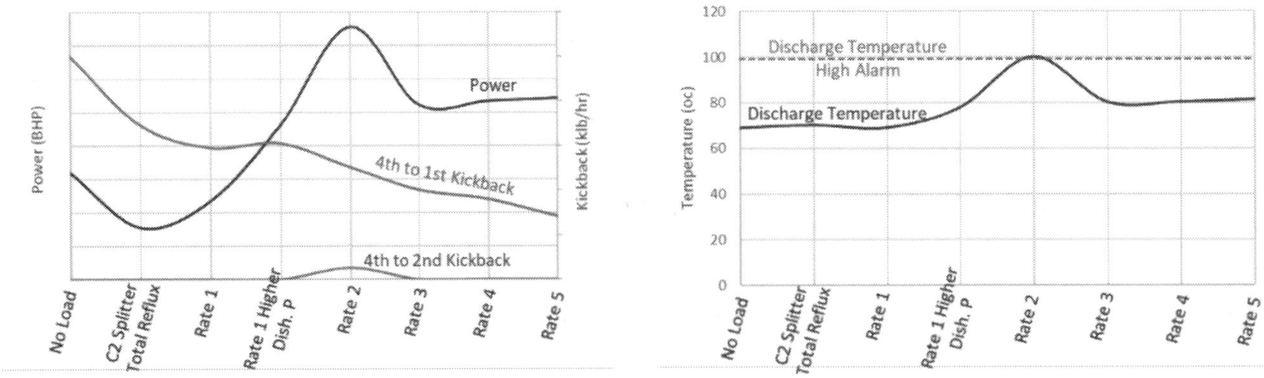

Figure 6: Propylene Refrigeration Compressor Power, Kickback Flows & Discharge Temperature

Those standalone models suggest a power reduction when transitioning the PRC from no load conditions to the C2 Splitter on total reflux. This reduction is expected as the heat pump circuit created by condensing C3R vapor from the 1st extraction in the C2 Splitter reboiler reduce the volume of C3R traveling through the 3rd and 4th stage of compression. The heat pump circuit is completed by evaporating C3R in the C2 Splitter Condenser which should results in a colder C3R volume when compared to the equivalent C3R volume required from the 4th to 1st kickback circuit at no load. Unlike the plot on Figure 6, a reduction in compressor discharge temperature would then be expected. Given the purpose of those standalone models, C3R liquid quenching was not rigorously modelled and explain the unexpected discharge temperature behavior. The increase power from Rate 1 and beyond is primarily caused by the standalone model simulated for a higher discharge pressure while maintaining same suction pressure. Figure 6 also shows the 4th to 1st kickback flow reduction as load increase on the system. Except for Rate 2, the 4th to 2nd kickback is not required for preventing the 2nd stage surge margin to fall below 10%. Note that Rate 2, assumes that C2's from the Acetylene Reactors is transitioning from the flare valve to the C2 Splitter. As such, a number of C3R exchangers intended to heat the ethylene product have not been commissioned yet and partly explain this small bump for the 2nd stage kickback flow.

Based on the standalone model results, no real surprise was initially expected for starting the PRC and C3R system. However, the first few attempts on the OTS resulted in high compressor discharge temperatures. The discharge temperature rise was also found to be aggressive and in a real startup could require tripping the compressor. The PRC has no high temperature trip automated mechanism but pass a certain value, process operators are required by procedure to trip the compressor. This behavior was also something never experience by the OTS test operators on the pre-revamp compressor system. While understanding that quenching was not rigorously model in the standalone models, this raise the questions as why is this behavior so different? The answer must obviously be related to changes when comparing pre and post revamp. Other than heat integration flowsheet alterations; the compressor re-wheel and conversion of the 3rd stage suction from side load to extraction are the main changes. Figure 7 compares the pre and post revamp polytropic head curves.

As illustrated, the compressor re-wheel design significantly pushes the surge line to the right on the 1st stage, hence resulting in higher volumetric rates for operation above 10% surge margin. The surge line on the 2nd stage is almost an extrapolation of the pre-revamp surge line at higher polytropic head. In the pre-revamp configuration both kickback circuits would be open to prevent the surge margin to fall below 10% on the 1st and 2nd stage. But the significant shift on the surge line to the right on the 1st stage in the post revamp configuration explains why little if no kickback flow is expected for meeting the 10% surge margin minimum on the 2nd stage. In the pre-revamp era, satisfying the 1st and 2nd stage surge margin would more than satisfy the surge margin requirement for the 3rd and 4th stage. While the head curve is dropping on the 3rd and 4th stages, the surge margin requirements are expected to be met by just satisfying the 1st stage 10% margin. In summary, in the post revamp configuration, the entire recycle flow required for meeting the surge margin low limit of 10% will now travel from the 1st stage suction all the way to the compressor 4th stage discharge. Preventing accumulation of the heat of compression would then depend solely on the 1st stage suction C3R liquid quench.

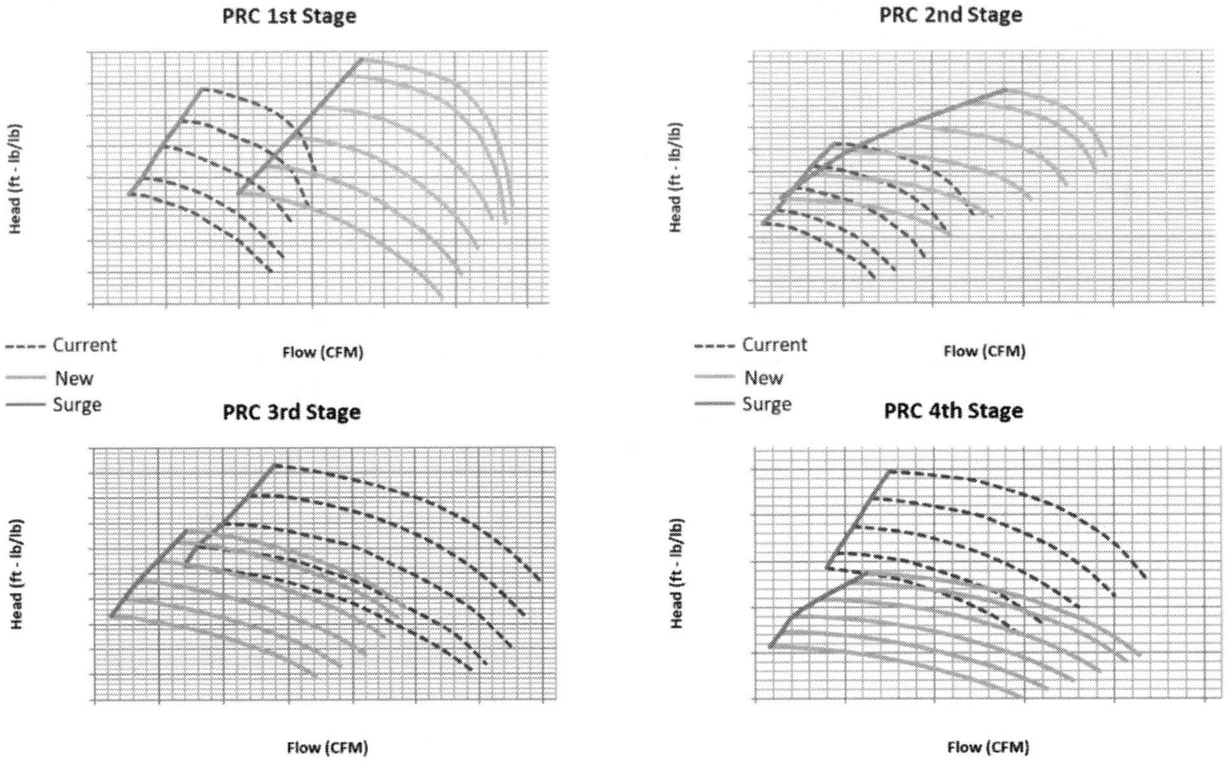

Figure 7: Propylene Refrigeration Compressor Pre & Post Revamp Polytropic Head Curve Comparison

Conversion of the 3rd stage suction into an extracting stage add on the volume of C3R carrying heat of compression. Figure 8 illustrates the differences between the pre and post revamp configuration. In the pre-revamp configuration excess vapor C3R dumped from the 4th to the 3rd stage drum along with liquid flash in the 3rd stage drum provides a sideload to the 3rd stage suction. In the post revamp configuration, this volume of C3R must now be dumped to the 2nd stage drum and supply the compressor through the 2nd stage suction; hence adding an additional stage of compression when compared to the pre-revamp configuration.

The anti-surge controls are configured to have the 4th to 1st and 4th to 2nd kickback control valves fully open when accelerating the compressor to minimum governor. As the compressor transition into run mode, the controllers will ramp the output down as they aim to surge margins set point of 10%. The change in head curve resulting from the compressor re-wheel required upsizing the kickback control valves. Therefore, the recycle flow to the 1st and 2nd stage would become larger in the post revamp configuration.

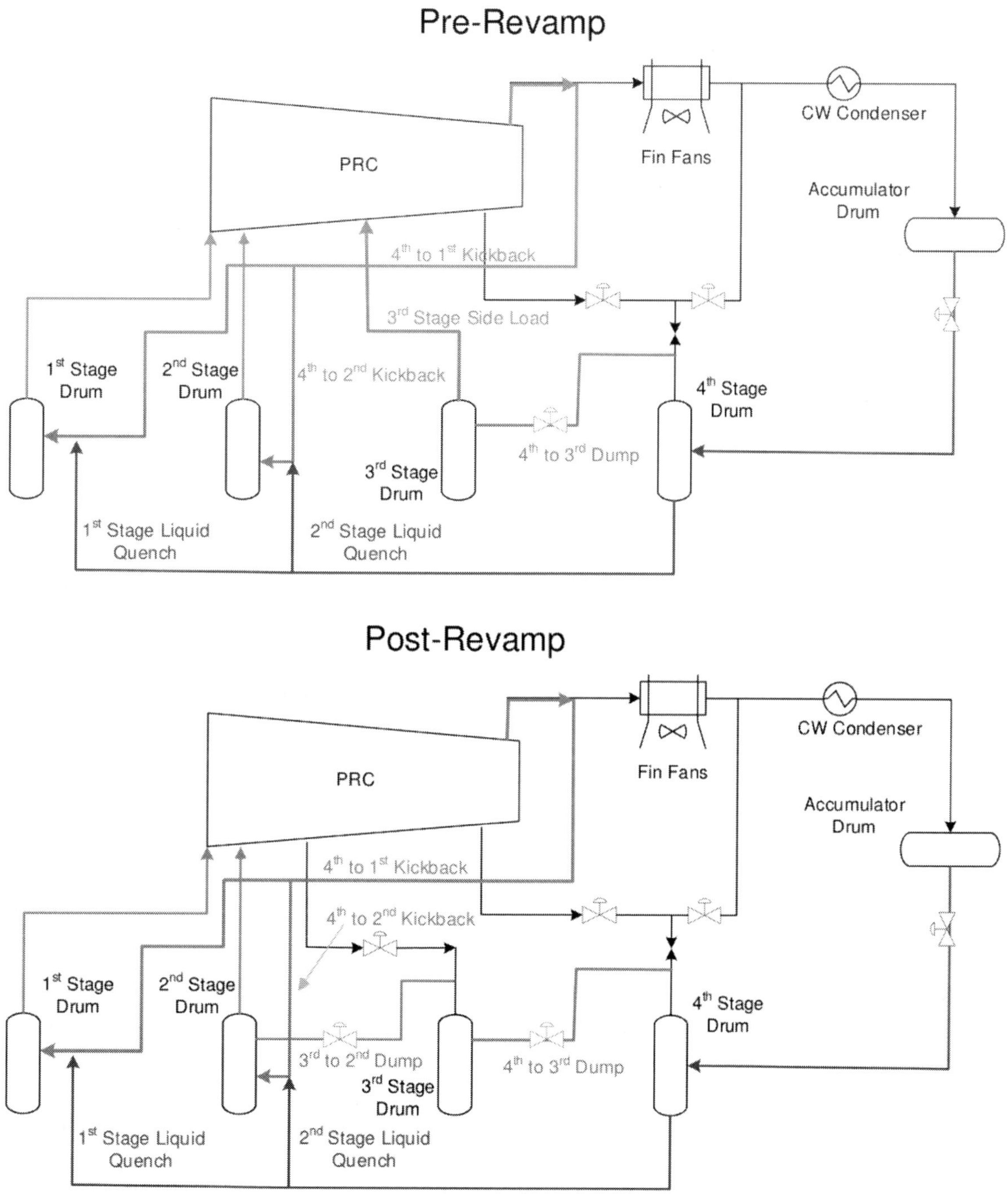

Figure 8: Propylene Refrigeration Compressor Pre & Post Revamp

As stated earlier, C3R liquid quench injection is used to prevent accumulation of the heat of compression when the anti-surge kickback circuits are open. C3R liquid quenching is regulated by a temperature controller, which used the temperature of the stream supplying the compressor as process variable (PV) and can be operated on automatic mode with set point (SP) input from the board operator. However, the intent is to operate in a remote set point fashion where the PLC supplies a temperature set point 20 °C above saturation. The remote set point configuration allows a constant temperature offset from saturation during pressure transient and prevent potential liquid accumulation in the knockout drums that could likely cause a compressor trip on high drum level. There are 2 liquid quench temperature controllers, one on the 1st stage suction and one on the 2nd stage side load stream.

After a few unsuccessful attempts on the OTS, it became clear that the quench controllers were too slow in preventing a high discharge temperature. The 20 °C offset from saturation was believed too high and the fluctuation of pressure during the acceleration and transition to run mode were also impacting the ability to maintain a reasonable discharge temperature. The configuration of the anti-surge and liquid quench controls were also discussed and challenged as the OTS was based on a PLC emulation rather than an actual PLC.

After discussions and review of the different factors influencing the high discharge temperature, the PLC emulation in the OTS was found accurate. The problem was then associated with the changes on the compressor and different strategies were brainstormed and tested on the OTS. The most successful strategy required to accelerate the compressor and transition into run mode with the anti-surge controls on manual and the liquid quench temperature controllers on either automatic with low temperature SP or manual depending on the ability of preventing excessive discharge temperature. The 4th to 1st and 4th to 2nd kickback control valves output were to be set at values predicted by the Aspen Plus Dynamics™ standalone models. The 1st and 2nd extraction control valves would also be open for minimizing the volume and number of compression stages followed by the C3R recirculating through the compressor. The anti-surge controllers would be switched to automatic control once establishing the desire compressor pressure profile and attaining stable operation.

C2 Splitter Total Reflux

Transitioning the main C2 Splitter to total reflux operation is not something new. The revamp project involves minor alteration on the tower that were not expected to change the methodology for establishing total reflux operation. The changes on the PRC were expected to allow operating the tower with vapor/ liquid traffic in the stable tray operating range at total reflux conditions. In the pre-revamp era, vapor/ liquid traffic was below tray turndown due to C3R supply limitations to the C2 Splitter reboiler until other process loads re-balanced the C3R system later in the startup sequence. That constraint goes away with the conversion of the 3rd stage suction from side load to extraction.

As for the PRC, Aspen Plus Dynamics™ standalone models of the C2 Splitter and C3R system were used for establishing operating targets. Transitioning the C2 Splitter to total reflux involve ethylene imports from the pipeline for inventorying the tower. As the tower pressure rises, ethylene will start condensing as it travels through the overhead condenser, with the liquid accumulating in the reflux drum. Once sufficient liquid accumulates in the reflux drum, some of the liquid inventory gets transfer to the tower sump via the reflux circuit. As the liquid level sufficiently rises in the tower sump, reboil heat is added. The heat duty to the reboiler will drive vapor/ liquid traffic on total reflux and is set for meeting the desire operating range of the trays. The heat duty ramp rate on the reboiler is critical as it must balance with the reflux rate. Aggressive increase in reboil rates may cause liquid vaporization rate that exceeds liquid supply to the sump and result in reboiler stall. As shown on Figure 9, the work done on the Aspen Plus Dynamics™ standalone model suggest a stair approach on reboiler duty step increase. The plot on the left shows the step increase in reboiler duty and the outcome on reflux rate. The plot is magnified on the right where liquid ethylene makeup to the sump is also plotted. As ethylene gets vaporized in the reboiler, it travels up the tower, condense in the overhead condenser and accumulates in the reflux drum. If the reflux drum level is regulated by a level to reflux flow cascade control loop, the reflux flow would be expected to increase proportionally to the increase in heat duty. However, there is a lag for the additional liquid volume collected in the reflux drum to travel down the tower and compensate for the loss of liquid inventory in the sump. To prevent stalling the reboiler in the standalone dynamic model, a small ethylene liquid makeup was added to the tower sump and as illustrated on the right plot of Figure 9 had to be commissioned at about 37% of a reboiler heat duty step increase to prevent the loss of liquid inventory.

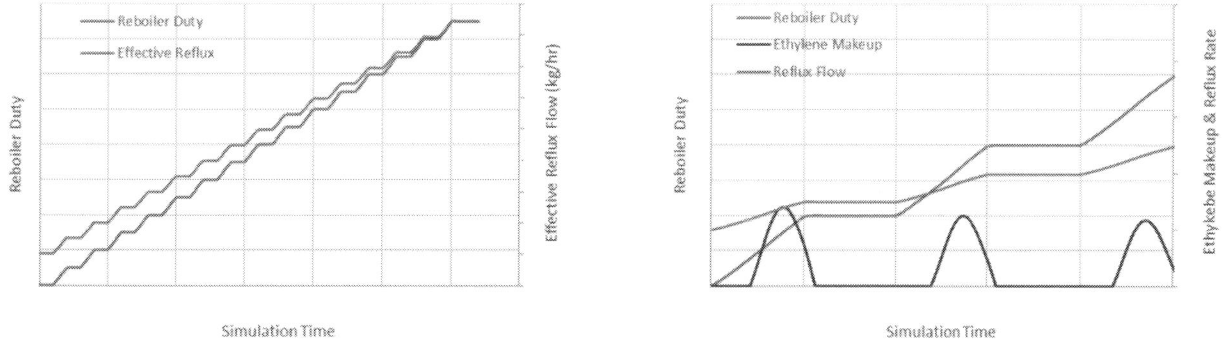

Figure 9: C2 Splitter – Total Reflux Transition Stair Approach

In the plant and in the OTS, this problem can be managed by providing sufficient liquid inventory in the reflux drum and the tower sump. As step increase are made on the reboiler duty, the board operator must pro-actively increase the reflux rate on flow control or on manual if the controller is too slow to maintain sufficient inventory in the tower sump. The next step increase on the reboiler must then wait for the reflux drum and tower sump level to stabilize. Ethylene inventory will be required at one point as once sealing the downcomers and the tray deck, liquid holdup will build on the tray requiring external makeup for keeping the reflux drum and tower sump inventory. Note that unlike the standalone model, ethylene makeup from the pipeline in the OTS is supplying the reflux drum.

As discussed earlier, the distillation block in the OTS performs vapor/ liquid equilibrium calculations. The trays are insensitive to flooding, weeping and lost of downcomer seal. As a result, the dynamic relative to building liquid holdup on the trays will be different then the plant. In reality, liquid typically weep through the perforation of the tray deck and vapor blow through downcomers until sufficient liquid seals the downcomers and the vapor velocity pass the weep point. Not meeting the process target for effective total reflux operation will therefore be inconsequential in the OTS. The process targets for this step relies on standalone models and tower internals rating tools.

Reboiler thermosyphon stability is more forgiving in the OTS and standalone model then in the plant. While the reboiler, capacity is affected by the liquid level, it is not impacted by flow stability phenomenon such as flow reversal and unstable flow regime.

The rehearsal session on the OTS dedicated to the C2 Splitter total reflux operation were associated with the need to aggressively increase reflux rate when stepping up the reboiler heat duty. Managing liquid inventories in the reflux drum and tower sump was a challenge as expected from the work on standalone models but manageable.

Difficulties in stabilizing the tower pressure and the C3R system prove challenging. Anti-surge controls on the C3R system were causing the compressor suction pressure to oscillate, which in turn was affecting operation of the tower overhead condenser and pressure. Given that the PRC suction pressure is strongly affected by the C2 Splitter condenser duty, a simulation run was conducted with the PRC anti-surge system on manual while transitioning the C2 Splitter to desire total reflux target conditions. This proof to be more stable and effective. The recommendation was therefore to try bringing the C2 Splitter on total reflux with the anti-surge controller on automatic but if similar oscillation were to be experience in the plant to switch them on manual until stabilizing the C2 Splitter.

Chill Train & Towers

Process targets for this section were calculated from steady state simulation models and distillation equipment rating tools. As previously stated, the tower in the OTS can be operated at conditions below turndown without impacting fractionation efficiency. It was important to emphasize this anomaly and provide the basis for the desire targets.

One of the challenges that arise when rehearsing the startup sequence of this section, was the need to execute a number of steps in parallel. It became clear that once the chilling sequence was initiated, that the Pre-Stripper bottom would eventually be ready to feed the Deethanizer, followed by the Demethanizer bottom while progressing the chilling sequence. In the first few rehearsals, the test operators got overwhelmed with material accumulating in the tower bottoms and had to liquid drain to flare. As shown on Figure 10, at this point in the startup sequence the process can be divided in two sections. The Chill Trains, which includes the chillers, knockout drums, cold box and turbo expander and the Pre-Stripper, Demethanizer and Deethanizers. The chill down sequence can be executed by one board operator who maintain focus on the chill train. A second operator focus on operating the tower without being distracted by executing the chill down sequence but by rather responding to its outcome on the towers. The Demethanizer overhead becomes a cross boundary between the 2 board operators but can be effectively manage if the startup objectives are understood. This strategy proved effective and accelerated the time required for steadying the towers while minimizing losses to flare.

While defining board operators' roles might be straightforward, communicating parallel activities is a challenge as written procedure are sequential by nature. To overcome this problem a procedure flow chart illustrating parallel activities and conditions required for moving from one step to the next was established as per the example illustrated on Figure 11.

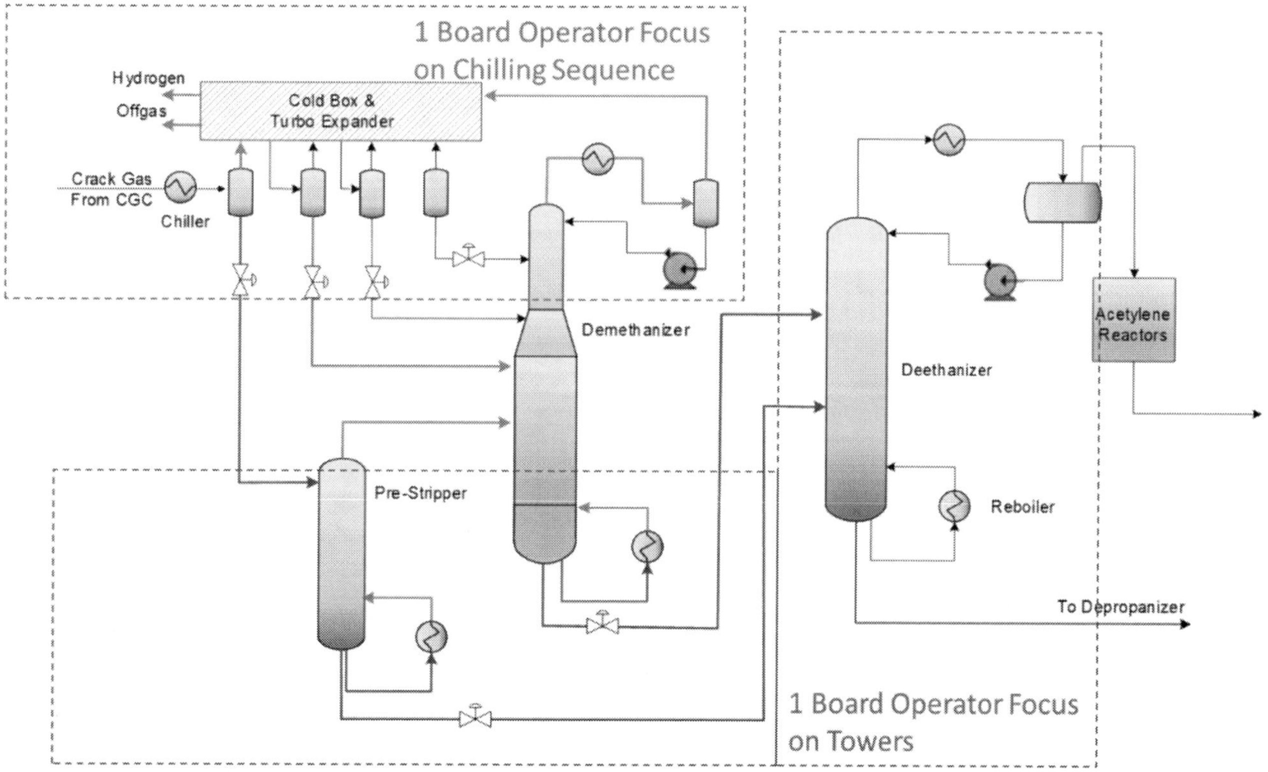

Figure 10: Dividing Board Operator Workload During the Chill Down Sequence

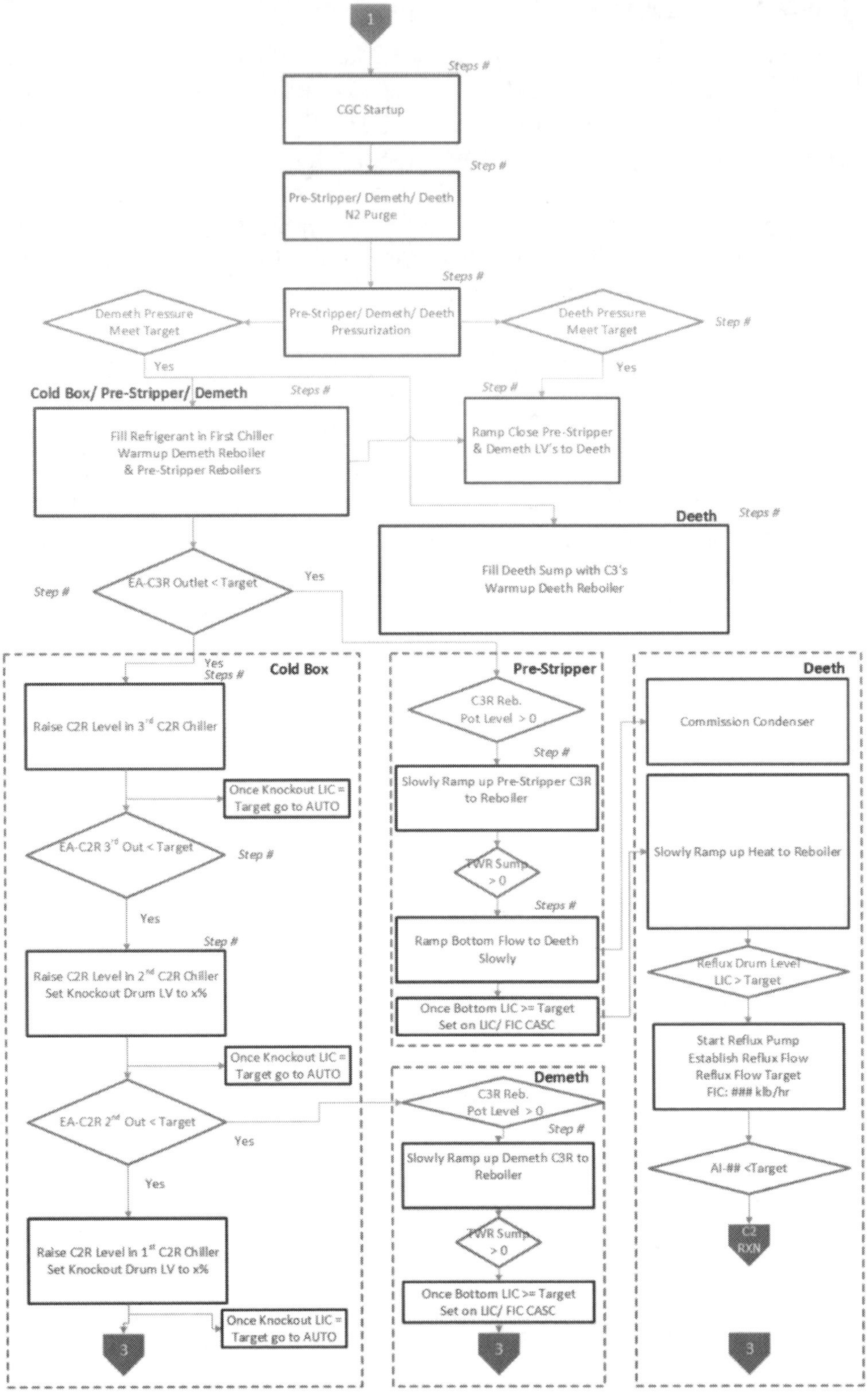

Figure 11: Procedure Flow Chart

Actual Startup vs Operating Training Simulator Learnings

Propylene Refrigeration Compressor (PRC) Startup

As discussed above, the conclusion and recommendations from the OTS rehearsals for starting the PRC was to operate the anti-surge controllers on manual. Kickback control valves openings had been calculated for meeting surge margin above 10% with additional safety margins. However, the presence of the anti-surge control system vendor field representative and unease from manufacturing leadership in operating this system on manual reverse the proposed plan. As such, the PRC was started with the anti-surge controls on automatic. Figure 12 summarizes the 1st 50 minutes of the PRC startup. While not shown on Figure 12 due to the speed scale, the compressor was at slow roll speed at time 0 minutes and acceleration to minimum governor was initiated at time 2 minutes on the plot. It is unclear why the output of the 4th to 2nd kickback circuit ramps down as the compressor accelerated. The controller was on automatic with a surge margin set point of 10% and a PV indicating 0%. The 4th to 1st kickback controller shows an expected output of 100% during acceleration. Figure 12 shows the compressor reaching minimum governor speed followed by a sharp decrease in speed. While the anti-surge controllers were operated on automatic, the board operator was instructed to take over C3R liquid quenching to prevent high compressor discharge temperature. As shown, the discharge temperature exceeded the high alarm value as the compressor reached minimum governor. The sharp decrease in speed is explained by the compressor power reaching the turbine maximum power due to excessive 4th to 1st stage kickback along with high volume of C3R liquid quench. Backing down on the C3R liquid quench is associated with speed recovery but at the expense of high compressor discharge temperature. The speed scale on the plot does not fully reflect the short margin from a critical speed trip. The first drop of speed came within about 100 RPM from the trip point. This situation put the board operator in a difficult situation, where managing high discharge temperature may result in low-speed trip, while managing the low-speed event could result in high discharge temperature potentially resulting in equipment damage.

This could be qualified as a moment where all the learnings from the OTS rehearsal kicks in and proper coaching of the board operator becomes crucial. As a first move, the board operator was instructed to open the 1st extraction valve with an output of 5%. The intent of this move was to bleed propylene off the 3rd suction and create a false load circuit between the 3rd and 2nd stage. The expectation is that it would save power and relief some heat of compression by short circuiting 2 stages of compression. While not shown on Figure 12, the 1st extraction valve was open at 5% around time 25 minutes. The move was not too aggressive and intended to validate the expected outcome. The benefits of this move can be assessed when comparing speed and discharge temperature at time 30 minutes. A 3.5 ^{o}C reduction in discharge temperature is observed when comparing speed and temperature from the last spike in temperature at time 22 minutes and the new spike at time 30 minutes. This also helped in sustaining minimum governor speed. While further increase in extraction would have help, a more aggressive solution was believed required. The 4th to 1st kickback circuit was still fully open and putting significant load on the compressor. All the process engineering work done in parallel to the OTS sessions and behavior observed on the OTS did not align with a 4th to 1st kickback circuit still fully open at this stage of the startup. The board operator was then instructed to start ramping down the output of the 4th to 1st kickback controller on manual as observed on the plot at time 33 minutes. The discharge temperature started dropping shortly after and went below the high temperature alarm point at time 44 minutes. Stable minimum governor speed was also sustained after those operating moves.

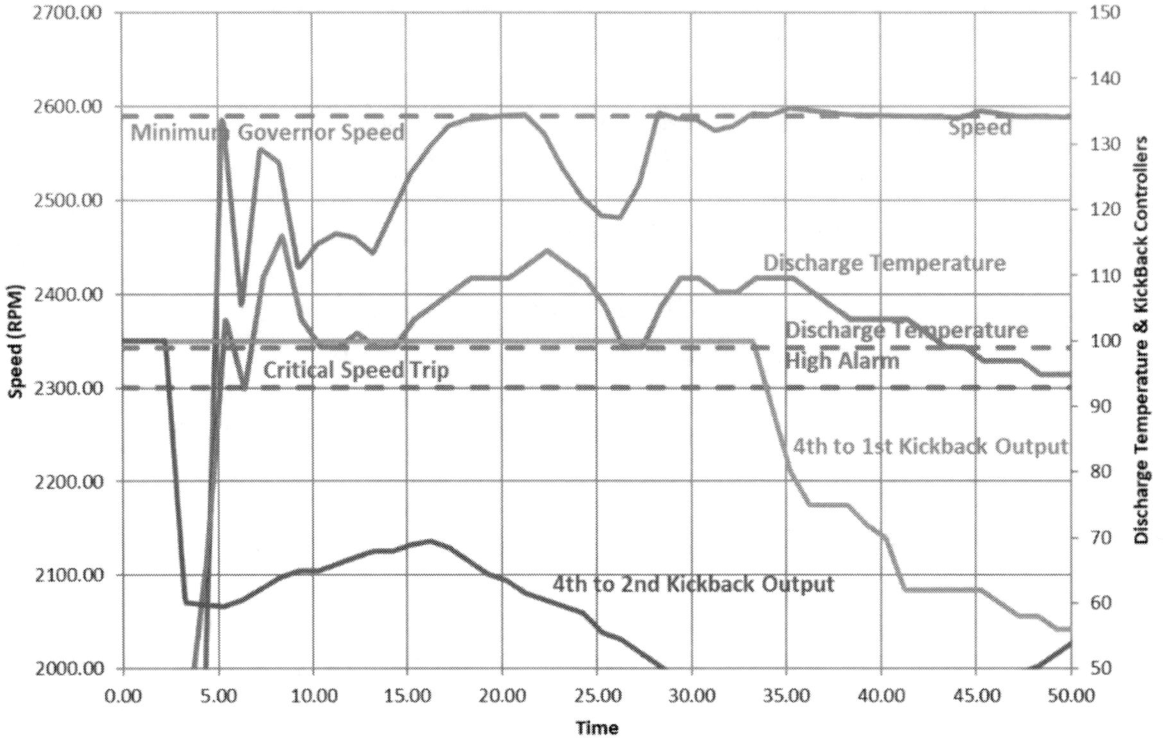

Figure 12: Propylene Refrigeration Startup

After consultation with process control, it was found that the 4th to 1st kickback anti-surge controller was holding a 100% output due to a bad temperature transmitter involve in the surge margin calculation.

While review of those events would suggest revisiting the anti-surge control strategy on startup it highlights the benefits from all the work conducted on the OTS prior to startup. Many of those events and behaviors had been experience on the OTS. Many alternatives had been tried and along with parallel process engineering work done, the startup team had a strong understanding of the system. The event described above occurred over a period of less than an hour and required quick decision making.

The importance of the startup team providing coaching to the board operators cannot be emphasized enough in this case. The startup team spend 1 year working on the OTS, while process operators had limited training session relative to the revamp plant startup due to their availability. Plant revamp context cannot be treated similarly as maintaining operators' proficiency in starting up an existing and unmodified facility.

C2 Splitter Total Reflux

Given the problems encountered on the PRC startup, the main C2 Splitter was brought up to total reflux operation with the PRC anti-surge controls on manual. Rehearsal on the OTS suggested an aggressive need for reflux when stepping up the reboiler heat duty to prevent a rapid rise in tower pressure. In the plant, the tower pressure remained in control throughout this sequence and as shown on Figure 13, the response was quite different than on the OTS. Figure 13 covers a period of about 24 hours (time scale in minutes) and can be divided in 3 periods. A first period, from time 0 to 700 minutes where the C3R flow to the reboiler is first ramped up to about 50 klb/hr, then ramps down in successive oscillation to about 30 klb/hr. During that period, the reflux flow experiences a few oscillations and is predominantly kept to zero from time 360 to 700 minutes. Frequent oscillation in the C3R condensate pot level between full and about 80% are observed and explain the reluctance of the board operator to increase the C3R flow. At this point, the problem is not fully understood, but the lack of reflux flow is believed problematic. It is suspected that the variation in C3R pot level should be manageable if the reflux flow is manipulated in a way that prevents a drop in the tower sump level. This potentially requiring external ethylene makeup if depletion of the reflux drum is experienced. As such, reflux and C3R flow to the reboiler are ramped up from time 700 to about 1,100 minutes. The reflux drum, tower sump and C3R condensate pot exhibit some significant oscillation between time 700 and 900 minutes. A magnification of the time scale between time 700 and 950 minutes shows that the reflux drum and tower sump level exhibit about 4 cycles of oscillations with their periods offset by 180°. This behavior suggest that the volume of ethylene vaporized in the reboiler travels up the tower and back down through the reflux circuit in an unsteady fashion. Some form of stability is obtained after time 835 minutes, where the amplitudes of those oscillations are significantly reduced and both reflux and C3R flow to the reboiler ramped up in a more control fashion.

The oscillations are believed caused by thermosiphon instability. It is suspected, that the thermosyphon was stalling at low C3R rate and low tower level. Higher sump level and C3R flow pass time 835 minutes became sufficient for sustaining circulation of the thermosiphon. Further increase in C3R flow brought operation above minimum thermosiphon stability criteria's and explain why the reboiler operation loss sensitivity relative to the tower sump level after time 1,000 minutes. The sump level would still influence the circulation rates and degree of vaporization, but the variations are believed to fall within stable conditions.

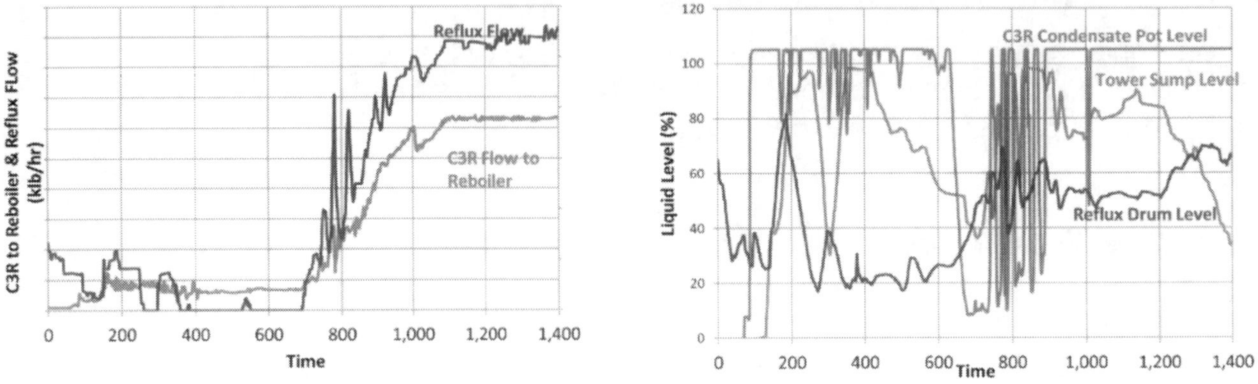

Figure 13: Reflux/ Reboiler Interactions

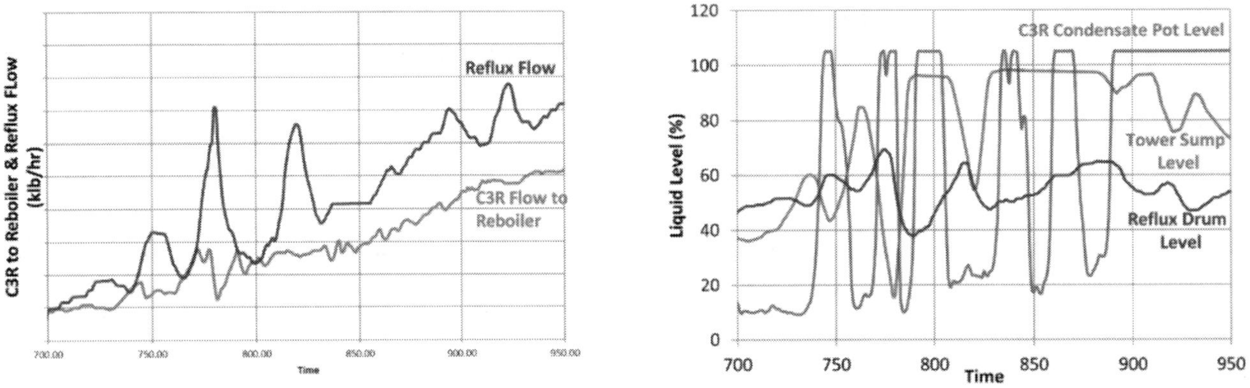

Figure 14: Reflux/ Reboiler Interactions - Time Scale Magnification

As stated earlier, thermosiphon reboiler are typically more forgiving in an OTS model and explain why this behavior or anything similar did not occurred during the rehearsal sessions. The focus on preventing the potential aggressive rise in pressure when starting the reboiler became a distraction and may have delayed troubleshooting the oscillations experienced. The rapid rise in pressure experience in the OTS is believed caused by a few factors. Unless done through some form of customized malfunctions, distillation block in OTS perform vapor/ liquid equilibrium calculations regardless of hydraulic. The tray hydraulic calculations are geared towards pressure drops and holdup and do not reflect weeping and vapor blow through downcomers. The total reflux operation in this tower involves pure ethylene, a single component for which boiling, and dew points are identical. Reality on the other hand differs from the OTS due to a transition from dry trays to weeping and vapor blowing through downcomers until the internal circulation rates becomes sufficient for sealing the trays. Unlike the OTS, large pool of liquid ethylene inside the reboiler, tower sump and reflux drum could be associated with temperature profiles and as such deviates from the homogeneous mixture assumed in a simulation model.

The oscillations experienced during this period were relatively inconsequential or manageable relative to the stability of the C3R system. Having the anti-surge controllers on manual proved effective in preventing oscillation originating from the C2 Splitter in migrating and causing severe and potentially undamped oscillation in the C3R system.

Transitioning the C2 Splitter to total reflux operation was not something new from a tower standpoint. However, the significant changes on the C3R system brought an element of caution and explain a slow transition that could have been done faster. The discussion so far emphasized on the use of the OTS and process engineering tools for setting up the startup procedure. However, this example emphasized the need to capture post startup learnings and evaluating enhancements that could potentially be configured in the OTS for reflecting a more realistic startup. In this case, the focus should be on meeting a more realistic pressure transient rather than emulating the reboiler oscillations. The procedure should be improved by requiring a shorter C3R ramp rate that meets stable thermosiphon conditions.

Conclusion

Updating the Operating Training Simulator (OTS) in a revamp project is primarily driven by change management for ensuring continuity in operator training. Properly scheduling the OTS update is also beneficial in the commissioning of control systems. This paper demonstrates the additional benefits of having the OTS available for developing a startup procedure in alignment with the revamp plant. Too often, in revamp projects, the startup procedure is an afterthought or is not given the appropriate level of attention due to time and resource constraint. "*I cannot imagine how we would have resolved those issues without all those rehearsal on the OTS*", has been a common statement expressed by team members. This paper describes only a few of the startup activities that took place. While the procedure developed was found adequate, response to equipment and instrumentation malfunctions required deviations and adjustments from the plan. The deep knowledge gained by the startup team in understanding the revamp plant and troubleshooting experience acquired during the many rehearsals on the OTS allowed the team to effectively adapt to malfunctions encountered while starting up the plant.

Understanding limitations of the OTS and supplementing the evaluation of process targets with other engineering tools is another key element for developing an effective and comprehensive procedure. Gathering learnings post startup, will fill the gaps experienced during a first execution and provides guidance for fine tuning the procedure and the OTS.

The numerous startup rehearsal executed by the startup team along with process engineering calculations and problem-solving sessions, granted the team with an elevated level of expertise. While the process operators got trained on the OTS, the limited number of sessions could not transfer the same level of knowledge. As such, some of the team members were assigned as board operator coaches. This strategy proved effective especially when face by challenging events as discussed in the PRC startup and C2 Splitter transition to total reflux.

The COVID 19 restrictions environment, while perceived as disruptive, demonstrated the capabilities in providing technical support remotely. By having process engineers on their computers in remote location, allowed them access to design data and process engineering design tools. It can be argued that this could also happen in the training simulator room. However, working from dual screen stations in an ergonomic office is more effective then on a laptop screen in an overcrowded room. The level of focus is also believed higher as less distractions occurred. People attending on a part time basis can take phone calls or come in and out without disturbing the team. Some face-to-face meetings may still be required, but this exercise proved effective by having some people in remote locations.

The skill set mixture of combining former experience process operators and process engineers together is an important factor contributing to the success of this work. An effective startup team must have a good balance of both.

Nomenclature

APC: Advance Process Control
C2R: Ethylene Refrigeration
C3R: Propylene Refrigeration
CGC: Crack Gas Compressor
DCS: Distributed Control System
ERC: Ethylene Refrigeration Compressor
FAT: Factory Acceptance Test
MV: Manipulated Variables
NGL: Natural Gas Liquids
OTS: Operating Training Simulator
P&ID: Pipe and Instrumentation Diagram
PFD: Process Flow Diagram
PHA: Process Hazard Analysis
PLC: Programmable Logic Controller
PRC: Propylene Refrigeration Compressor
PV: Process Variable
SAT: Site Acceptance Test
SP: Set Point

References

1. K. Doe, D. Coppard, "Process Simulations for Operations Training and Other Benefits", 2001 AIChE Spring Meeting, Houston, TX
2. R. Besuijen, R. Young, M. Guaraldo, "Operator Training Simulators – NOVA Chemicals' Experience for training, certification and other applications", 2012 AIChE Spring Meeting, Houston, TX
3. A. Bernard, "Distillation Tower Startup Considerations in Ethylene Plants", 2021 AIChE Spring Meeting, Virtual
4. J. Kendall, D. Dutschmann, J. Bethea, "Utilization of Dynamic Simulations for New Steam Cracker Start-Up Activities", 2022, AIChE Spring Meeting, San Antonio, TX

Trademarks

Aspen Plus Dynamics™ is a registered trademark of ASPENTECH.

124f Low Cost Debottleneck of a C2 Splitter

Carolyn Granade

This paper will discuss an example of a low cost solution for a C2 Splitter debottleneck project. The project leveraged a first-time-in-service application of an enhanced heat transfer heat exchanger to overcome a system constraint with the propylene refrigeration system and more completely utilize the hydraulic capacity of the C2 Splitter. There will be a summary of the evaluation of constraints on the tower and the supporting refrigeration system. Then, this paper will detail the selected constraint-removal solution and associated technology and operational risks, including an enhanced heat transfer application on the overhead condensers. Finally, there will be a post-start-up analysis of the performance of the new equipment and overall tower capacity increase for project prove-out.

Steam Cracker Thermal Imaging and Automated TMT Monitoring

James Cross

Thermal imaging devices are commonly used for periodic inspection and measurement on steam crackers, but less so for continuous thermal imaging and temperature data collection and analysis. Continuous monitoring and repeatable, reliable measurements allow operators to extend furnace runtime closer to the material limits. Automated detection of hot spots, together with temperature-controlled decoking, avoids accidental overheating of the coils, increasing their lifetime and reducing maintenance costs.

A pilot installation in Gelsenkirchen, Germany highlights some key advantages and considerations when implementing the technology. With real-time remaining coil lifetime prediction algorithms, software can predict the risk of coil rupture before it becomes a substantial risk, giving operators the insights they need to optimise replacement intervals while mitigating the risk of rupture.

SPRING23
+19TH GCPS
A Joint AIChE and CCPS Meeting

Deploying Proven Turbomachinery Solutions to Assist Plant Owners with Energy Savings & Lower Carbon Emissions

Yuya Tanaka
Mitsubishi Heavy Industries Compressor Corporation
4-6-22, Kan-on-shin-machi, Nishi-ku, Hiroshima 733-8553 Japan
yuya.tanaka.3e@mhi.com

Prepared for Presentation at
American Institute of Chemical Engineers
2023 Spring Meeting and 19th Global Congress on Process Safety

Deploying Proven Turbomachinery Solutions to Assist Plant Owners with Energy Savings & Lower Carbon Emissions

Yuya Tanaka
Mitsubishi Heavy Industries Compressor Corporation
4-6-22, Kan-on-shin-machi, Nishi-ku, Hiroshima 733-8553 Japan
yuya.tanaka.3e@mhi.com

Abstract

Historically oil, gas, and petrochemical plants would revamp and/or replace turbomachinery in order to increase plant capacity or replace end of life machinery. Mitsubishi Heavy Industries Compressor Corporation (MCO) considers itself a world-class provider of brownfield equipment due to their vast experience, knowledge, and custom-built design philosophy.

Recently many plant owners have started their investigations into how they can reduce the carbon intensity of their existing assets, as well as new grassroots developments, as “Energy Transition” becomes more mission critical to their respective companies.

Turbomachinery manufacturers have been receiving new study requests regarding low carbon and carbon neutral solutions. Specifically in this paper, we will highlight our equipment replacement and revamp solutions that can significantly improve plant’s efficiency & reduce their CO2 emissions by utilizing proven well-referenced turbomachinery technology.

As the result, plant owners can achieve “Energy Savings” by utilizing Mitsubishi Heavy Industries Compressor Corporation (MCO) compressor & steam turbine solutions without having to introduce “first of its kind” technologies.

1. Introduction

The reduction of greenhouse gas (GHG) emissions has become a major issue and has created an urgent need for solutions to reduce the carbon dioxide emissions of industrial facilities. As governments around the world set increasingly ambitious targets for reducing CO_2 emissions, the transition to a low carbon or zero carbon economy has become a pressing issue for industry. In particular, emission-intensive industries such as coal-fired power generation, petrochemical manufacturing, steelmaking, cement, and liquefied natural gas production facility need to implement concrete objectives for reducing emissions. Therefore, both reduction of CO2 emissions and carbon capture and storage (CCS) must be promoted to achieve a low-carbon or zero-carbon economy.

MCO has rich experience of compressor & steam turbine revamp/replacement services and contributes to the reduction of CO2 emissions by improving energy efficiency through modernization, reliability improvement and capacity increase, which have been accumulated through many years of experience. Shown in **Figure 1.1** is the upgrading history of a certain company's ethylene plant in Thailand. This original plant was constructed 20 years ago with a nameplate capacity of 600KTA. Over time, this plant has carried out three separate expansion projects. As the result, 156% capacity-up from original plant was achieved by replacement of the compressors and turbines through footprint replacements. In this way, plant capacity can continuously be increased while improving the turbomachinery efficiency. In many cases it's more cost effective (CAPEX) than that of a grass root project. This is the typical case that customer can obtain large amount of production by replacement on same footprint as existing.

In addition, MCO has well-proven achievements in CCS (Carbon Capture) applications, employing both CO2 integrally geared compressor and inline compressors. Both type compressor can contribute to large capacity CCS application.

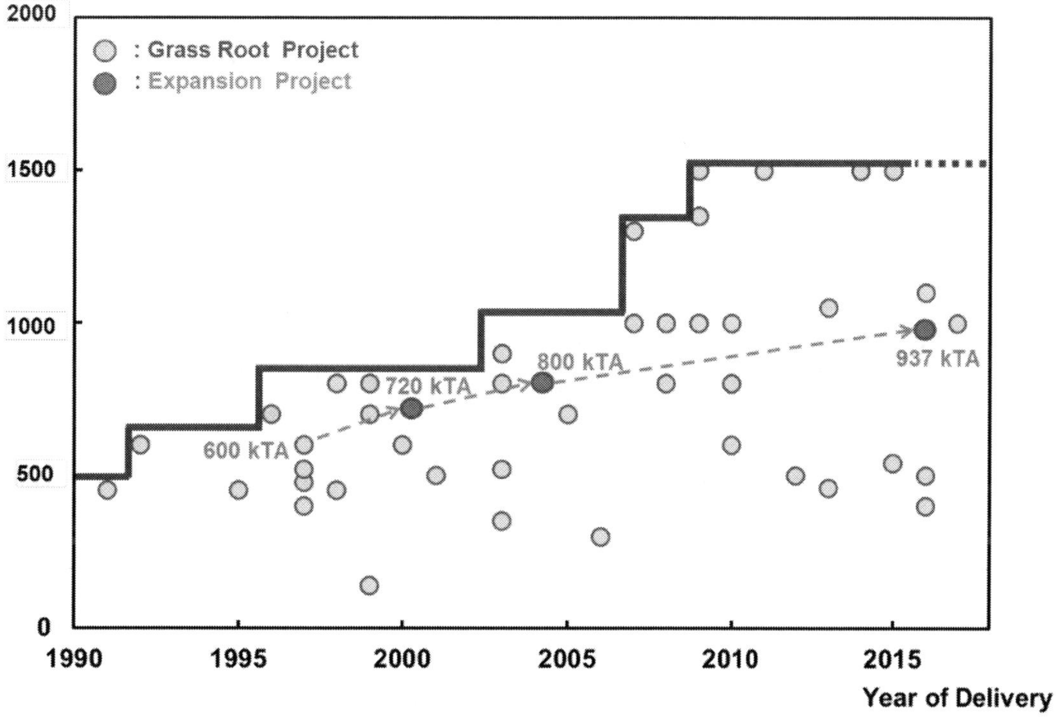

Figure 1.1 Trend of Ethylene Plant Capacity

2. MCO's Features

2.1 Proven Technologies

Effective improvement solutions with proven technologies should be proposed when looking at these projects. A 3D impeller for a compressor, for example, would be radically more efficient if a 2D impeller were applied to an existing compressor. As for steam turbines, it is also expected that the new advanced shape of low-pressure stages will be much improved from the existing older ones. These technologies, as well as other modern designs involved in modernization, could make a significant contribution to efficiency. Capacity up of compressor and steam turbine can be achieved by optimal design casing. If the optimal design is applied for casing external, larger machines for capacity up suit existing foundation which size was adjusted for existing machines. Such design has been applied for many revamp projects and modification of existing foundation was avoided for almost all of projects. It is beneficial for customers because modification of existing foundation should be checked with various viewpoints.

2.2 Advanced Technologies

Not only proven technologies, but also advanced technologies can be included in revamped machines. As for charge gas, cracked gas or process gas compressor, it is important to take measures to inhibit the fouling of compressor. Oil injection system at return bend is well known method. As you can see in **Figure 2.2.1**, "Whole wash oil injection" was developed. The whole wash oil injection type is expected to prevent bias of oil and maximize the effect of oil injection.

As other advanced technology, remote monitoring system is developed as show in **Figure 2.2.2**. Monitoring by remote service of compressor operations, using artificial intelligence, enables to establish optimum operating conditions. In the unlikely event of a malfunction, the system displays guidance on what action to take. The system enables operators to better visualize plant status and is able to propose an optimized maintenance schedule, thereby reducing maintenance costs. This system can be applied existing plants additionally.

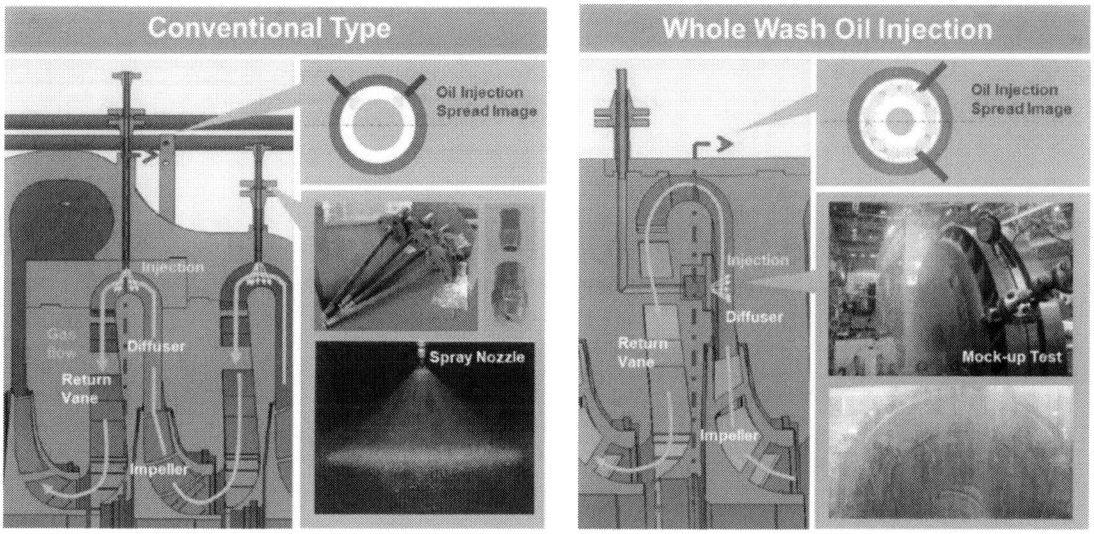

Figure 2.2.1 Wash Oil Injection

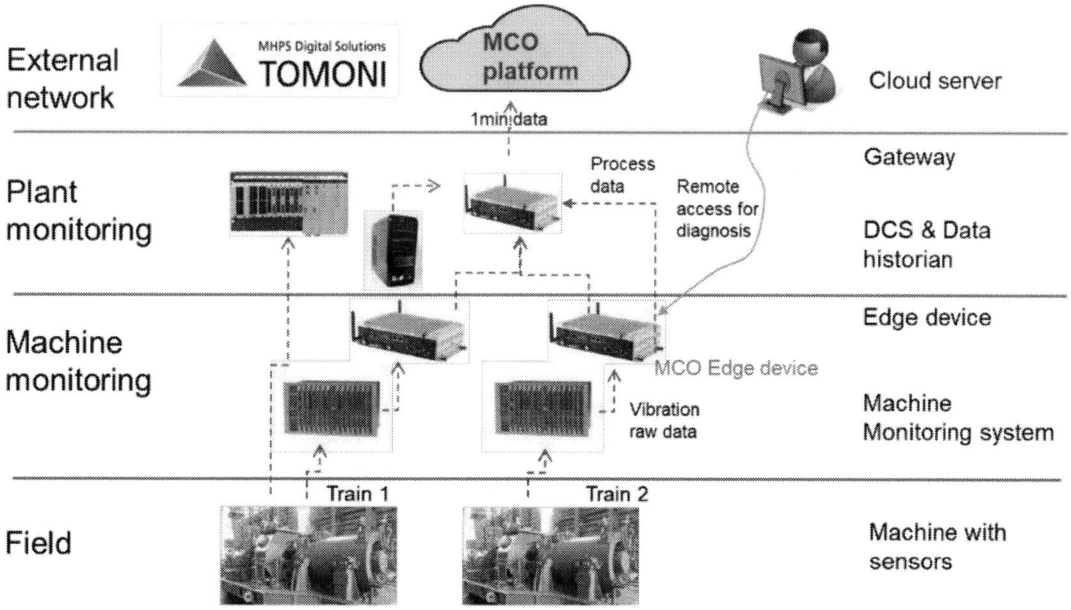

Figure 2.2.2 Monitoring by remote service of compressor operations

2.3 Reuse Items

Minimum modification also contributes to energy savings because it prevents unnecessary modification. Existing equipment will be checked by design data and site survey at project development stage. Shown in **Figure 2.3**, marked equipment has a possibility of huge impact of cost and turn around period if modification is required. By checking reusability of such equipment in advanced, customer can estimate total cost more accurately. In addition, by minimizing modification items, it will contribute to energy conservation in several aspects such as reducing new material, less waste, work, transport, and so on. The revamp is based on not only various technologies but also many kinds of experience. Therefore, the revamp can suit to each customer's request.

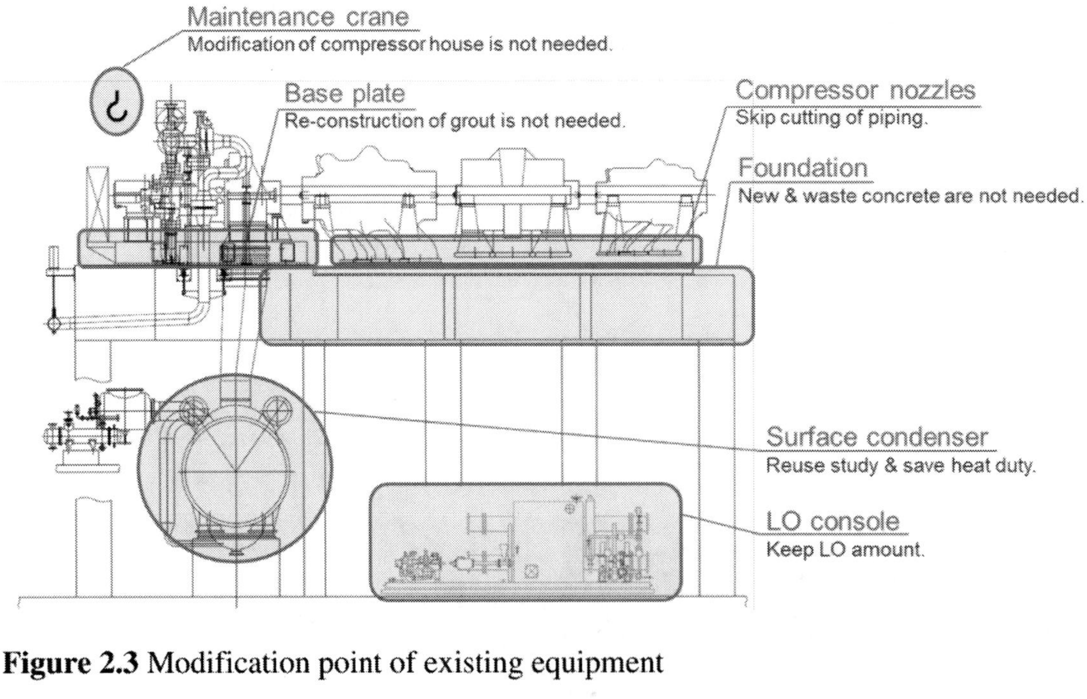

Figure 2.3 Modification point of existing equipment

2.4 CCS

High-pressure CO2 compression systems are 'transcritical', i.e. with subcritical low-side and supercritical high-side pressure. Supercritical CO2 has a density as high as that of its liquid phase while the viscosity remains closer to its gaseous phase. Knowledge of real gas behavior in the supercritical region plays a key role in designing transcritical CO2 compression systems. Thermo-physical and transport properties of CO2 abruptly change at thermodynamic conditions slightly above the critical point. The design and performance of fluid machinery can be significantly impacted by the impurities in the CO2 mixture in supercritical phase. For example, stainless steel grades (e.g., 304 and 316L) have been successfully used for wet CO2 compressor stages. Even though carbon steel can be used for dry stages (i.e., downstream of dehydration unit), the use of stainless steel on dry sections may improve the robustness of the system, as indicated by end-users' operating experience. Stainless steel overlay and anti-corrosion coating such as ENP coating are also used for wet CO2 compression applications. [1]

Taking advantage of our proven track record of both Inline and integrally geared compressor as large CO_2 compressors as shown in Table 2.4, and the ability to manufacture them in the same factory, the most suitable combinations for the target plant can be applied. There are pros and cons for inline compressor and integrally geared compressor. In general, inline compressor has easy maintenance and wide operating range but integrally geared compressor has higher efficiency than inline compressor. In addition, capital expenditure associated with CO_2 compressors for CCS applications can be reduced by making ancillary equipment more compact and employing a modular design that integrates compressors and associated systems. The modular approach to installation enables operations to save time and costs on site.

Table 2.4 Line up of CO2 compression

Type	In-line Compressor	Integrally Geared Compressor (IGC)
	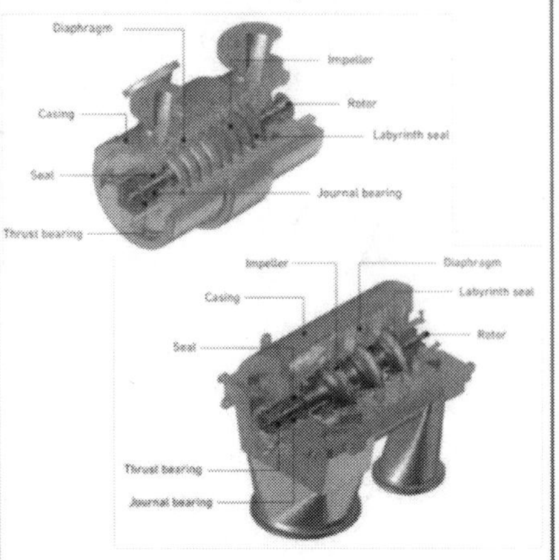	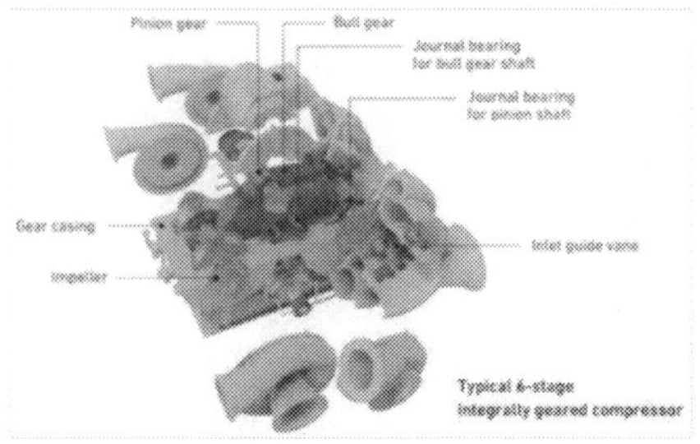
Capacity	Currently 1,000 TPD – 6,000 TPD (0.37 MTA – 2.2 MTA)	
Discharge Pressure	Up to 700 BarA	Up to 200 BarA
Service	Urea, EOR, CCS	Urea, CCS

3. Experience

3.1 Revamp/Replacement

Shown in **Figure 3.1.1** is one of our experiences of replacement project for ethylene plant in Singapore. The purpose of revamp is modernization and value up (overhung damper, swirl canceller, etc...) of compressors which were manufactured by another OEM. New compressors were installed on existing base plate by footprint replacement. Not only existing base plate but also other existing equipment were reused or applied minor modification. By improving efficiency, required power was reduced by 14%. Required modification and reusability of existing equipment was checked at early stage of project. Therefore, customer could proceed with this revamp project as planned and achieved in scheduled turnaround window. Also, many technologies were included in new compressors with considering trouble of existing compressors. Customer could achieve energy saving by reducing required power and prevention of waste caused by troubles. According to the reduced power the steam consumption is also reduced. The energy cost was approximately 4.3 Mil. $ per year (= Total -27 t/h x 20 $/ton x 8000 h/year). Not only could the customer achieve OPEX savings, but also the reduction of CO2 emissions. We believe that footprint replacement with the latest technology can improve OPEX, in the long run and initial investment cost will be collected in the early stages.

Shown in **Figure 3.1.2** is another experience of replacement project for ethylene plant in Philippine. The purpose is not only modernization and 148% capacity upgrading from original plant, which were manufactured by other OEM, but also future expansion (+22%) after capacity up. Therefore, all casings of compressor and steam turbine are designed to have the capability of future capacity-up (+22%) by Footprint Replacement. If the future expansion is carried out, the only internal parts will be changed, and field work schedule can be shortened. This leads to the energy saving of capacity up with minimum cost.

Shown in **Figure 3.1.3** is other experience of revamp project for LNG plant in Brunei.

The purpose of revamp is value up and capacity up of steam turbines which were manufactured by MCO. Internal revamping of steam turbine was applied and required capacity up was achieved. In spite of rate of capacity up was different for each turbine, identical internal parts could be applied for all turbines. As the result, 1 spare rotor can be used as common for all turbines. In addition, some countermeasures against customer's concerns caused by actual operating condition were included. These are contributed to value up of actual operation. Customer could achieve capacity up with minimum modification and value up with countermeasures for actual operation.

Service		Charge Gas Comp.	Propylene Ref. Comp.	Ethylene Ref. Comp.
Replacement Summary (Ship date : 2018) New / Modify / Re-use		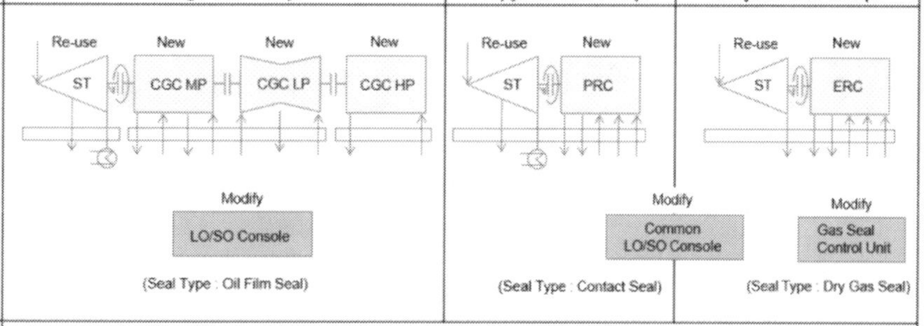		
Category		Modernization & Value-up		
Required Energy	Existing Machines (Other OEM)	100%		
	New Machines (MCO)	86%	88%	79%
Customer's Benefit by Revamp with MCO		- Totally, -13.7% was improved. - MCO proposed above details at early stage of project and customer could plan concretely. - Many technologies are applied for value-up. (Overhung damper, Swirl canceller, etc...)		

Figure 3.1.1 Summary of FPR Study for other OEM

Service	Charge Gas Comp.	Tertiary Ref. Comp.	Hydrogen/Methane Comp.
Replacement Summary (Ship date : 2019)			

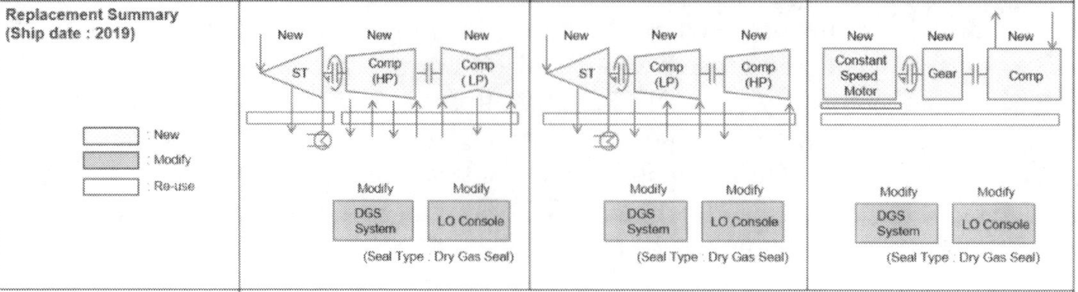

Category	Modernization & capacity up
Turbine/Motor Rated Power	+30.5% ~ +69.7%
Customer's Benefit by Revamp with MCO	- The higher improvement of OPEX would be expected than OEM revamp. - All casings of compressor & steam turbine will be designed to have the capability of future capacity-up (+22 %) by Footprint Replacement.

Figure 3.1.2 Summary of FPR Study for other OEM

Service	MCR LP	MCR-HP	Propane
Replacement Summary (Ship date : 2020)			

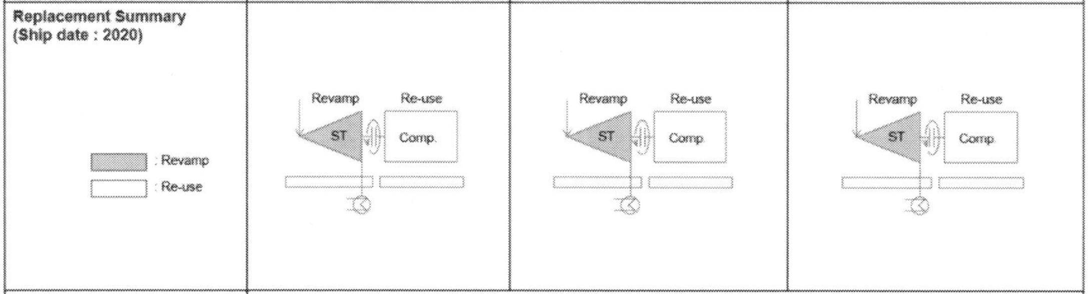

Category	Capacity-up & Value-up
Turbine Rated Power	+20.0 ~ 44.9% (All turbines are identical design)
Customer's Benefit by Revamp with MCO	- Capacity-up by identical internal parts for 3 turbines. (1 spare rotor is common for all) - Many technologies are applied for value-up. (PTA welding, Boronizing treatment, etc...)

Figure 3.1.3 Summary of FPR Study for MCO machine.

3.2 CCS

MCO has significant experience with continuous duty CO_2 compressors used in urea manufacture, CCS, and EOR (Enhanced Oil Recovery), enabling the safe operation of carbon capture systems.

Shown in **Figure 3.2.1** is an inline CO_2 compressor delivered to a world pioneering CO_2 capture and storage project in 2003. The CO_2 compressors is an Industrial-scale CO_2 Storage project that have been in operation since 2004. The highest-class discharge pressure in the world, in CCS application.

Shown in **Figure 3.2.2** is an integrally geared CO_2 compressor delivered to world largest 5000 t/d CO_2 recovery project using the CCS technology of MHI. CO_2 is compresses by 1,885 psi (130 bar) and the largest class capacity in the world for CCS/EOR compressor.

Figure 3.2.1 Inline Compressor for CO_2 compression

Figure 3.2.2 Integrally-geared Compressor for CO_2 compression

4. Conclusion

(1) Revamp solution can help with energy savings. Required power and steam consumption can be improved by revamp.

(2) MCO's features focus on minimum modifications. Minimum modification also contributes to energy saving because it prevents unnecessary modification.

(3) MCO has the Know-How and vast experience of footprint replacement to minimize site work by re-using existing foundation without modification.

(4) MCO can propose proven & advanced technology. Customer can obtain countermeasure for concerns of existing machines and latest technology by revamp.

5. Reference

[1] Matt Taher et al. "Carbon Capture: CO2 Compression Challenges", 2021.

96b A Survey of Compressor Wash Oils

Michael SMITH

A Survey of Compressor Wash Oils

Dave Smith, Senior Marketing Manager, Nalco Water, Sugar Land, TX

The Process Gas Compressor (also known as the Cracked Gas or Charge Gas Compressor) is a key unit operation in the production of ethylene via the steam cracking process. It is a complex and expensive piece of equipment, and its operation often determines the run length of the unit – a compressor malfunction can result in the entire unit being shut down. Declining performance of the compressor can negatively impact energy consumption and carbon emissions, product yield, and throughput, can cause damage to the machine, or unplanned unit outage. To prevent these costs, ethylene producers have implemented a number of mitigation measures. One such measure is the use of a wash oil on the compressor.

Wash oil is a liquid hydrocarbon used to remove polymer that is formed in the compressor. It acts as a scouring agent and a solvent for polymer that is formed. There are several qualities that comprise a high-quality wash oil – that is, one likely to perform efficiently at keeping the compressor clean and functioning efficiently.

The purpose of this paper is to discuss some of the qualities of a high-quality wash oil, the tests that can be used to determine those qualities, and a survey of various wash oils used by ethylene manufacturers. In addition, the paper will explore some of the changes in wash oil quantity and quality that have occurred over the past decade, particularly in North America, as manufacturers have moved towards cracking lighter feedstock slates.

96c Performance Optimization of a Charge Gas Compressor - a Story of Success

Serge Staroselsky, Ricky Lim, Alex Biggs

As most of the refineries and petrochemical plants were built 30 or more years ago and an increasing number of plant owners are searching for a new approach to maximize production throughput and reduce operational costs, the low reliability and poor performance of their equipment is a real problem that production managers are facing daily.

This article describes how compressor, steam turbine and heat exchanger equipment maintenance can be improved using effective diagnosis of the root causes that are generating performance degradation.

A case study is presented for the Charge Gas turbo-compressor train of ENI Versalis Ethylene plant located in Priolo, Italy, where recently implemented performance monitoring software "Total Train Performance Advisor" has been successfully used to diagnose compressor fouling due to polymerization.

The paper provides an overview of the performance monitoring calculation and demonstrates how such systems can be integrated with the plant DCS and Historian, so that operating personnel can easily monitor Key Performance Indicators, such as steam consumption efficiency, and plan and execute corrective actions

AIChE technical Paper reference Number: 96D

Title: Designing for reliability when using electric drivers to reduce carbon footprint in petrochemical plants

Author:

David Anderson, Chief Specialist Electrical Control and Process Automation Systems has 16 years experince at TMEIC implementing Electric drive and motor systems in Petrochemical, Mining, Cement, and water/wastewater applications. He holds a BS in Mechanical Engineering from Virginia Polytechnic Institute and State University in Blacksburg Virginia,

Abstract:

The application of electrical motors and drives for high-power compressors is steadily increasing in both prevalence and capacity, due to pressures for point emissions and carbon footprint reduction as well as the energy savings process improvement available in motor speed control. Large electric drivers have been in use for more than 40 years and even the latest and largest topologies have more than 10 to 15 years of experience in industrial and petrochemical applications. To users more familiar with mechanical drivers, the switch to an electric driver, particularly an electric driver with a variable frequency control, involves re-learning a new set of risks and requirements.

The process outlined in this paper will help to connect the application and specification of electrical equipment to the desired result of high system reliability. The reliability of any system involves considering the appropriate application of each component in a system. In addition to general application parameters the applied environmental conditions such as in the case of an electrical system temperature, current, vibration, voltage must be within the limits of that component.

A typical electrical drive system is made up of switchgear, utility side impedance (reactor or transformer), switching power devices for frequency conversion, a cooling system, a motor and the driven equipment.

The operating conditions for each of these components and appropriate design for those conditions can be understood and optimized through careful study of the application and use of appropriate standards and specifications. When this assessment is done as outlined it contributes to the extension of the system's useful life, increases reliability, and provides value to the process operator.

Introduction

A plan to maximize system reliability must include an evaluation of the application and operating conditions, auditing of the manufacturing process, effectively specifying the equipment to meet installation and regulatory requirements as well as a plan for routine maintenance.

Understanding the application

The first question involved in ensuring the reliability of process machinery in any industry is to ensure that the purpose of that machinery is understood. The required, load speed and duration of operation must be confirmed, including the expected method of operating the equipment.

From the desired operating characteristics a mechanically driven, fixed speed electric or variable speed electric driver can be selected to meet the process requirements. If equipment is selected that cannot deliver the desired operating envelope or performance, the effort to obtain a satisfactorily functioning and reliable system will be undercut at the very start. Performance priorities must be categorized and ranked. Then rotating equipment must be selected that meets the process performance requirements. However, since one of the greatest factors in reliability and user satisfaction is having experience with a particular method, priority is often given to equipment with which plant personnel already have experience. However, a careful selection process, followed by investment in engineering and maintenance, can greatly reduce the risk associated with adopting advances in technology or control.

When this process is applied to an electric driver system the first consideration should be that an electric driver system is most vulnerable to variations in installed environment, electrical utility variation, and load requirements. Collecting data to understand these parameters in a particular application and communicating that data to the equipment designers is critical to ensuring that equipment will operate reliably. This information can then be communicated to the manufacturer utilizing the datasheets that accompany standards such as API 541 or IEEE 1566. These incorporated datasheets prompt the user to provide key site characteristics including utility data such as available voltages, expected voltage variation, available short circuit, required harmonic performance to the electric equipment vendor.

When selecting a variable speed or constant speed electric driver, the key deciding factors are the need for variable speed process control, the amount of load required during a start and the impact of starting on the power grid. When a DOL (Direct on line) start of a motor is performed the current drawn from the power grid may be as much as 6 times rated current. The resulting momentary load increase causes a voltage drop throughout the power system while the motor is being started (Figure 2). When started on a power grid with reduced overall capacity, such as one that has a longer power line between the load and the substation, the voltage drop can be enough to negatively impact other loads at the plant or other customers that share the utility. In this case depending on the required load during a start a variable speed drive may be the only way to start the load. While the increase in total number of components required by the VFD(Variable Frequency Drive) may appear to be a downside there are significant benefits that must also be considered. The significant voltage variation that is experienced during a DOL motor start is a significant risk to overall plant reliability. Additionally the use of a VFD reduces the mechanical forces experienced in a motor winding during a start and the overall stress on the power system from inrush current. The VFD is able to operate the motor at its ideal slip frequency from the beginning of a start, reducing stress on many components of the system. These advantages can result in a net reliability gain if the correct steps are taken to ensure reliability of the VFD system.

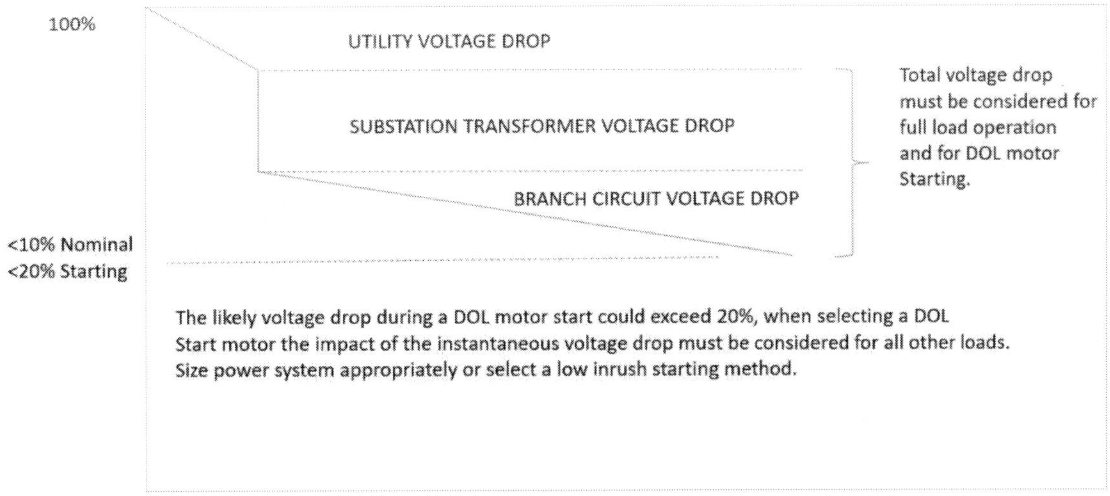

Figure 1: Voltage drop evaluation for a branch circuit in an industrial facility

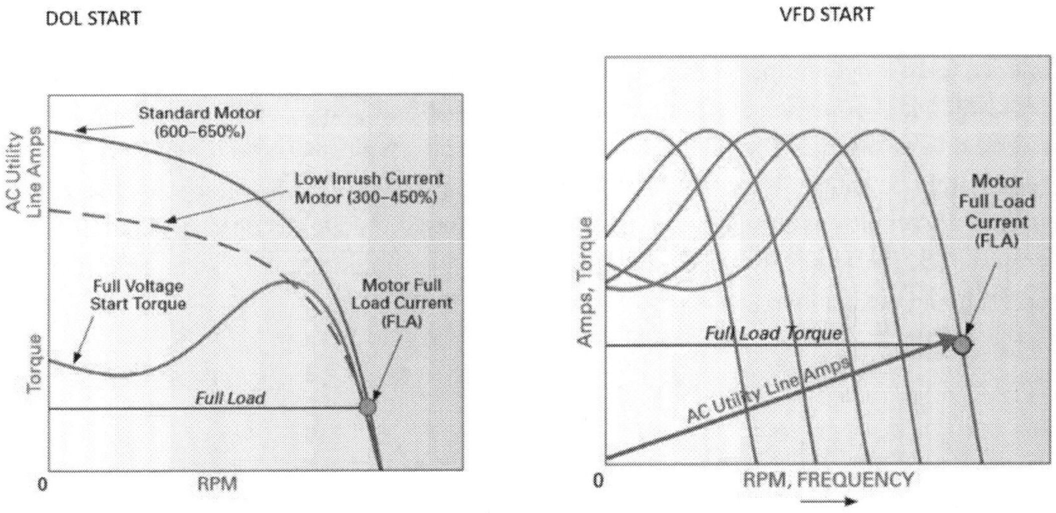

Figure 2: Available Torque and Current comparison for motor starting.

Using Equipment Standards to maximize reliability.

A variable speed electric driver is typically made up of the following key parts, each of which can be defined effectively utilizing the appropriate industry standards.

1. Input disconnection for isolation and protection of the equipment.
2. Inductor or transformer used to condition the power quality and isolate the VFD from power system disturbances.
3. Frequency converter to provide variable frequency power to the motor.
4. Motor to rotate the driven equipment.

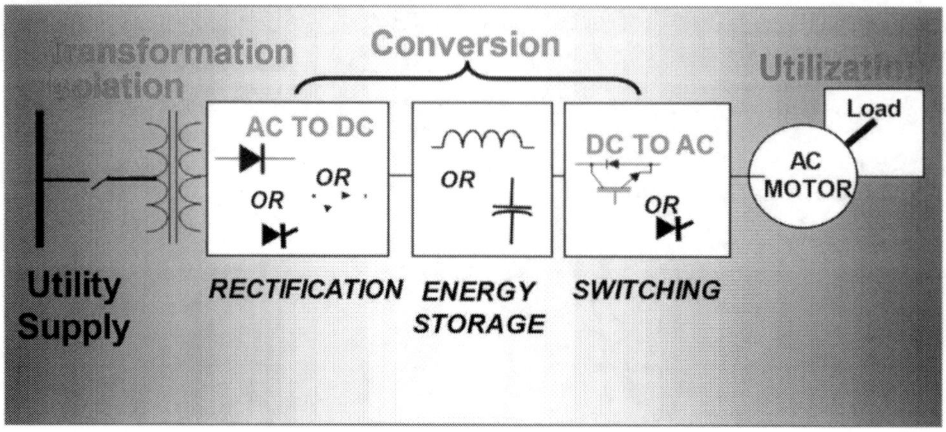

Figure 3: VFD System

The standards developed by IEEE, API, and NEMA are important tools to ensure the reliability of an electrical system. They provide the building blocks to understand the application of the equipment, understand the constraints due to various designs or topologies, and apply conservative principles for reliability and safety.

Input Disconnect

The maintenance team, engineering team and vendor all need to understand the application of the switchgear for isolation. For instance a medium voltage circuit breaker may be rated for only a few thousand operations before being rebuilt, however a contactor is typically rated for several hundred thousand cycle before a rebuild. If electrical equipment for a process must be energized and de-energized frequently provisions should be made to utilize a medium voltage controller that disconnects the load using a contactor. In all cases the expected usage of the disconnecting means and the required service interval should be included in the maintenance plan.

Application related reliability considerations for input disconnecting switchgear

a. Consistent testing protocols
b. Correct application of switchgear:
    i. What is the designed number of switching cycles?
    ii. What is the required yearly number of switching cycles for the application?
    iii. What is the designed breaking and closing current?
    iv. What are the required breaking and closing currents for the application?
    v. What is the expected maintenance performance, how often with the equipment be accessed by personnel, and what is required to utilize the equipment safely?
    vi. What is the expected switching surge when the device is opened, and what mitigation is required to protect other electrical equipment in the power system from that switching surge?
c. Correct installation of the switchgear
    i. Are the capabilities and characteristics of the switchgear suitable for the installed environment.

Input isolation transformer

The following IEEE standards are used to ensure the isolation transformer is applied correctly for the required service.

C57.12.00/01 define the general requirements for power and distribution transformer, C57.12.90/91 define the testing methods for transformers. C57.18.10 defines the requirements for transformers used with power conversion equipment.

C57.18.10: Practices and requirements for Semiconductor Power Rectifier Transformers

To get the best life out of a VFD service transformer, the designers must understand the impact of harmonics on insulation aging from a thermal perspective and with regards to the voltage stress applied to the insulation. Applying an insulation system suitable for the observed voltages and temperatures means a longer than 20-year service life can be obtained. Section 8.7 outlines a method for analyzing the expected load losses that are observed in the operation of a rectifier transformer used on the input or output of a VFD. Discussing the expected harmonics, and the calculation of the losses based on that spectrum with the associated engineering team is the first step in ensuring that the transformer can support reliable operation for in the user's process for the life of the equipment. Without this detailed analysis the user runs the risk of an early failure due to under specifying or an excessive upfront cost due to over specifying the equipment.

C57.12.90 Test code for liquid immersed Distribution Power and regulating transformers

&

C57.12.91 Test code for dry type distribution and power transformers

These standards define the suggested testing methods for general application of dry type and liquid immersed power transformers. The test plan for a critical service transformer must be reviewed against the recommendations in these standards. Load loss testing, (section 9) dielectric testing (section 10), and temperature rise testing section 11, should be carried out based on accurate understanding of the service conditions for the transformer. This is critical to insuring reliable operation of the transformer throughout its service life.

C57.12.00 General requirements for liquid immersed Distribution Power & regulating transformers

&

C57.12.01 General requirements for dry type distribution and power transformers

In addition to specifying minimum construction standards for the Vendor, these specifications also define how the transformer is expected to be applied. For instance, Section 4 defines the expected design constraints for "usual" and "unusual" service conditions. A defined temperature operation range is listed as a maximum air temperature of 40C and an average temperature over 24 hrs. of 30C. If either the average or maximum temperature is exceeded during normal use of the transformer, then C57.91 must be referenced to determine the expected impact on transformer life and reliability based on the operating conditions and determine the appropriate corrective path to maintain reliable operation in critical service applications.

The definition of operating voltage, utility connection and dielectric insulation rating, supplied in C57.12.00 and C57.12.01, is critical to attaining long transformer service life. Attention should be paid to the type of utility grounding system and risk of transient voltages at a particular site. These factors must be used to determine the transformer insulation system rating. At a 15 kV nominal system voltage the transformer insulation must be rated for 34 kV with an ungrounded or resistance grounded system, but only 26 kV for a solidly grounded system. Accurate communication with the transformer vendor regarding the applied utility voltages can ensure the insulation system of the transformer is specified in a reliable manner. Standard and optional peak voltage insulation ratings are also listed for each voltage class. If a user understands their power system characteristics, including risks of switching surge and lightning events at a particular site, they can use this information to select a peak voltage rating that will increase the likelihood of reliable service throughout the life of the equipment.

Questions to ask when defining the insulation system requirements for a VFD service transformer.

1. What is the plant grounding system?
2. What is the expected extent of voltage variation?

3. What is the expected risk of lightning strike or switching surge?
4. What are the expected harmonics produced by the drive, and already existent on the utility?

Defining VFD characteristics using IEEE 1566

The IEEE 1566 VFD standard provides guidance in the specification of variable frequency equipment which can contribute to reliable operation of power equipment on critical loads. It is intended to be applied for variable frequency drives greater than 375 kW and 750 Volts, such as are used throughout the petrochemical industries. The purpose of the standard is to help the user define the performance requirements of a VFD system in a manner that improve the clarity and quality of communication between the user and manufacturer.

As with all electrical equipment, the primary risks to a VFD are ambient temperature and other environmental considerations, over voltage (either continuous or transient), and exceeding the designed current. Utilizing IEEE 1566 to clearly communicate the expected operation environment with the vendor is a major contributor to successful implementation of a VFD system.

The IEEE 1566 Standard is comprised three main parts. The first is the datasheet which helps the user communicate the expected application and performance requirements to the manufacturer. The datasheet is made up of three pages intended to be filled out by the user, and three pages intended to be filled out by the manufacturer.

The purchaser section includes details pertaining to the supplied utility and grounding characteristics, the installation physical characteristics, the expected performance and configuration of the VFD, the required testing, and the required special services. Effective use of the utility section, performance sections, and testing sections helps to ensure effective communication needed to specify a system that can operate reliably in the intended service.

The portions of the datasheet meant to be filled in by the manufacturer includes the rating and mechanical characteristics, the expected performance, the expected impact on the motor insulation, as well as a basic summary of the VFD design and component ratings.

A complete understanding of each of these items is important to ensuring clear user-manufacturer communication. A critical part of this discussion with the manufacturer regards the voltage rating of the power devices in a drive. Selecting a power device with a conservative voltage rating is a key contributor to VFD reliability and an important part of evaluating the FIT rate (Failure in time) for each component [1]

Importance of Equipment preservation during shipping and storage

Storage conditions during shipment may have negative impacts on the life of the equipment.

- Humidity during storage
- Vibration
- Rodent/insect damage
- Aging of components

Inspection should take place when receiving and during storage to ensure that environmental conditions do not impact equipment reliability. Check with vendors to confirm any storage life limitations on reserved sparts to be sure they are ready for use when needed.

Importance of a Maintenance Plan

What actions are required to maintain environmental conditions?

- Cleaning
- Filter replacement
- HVAC system maintenance

Regular in-person inspection of the electrical installation is recommended to observe abnormal temperatures, odors, vibration and other physical conditions such as corrosion or mechanical abnormalities [3].

Which components have a wear out period and what is the timeline for that wear out period? (Figure 4, a)

- Anti-friction Bearings
- Electrolytic capacitors
- Mechanical seals
- Mechanical switching devices

Which components do not have a wear out period(Figure 4, b).

- Semiconductor power devices
- Film type capacitors

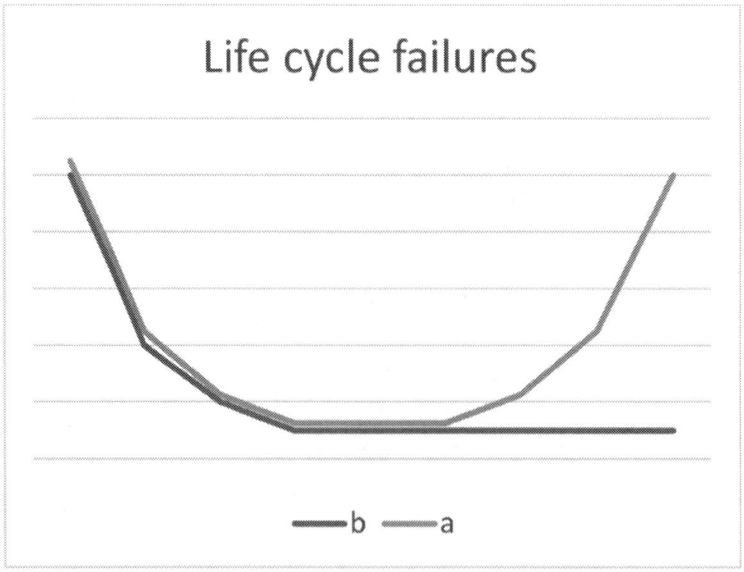

Figure 4: Comparison of expected failures of components subject to wear (a) and components not subject to wear (b)

Redundancy as part of a maintenance plan

While a component or system failure can occur in any method of power delivery, dependence on electric power as the main supply for an industrial process adds an external vulnerability. To alleviate this risk the user can take the following steps. Historical data for the power system should be collected. Reports on the yearly experienced outages should be created and evaluated. The expected performance of the electrical systems should be evaluated for short periods of power loss (less than 1 second) as well periods of power loss greater than 1 second. The operator can then plan for the impact of utility reliability on the process and make provisions in the electrical control system to minimize disruptions. Potential strategies could include redundant power feeds, energy storage for key elements of the process, or local generation capability.

An evaluation of the technology available, the expected reliability of the equipment, the importance of maintaining process operation and the desired method of returning to service all figure into choosing the appropriate redundancy for a given process. This evaluation helps to select one or more of three different approaches to redundancy.

Component redundancy involves providing redundancy of individual parts in a system such as control power supplies, PLC controllers, or power modules in a variable frequency drive. This redundancy improves the likelihood that equipment can continue to operate after a failure, however, it typically requires the equipment to be shut down to replace the failed component. Additionally depending on the design, it may increase the risk of a cascading failure. However, it does allow the system to continue operating until downtime can be scheduled.

Assembly redundancy involves arranging duplicates of entire portions of the power system redundant transformers, redundant VFD banks, motors or controller systems. This increases the likelihood of continued operation in the event of a more widespread failure. Additional this configuration makes it easier to work on a portion of the equipment while the process continues running. Cost and space requirements are typically increased and depending on the arrangement of the system, switching components on or off-line during a failure may present an additional challenge.

Process or capacity redundancy involves duplicating the entire process so that no matter what fails in the system the rest of the system can continue operating, typically at a reduced capacity. This type of redundancy provides the greatest capability to continue operating during a failure, and ease of servicing equipment, but also has the highest cost, because the driven equipment must be duplicated.

With planning for safety, redundancy can be used to allow maintenance during operation. The upfront cost of redundancy must be evaluated against the risk and cost of downtime.

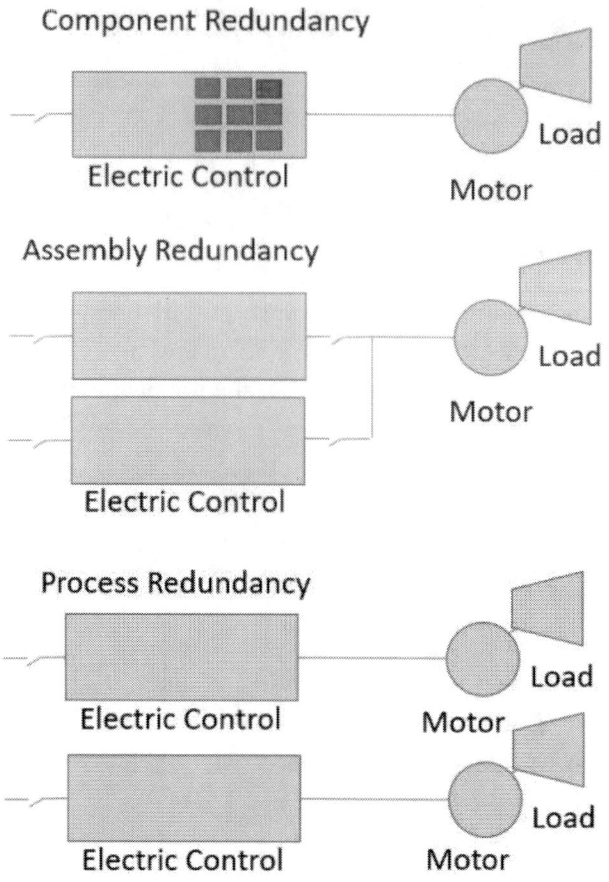

Figure 5: Electric driver redundancy configurations

Conclusion

Process requirements, available utilities, and required regulatory compliance may limit the available power sources in a particular application. However, high reliability is possible with careful engineering throughout the design and commissioning process with any of the commonly used methods for power conversion.

A reliable system with a long service life depends on investing the resources required to understand the installed conditions and application requirements of the equipment.

Utilizing the appropriate standards provide a framework for clear communication throughout the engineering and manufacturing stages of a project.

After installation, a well planned and executed design can still cause unnecessary and costly downtime unless the requirements for maintenance are followed proactively. There must be continued engagement on the part of the owner and manufacturer to keep the equipment running as intended.

Acknowledgements:

The author would like to thank Elliott Group and TMEIC for the support in publishing this paper. In addition, the contributions from Douglas Phares and Michael Baldwin are greatly appreciated.

References:

[1] Manish Verma, "Application Edge Volume 1 Issue 6 Variable frequency drive reliability"

[2] Hiromi Hosoda, Mostafa Al Mamun and Teruo Yoshino "Trends in MW-Rated VSI Technology and Reliability for Adjustable Speed Drives"

[3] Albert Sijm and Manish H Shah "The road to a fully integrated asset management strategy applied to rotating equipment"

Utilizing Simulation Tools to Ensure Successful Motor Startup of Centrifugal Compressors in Ethylene Cracker Applications after Switch from Turbine to Electric Drive

AIChE technical paper reference number: 656291

Justin Iwinski
Elliott Group, Systems Engineer
Jeannette, PA
jiwinski@elliott-turbo.com

Mario Valdivieso
Elliott Group, Application Engineer
Houston, TX
mvaldivi@elliott-turbo.com

Keywords: dynamic simulation, electrification, ethylene, motor drive

Abstract

Traditionally, critical compression processes in ethylene plants have been steam turbine driven due in large part to the high power demand. However, current industry and global trends to curtail emissions and increase efficiency have highlighted the benefits, of electrification (or motorization) of these critical processes. These potential changes result in a significantly different set of requirements and constraints in order to start and operate compressor trains. To ensure the motor is sized properly for the process conditions and operational expectations of the end user, simulation tools can be utilized to evaluate the torque requirements for various electrical configurations, including direct-on-line (DOL), variable frequency drive (VFD), or soft-start current limiting devices. This paper will explore the system details relevant to developing a speed vs. torque curve, the sensitivity of the output to these variables, and the operational adaptations that may be required to support an electrification initiative.

1 Introduction

Centrifugal compressors are used in the ethylene production to move feedstock through a steam cracking process (cracked/charge/process gas compressor), and to provide the cooling integral to the separation process (propylene and ethylene refrigeration compressors). Steam turbines drivers have been a natural choice to run these compressors due to the availability of steam from the cracking process, but the generated steam could instead be utilized to recover power via electric generators to support a transition to electric drives for critical compressor applications.

The benefits of electrification in ethylene production are well understood in terms of efficiency and ease of maintenance. It is an enticing notion to "drop" an electric motor into a functional process and reap the benefits. "Drop in" solutions become more complicated during the motor selection process, however, due to the array of possible motor types and driver configurations. Will the motor be fixed speed direct-online (DOL) or will a variable speed/frequency drive (VFD) configuration better suit the functional needs of the service in question? A DOL configuration may be attractive in terms of capital expense, but can the local power grid support this type of load? Perhaps a soft-start device to minimize in-rush current is needed due to local grid stability. In any case, the familiar acceleration schedule of a steam turbine, which is friendly in terms of acceleration torque demands, will likely not crossover as part of the electric motor startup procedure. The last thing an end-user wants is a compressor that will start only after flaring costly hydrocarbons.

The same considerations apply for new construction where previously long, high power compressor trains can be split into shorter, electric motor-powered trains operating in series.

Simulation tools can be used to establish torque requirements during acceleration for various starting configurations while considering the operational/process needs of the plant. However, a major hurtle for effective analysis is the general understanding of how certain parameters affect the starting load torque of a compressor. The focus of this paper is to identify the critical parameters and demonstrate the sensitivity of each parameter on the simulation output.

Figure 1. Electric drive double-flow large frame compressor, typical for stage 1 cracked gas service.

Typical centrifugal compressor services in ethylene production include multiple stage iso-cooled arrangements (charge gas/cracked gas) and sideload refrigeration arrangements (ethylene/propylene/propane). For simplicity, this paper will use a single-section compressor to demonstrate parameter sensitivity.

2 Critical Startup Simulation Parameters

The parameters required to accurately determine the compressor gas load and required acceleration torque can be separated into two categories: 1. compressor design details and 2. loop configuration details. The compressor design details will be available from the compressor original equipment manufacturer (OEM) and the loop configuration details will need to be obtained from the site and/or operations. This makes the simulation a collaborative effort between the end user, engineering procurement contractor, and the compressor OEM. Therefore, if a current (or past) customer approaches a compressor OEM for an updated speed vs. torque curve, the OEM will likely not have all the relevant details to generate a curve that reflects the actual dynamic performance of the machine.

What is a speed vs. torque curve?

Often referred to as an N-T curve, a compressor speed vs. torque curve illustrates the torque required from the motor to accelerate the rotating equipment to a speed where performance control can be enabled. This speed may be the low end of the continuous operation speed range for a variable speed drive, or it may be motor rated speed in the case of a fixed speed drive. Usually presented on a 'per unit' scale, the curve will indicate the expected motor torque (as a percentage of motor rated torque) required by the system through the speed range (as a percentage of compressor rated speed). Note that motor speed and compressor speed is equivalent when presented as a percentage of rated speed (the actual shaft speed, low vs. high, will be offset by the gear ratio employed).

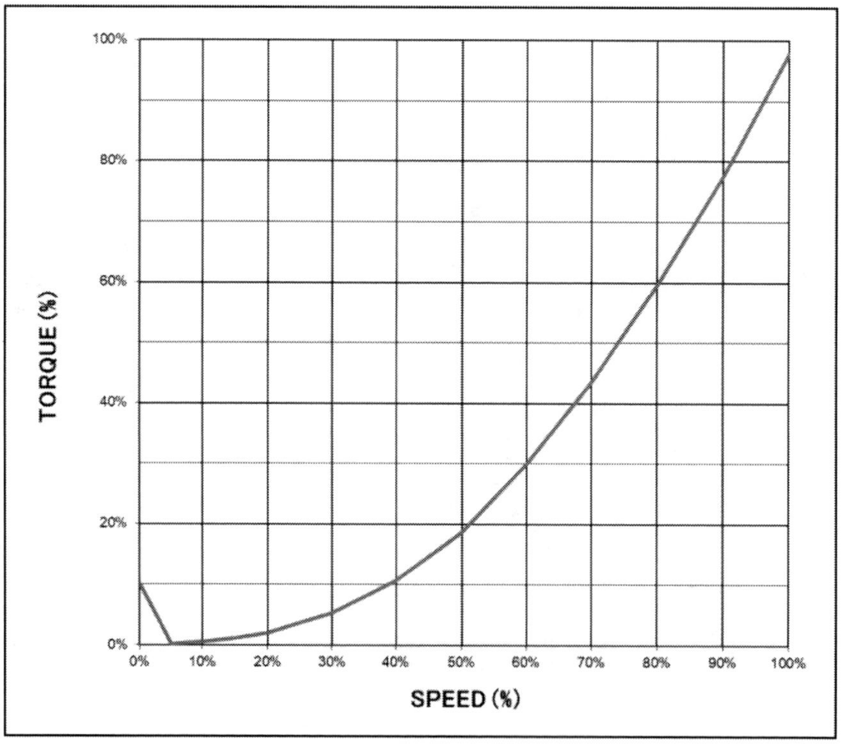

Figure 2. Sample speed vs. torque curve. The curve illustrates an estimated break away torque at 0 RPM (10% of motor rated torque) and proceeds to plot the driven equipment load torque at 10% increments up to 100% motor rated speed.

2.1 Compressor Design Details

The compressor design details can be obtained from the compressor selection and performance information, and therefore, it can be assumed that the OEM will have access to these details. The details are based on internal design characteristics of the compressor, which map the gas conditions at the inlet to expected discharge conditions given the head/efficiency curves at a given speed. Once these details are established, and assuming that the process gas composition cannot be varied during startup, the only parameter that can be realistically adjusted is the final compressor speed. Minimizing the compressor speed during the startup transient will certainly affect the startup load torque.

Table 1. Compressor Detail Input Table

Data from Performance/Output		
Design compressor speed (map speed)	(rpm)	
Final compressor speed	(rpm)	
Bearing loss @ design speed	(hp)	
Mol wt		
K (Cp/Cv)		
Gas compressibility (Z)		
Critical pressure	(psiA)	
Critical temperature	(degR)	
Sidestream flange volume flow	(cfm)	
Design inlet volume flow	(cfm)	
External Design head	(ft-lb/lb)	
External Design efficiency		
Section design discharge pressure	(psiA)	
Section design discharge temperature	(degF)	
Data from Map		
External Stonewall inlet volume flow	(cfm)	
External Stonewall head	(ft-lb/lb)	
External Stonewall efficiency		
External Surge inlet volume flow	(cfm)	
External Surge head	(ft-lb/lb)	
External Surge efficiency		

2.2 *Loop Configuration Details*

The loop configuration details relate to parameters outside of the compressor itself. These details may be found on equipment datasheets, P&IDs, process licensor documentation, or may rely on operator experience/trend data from the site. Included in the configuration details are assumptions related to the operation of the motor, such as acceleration time/schedule and available torque. When considering the switch from a steam turbine to electric motor, these assumptions will be entirely new. Clever application of the starting assumptions can result in efficient torque usage from the motor, potentially allowing startup at higher loop pressures.

Table 2. Loop Configuration Details Input Table

Loop Configuration Details		
Discharge check valve downstream pressure	(psiA)	
Anti-Surge valve Cv		
Does recycle line return downstream of throttle/isolation valve?		
Throttle valve stroke fraction		
Acceleration time	(sec)	
Motor full load speed	(rpm)	
Motor full load power	(hp)	
Suction volume	(ft3)	
Discharge volume	(ft3)	
Initial loop pressure	(psiA)	
Initial loop temperature	(degF)	
Cooler outlet temperature	(degF)	
Total Train Inertia WR^2 (ref. to LSP shaft)	(lb-in2)	
Gear Losses	(hp)	

3 Parameter Startup Torque Sensitivity

In this section, a sample of the relevant parameters will be explored with simulated examples that illustrate their effect on the startup load torque. Because the effects may be amplified or muted by the drive type selection, those elements will also be considered for each item. Each motor type has advantages that can be quantified in a startup simulation.

The primary value that will be used to compare torque between cases is 'torque margin'. Torque margin (TM) is defined as the difference between the available torque from the motor and the compressor load torque expressed as a percentage of the available motor torque.

$$TM(\%) = (\tau_{motor} - \tau_{load}) / \tau_{motor}$$

In accordance with American Petroleum Institute (API), conventional sizing of electric motors is based on 110% of the greatest power required (including gear and coupling losses) for any of the specified operating conditions. API further specifies that a 10% torque margin (based on motor full rated torque) must be maintained between the driven equipment load torque and the motor developed torque during acceleration. Evaluation of this margin is especially critical on DOL configurations where the motor torque capability, already limited in the mid-speed range, must consider potential voltage drop to 80% rated voltage; soft-start capability must consider the current-limited torque curve. Despite these limitations, the TM during startup should not be permitted to influence the motor rating (unless in extreme circumstances) because of the potential for a negative impact to efficiency during normal operation, negating a portion of the electrification benefits. Therefore, it is optimal to find a starting sequence that fits within the power constraints of a conventionally sized motor.

Figure 3 illustrates a soft-start motor capability curve and simulated compressor load torque curve, with torque availability reduced through a current-limiting soft-start device. Note that the minimum torque margin occurs near the middle of the speed range in this example. Despite lowering motor torque availability, limiting current may also benefit the starting load torque profile by enabling a longer acceleration time, thus minimizing the acceleration torque required (lower in-rush currents generally allow for extended motor overheating intervals).

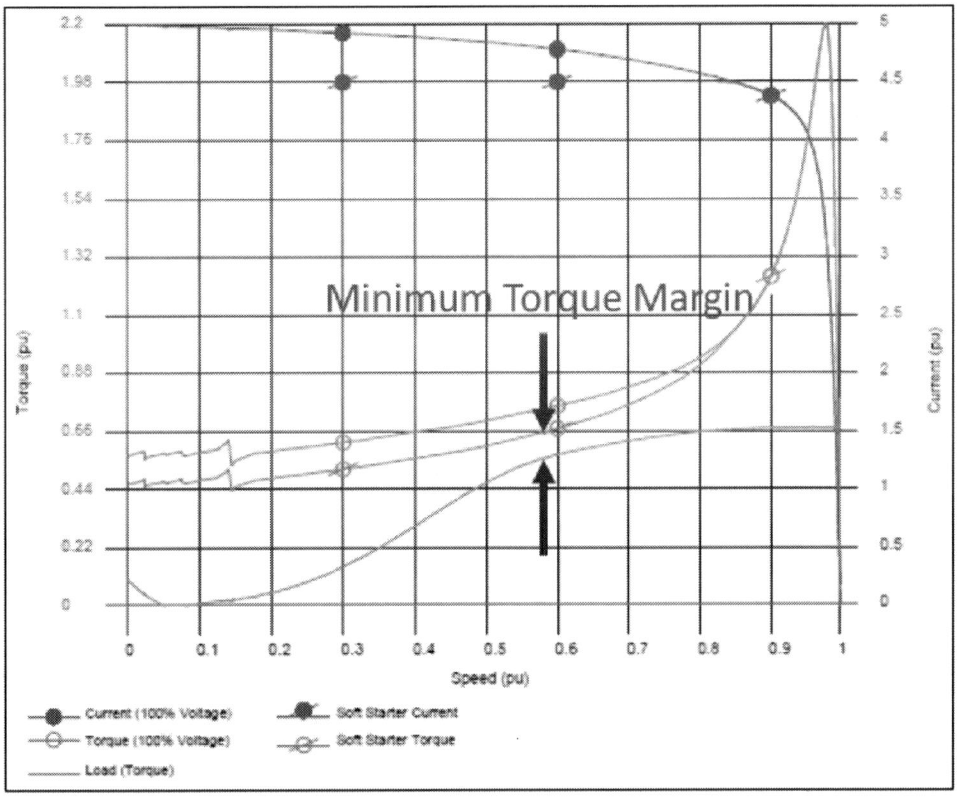

Figure 3. Induction Motor Capability Curve (Soft-Starter) and Compressor Load Curve

3.1 Speed and Acceleration Schedule

A steam turbine acceleration schedule is typically long; the acceleration schedule may exceed 1 hour in duration. The acceleration time for VFD can be adjusted based on the system needs, but 120 seconds (2 minutes) is typical. A Direct-on-Line acceleration must be very fast due to the heating associated with the large current draw; overheating risk can manifest in less than 30 seconds. A notional overlay on the steam turbine acceleration schedule highlights the time scale differences (Figure 4).

In terms of torque, the slower acceleration of a VFD and the ability to limit the acceleration to the minimum continuous speed of the compressor is a distinct advantage over DOL, albeit at additional capital expense. When developing the electric motor startup sequence for the compressor, the final speed and even the ramp rate can be manipulated to minimize the torque demands of the motor.

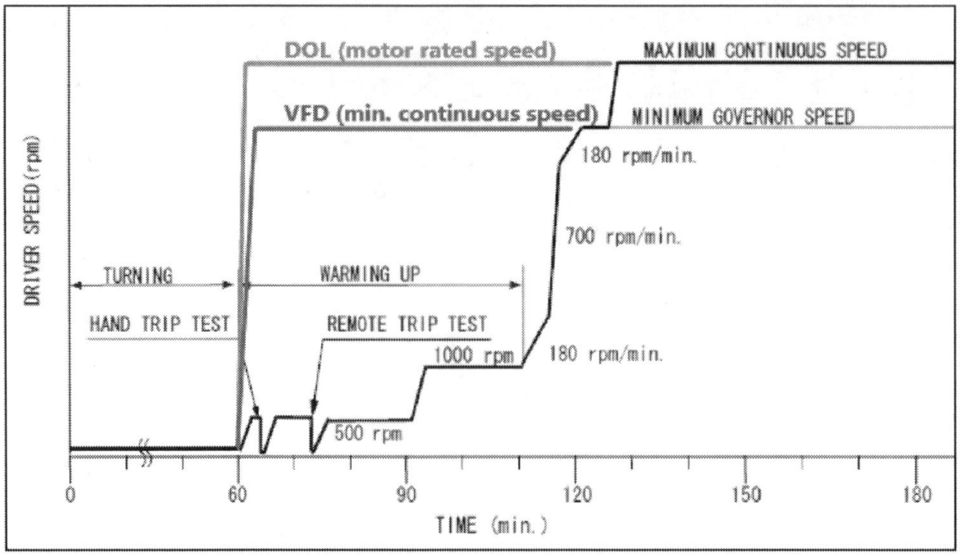

Figure 4. Steam Turbine Acceleration Schedule with VFD-motor and DOL-motor Overlay

The example below (Figure 5) illustrates a comparison of a VFD acceleration for a single-section mixed refrigerant compressor assuming a 120-second linear ramp with equivalent starting conditions. The only difference is the final speed: 80% rated and 100% rated speed.

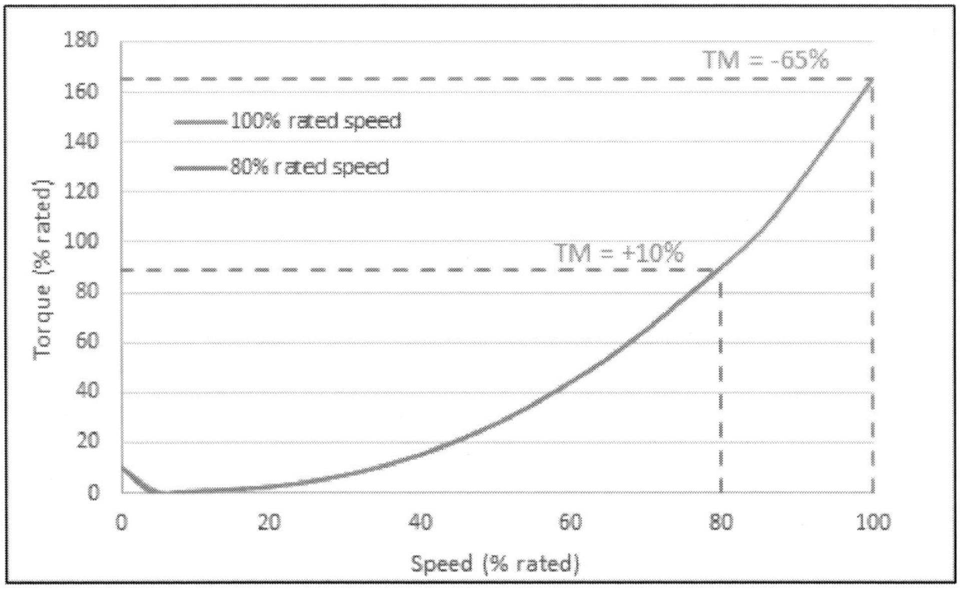

Figure 5. Torque Comparison, varied Final Speed (VFD)

Consider now that a fixed-speed drive must accelerate to 100% rated speed. It is not practical to size the motor to accommodate 65% more torque than normal sizing conventions suggest, therefore, the initial starting conditions (3.2) will need to be adjusted or a throttle valve (3.3) may need to be added to reduce the pressure at compressor inlet flange.

3.2 Initial Conditions (Pressure/Temperature)

The initial conditions can be a major contributor to startup load torque for services in ethylene production, especially in refrigeration loops that can see vapor pressures much higher than compressor design. Flaring is a costly, but effective, solution for starting against high vapor pressure with any driver type. Higher pressures are exceptionally difficult for a DOL configuration, but can be accommodated with additional equipment (3.3).

The example below (Figure 6) illustrates a comparison of a VFD acceleration for a single-section mixed refrigerant compressor assuming a 120-second linear ramp to 80% rated speed with different starting loop pressures. This parameter may be difficult to change given that it is largely based on gas inventory, loop volumes, and process/ambient temperatures, but the initial conditions of the desired startup must be specified for accurate torque modeling.

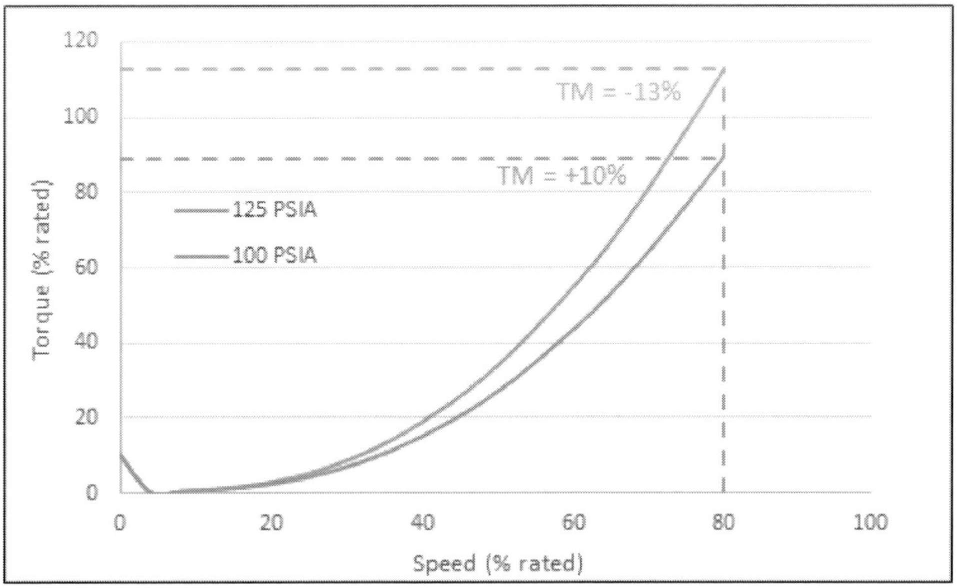

Figure 6. Torque Comparison, varied Initial Pressure

3.3 Throttle Valve Effects (Process Control Method)

The variable speed operation of a steam turbine also provides the process control method for the compressor train. A VFD configuration can mimic that type of operation, but a DOL configuration may need some other type of process control because the speed is not variable. Any process control change will likely require significant changes to the overall performance control system. A common type of control for fixed-speed configurations is suction throttling. Suction throttling can also be employed on variable speed configurations specifically for the purpose of reducing starting torque when accelerating with very high system loop pressures (as with warm, saturated refrigerant in a propane or ethane refrigeration unit).

A suction throttle valve provides a controllable mechanism to reduce the flange pressure at the compressor suction. As the compressor accelerates, a fixed throttle position will result in an increasing pressure drop. Minimizing the increased torque effect caused by high settle out pressures. The example below (Figure 7) illustrates the effect of throttling from a given initial pressure.

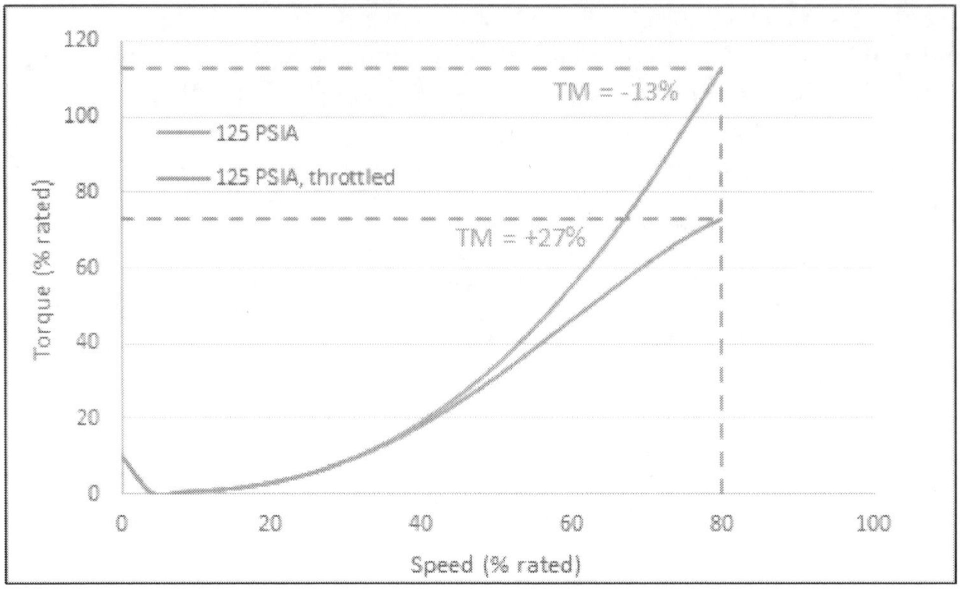

Figure 7. Torque Comparison, Throttled vs. Unthrottled (VFD)

3.4 Downstream Pressure

The system resistance downstream of the compressor discharge check valve should be understood and accurately captured in a startup simulation. In some processes, the compressor train system may need to be started against a closed discharge block valve. In that case, there will be no opportunity to allow mass to leave the loop during acceleration. However, if the conditions downstream support mass flow out of the compression loop, to a downstream condenser/accumulator for example, a slow acceleration via VFD can move mass out of the compressor loop, ultimately reducing the flange pressure for when the compressor reaches a high gas load speed range.

The example below (Figure 8) represents an unthrottled VFD acceleration in 120 seconds with discharge block valve closed vs. 240 seconds with discharge check valve modeling (i.e., the check valve opens when the system reaches normal discharge pressure). Each curve starts with an initial loop pressure of 175 PSIA; downstream pressure modeled at 250 PSIA.

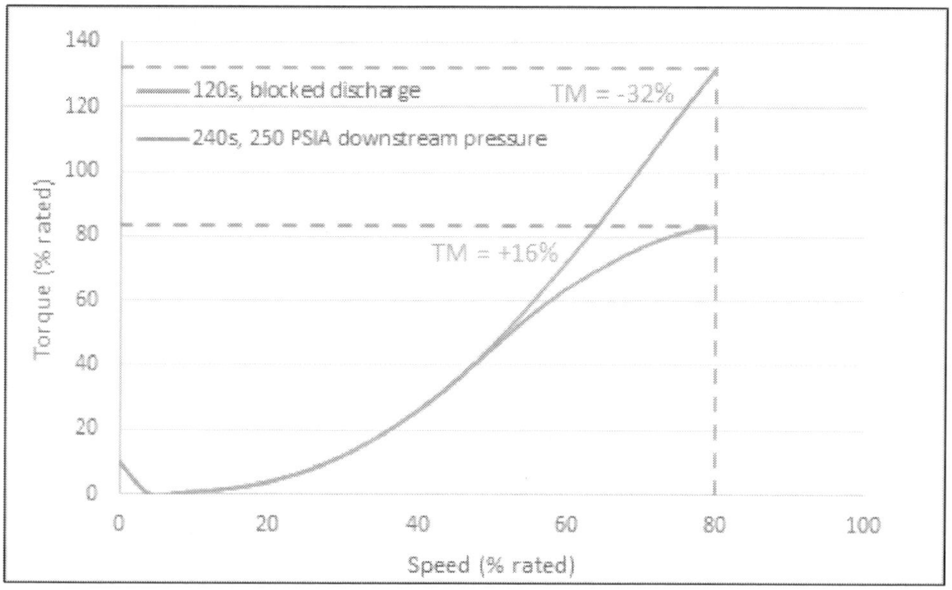

Figure 8. Torque Comparison, Discharge Blocked vs. Check Valve (VFD)

3.5 System Volumes (Suction/Sideload/Discharge)

System volumes directly influence the pressure rate of change around the compressor loop. A very large suction volume will resist changes in pressure, and therefore, a large suction volume will result in higher load torque for a given initial vapor pressure. Likewise, a larger discharge volume may allow a more torque friendly redistribution of mass throughout the system, thus enabling an effective pressure drop at the suction.

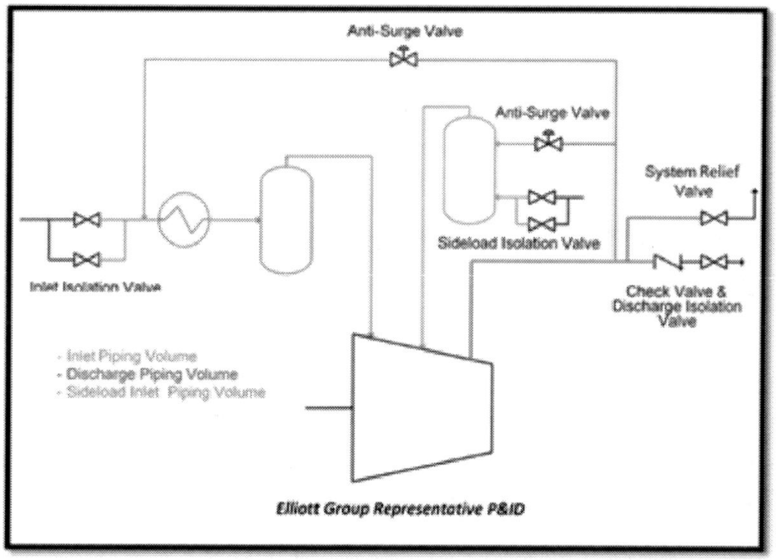

Figure 9. A typical 2-section sideload refrigeration loop, are outside of the compressor vendor's scope but must be estimated reasonably well or calculated from piping isometric drawings.

The volume effect is why suction throttling reduces torque when the valve is properly located. If a suction throttle valve is placed upstream of the suction drum/recycle tie-in, it will be ineffective at reducing the pressure at the inlet flange because: 1. the volume will be too large. 2. The recycle flow will be unthrottled.

The example below (Figure 10) represents a single-section throttled DOL acceleration (10 seconds to 100% speed) with a given throttled volume compared to the same acceleration with the throttle valve located farther from the suction flange (10x the suction volume). The volume increase may seem unreasonable, but consider the volume that a misplaced suction knockout drum could add to the throttled suction volume. Reducing the load torque back to +10% margin will take a significant pressure reduction via flare or other means.

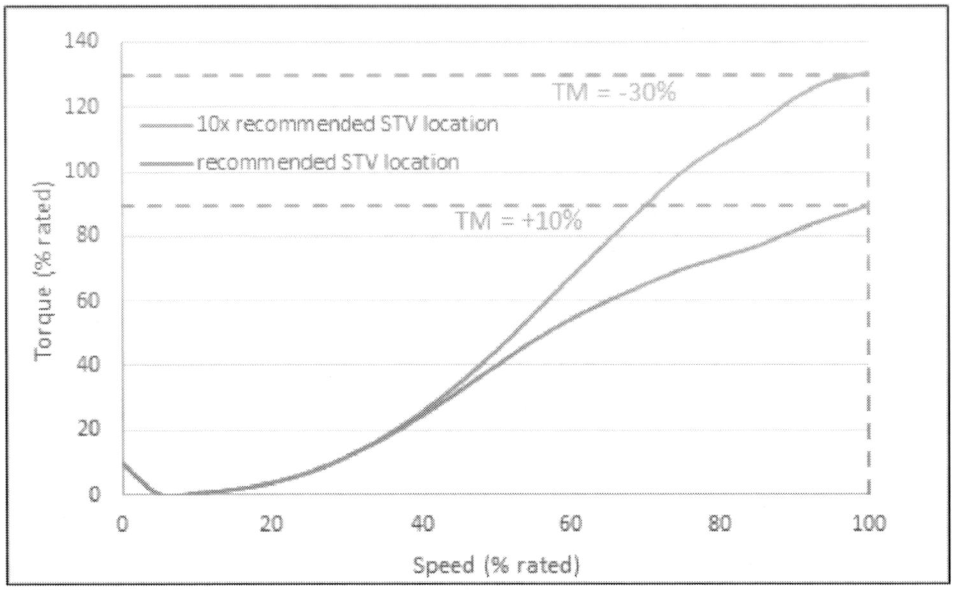

Figure 10. Torque Comparison, 10x Larger Throttled Volume (DOL)

3.6 System Inertia (Gear Ratio Effects)

The last parameter that this report will consider is the system inertia. If included in the model, the inertia will manifest in the speed vs. torque curve as the acceleration component of the total load torque (faster acceleration intervals will require more acceleration torque). Please note that if a speed vs. torque curve is being developed for a motor requisition, the inertia is typically not included because the motor inertia may not be available from the vendor at that stage. Instead, the driven system inertia (couplings, gear rotors, and compressor rotor) is provided to the motor vendor along with the expected speed vs. torque representing the gas load. For DOL configurations, the motor vendor will compute the acceleration time based on motor capability and total string inertia to ensure that the acceleration occurs within the motor's thermal limitations.

The important concept to consider when switching from a steam turbine is the inertia effects of a speed increasing gear. When inertia is added to a simulation model where the dynamic loads are computed on the low-speed shaft, the compressor rotor inertia (and other high-speed components) must be multiplied by the gear ratio squared. This is referred to as reflected load inertia. Preserving some torque margin, as was demonstrated in the examples above, is critical to ensure that there is motor torque available to accommodate the required acceleration torque, a concern that is nearly negligible in steam turbine drives.

4 Other Considerations

Beyond the increased starting torque associated with faster acceleration schedules, other torque related effects can lead to compressor design implications. For example, the criteria for shaft end sizing is very different between steam turbine and motor drive applications. Analysis for motor drives must consider steady-state and transient torsional excitations, as well as, short circuit air gap torque on the order of 3 to 15 times the full motor rated torque. Therefore, it is likely that a larger compressor shaft end will be required when switching from a steam turbine to an electric motor driver.

Furthermore, increasing the compressor shaft end size can force an increase in the size of the journal bearings, which are limited to the size of the bearing housing. Once the largest shaft end/bearing journal combination is exceeded, the only option on the compressor design is to go to the next compressor frame size. Larger shaft end sizes add to the overhung weights of the compressor, which can result in rotordynamic complications. Changes to the compressor frame size could also affect the design head, efficiency, and overall performance of the compressor.

Various types of shutdown/startup scenarios that may be encountered should be shared with the simulation vendor as well. Air run dryout or nitrogen purge operations can see molecular weights that are different from design, resulting in varying flow rates and load torques. Condition variations with respect to cold/hot restarts, seasonal process gas changes, or seasonal ambient temperatures can also influence the speed vs. torque curve. A comprehensive startup analysis should include these effects so that the highest load torque case will be captured.

These considerations, though not specifically included in the Section 3 parameter list, can affect the details that are provided as simulation inputs.

5 Conclusion

Dynamic simulation tools provide a method to evaluate the configuration and conditions surrounding a centrifugal compressor in order to optimize the startup sequence for efficient motor sizing. The optimization process can include adjustments to a variety of configuration parameters, introduced in Section 3, that each play a role in the dynamic behavior of the system. Though some parameters may be difficult to adjust without sacrificing process efficiency or increasing operational expenses, changes to other parameters may provide benefits with little effort. Consider potential changes to the initial starting conditions (pressure/temperature), where the feasibility of venting may depend on the type of process. For example, venting is likely not possible in a closed-loop refrigeration system which may be a determining factor in sizing a driver.

Simulations also provide the flexibility to experiment with alternate configurations/sequences in order to meet the existing electrical conditions available at the plant or to identify the scope of utility upgrades that may be necessary to support an electrification initiative.

Effective communication between all parties involved is the key factor in effectively applying assumptions for the startup dynamic simulation. Understanding which parameters are fixed and which parameters may be varied on a particular project will help to identify the envelope where a successful startup configuration can be engineered for the centrifugal compressor.

6 References

[1] Kalyan Malnedi, Shane Harvey, Axel Rauber, "Mechanical and Electrical considerations of electric motor/Drivers for large centrifugal compressors in Ethylene cracker applications" AIChE Spring Meeting, Ethylene Producers' conference, 2022

SPRING23
+19TH GCPS
A Joint AIChE and CCPS Meeting

Case Study: Ethylene Product Pump - Mechanical Seal Design Enhancements for Improved Reliability

Brian Kalfrin
John Crane
4001 Fair Drive, Pasadena Tx
bkalfrin@johncrane.com

Prepared for Presentation at
American Institute of Chemical Engineers
2023 Spring Meeting and 19th Global Congress on Process Safety
Houston, TX
March 12 – March 16, 2023

Case Study: Ethylene Product Pump - Mechanical Seal Design Enhancements for Improved Reliability

Brian Kalfrin
John Crane
4001 Fair Drive, Pasadena Tx
bkalfrin@johncrane.com

Keywords: mechanical seal, ethylene, pump, reliability, design, CFD, FEA

Abstract

One of the largest ethylene and polyethylene production complexes in the world located in Alberta, Canada was dealing with marginal reliability of the mechanical seals on their ethylene product pumps. The ethylene product pumps are classified as API 610 BB5 types and the duty conditions for the mechanical seals are arduous, experiencing seal chamber pressures continuously above 1000 PSIG at sub-zero temperatures. The pumps operate at shaft speeds above 5,000 RPM providing additional challenges for the mechanical seal to overcome. The heritage mechanical seals utilized in this application were a dual unpressurized configuration supported by an API Piping Plan 11 and 52. Through close collaboration with the end user, the supplied seal provided by the manufacturer was able to achieve an 18 – 24 month Mean Time Between Repair (MTBR). Known failure modes of the existing mechanical seal identified during failure analysis activities were addressed through modification of the seal to the current iteration. The most recent limiting factor in seal performance was attributed to breakdown of the dynamic secondary sealing element in both the inner seal and containment seal. The dynamic secondary seal, or o-ring, experienced a high degree of abrasion and breakdown due to suspected axial movement of the pump rotor. Additional analysis of the pump further supported these findings. It was the seal manufacturer's recommendation to address the root cause of seal failures by using an alternative configuration. The alternative seal configuration utilized active seal face features for optimized sealing of the ethylene while incorporating a non-pusher secondary seal (NPSS) to both the inner seal and containment seal to address the dynamic o-ring concerns. Additionally, improvements to the circulation and cooling of the Plan 52 loop were identified and optimized through CFD modeling of the internal circulation device and support system piping. All modifications were incorporated into a redesigned seal cartridge that was installed during a unit outage in September 2021. Performance of the new design to date has been extremely satisfactory despite several documented upsets in the process stream.

1 Introduction

The application is described as a Reactor Feed pump where the process fluid, ethylene, is delivered by a 12-stage barrel pump, API designation BB5. The process fluid conditions are as follows:

Process Fluid:	Ethylene
Temp:	-25 °C / -13 °F
Seal Chamber Pressure:	1,102 psig (75.9 bar)
SG:	0.446
Vapor Pressure @PT:	322 PSIA (21.2 bara)
Speed:	> 5,000 RPM

The original mechanical seal is classified as a balanced, pusher, dual unpressurized mechanical seal supported by an API Piping Plan 11 / 52. The seal utilized antimony-filled carbon and sintered silicon carbide seal faces, low temperature FFKM elastomers, and an integrated circulation device (pumping ring) on the containment seal to facilitate lubrication of the buffer fluid in the Plan 52 loop. The inner seal receives cooling and lubrication from the process fluid, supplied from discharge regions of the pump through the Plan 11 piping. The original mechanical had undergone several design adjustments to internal geometry to help optimize performance in terms of heat generation and leakage. In its current iteration, it was achieving a Mean Time Between Repair (MTBR) of between 18 – 24 months. The equipment owner was seeking to improve the MTBR of these pumps to meet scheduled outage intervals of at least 60 months.

The original mechanical seal is shown below:

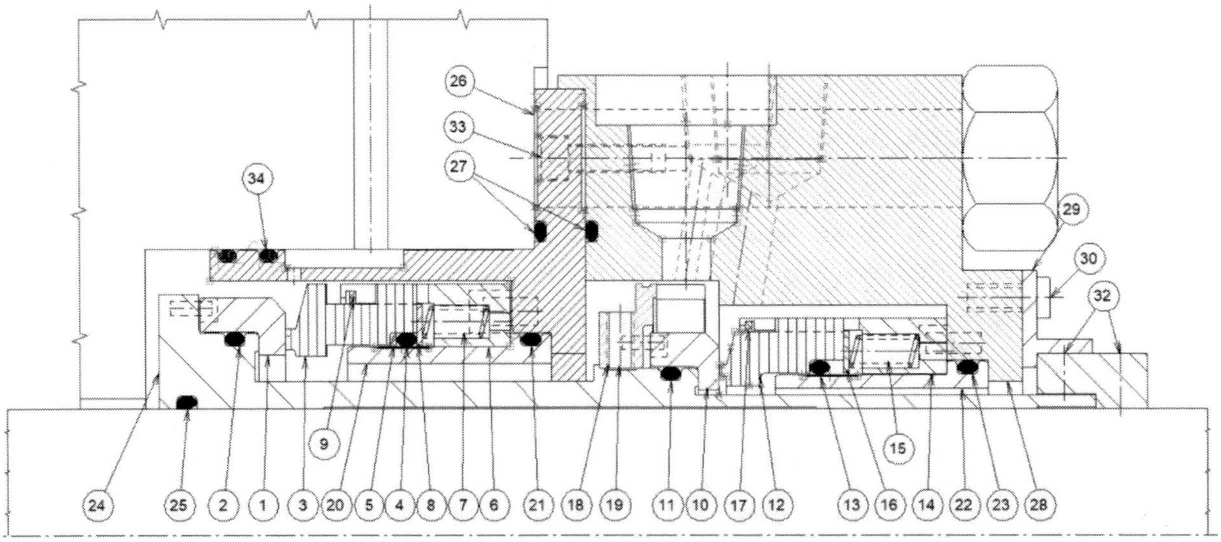

Figure 1: Original Mechanical Seal

2 Application Failure History, Analysis, and Resolution

2.1 Application Failure History

The historical mechanical seal design achieved between an 18 to 24-month MTBR as noted. The failure symptoms observed during seal component condition analysis were noted as follows:

Inner Seal:

- Excessive inner diameter (ID) wear and edge chipping
- Dynamic o-ring fretting and hang-up damage
- Polymerization of the process fluid at the seal faces

Containment Seal:

- Dynamic o-ring hang-up and fretting damage
- Mating ring o-ring heat damage
- Primary ring face wear, grain pullout, and ID edge chipping

On one occasion, during pump shut down after an inner seal failure the secondary sealing element (o-ring) suffered severe hang-up. This hang-up, combined with pump rotor axial movement (~ 0.0035" / 0.889 mm) translated to a gap variance between the seal faces in a static condition that allowed ethylene to leak to the atmosphere.

The primary ring damage identified as excessive wear, chipping, and blistering are indicative of elevated seal face temperatures, particularly in contact regions between the primary ring and mating ring surfaces. Microscopic examination of the seal faces found evidence of polymerized deposits on the face surface. It is not uncommon to develop localized areas of elevated temperatures at the seal faces that could support ethylene polymerization (1). Additional primary ring wear related to heat is consistent with the damage identified in the form of ID chipping and blistering.

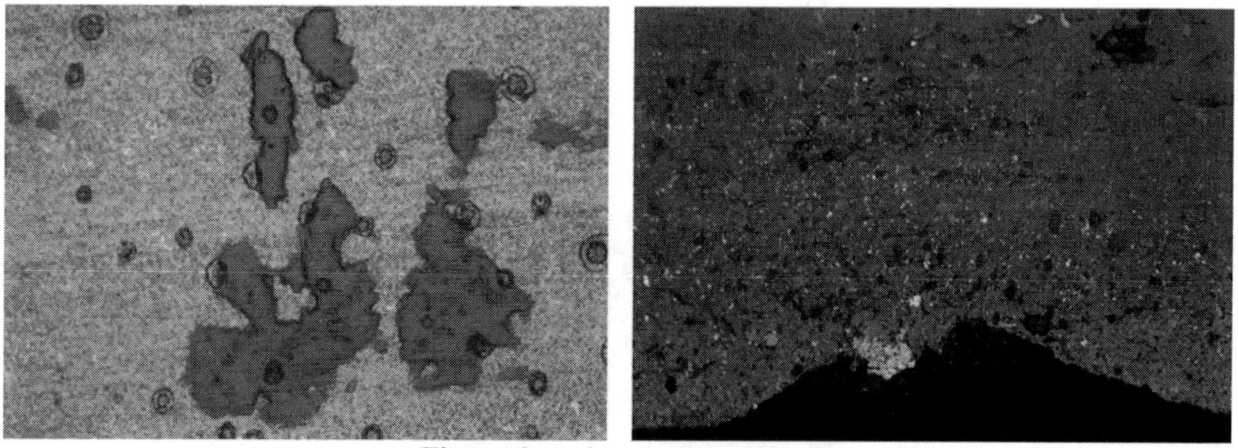

Figure 2: Primary Face Wear

The dynamic o-ring experienced severe fretting damage, even though the o-ring was sliding against a sleeve with a High Velocity Oxygen Fuel (HVOF) tungsten carbide coating. In many cases, application of a hard coating to adjacent sliding surfaces for a dynamic o-ring has been considered a preventative measure against fretting damage. The coating only masks the problem for a prolonged period, requiring the root cause of the fretting damage to be addressed. In this instance, shaft axial movement and subsequent seal gap instability due to vaporization of the process fluid at the faces were the likely contributors to the degradation of the o-ring during the operation of the mechanical seal.

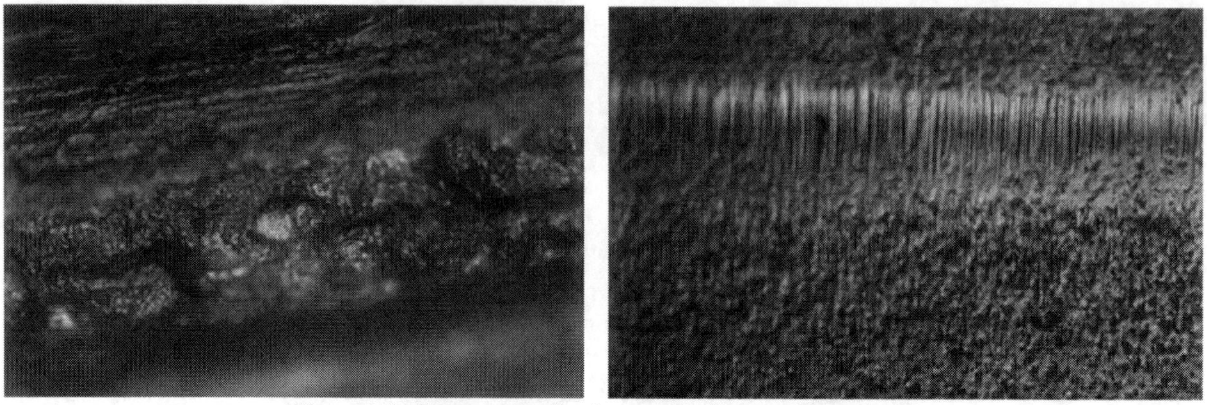

Figure 3: Dynamic O-ring and Sleeve Fretting

2.2 Analysis of Inner Mechanical Seal

2.2.1 Face Geometry and Secondary Seal Modifications

The inner seal face inspection highlighted that there was excessive face contact, more heavily concentrated at the seal face ID. This damage indicated that there was inadequate fluid film formed in the sealing gap (between the two faces). A sealing interface is subject to a closing force which is the sum of the hydraulic pressure and the spring force. This is resisted by an arithmetically equivalent opening force which is the sum of the hydrostatic pressure (of fluid in the interface), hydrodynamic pressure (caused by rotation) and contact pressure, usually just at the asperities. If the fluid film is not formed adequately, then the contact pressure must increase to provide equilibrium, and this leads to a rapid increase in heat generation, vaporization, and material degradation. Figure 4 illustrates this concept.

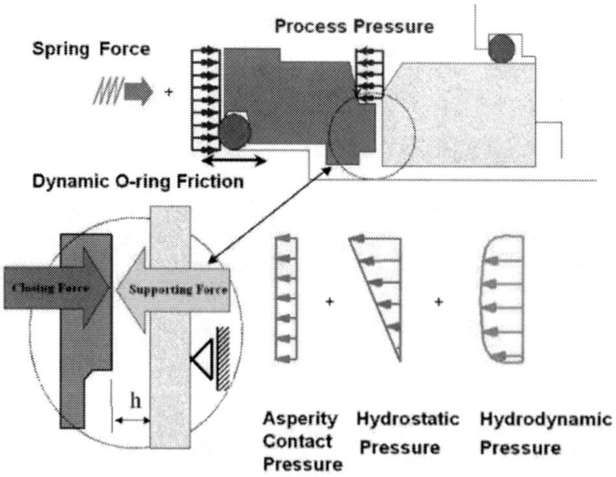

Figure 4: Seal Face Equilibrium Equation

Low molecular weight hydrocarbons such as ethylene provide very little in the way of hydrodynamic load support between the mechanical seal faces. The faces are likely to operate in a regime of solid friction to mixed lubricity in which the face materials can experience potentially higher wear rates. The elevated temperatures from rubbing friction and potential break down of face material due to hydrostatic loading are evident in the condition of the faces post failure in this case. Comparison analysis of performance using Finite Element Analysis (FEA) software has proven useful in these applications to evaluate face loading, contact pressure, and wear potential.

The existing seal configuration used a passive face technology in the form of hydro-pads to overcome face friction and reduce wear. The benefits of implementing active seal face features into low molecular weight hydrocarbon applications is well documented [2].

To date, most solutions have involved passive control of the sealing gap, with the emphasis on boundary lubrication or hydrostatic fluid film formation. There have also been attempts to introduce hydrodynamic seal face features (such as hydro-pads, hydrodynamic micro-grooves, or thermo-hydrodynamic circulation grooves) to overcome the excessive wear that reduces reliability, particularly in high pressure, high velocity applications. However, many such seals perform inefficiently; those which feature enhanced lubrication produce high leakage due to the fluid film at the sealing interface not being effectively controlled, while the lubrication regime of low-leakage seals often limits both pressure and velocity performance.

The successful solution lies in a technology that offers the full fluid interface lubrication required for increased reliability, but which also incorporates active leakage reduction. This is accomplished using a combination of enhanced lubrication features and an interface pumping arrangement that returns fluid from the sealing gap to the pressurized fluid space. Together with modern high-precision micro-machining techniques this evolved into a patented technology which offers a highly effective sealing solution for volatile liquid process applications.

The active face technology centers on a series of carefully positioned seal face grooves, each just a few micrometers deep. Square 'inlet grooves' on the seal face, which are linked to the pressurized fluid area, allow the process fluid to penetrate deep between the seal faces where it is dragged by liquid shear in a tangential direction from the sliding counter face. When the fluid reaches the edge of the inlet-groove, it generates a strong hydrodynamic pressure that lifts the seal faces and stabilizes the sealing gap. This allows full fluid film lubrication to be achieved at quite low shaft speeds, thus guaranteeing low friction and wear. Even at faster shaft velocities the seal operates with less interface friction when compared to conventional plain face seals thanks to an increased interface gap (see Figure 5).

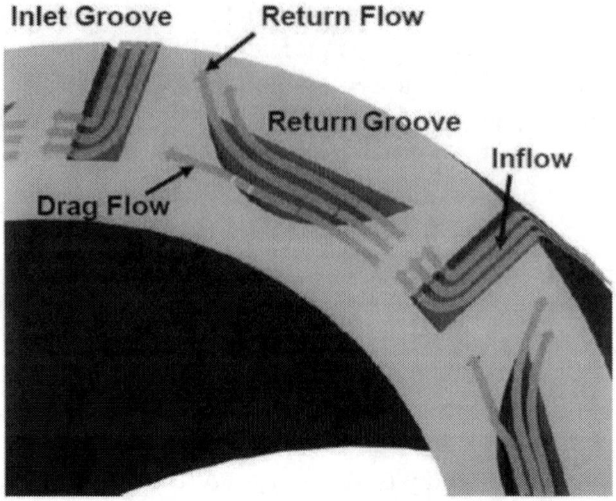

Figure 5: Active Seal Face Technology Features

Comparison of the existing inner seal with a proposed solution utilizing improved active face features predicted substantial improvements. At the desired operating conditions of 1102 PSIG seal chamber pressure at 6200 RPM, the FEA results comparison is shown below:

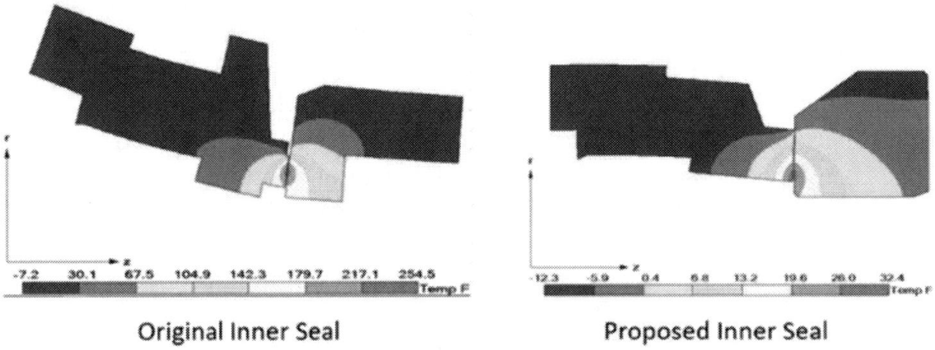

Percent Reduction with Proposed	
Heat Generation	74%
Contact Pressure	88%
PcV	88%
Pressure Rotations	77%
Avg Face Temp	85%

Figure 6: Performance Comparisons

The benefits of the proposed design are numerous, but overall, there is improved film stiffness and face separation with the active face features incorporated to the mating ring. The net reduction in heat generated is directly attributed to a lower hydraulic balance with the proposed design. The active face features provide improved leakage control, therefore mitigating the need for increased face loading currently utilized with the passive method.

The other predominant failure mode of the original seal configuration was the degradation of the secondary sealing element in the seal head. The original seal configuration is defined as a pusher seal. By definition, a pusher seal is a seal that moves forward (with the O-ring) as the seal faces wear. In normal operation, the O-ring will be pushed forward by the spring load and hydraulic forces. In operating scenarios where the pump experiences repetitive axial movements where the shaft moves towards the drive end or non-drive end, the O-ring will move across the metal sleeve surface. In addition, instability at the seal faces due to vaporization and expansion of the process fluid changing state can exacerbate the issue with continued wearing action on the O-ring, resulting in seal failure. Any pusher type secondary sealing element (wedge, spring energized polymer seal, etc.) would be susceptible to this wear pattern. The only way to eliminate the secondary sealing element breakdown from this mode of wear is to utilize a non-pusher secondary seal.

Figure 6 shows a single non-pusher single seal using a unique secondary sealing technology to prevent hang up. This design has been developed to extend reliability in high duty applications particularly where dynamic O-ring degradation is a concern. The unique feature is a non-pusher secondary seal elastomeric element that stretches and compresses to take up shaft/axial movement or wear. The sealing faces of a non-pusher seal track each other while in operation, but the secondary seal remains in place. This eliminates the wear or fretting caused by the constant shuttling or movement of the pump. Replacing a bellows that has a convolution with a flexible sealing membrane (non-pusher secondary seal), that lacks a full convolution, eliminates bellows collapse and axial rigidity. As shown in Figure 7 the flexible sealing membrane includes a flange portion, a coaxial portion, and a connecting portion extending between the flange portion and the coaxial portion. Each of the flange portion, connecting portion, and coaxial portion lie against a portion of the axially shiftable primary seal ring, an anti-extrusion ring, or a stub sleeve and in this respect are supported against collapse due to high process chamber pressures [3].

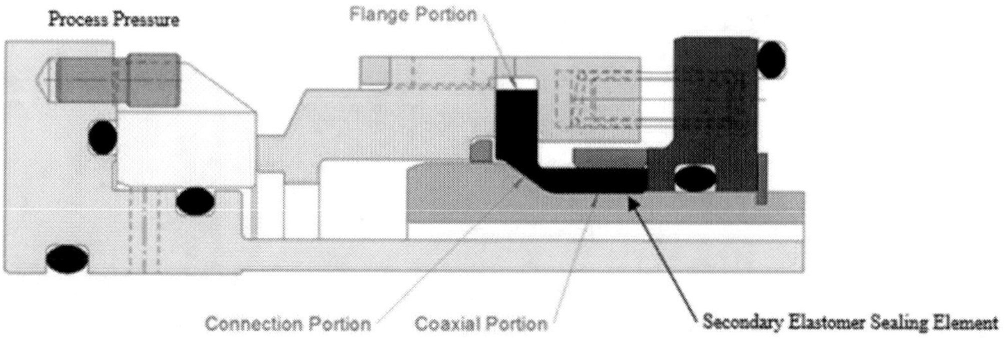

Figure 7: Non-Pusher Secondary Sealing Element

The implementation of the improved face lubrication features in addition to the inclusion of a non-pusher secondary sealing element (NPSS) were recommended to address the two primary causes of inner seal component wear in this configuration.

2.2.2 Piping Plan Adjustments

While considering the recommended improvements to the inner seal geometry, it was important to additionally validate the effectiveness of the primary seal flush. The inner seal is lubricated by the process fluid supplied by an API Plan 11 (circulation from discharge region of the pump to the seal chamber). To evaluate both the existing and proposed seal performance in terms of the effectiveness of the Plan 11 to remove seal generated heat, an analysis of the piping and integral seal flow diverter was conducted using Computational Fluid Dynamic (CFD) analysis.

The provided operating parameters and boundary conditions considered in the CFD model are as follows:

Fluid properties based on ethylene at -13 °F (-25 °C) temperature.

- SG: 0.446
- Viscosity: 7.07E-5 Pa·s
- Specific Heat: 0.868 BTU/lb·°F

Boundary conditions

- Plan 11 flush from 1st stage discharge, 1223 psig (84 Bar) (assumed 1/12th pump differential)
- Seal chamber outlet = suction pressure; 1102 psig (76 Bar)
- Differential pressure across inlet and outlet: 121 psi (8.3 Bar)
- Seal chamber throat bushing: 0.010" (0.254 mm) radial clearance
- Shaft Speed: 6200 RPM

The initial analysis was conducted on the existing inner seal arrangement, with a calculated heat generation value of 20,546 Btu/hr applied to the cylindrical nose of the primary seal face and radial mating ring surfaces as a uniform heat flux. See figure 8 below:

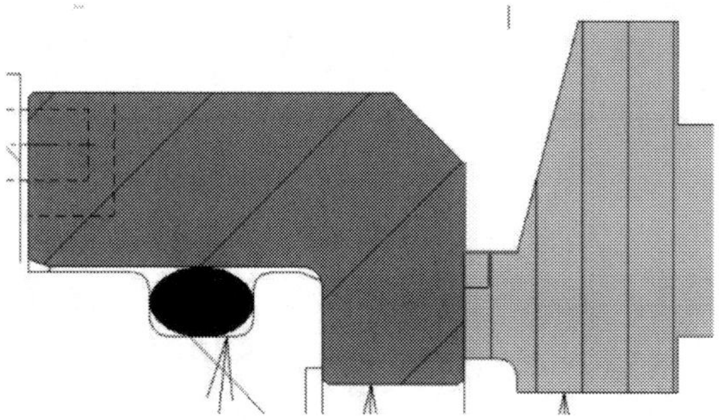

Figure 8: Heat Input for CFD Model – Existing Seal

Additional considerations in the CFD model were the piping size, drill through diameter in the pump casing, existing restriction orifice diameter in the flush line, and seal chamber throat bushing dimensions. All details were confirmed with the end user through measurement and pump maintenance record documentation. After a few simulations of the existing flush porting, model results showed generally good performance with equal flush distribution through the seal flow diverter porting (velocity end view, Figure 9). Sectioning the model geometry through one of the diverter holes, it is noticeable how the fluid flow tends to hug the radial face of the seal mating ring, impacting the internal bore inner diameter of the flow diverter prior to exiting the seal chamber. This phenomenon at each hole explains the localized zones of increased velocity also evident in the end view. See figure 10.

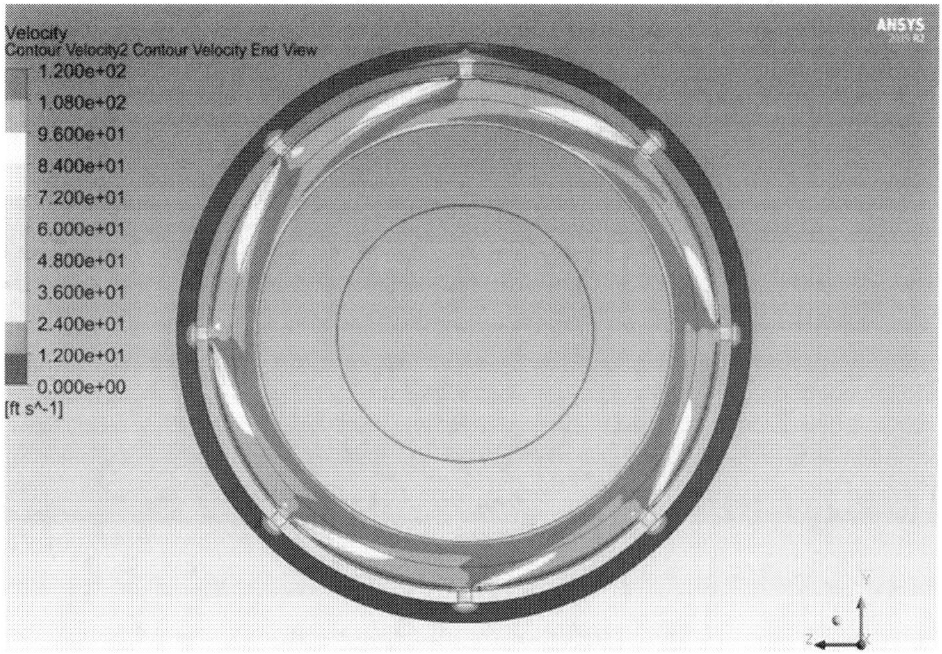

Figure 9: Velocity End View

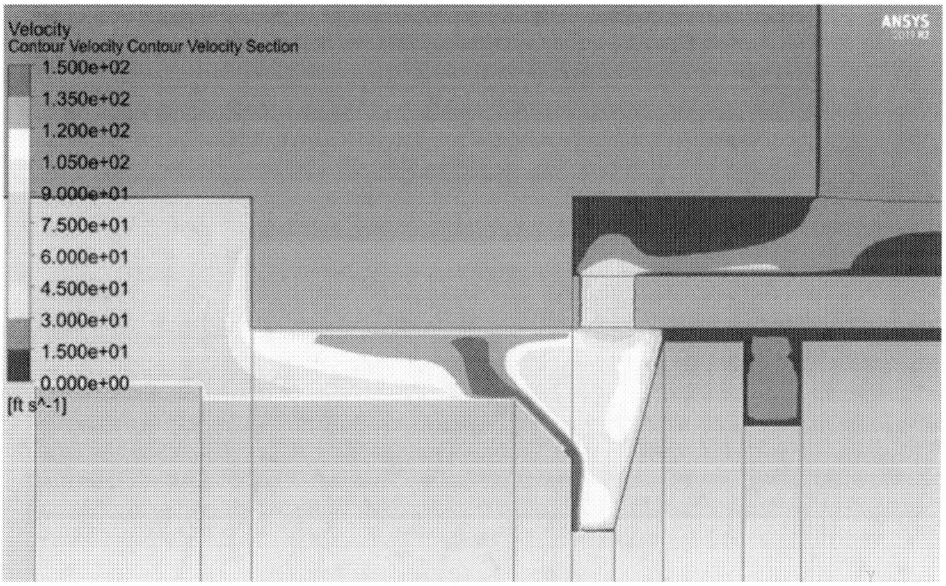

Figure 10: Velocity Section View

Overall, considering the model inputs, the calculated flush flow rate through the piping arrangement per seal was 16.1 GPM (60.9 LPM). The target flush flow rate for this application was 18 – 20 GPM (68.1 – 75.7 LPM). Once the analysis was completed for the existing inner seal arrangement, model geometry was generated for the proposed design. Fluid properties were maintained from the analysis of the existing cartridge design and boundary conditions kept consistent except for the seal face heat generation. The proposed seal design had 18,460 Btu/hr applied to the cylindrical primary ring nose and radial mating ring surfaces as a uniform heat flux (see Figure 11). The reduction in heat generated by the seal is a direct correlation between the hydraulic balance reduction, increased film thickness, and hydrodynamic load support generated by the active seal face features.

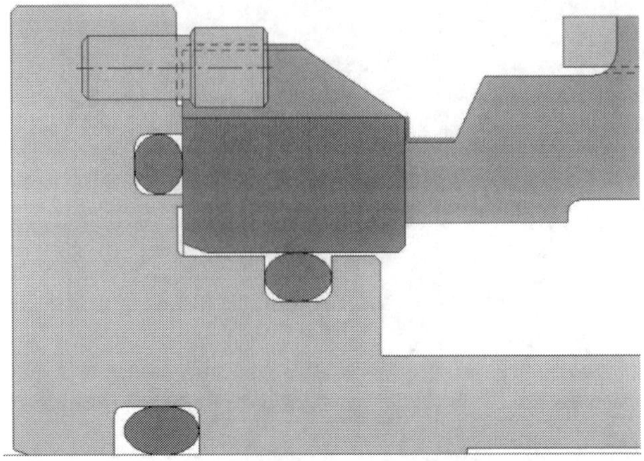

Figure 11: Heat Input for CFD Model – Proposed Seal

The proposed seal design utilized a different mating ring configuration with OD slots and drive pins oriented at the OD. This mating ring configuration is conducive for high speed and pressure applications and yields low pressure and thermal distortions due to optimized hydraulic loading. As the CFD model would show, there were additional benefits yielded from implementing this design. For computer model processing efficiency, the rotating mating ring pin and pin slot were simplified as a simple slot as shown in Figure 12.

Figure 12: Mating Ring Drive Slot

As with the previous analysis. the model outputs yielded good results with equal flush flow distribution through the flow diverter holes. Sectioning the geometry through one of the diverter holes, at the mating ring pin slot location was of particular interest. The section view indicated that the slots on the OD of the mating ring aided in channeling fluid flow away more efficiently out of the seal chamber. This contrasted with the inefficiency shown in the previous model with the flow pattern radially projected outward. The calculated flush flow rate considering the revised seal design was 18.3 GPM. The combined effects of the optimized face geometry and flush effectiveness predicted significant improvements in seal flush flow efficiency overall.

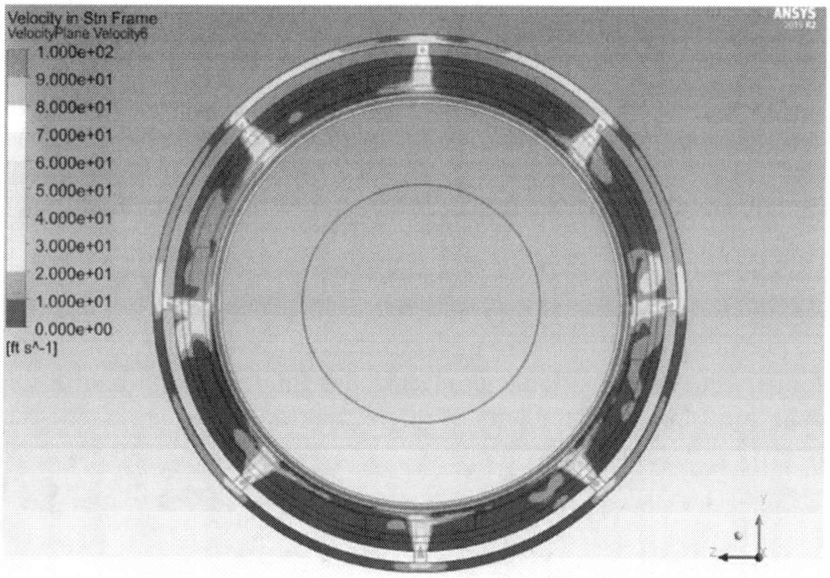

Figure 13: Velocity End View

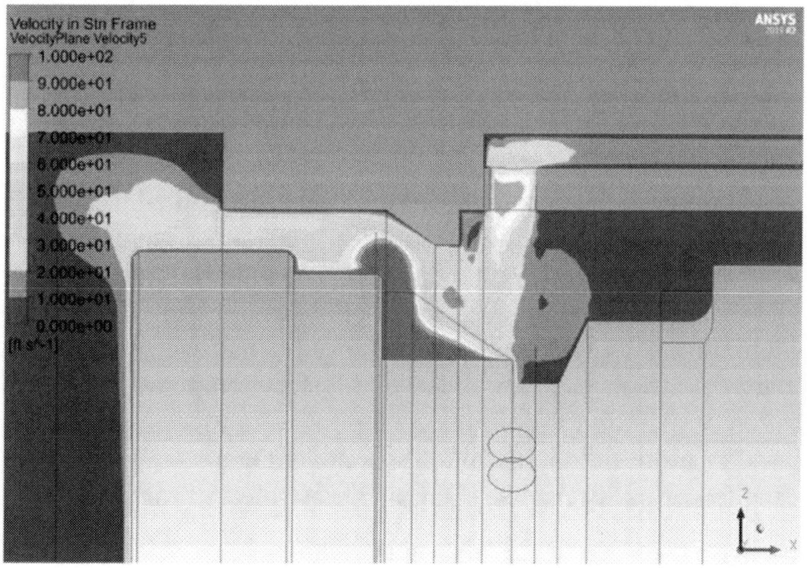

Figure 14: Velocity Section View

2.3 Analysis of Containment Seal

2.3.1 Face Geometry and Secondary Seal Modifications

Concerns associated with the performance of the inner seal aside, the overall performance of the containment seal was considered very good. There were several design modifications over the years implemented to the containment seal in terms of face geometry adjustments and pumping ring modifications that were implemented to improve performance. The result was a very reliable design. Recent failure observations, however, found that the dynamic O-ring in the containment seal was suffering from the same degradation and wear as the inner seal dynamic O-ring. Due to this issue, it was important to consider utilization of a non-pusher secondary sealing element in the containment seal as well.

To assist in the analysis, FEA modelling was again utilized. Several iterations were considered to first compare the performance of the original containment seal, a proposed containment seal with a non-pusher element and identical face geometry, and finally a version utilizing improved seal face materials. The summary of the results of the model outputs is shown below:

Outer Seal	Original (Carbon vs SiC)		Option 1 (Carbon vs SiC)		Option 2 (SiC-G vs Diamond)	
	Startup	Steady-State	Startup	Steady-State	Startup	Steady-State
Heat Est (btu/hr)	2821	2918	3534	3631	4643	4661
Avg. Face Temp (F)	305	310	305	310	263.7	264
Taper Sum (min)	1.4	1.6	1.7	1.8	1.5	1.5
PcV (PSI-FPM)	42,858	0	0	0	5,776	0
Min Film (µin)	14.5	25.7	22.8	38.1	42	49
Contact Press (PSI)	9.0	0.0	0.0	0.0	1.0	0.0
Leakage (g/hr)	0.11	0.05	0.10	0.17	0.26	0.22

Figure 15: Containment Seal Summary Results

Option 1 considered only seal face geometry modifications identical to the original containment seal and the non-pusher secondary element. Option 2 considered identical face geometry as Option 1, but with the inclusion of an enhanced seal face material pair incorporating a Silicon Carbide – Graphite composite face running against a diamond treated silicon mating ring. Utilizing diamond treatment to hard face materials is advantageous in many high duty applications primarily as it will yield much lower interface temperatures due to the very low coefficient of friction of diamond. Additionally, the inherent topography of the hard face material pair allows for more fluid in the interface, advantageous in this instance as it ensures a stable fluid film and low asperity contact. Option 2 was selected for the improvements in interface temperature reduction and reduced PV (pressure-velocity) values over the original. A comparison of the options generated from the FEA model is shown below:

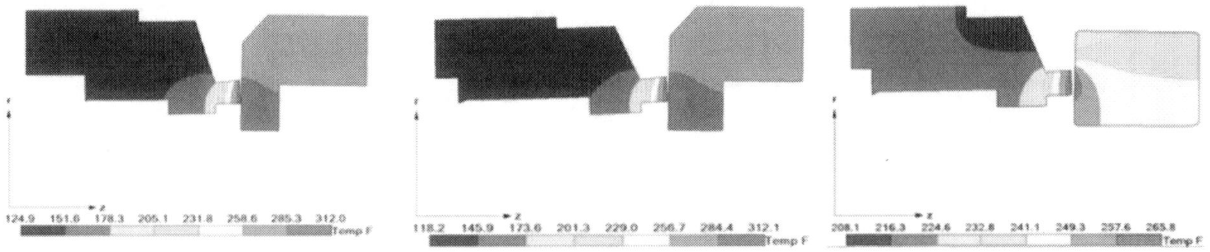

Figure 16: Containment Seal Comparison

2.3.2 Pumping Ring Modifications – API Plan 52 Analysis

The final component of the system requiring evaluation was the performance of the API Plan 52 support system and the internal circulation device, or pumping ring, within the mechanical seal. To evaluate the performance of the pumping ring, the interconnecting piping parameters and support system details were obtained. Considering the information below, a system resistance curve was generated.

System Resistance Components:

- 4 x 90° Short radius bends
- 2 x Seal plate connections (reduction)
- 1 x Tangential seal plate hole
- 1 x Expansion connection
- 1 x Reduction connection
- 13.33 ft ¾" x 0.065" wall tube

Buffer Fluid

- Light Synthetic Oil
- Specific Gravity: 0.779 @ 50°C
- Viscosity: 4.8 cP @ 50°C

After establishing the system resistance curve, several iterations of pumping ring design were considered. The first iteration was the original pumping ring design from the existing mechanical seal. The iteration was simulated using the CFD software to establish a baseline and validate calculated circulation rate estimates. Once the original pumping ring was validated, multiple iterations were reviewed to identify improvement areas. The goal of the analysis was to achieve comparable or better circulation rate as the original, while seeking to improve overall heat dissipation and efficient heat transfer. Pumping ring design iterations are shown in Figure 17.

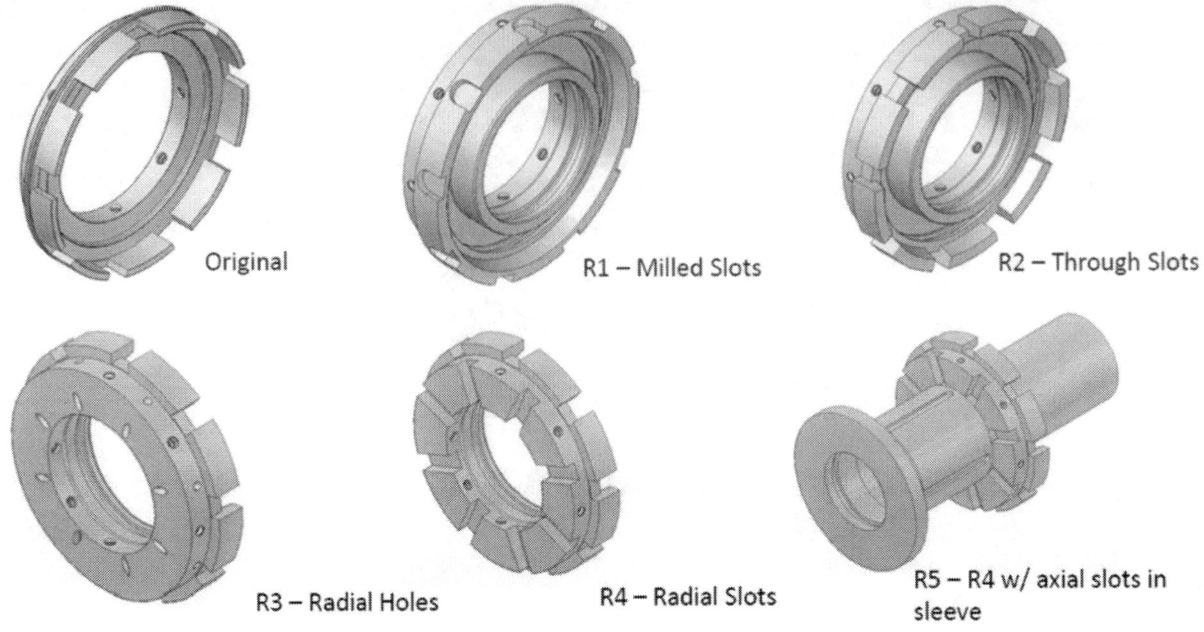

Figure 17: Pumping Ring Design Iterations

The Pumping Ring iteration results are summarized below:

- Original pumping ring design yields results of 35.2 l/m (9.3 GPM) considering system resistance curve.
- R1 design of milled slots produced significantly lower flow rate when overlaid on the existing system resistance curve and was discounted.
- R2 design improved circulation rate slightly to 39.7 l/m (10.5 GPM); CFD predicted model shows room for improvement with fluid circulation throughout the containment cavity.
- R3 design iteration has comparable circulation rate closer to the original at 36.0 l/m (9.5 GPM); addition of radial through holes improve fluid circulation in adjacent containment cavity regions.
- R4 design iteration has comparable circulation rate to R3 at 35.2 l/m (9.3 GPM); CFD predicted model shows improvement of fluid circulation further into the containment cavity, beneath inner seal faces.
- R5 design iteration has comparable circulation rate to R4 at 35.2 l/m (9.3 GPM); CFD model predicts the best performance in terms of matching desired circulation rate with overall improvement to containment cavity circulation.

The CFD predictions of each pumping ring iteration aided in illustrating advantages of each modification in terms of increased fluid distribution throughout the containment cavity (cavity between the inner and containment seal in the cartridge). For comparison, a CFD velocity profile of R2 and R5 is shown below. The velocity gradients for R5 permeate through the containment cavity, including the area beneath the inner seal faces, minimizing stagnation. The improved velocity gradients would result in more efficient heat transfer and reduced temperatures in the buffer system.

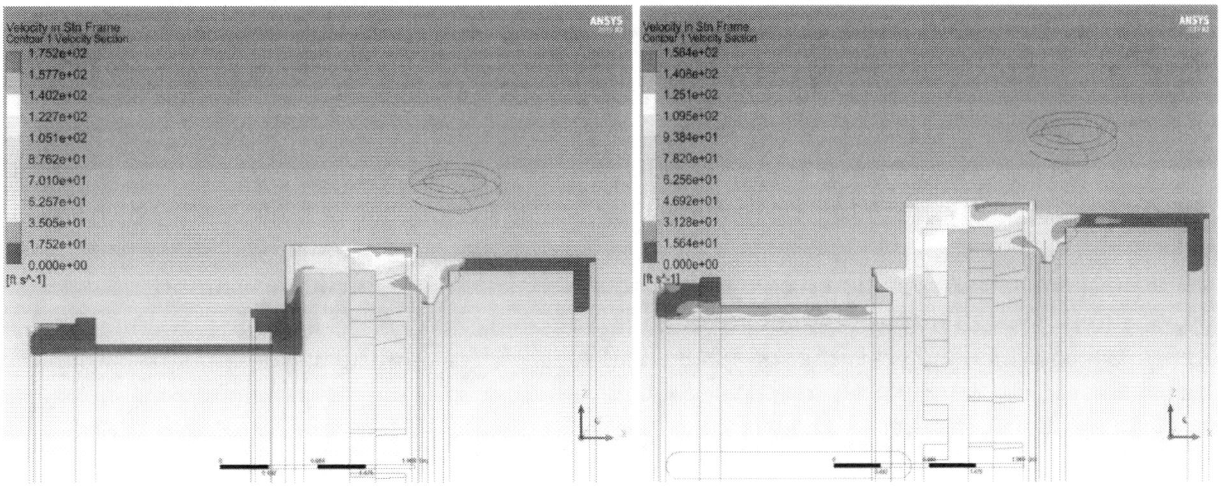

Figure 18: R2 vs R5 Velocity Comparisons

2.4 Summary of Results

The finished proposed seal design is shown below:

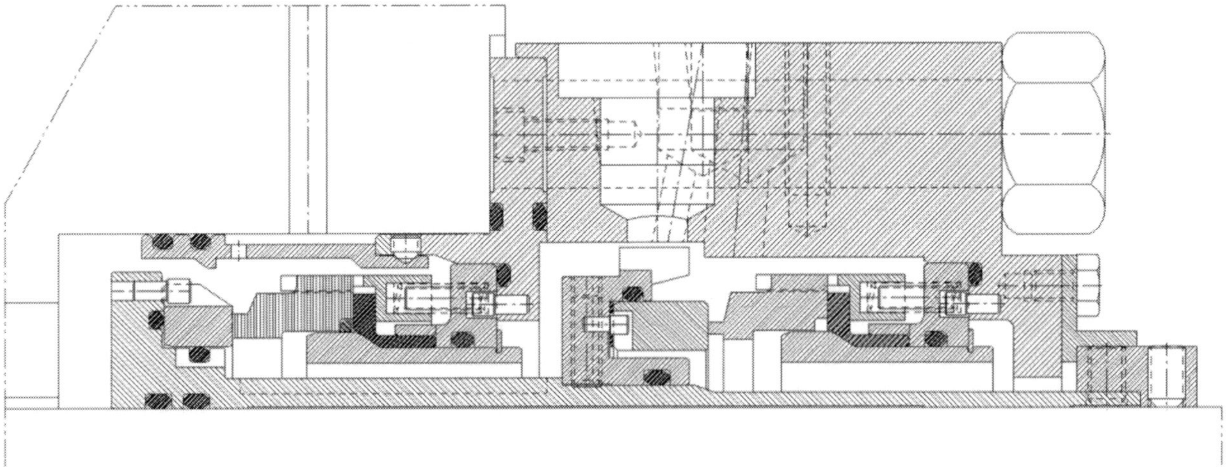

Figure 19: Final Design

The proposed mechanical seal was installed in September 2021. To date:

- There is approximately a 10 – 20 °C reduction in the API Plan 52 reservoir temperatures observed on both drive end and non-drive end seals.
- Inner seal leakage results:
    - Measured by seal buffer reservoir pressure transmitters.
    - No change in performance observed during multiple pump stops/starts.
    - Leakage rate is slightly higher than original seal as anticipated (hydraulic balance reduction and enhanced fluid film).
    - Leakage rate has decreased over time – seal profile wear in is consistent with FEA results.

- Pro-active mechanical seal inspection is planned for Fall 2023 outage to confirm modifications resulted in the desired outcomes in terms of performance.

Conclusion

The proposed seal design has provided as expected performance without experiencing any detrimental impacts even during operational upsets. The performance of the improved seal design has further substantiated the benefit of advanced analysis tools coupled with documented application expertise as a good recipe for success when attempting to address problematic applications. Design features highlighted in this case study are worthy for consideration to improve and extend the reliability of mechanical seals in low molecular weight hydrocarbon services.

3 References

[1] Kalfrin, B., Dwyer, T., Choma, J, 2019, "Ethylene Product Pump Mechanical Seal Upgrade Lessons Learned," *AiChE Paper T4-550804,* 2019 Spring National Meeting, Ethylene Producers Conference, New Orleans, La

[2] Kalfrin, B., 2020, "Utilization of Laser-Etched, Active Seal Face Surface Features: Maximizing Reliability in Ethylene Applications," *AiChE Paper 588699,* 2020 Spring National Meeting, Ethylene Producers Conference, Houston, Tx

[3] Rasmussen, D., Morton, J.L., Wasser, J., 2018, "Crude Oil Non-Pusher Secondary Seal". *Texas A&M, 34th Pump User Symposium*, USA.

Combined Experimental and Kinetic Modeling Study of Thermal Decomposition of Deha: Formation of NO and HCN

Cato A.R. Pappijn, Aleksandar Bojkovic, Ruben Van de Vijver, Georgios Bellos, Marie-Françoise Reyniers, Guy Marin, Kevin Van Geem

Thermal decomposition of the steam cracking additive diethyl hydroxylamine (DEHA) leads to formation of the small nitrogen-containing compounds HCN and NO, which can have an influence on the process safety. In this contribution the thermal decomposition of DEHA is studied both experimentally and theoretically under typical steam cracking conditions. The experimental investigation is conducted in a 90 cm long annular quartz flow reactor, having outer and inner tube diameters of 2.1 and 1.5 cm respectively. Steam cracking experiments of propane with addition of 1000 ppmw/HC DEHA are performed over a temperature range of 700 – 850 °C at a fixed pressure of 1.5 bara. This enables to investigate the influence of temperature on the distribution of the nitrogen-containing products, as well as the interaction with the propane steam cracking process. The reaction effluent is sampled on-line at 300 °C. The unique on-line sampling system consists of valve-based sampling manifold and uniformly heated transfer lines connected to a comprehensive two-dimensional gas chromatograph (GC × GC) equipped with a flame ionization detector (FID) and a nitrogen chemiluminescence detector (NCD). This set-up allows to detect and quantify both pure (with the FID) and nitrogen-containing compounds (with the NCD) using a fixed flow of 3-chloropyrdine as an internal standard (added to the effluent right after the exit of the reactor before the sampling point). The GC × GC-FID/-NCD settings are optimized for trace analysis of nitrogen-compounds in olefin-rich hydrocarbon matrices produced during steam cracking. There is a complete conversion of the reactive DEHA over the complete temperature range 700 – 850 °C, with HCN as the main nitrogen-containing product. Acetonitrile (CH_3CN), 2-propenenitrile (CH_2CHCN), NO, diethylamine and pyridine are minor nitrogen-containing products. To gain understanding of the underlying decomposition mechanism of DEHA in the presence of a hydrocarbon matrix, a kinetic model is constructed with the in-house automatic kinetic model generation software tool called Genesys. Genesys makes use of user-defined databases containing thermodynamic and kinetic parameters for hydrocarbons and nitrogen-containing compounds calculated with the ab initio CBS-QB3 composite method. If data is not available for certain species or reactions, the group additivity method based on these high-level ab initio calculations is applied. No adjustment of the kinetic parameters to the experimental data is performed. The kinetic model succeeds in capturing the main experimental trends for the decomposition of DEHA in a propane matrix. Rate of production analyses reveal the important pathways for HCN and NO_x formation. Decomposition of DEHA is initiated both via hydrogen abstraction from Cα, i.e. the carbon next to the nitrogen atom, as well as via homolytic scission of the weak N-O bond. A series of hydrogen abstraction and β-scission reactions leads to formation of HCN as the main nitrogen-containing product, which is a stable product species in the studied temperature range.

Near-Misses Temperature Excursions below MAT Identified during AR/BF Studies on Unexpected Equipment.

Giorgio Franceschetti, Leila Lisboa Santos, Jussara da Rocha Duarte, Priscila Auad, Thais Helena A. B. de Camargo

Braskem has implemented a preventive program to analyze, evaluate and mitigate the risk of auto-refrigeration/brittle fracture in its existing plants and units according to 'API 579 Recommended Practice for Fitness-for-Service, Assessment of Existing Equipment for Brittle Fracture'. This program includes a level 2 assessment of a complete section of a plant or unit, including some equipment that, in a level 1 assessment, was not considered a low temperature candidate and therefore would not be susceptible to a brittle fracture scenario. The result of this level 2 analysis, using historical data, identified low temperature excursions close to the MAT caused either by the phenomenon of auto-refrigeration, as by the loss of heat or contamination with lighter HC. Although these temperature excursions did not result in operation below the MAT, they reached very close values and therefore we decided to include specific recommendations to reduce the frequency of these scenarios reoccurring.

Understanding Heat Integration within Ethylene Facility to Correctly Size Flare Headers

Shailesh Saraykar

If there is any processing plant with tremendous amount of heat integration, it's an Ethylene plant. It's important to understand how these heat exchanging systems are linked to each other for correct sizing of flare sub-headers and headers. A case example if presented where heat integration detailed study was performed.

148a Safe Hot Tapping and Line Isolation on Ethylene Systems

Matthew Guest, Ian Littlepage

Performing hot tapping and or in-service line isolations on ethylene systems carries a high level of risk. Proper assessment and management of the risk is vital to the safe execution of such work. Is an understanding of ASME-PCC-2 and/or API-2201 enough?

From gaining understanding of the scope through project completion, hot tapping and line isolations on critical systems should be handled as a joint venture, not a transaction.

Best practice for safe execution begins with data acquisition. It is imperative that a complete understanding of all characteristics of the carrier pipe and the desired outcome of the project are understood. This information supports fitting design and manufacturing, and development of the in-service weld procedures.

From there, development of the field execution procedures and a thoruogh risk assessment can be carried out. Assessment of the equipment inspection and maintenance programs, including documentation, along with technician training and certification processes then set the standard to ensure the utmost in safety and quality.

148b In-Service Welding on Ethylene Piping: Defining and Demonstrating Safe Conditions

Rick Noecker II, Matthew Boring, Chris Vrolyk, Timothy Burns, Zhiyan Wang, David Futch, Vinay Naik

In-service welding is a common practice used to perform repairs and install hot taps to both piping and pipelines. It involves arc welding on a systems containing fluids that are under pressure in either flowing or static conditions. Safety considerations for in-service welding are well defined for many hydrocarbon fluid streams, but there have been reservations to perform in-service welding on high purity ethylene systems due to concerns of exothermic decomposition reactions leading to failure of the pipe. This work examined existing literature, including past industry ethylene incidents, and experimental data to develop an analytical model that may be used to define safe in-service welding conditions on high purity ethylene systems.

The test vessel consisted of commercially pure ethylene (<5ppm O_2) contained in a thin walled pipe section (<0.203") at high pressure (1,200 and 1,500 psi) under static conditions. Nine in-service welds were completed using various welding heat inputs with the intent of producing the highest possible pipe wall temperatures to facilitate conditions that may lead to failure due to self-propagating ethylene decomposition. Piping failure was not observed; however, all welds did result in localized ethylene decomposition that produced a carbon deposit on the inner diameter pipe surface, local diffusion of carbon into the pipe wall resulting in the development of carburized layer on the inner diameter pipe surface of all welds, and eutectic iron forming adjacent to the carburized layers on the ID surface on several welds. The experimental results were used to validate the models. The experiments also provided evidence that limiting inner diameter surface temperature to ≤1,800°F, an already common practice for mitigating the risk of burning through the pipe wall during in-service welding, is also suitable when performing in-service welding on ethylene piping and pipelines to prevent ethylene decomposition related failures and eutectic iron formation.

This paper provides an overview of ethylene decomposition, a summary of the work and an approach that maybe used to define the safe in-service welding conditions on high purity ethylene systems.

148c Hot Restart Following a Furnace Trip

Matthew Boian, Bryan McVicker

Hot Restart Following a Furnace Trip

Authors

Bryan McVicker

Matt Boian

Abstract:

A hot restart is the result of an unplanned shutdown or protective system "trip" of an online furnace. Coke deposited in the radiant coils, inherent to steam cracking reactions, leaves the coils vulnerable to brittle fracture damage as the furnace cools. Radiant coil damage can lead to higher consequence events including but not limited to a hydrocarbon loss of containment and firebox flooding. A safe, rapid restart is therefore desirable to protect against event escalation and must be balanced with the risk inherent to furnace startup operations.

This presentation intends to summarize the key items that should be considered when attempting to restart a furnace following an online trip. These items rely in part on successful execution of observations and tasks by responding personnel on the DCS console and in the field and therefore necessitate the availability of critical information and training.

Proper management of hot restart risk increases in difficulty during an upset involving multiple furnaces. A method for quickly establishing furnace restart priority will be discussed including an example of integration into a site DCS interface.

Furnaces in the context of this paper are referring to Pyrolysis Furnaces, also known as Steam Crackers.

148d Causes and Future Prevention of a Firebox Overpressure Event

Hassan Al-Ameer, Ryan Snell, Diaa Ghonaim, Adam Ashworth

During normal cracking operation of an ethylene furnace in one of Saudi Chevron Phillips plants, the induced draft inlet damper suddenly went to the closed state at full firing rate. There was no furnace trip for this condition by design. This led to high firebox pressure with hot flue gas coming out of the peep doors and resulted in damage to the firebox structure and surrounding instrument cabling and lighting. This paper will go through the details of the incident, the learnings from the incident investigation and some corrective actions implemented.

148f NOVA Chemicals NORM Experience during Plant Maintenance Turnarounds

Heidi Ladanchuk, Kellie Lines

Contamination from naturally occurring radioactive material (NORM) at our manufacturing facilities in Alberta and Ontario is introduced in the form of Radon-222 from our ethane feedstock. The Radon-222 concentration in ethane entering each manufacturing location varies for many reasons such as well depth, travel time, and storage time prior to arriving on site for processing. The Radon-222 decays to Lead-210 which deposits inside piping and equipment. Since Radon-222 progeny emit alpha, beta, and gamma radiation, protecting workers and the environment can be challenging.

NOVA has many years of experience in NORM identification, evaluation, and control during normal operation, routine plant maintenance, and turnarounds; with turnarounds being much more difficult to control due to quantity of NORM contaminated equipment being opened at same time, and number of workers present. This presentation will discuss NOVA Chemicals NORM experiences with identification, evaluation, and control of NORM as it relates to protection of our workers and the environment to levels ALARA during plant maintenance turnarounds.

148e *Naturally Occurring Radioactive Materials Experience in the Ethylene Plant Turnaround*

Gurwinder Singh, Kurtis M Bergman, Norman Wilhelm

A recent LyondellBasell turnaround observed an increase in NORM (Natural Occurring Radioactive Materials) findings in process equipment. NORM is a known health hazard due to low levels of radioactivity from the elements of natural origin. As the health and safety of our employees and contractors is a top priority of LyondellBasell, NORM is monitored closely when opening equipment. In addition to detailed upfront planning, proper measuring techniques are used to safely handle NORM and protect personnel. This paper will review the NORM history at the LyondellBasell gas cracker located in La Porte, Texas (U.S. Gulf Coast Area) and present the planning, findings and safe handling practices used during the turnaround.

AIChE Paper Number: 61A

Development of the Rotating Olefins Cracker for Decarbonized Production

Silvano Saretto, Ph.D.
Head of Turbocracker Technology
Siemens Energy

Eric Wagner, Ph.D.
Director Technology Development
Technip Energies

Cristiano Balestrino
Portfolio Manager of Turbocracker Technology
Siemens Energy

David Ransom
Head of Technology and Systems
Siemens Energy

Joel Guillaume
Director of Process and Technology
Technip Energies

Prepared for Presentation at the 2023 Spring National Meeting

Houston, TX, March 12-16, 2023

Development of the Rotating Olefin Cracker for Decarbonized Production

Dr. Silvano Saretto[1], Dr. Eric Wagner[2], Cristiano Balestrino[1], David Ransom[1], Joel Guillaume[2]

[1] Siemens Energy

[2] Technip Energies

Abstract

Siemens Energy (SE) and Technip Energies (T.EN) are jointly developing the Rotating Olefins Cracker (ROC) technology to decarbonize the olefin production process. The ROC technology employs turbomachinery to supply the energy input for steam cracking via direct fluid heating of the process flow in place of fired heating in conventional furnaces, thus eliminating the reliance on combustion of fossil fuels for the manufacturing of light olefins.

The ROC technology offers driver flexibility. When driven by electric motors or hydrogen-fired gas turbines, it provides a pathway for decarbonizing light olefins production. The decarbonization impact is even more significant when the electric power or hydrogen fuel is derived from renewable sources. The ROC process, once fully commercialized, is also expected to have better first pass olefins yields, with similar operating costs, compared to current commercially available technologies.

The companies have already validated the fundamentals of the reactor technology in laboratory testing, as well as performance testing of the first turbomachinery prototype with inert gas. The combination of laboratory testing and turbomachinery testing has validated the two essential building blocks of the technology. This paper presents an overview of the ROC technology, the potential for process decarbonization, and test results from both rounds of validation testing.

Introduction

Ethylene is the most widely produced organic chemical in the world and is the primary building block for many manufactured plastics. While sustainable pathways of ethylene production (such as ethanol dehydration) continue to show promise, the near and mid-future for production of olefins and other High Value Chemicals (HVC) will remain dependent on steam cracking of fossil feedstocks, such as ethane and heavier feeds. And since most of the carbon in the fossil feedstock ends up in a permanent solid product, such as plastic, the CO_2 produced from hydrocarbon combustion to supply the process energy input in ethylene production is the major contributor to the Scope 1 and Scope 2 emission footprint for olefin producers.

Like many commodity chemicals, ethylene production is a mature technology that has been historically focused on the largest practical scale and lowest cost of production. Its production cost is largely dependent on the cost of feed and fuel. While ethylene plants are located around the world, their construction location and timeline is often loosely tied to the availability of low-cost feedstocks. In recent years, for example, the abundance of low-cost light hydrocarbons on the U.S. Gulf Coast has led to a number of crackers being built in the region. The future growth of ethylene, however, is critically dependent on solutions and technologies to reduce CO_2 emissions from the production process. Many encouraging activities devoted to decarbonizing olefins production are already underway.

Since the chemistry of hydrocarbon pyrolysis requires high temperatures, the largest contributor to CO_2 emissions in the ethylene plant is the cracking furnace. Here, the combustion of hydrogen and light hydrocarbons provides the heat of cracking for the ethylene production process. In this, it is assumed that future ethylene production processes will continue to inherently require the supply of high amounts of high-temperature heat to drive the endothermic cracking process.

One way to reduce CO_2 emissions is to replace the furnace fuel with pure hydrogen. This method will likely gain in popularity. However, the amount of hydrogen produced in an ethylene plant is not sufficient to provide all the H_2 fuel required for the furnaces. So supplemental hydrogen must be produced, either via a blue H_2 process, or (green) electrolysis. Blue H_2 production will require the capture and sequestration of the CO_2 produced in the process, and green hydrogen production has its own challenges with scalability, cost, and electrical consumption (energy requirements for green hydrogen are several times greater than for grey or blue H_2 production).

Another option is the wholesale capture of post-combustion CO_2 in existing and future furnaces. While the technology exists for this approach, the associated costs are significant and the question remains as to the suitability of each production site for near-site sequestration.

Direct or indirect electric heating of the cracking coil is another option that has garnered significant interest across the industry. The greenness of the electricity is, of course, an important consideration when it comes to both Scope 1 and Scope 2 emissions. In such a scenario, the cracking coil would remain largely as it is today, although there would be no combustion or convection section.

The Rotating Olefins Cracker (ROC) technology is a radically innovative concept that employs turbomachinery to convert mechanical work (potentially provided via carbon-free energy, e.g.

green electricity) directly into high-temperature heat to sustain the pyrolysis process. Unlike the electrical furnace concept, however, the ROC does not require the same reactor coil configuration. Therefore, it allows for a compact and efficient design compared to radiant coil designs, regardless of the heat source.

Siemens Energy and Technip Energies are cooperating for the development, commercialization, and licensing of the ROC - to bring together critical know-how and expertise as market leaders in high-barrier-to-entry technology fields, such as critical Rotating Equipment (Gas Turbines, Compressors, etc.) and licensed Process Technology.

The integration of these capabilities, brought together into a joint development effort, is seen as a critical success factor to bring this transformational technology to market.

Under the cooperation, the companies have progressed the technology through fundamental validation and cooperated in design activities of a full system prototype (ROC turbomachinery and integrated process systems) to be built and operated as a demonstrator unit at an operating steam cracker site.

The two companies have signed a non-exclusive Memorandum of Understanding (MOU) with the Cracker of the Future (COF) Consortium, expressing the intent of the parties to negotiate a contract to install a hydrocarbon demonstration unit utilizing the ROC technology in a plant operated by one of the COF members [Ref. 2].

ROC Technology and Process Decarbonization

ROC operating principle

The fundamental operating principle of the ROC machine is to increase the temperature of a working fluid through the process of a shock wave, to the point that the process fluid begins to undergo the endothermic process of chemical cracking. Every feature of the machine and supporting furnace island is built around this principle. It starts with a driven machine that takes in the process flow, which in this case is a steam/hydrocarbon mix. Since the interface to the driven machine is simply a shaft end, it can be driven by anything which can meet the speed and power requirements (e.g. electric or turbine drive).

Figure 1 is a sketch from the awarded patent for the technology and shows the machine concept starting from the driver. While each and every element is described in the patent, only select ones are described in the following paragraphs.

The feedstock is first drawn into an aerodynamic section, which consists of an inlet region that collects the flow and prepares it for entry into the axial impulse impellers (shown as Item 50 in the figure). Pure impulse impellers are necessary for the application, since the sole function of the aerodynamic section is to accelerate the flow without pressure rise. The flow is accelerated to supersonic (greater than Mach 1) conditions and is then carried from the impeller section to the diffuser section through an annular passage. The flow is then decelerated in the static diffuser section (Item 70) where the standing shock wave is formed (Item 90). The diffuser section consists of an array of circumferentially spaced, axial diffuser channels. Item 20 provides the pressure containment barrier necessary for process safety and integrity. Finally, the flow is collected through a transition duct, (Item 92) to recombine into a pipe flow for further downstream processing. The time of flight through this entire process is on the order of 10 milliseconds (ms). This is in contrast to state-of-the-art furnaces which have heating times of the order of hundreds of milliseconds. It's also important to note that the ROC technology does not depend on external heating of the process fluid, which leads to an increase in energy efficiency by avoiding losses between the heat source and the operating fluid. Not shown in Figure 1, the flow leaving the ROC machine then proceeds to the quench process and other typical cracked gas process steps, just as any other furnace technology.

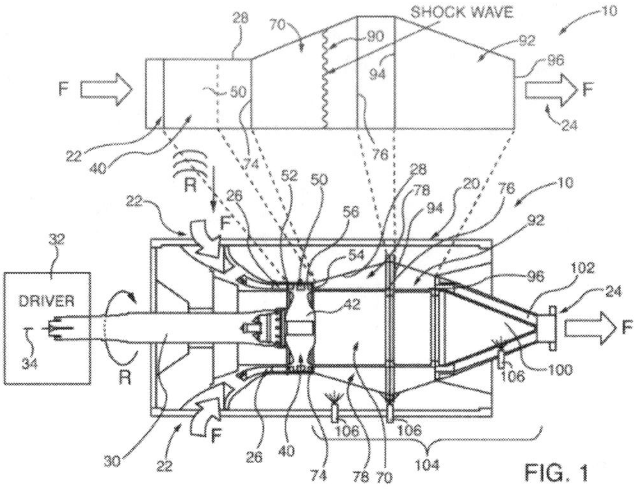

Figure 1. Patent Image of ROC Concept

Figure 2 provides an example of the evolution of the static temperature and the static pressure within the ROC. As seen from this graph, the static pressure and temperature remain approximately constant in the aerodynamic section of the machine, where the flow is accelerated from low subsonic velocities to supersonic ones. In the diffuser, where the flow is decelerated through the shock, a sudden increase in both static temperature and pressure occurs. The diffuser geometry is tailored to achieve the process target pressure and temperature after the shock.

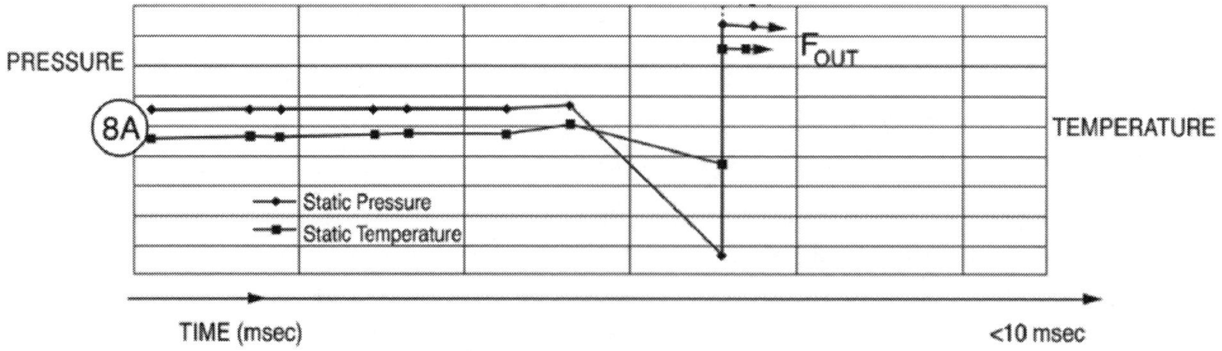

Figure 2. Static pressure and temperature evolution in the ROC.

The ROC can be designed to crack the same feeds as conventional cracking furnaces, ranging from ethane to gas oils, with the only requirement being that the hydrocarbon-steam feedstock be fully vaporized prior to entering the turbomachine. For example, the ROC hydrocarbon demonstrator system, introduced later in this paper, is intended to be operated on propane and naphtha across a wide range of severities.

Benchmarking and value-add of ROC technology vs. current state-of-the-art

Hydrocarbon steam cracking in fired furnaces is a well-established and widespread production process for olefins. Over the last several decades, incremental technology improvements have been introduced to increase furnace throughput, enhance product yields, and improve energy efficiency. On the other hand, the conventional process remains reliant on the combustion of fossil fuels for its primary energy input - mostly because the high temperature and energy intensity of the endothermic process limits alternatives to fired heating. For these reasons, olefins production via steam cracking is typically considered a "hard to decarbonize/abate" industrial sector.

The ROC technology introduces an innovative and transformational pathway to decarbonization for this industry, as it utilizes a different operating principle to meet the demanding process requirements without the combustion of fuel. Rather, it derives the process energy input from the direct conversion of mechanical work, which in turn can be produced with carbon-free energy sources, such as renewable electricity.

This fundamentally alters the energy balance of the steam cracker island, as illustrated by Figure 3 below. A much larger proportion of the energy input is applied by the ROC directly to the process flow to sustain the reaction endothermicity, and without the indirect heating mediated by transferring heat from hot combustion products into the process flow in the tubing coils of a conventional fired furnace.

As a result, the energy input is directed much more efficiently towards the conversion of feedstock into the desired high-value products, and less of it is left as a "by-product", in the form of sensible heat of flue gases to be recovered by generating high-pressure steam.

This, in turn, enables and supports the electrification of the downstream recovery process, as energy users (e.g. process gas compressors, etc.) can now be driven with electric motors rather than steam turbine drives, which come with an associated energy conversion penalty.

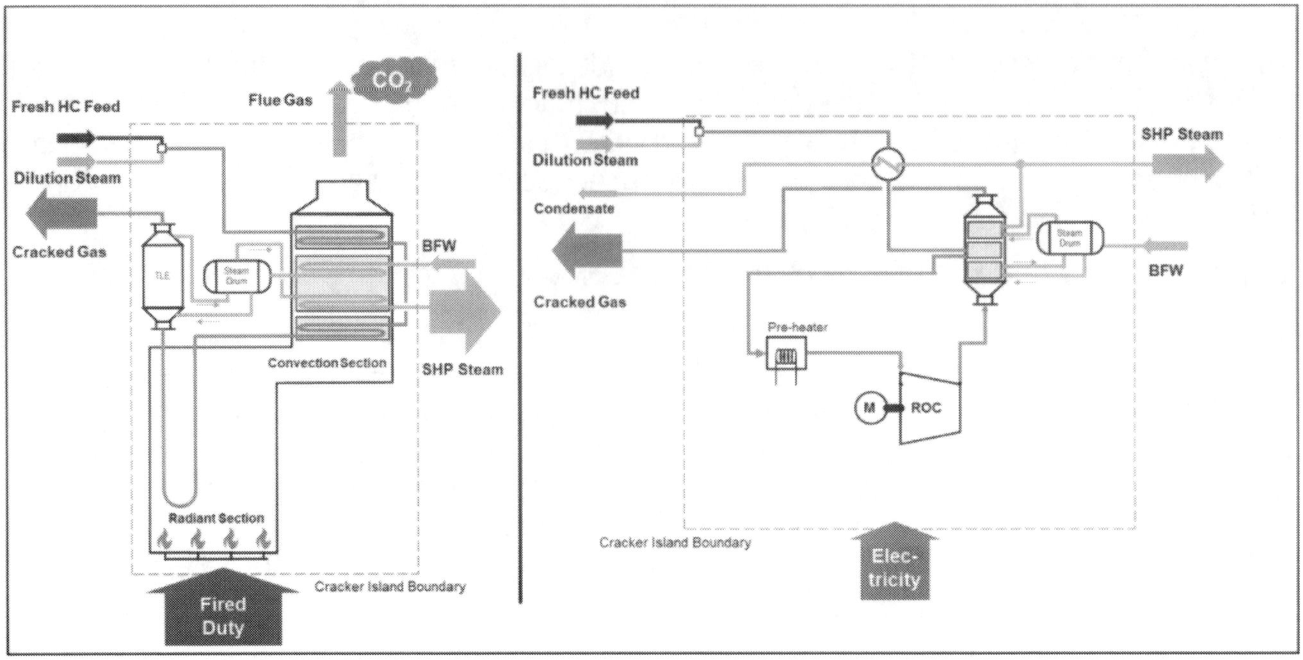

Figure 3. Energy balance of conventional Naphtha cracker (left) vs. ROC (right).

With all this taken into account, the ROC technology is anticipated to enable substantial improvements in the overall specific energy consumption of steam crackers.

Most importantly, the primary energy input to the process can be carbon-free. Figure 4 illustrates how, with the increasing penetration of carbon-free sources within the electricity generation mix, the ROC technology can enable substantial improvements in the GHG footprint of olefin production. When electrification with 100 percent green electricity is feasible, virtually all Scope 1 & 2 CO_2 emissions can be eliminated from the cracking process.

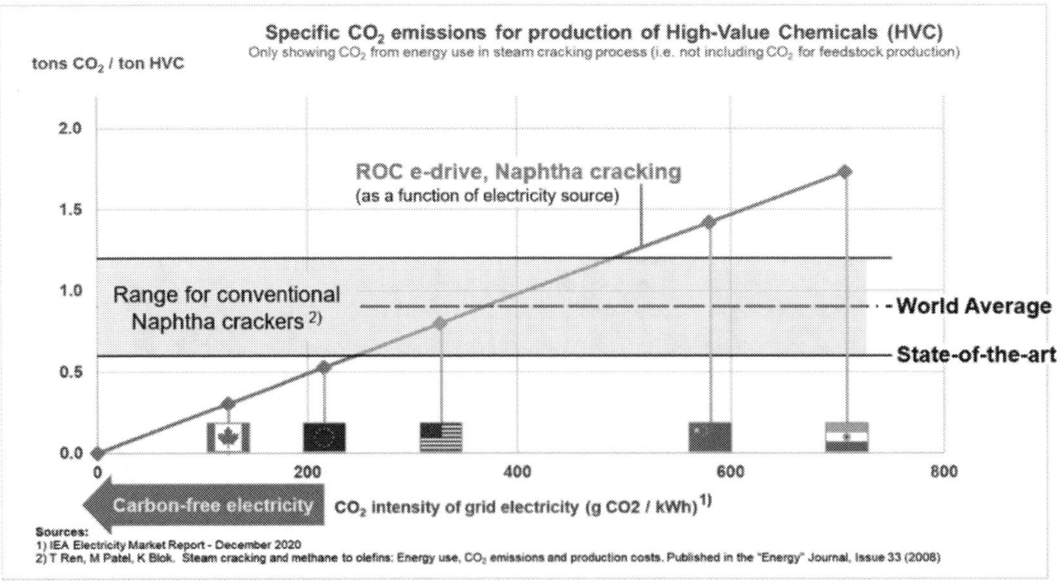

Figure 4. Specific CO_2 emissions of olefins production from ROC as a function of electricity source.

In addition, the operating principle of the ROC (see Figure 2 above) produces a near-instantaneous temperature rise in the process flow and eliminates heat transfer through process coils.

These intrinsic features have potential to translate into substantial gains in the process performance and economics, as they overcome the fundamental limitations of current state-of-the-art technology. Since furnace coils require heat transfer through tube walls, there is a physical rate limit of the heat transfer to the process fluid through the heating coils, with process performance constrained by physical size and tube metallurgy.

First, the near-instantaneous temperature rise generated by the ROC allows optimization of the reaction volume to better control chemical kinetics and maximize the yield of the desired high-value products. This is in comparison to the gradual heating and longer residence times dictated by the bulk volume of long tube bundles in furnace heating coils.

Second, a key challenge of conventional steam cracking furnaces is the accumulation of carbon deposits (coking) within the heating coil passages, which is determined by the large temperature differential required to transfer heat through the tubing coil into the process flow. A de-coking procedure is required on a periodic schedule to eliminate these carbon deposits, which interrupts production and drives significant operational burden and maintenance costs, as well as the need for additional capital investment in "standby" units (n+1 concept).

In the ROC, reliance on surface heat exchange in tube coils as the means for process energy input is eliminated altogether, and metal wall temperatures remain below the process flow temperature throughout. This is expected to result in a lower propensity for coke formation, and the potential for longer run times.

Lastly, turbomachinery is known for being highly compact equipment, and capable of converting large quantities of energy in a small footprint. This makes the ROC inherently capable of scale-up, and can potentially overcome the constraints to economies of scale in traditional steam cracking processes, which are currently dictated by the maximum physical size of fired furnaces.

In summary, the ROC technology is expected to enhance end-user value in multiple areas, as illustrated in the table below. It should be noted that several of these advantages also are relevant when compared to other decarbonization options for steam cracking, such as electric heating or the use of low-carbon fuels in fired furnaces, as these will still be subject to the challenges and physical limitations of high-temperature heat transfer.

Impact	State-of-the-art crackers	Rotating Olefins Cracker (ROC)
GHG Emissions	Reliant on fuel combustion for process energy input. Fossil fuel use in steam cracking accounts for about 30 percent of GHG emissions from the entire chemicals industry [Ref. 1].	Carbon-free energy can be used to supply the ROC rotating machinery with a) electric motor drive or b) hydrogen fueled gas turbine drive.

Impact	State-of-the-art crackers	Rotating Olefins Cracker (ROC)
Product Yield	Heat transfer rate limitations in tubing coils causing gradual temperature rise and larger than optimal residence times for the cracking reactions.	Near-instantaneous temperature rise, with opportunity to optimize reaction volumes and residence times towards chemical kinetics that maximize yield of the desired high-value products over undesired secondary reactions.
Coking	High wall temperatures in cracking tube coils are required to create heat flux; contact with hydrocarbons causes accumulation of coke. Required de-coking procedure interrupts production, driving substantial operational cost and need for standby assets.	Different heat generation mechanism ensures metal wall temperatures remain below flow temperatures throughout. Very high flow velocities ensure uniform mixing and temperature distribution, and discourage formation and accumulation of coke.
Equipment Size	Furnace output capacity is proportional to the surface area (length) of tubing coils, requiring large quantities of high-grade alloys and driving the overall size of cracker furnaces.	Flow heating, not limited by convective heat transfer rates, make the ROC inherently much more compact for the same output. This drastically reduces the use of high-grade materials as well as the overall size and footprint of the process island, facilitating modularization and pre-assembly to reduce on-site construction and project cost.
Scale	Structural limits in furnace construction constrain the economies of scale of the furnace island in steam cracking plants, with multiple units typically operating in parallel to feed the downstream process.	Turbomachinery is inherently capable of processing high throughput, and scales very favorably by doubling its output in much less than double the overall equipment size. ROC has potential to overcome several constraints to economies of scale.

ROC technology maturation program

Development of machinery solutions like the ROC requires a stepwise approach, with each step intended to validate operating principals, design methods, and integrated system performance. For this specific technology, there are two fundamental building blocks that are critical: 1) use of a shock wave to induce the endothermic reaction, and 2) the ability to generate a standing shockwave in the turbomachine design. Both must be demonstrated before expending the effort and resources on a number of design and engineering challenges, which must be overcome for final product development. Figure 5 and the following paragraphs describe the development program, including both the validation of the operating principle and the ensuing detail design/validation of the integrated system.

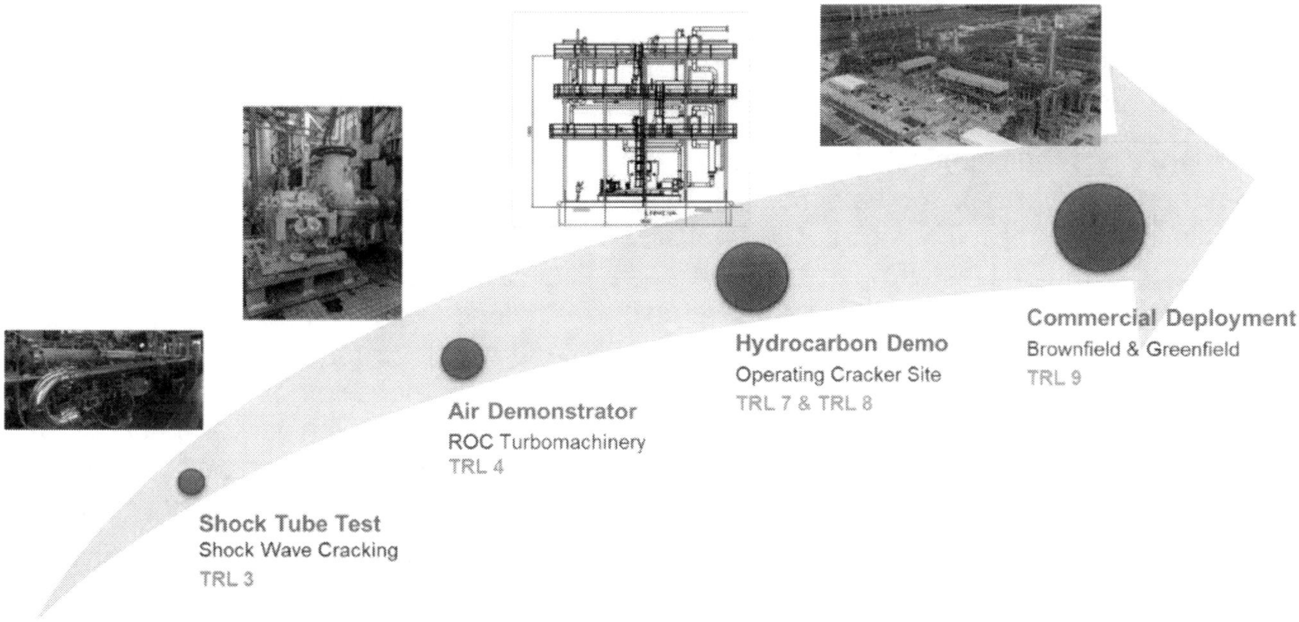

Figure 5. ROC technology maturation roadmap.

Prior to embarking down the path of generating a steady shock wave process, it is important to validate that a shock wave can create the necessary conditions for cracking hydrocarbons, and that this process can be predicted with reasonable accuracy. The Siemens Energy / Technip Energies' team enlisted the expert help of Dr. Ron Hanson and his team at Stanford University to support this effort. The team conducted a series of shock tube experiments with varying mixtures of hydrocarbons and inert gases. Each experiment consisted of a single shock wave passing through the premixed working fluid, setting off a chemical reaction to crack the relatively heavy hydrocarbons into lighter hydrocarbons. Resulting gas species were monitored using optical methods, tuned to identify specific species, such as propylene and ethylene. Results of this work are presented later in this paper.

Following validation of the shock wave cracking process, and the ability to predict the yield of such a process, the next part of the development program involves the generation of a steady shock wave environment for the cracking process. For this validation, Siemens Energy designed, built, and tested an aerodynamic prototype machine to operate on inert gas (air). Leaving out the cracking process from this demonstrator enabled the team to focus on the aerodynamics and overall mechanical design of the ROC without having to yet deal with the elevated temperatures required for hydrocarbon cracking, or the complications and expense of processing massive amounts of hydrocarbon/steam mixture. Results of this aero demo test are also presented later in this paper.

With these two fundamental building blocks complete, the focus of the development program moves on to design, build, and validation of the entire ROC turbomachine, and then the validation of the ROC furnace island.

As mentioned previously, demonstrating the actual machine with a steam/hydrocarbon mixture requires a massive amount of feedstock, certainly more than is practical for existing experimental facilities. This is mainly due to the fact that there is a minimum practical scale for the ROC turbomachine, below which the aerodynamic design becomes dominated by losses in the clearance region between the rotor and the stator. While the technology scales up quite easily, it does not scale down from the aero demo prototype size. For the ROC, the minimum practical scale for an operational prototype is a system with a capacity of 25-30 kTA of ethylene.

In addition, since the chemical composition of the feedstock changes through the ROC machine (it is a chemical reactor), the test must be performed in an open loop, and viable high-value chemicals are produced. Therefore, the final step in the development plan is to build an entire ROC furnace island and install/operate the unit as a pilot demo at an existing olefins plant where the output from the ROC can be processed in the ethylene plant recovery section. Figure 6 provides elevation views of the pilot demo ROC island and a traditional furnace of similar capacity: for the same capacity, the ROC island is significantly more compact than a traditional system. The pilot demonstration will provide valuable feedback for design validation of the entire system, and certainly for many aspects of the ROC turbomachine. All of this feedback will influence final commercial design and aid in achieving program goals for performance and reliability.

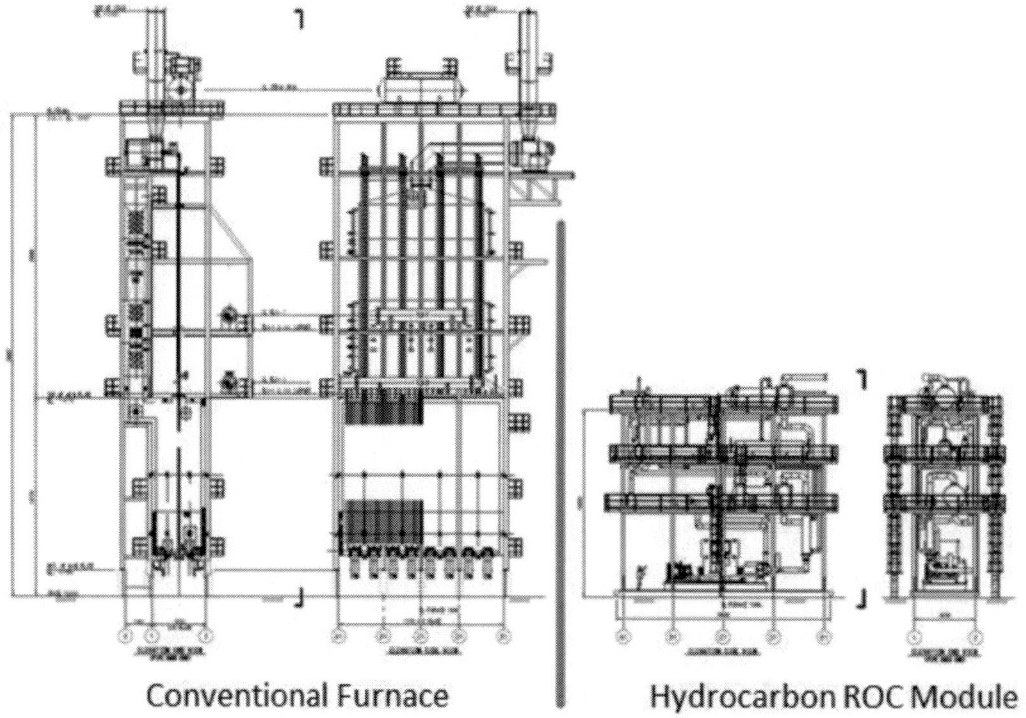

Figure 6. Elevation views of similarly sized traditional furnace and ROC islands.

Shock Wave Pyrolysis Chemistry Validation

Validation approach

The design of conventional cracking furnaces is performed at Technip Energies through the use of its proprietary simulation package SPYRO®. Known primarily for its yield prediction capabilities, the SPYRO Suite also contains modules simulating the furnace combustion side

(FIREBOX®), the convection section heat recovery (CONVEC), as well as the transfer line exchangers (TES®). The use of the four models together means the entire cracking furnace heat and energy balance can be performed in a single simulation (see Figure 3.)

The impact of the combustion side on the reaction kinetics is accounted for, because the coupled simulation between SPYRO and FIREBOX ensures that the time-temperature profile of the process gas is accurately predicted as it flows through the radiant coil. The impact of coke formation is predicted over time by providing the increase in tube metal temperature (TMT) and pressure drop.

Modeling of the ROC requires a different approach, compared to conventional cracking, primarily because there is no coil per se, and because the reaction heat is not supplied continuously from external combustion, but rather instantaneously via the standing shock wave. Nevertheless, the SPYRO model was adapted to simulate the exact conditions envisaged inside the ROC and thereby can be used to design such a device for demonstration purposes, or more importantly, for industrial applications.

For more than forty years, SPYRO has been shown to accurately predict the pyrolysis process for a wide range of operating conditions, feedstocks, and reactor configurations. This is because it has a rigorous, thermo-chemical basis. So, while conventional cracking occurs at temperatures in the range of 800 – 900 °C, and residence times of ~100 - 500 ms, the model will accurately predict the pyrolysis chemistry at temperatures much greater than 1000 °C and residence times of just a few milliseconds, and everything in-between those two extremes.

Test campaign and results

In order to obtain some initial validation of the SPYRO model vs. experimental data, a laboratory test campaign was planned. Conducting bench-scale, steady-state laboratory experiments under conditions expected to take place in a commercial ROC has practical challenges related to the aerodynamic requirements of the tests. Hence, it was decided to conduct cracking experiments under a wide range of conditions, but in a single shock reactor. These experiments were performed at Stanford University using shock tube reactors that have a long history of being used for chemical kinetics measurements under conditions of very short residence time and very high temperatures. Various diluted mixtures of hydrocarbons and argon, containing both paraffins and olefins of varying concentrations, were cracked at a wide range of temperatures, pressures, and residence times (representative of the conditions expected in the ROC). SPYRO was adapted to model the exact time-temperature shock experience.

In its simplest form, a shock tube is an apparatus consisting of a long pipe closed at both ends in which a high-pressure driver section is separated from a low pressure driven section by a thin diaphragm. When the diaphragm is ruptured, the instantaneous release of gas from the high-pressure section generates a shock wave that travels through the low-pressure section containing the test gas. The shock wave reflects off the closed, low-pressure end of the tube, increasing the temperature of the test gas almost instantaneously. The initial pressure in the test section and the strength of the shock wave are tailored to achieve the target post-shock pressure and temperature. The evolution of key chemical species and/or temperature near the closed end of the shock tube is then monitored using a time-resolved diode laser absorption system. [Ref. 5]

As illustrated by Figure 7 below, laser-based measurements of the reaction products from the pyrolysis reactions in this campaign showed good alignment with predictions made by the SPYRO software.

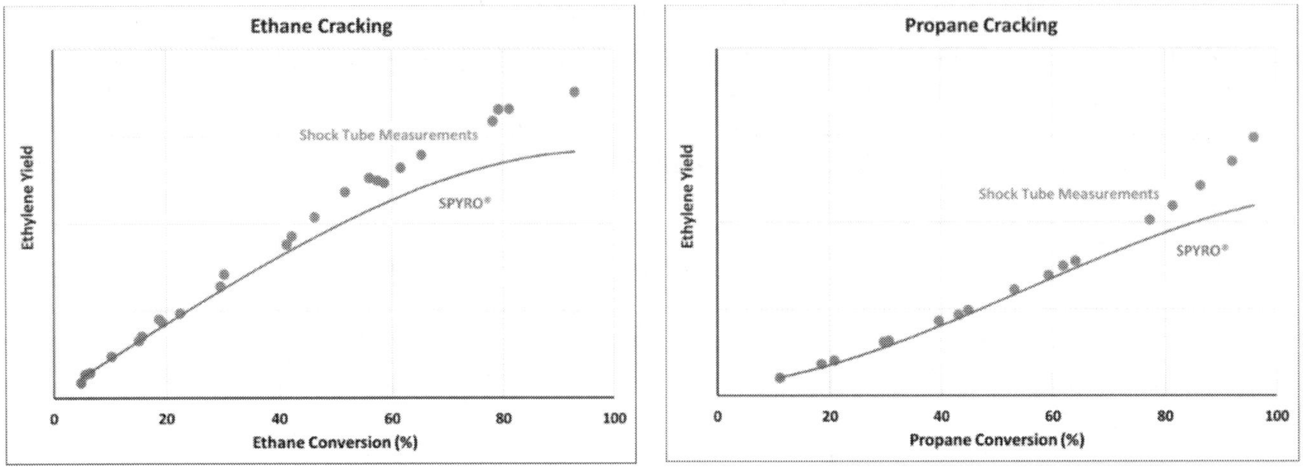

Figure 7. Shock Tube measurements for Ethane and Propane cracking experiments.

The initial shock wave tests at Stanford were set up to assess the gas phase kinetics of hydrocarbon pyrolysis after a shock wave. While the carbon balances of those experiments were excellent, they were not set up to quantify, or even detect the formation of coke. The environment for coking in the ROC will be completely different compared to a conventional cracking coil where there is a steep temperature gradient from the combustion chamber, across the reactor tube wall, across a film temperature gradient, and into the bulk gas. So while the location of coke accumulation in a radiant coil is seen on the reactor wall, there could be gas-phase carbon formation in the ROC. As a result, a series of shock wave experiments were carried out at Stanford with a modified experimental rig that would quantify the formation of solid soot particles in the reacting gas. These experiments were largely uninteresting because soot was only detected at conditions many hundreds of degrees centigrade higher than those expected for ROC operation.

Concurrently with the test validation, SPYRO was also used to predict the expected yields in the ROC hydrocarbon demonstrator for propane and naphtha feeds and compare them with those obtainable with a state of the art conventional, cracking furnace. The results of this analytical effort are presented in Table 1. As seen from the table, the ROC provides a significant improvement on the yield of high value chemicals produced by the pyrolysis process at the same relative severity.

Table 1. ROC to conventional furnace yield ratios for propane and naphtha.

	90% Propane Conversion Yields Ratio	Naphtha P/E=0.42 Yields Ratio
Ethylene	1.09	1.06
High Value Chemicals	1.07	1.04

Notes:

1. High value chemicals defined as Hydrogen, Ethylene, Propylene, Butadiene and Benzene.
2. Results based on ultimate yields.

ROC Aerodynamic Demonstrator Validation Testing

To validate the aerodynamic performance of the ROC, Siemens Energy designed, built, and tested an aerodynamic demonstrator test unit. The demonstrator is designed to operate by intaking ambient air, accelerating it to supersonic speeds, and creating a standing shock wave in the diffuser, which converts the fluid kinetic energy into temperature rise.

Turbomachine aerodynamic and mechanical design

A schematic of the demonstrator aerodynamic design is provided in Figure 8. The ROC aerodynamic demonstrator comprises a radial inlet, equipped with a set of scoop vanes, that uniformly redistribute the inlet flow around the inlet annulus of the turbomachine. A series of inlet guide vanes is then used to properly direct the flow into the first rotor. The architecture of the turbomachine is axial and comprises multiple rows of rotating blades and of stationary vanes. The stationary vanes redirect the flow exiting each rotor so that it is properly aligned with the following rotor blade. The succession of rotating and stationary vanes compounds the velocity of the fluid which is accelerated to supersonic speeds prior to entering the diffuser without a significant increase in static pressure. In the diffuser, the flow is de-swirled and a standing shock train is created, which decelerates the flow from supersonic to subsonic velocities, thereby converting the flow kinetic energy into a temperature rise.

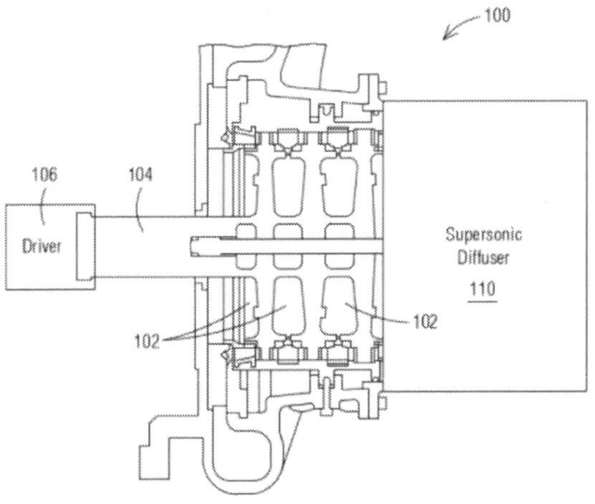

Figure 8. Schematic of ROC aerodynamic demonstrator

As the ROC aerodynamic demonstrator operates with air, rather than a high-temperature mixture of hydrocarbons and steam, the turbomachine is designed to operate in aerodynamic similarity between the two different sets of inlet conditions and fluids. Therefore, the results obtained from testing can be readily scaled and provide a direct experimental validation of how the turbomachine would operate at the cracking process design conditions. The ROC aerodynamic demonstrator is designed to achieve a ratio between outlet and inlet temperatures larger than 3.2 when operating with air. At the design condition, the power required to drive the turbomachine is more than 1.6 MW.

The main flow path of the turbomachine is heavily instrumented with pressure and temperature sensors which are used to verify adequate aerodynamic operation.

Figure 9 (a) and (b) provide a view of the CAD model of the demonstrator test article and a picture of the same installed in the test bed, respectively. The mechanical arrangement of the turbomachine is a standard overhung design, with tilt pad journal and thrust bearings located in the pedestal, connecting the ROC demonstrator to the base-frame. The bottom half of the radial inlet is connected to the same pedestal and provides the support to ground the outer case and the conical reducing connector to the exhaust stack. A separation is maintained between the radial inlet and the pedestal to minimize heat soak back into the pedestal itself and the bearings.

(a) *(b)*

Figure 9. (a) Three-dimensional CAD model of the ROC aerodynamic demonstrator. (b) ROC Aerodynamic demonstrator installed in the test cell.

The rotor system comprises multiple discs with Hirth serrations connected to the rotary shaft via a single tie bolt and tie nut. The airfoils for each stage are integrally machined with the disc, a configuration that is referred to as "blisk".

In addition to the main flow path instrumentation, the aerodynamic demonstrator includes a suite of instrumentation to monitor the health and safety of the turbomachine.

Test facility

Testing of the aerodynamic demonstrator was conducted in the Lincoln, UK, Siemens Firth Road test facility. A schematic of the drive system is presented in Figure 10. This consists of a TB 5400 gas turbine engine rated at 3.4 MW and with a rated output shaft speed of 7,950 rpm. The gas turbine is connected to two gearboxes providing a total gear ratio of 1.727:1. This system allows the speed of the ROC demonstrator to be continuously varied between 6,638 rpm and 15,067 rpm. A torquemeter is inserted between the Allen 11 gearbox and the ROC demonstrator, which allows the operator to view the input power to the turbomachine.

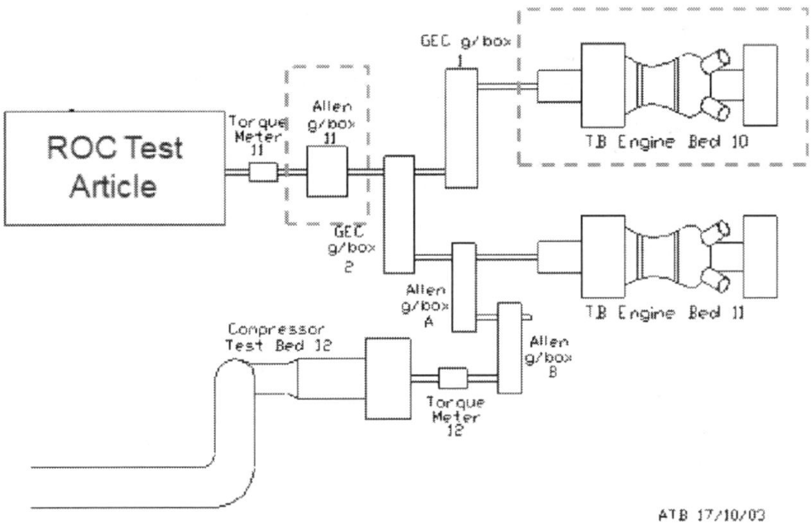

Figure 10. Driver and gearbox arrangement of the Lincoln, UK, Firth Road test facility.

Figure 9 provides a schematic of the test setup for the ROC air demonstrator and Figure 10 is a picture of the test bay. An air intake with filter house is located outside of the test cell. An inlet air heater is used to provide a constant inlet air temperature independent of the ambient conditions. An orifice plate is placed in the inlet duct system to measure the mass flowrate of air processed by the unit. The inlet ducting is then connected to the radial inlet of the ROC aerodynamic demonstrator via flexible bellows. The high temperature exhaust from the unit is ducted to the outside through an exhaust stack equipped with a valve, which can be used to vary the system discharge pressure.

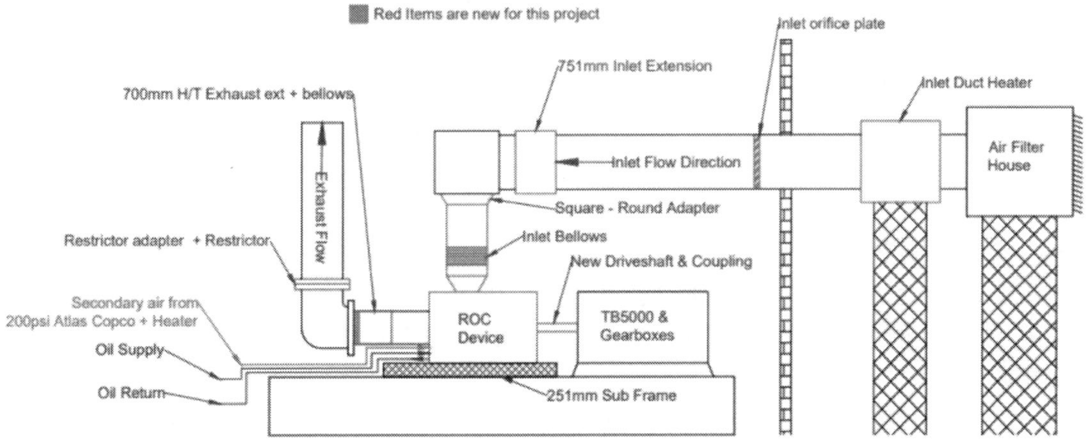

Figure 11.. Schematic arrangement of the ROC test cell.

(a) (b)

Figure 12. (a) Picture of the test bay. The facility control room is located in the right portion of the picture. (b) Picture of the ROC aerodynamic demonstrator test bed.

The test facility also provides all the auxiliary services and system required to run the turbomachine, such as a lube oil system, a heated purge flows air supply system, a control system for the gas turbine and the auxiliary systems, low frequency and high frequency data acquisition systems, a pressure scanner measurement system, a strain gage signal conditioning module, and a Bently Nevada bearing vibration monitoring system.

Test Program and Test Results

The ROC aerodynamic demonstrator test campaign began on Aug 8th, 2022 and concluded on Sept 1st, 2022. The test campaign explored the operation of the turbomachine for a wide range of operating speeds varying from 60 to 104 percent of design speed. At each speed, the seal flow used to purge the wheel space cavities were varied to assess their effect on the aerodynamic operation and performance of the turbomachine. Furthermore, testing with a varying degree of backpressure was conducted to experimentally determine the turbomachine stall line.

For any given operating point, the desired speed and desired level of backpressure is set, and the turbomachine is operated at this condition until pressure and temperatures across the machine no longer vary. Once the operating point is stabilized, one minute of data is recorded and averaged to calculate the performance data presented in the figures.

Once the aerodynamic testing was completed, a four-hour API mechanical test was conducted to demonstrate stable operation across a range of bearing lube oil flows and temperatures.

Figure 13 provides an overview of the turbomachine performance across a range of speeds and wheel space purge flow setting. The performance of the turbomachine is represented by the outlet-to-inlet total temperature ratio and the corrected mass flow processed by it (defined as the inlet mass flow divided by the ratio of the actual inlet speed of sound and the speed of sound at 15 °C), which are plotted against the machine Mach number (defined as the ratio of the rotor tip speed to the inlet speed of sound). Each point on the plot represents a different operating condition and the solid lines represent the pre-test prediction obtained from Computational Fluid Dynamic (CFD) simulations. It can be seen from these figures that there is excellent agreement between the predicted and measured total temperature rise across the entire operating range. At design point, the turbomachine is capable of heating air from ambient temperature to temperatures in excess of 700 °C.

The measured corrected mass flow rate indicates a four percent shortfall, 2.18 kg/s in test vs. 2.27 kg/s predicted, at design point operation. It is hypothesized that the discrepancy is caused by unforeseen internal leakages form the flow circuits supplying air to purge the wheel space cavities.

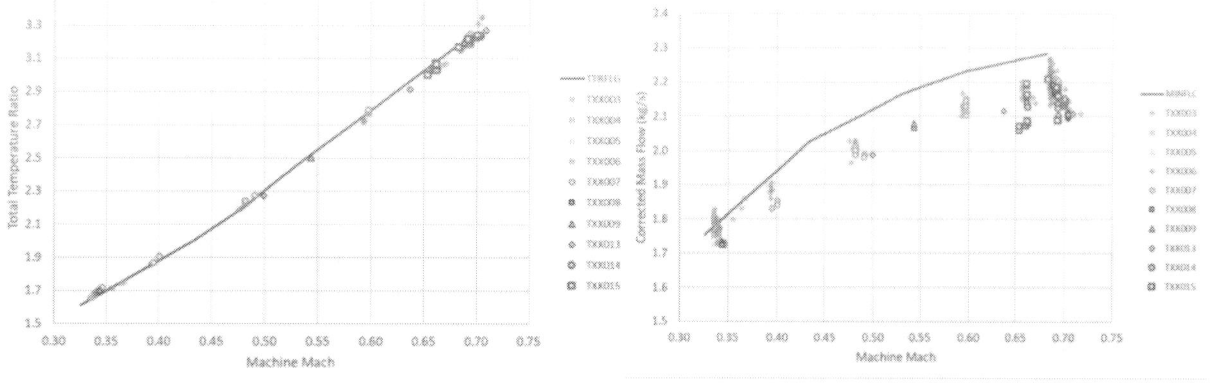

(a) (b)

Figure 13. Total temperature ratio and corrected mass flow as a function of the machine Mach number.

Rotordynamic performance of the aerodynamic demonstrator was very good and vibrations remained well below the API allowable levels. The data collected showed excellent agreement with predicted values of critical rotordynamic parameters.

A four-hour mechanical running test was also performed during which rotor speed, bearing oil flowrates, and bearing oil inlet temperatures were varied per the API test protocol. Peak-to-peak recorded vibration levels throughout were of the order 7-8 μm for the DE bearing and 3-4 μm for the NDE bearing, and remained well below the allowable API test limit of 23.6 μm. Similarly, bearing temperatures were 80-90 °C for the thrust bearing, 77-85 °C for the NDE journal bearing, and 67-77 °C for the DE journal bearing, again well below the 110 °C allowable limit. The successful completion of the mechanical running demonstrates the robustness of the rotordynamic design of the test unit.

Blade vibration and high cycle fatigue are one of the main concerns for the design of a turbomachine. During the ROC aerodynamic demonstrator Finite Element Analysis (FEA) tools were used to calculate modal responses of the airfoils across the speed range. If a resonance was observed, either the blade geometry was modified to move the resonance out of the operating range, or a forced response analysis was used to determine that stresses remained below allowable limits.

During testing, a BVM system was used to monitor aerofoil vibration using industry best-practice of 8 probes per stage. Vibrations were monitored across the whole speed range throughout the test program. Figure 14 provides an example of the data collected. The measured vibrations at various operating conditions matched frequency predictions and in every case the amplitudes were acceptable and below the established limit. Based on the data collected, it can be concluded that the aerofoils have sufficient strength for continuous operation at design conditions without risk of fatigue failure.

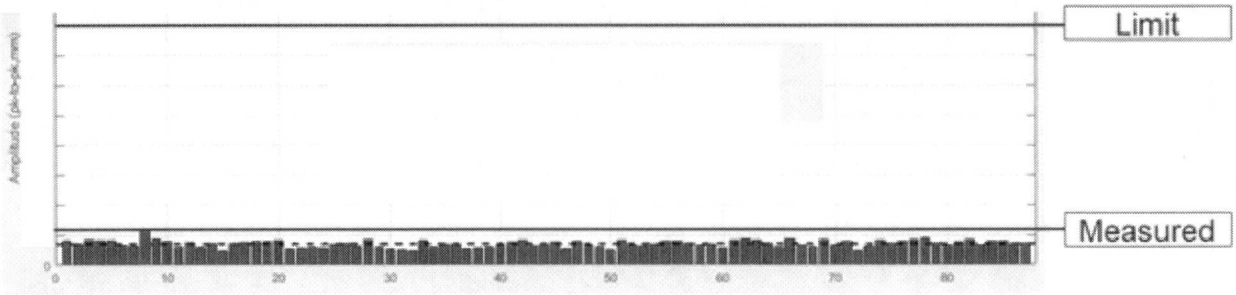

Figure 14. Measurement of blade vibration amplitudes.

Figure 15 provides a summary of the tip gaps measured throughout the test campaign. The dashed line represents the target design hot running tip gap. It can be seen that for rotor two and three the measured tip gaps are very close to the design target, whereas the tip gap is considerably larger for the first rotor. The reason for the discrepancy between expected and measured tip gap is currently being investigated. The data collected also suggests that no rub occurred during testing which has been confirmed both by monitoring of the blade vibration and, for the third rotor, by borescope inspection.

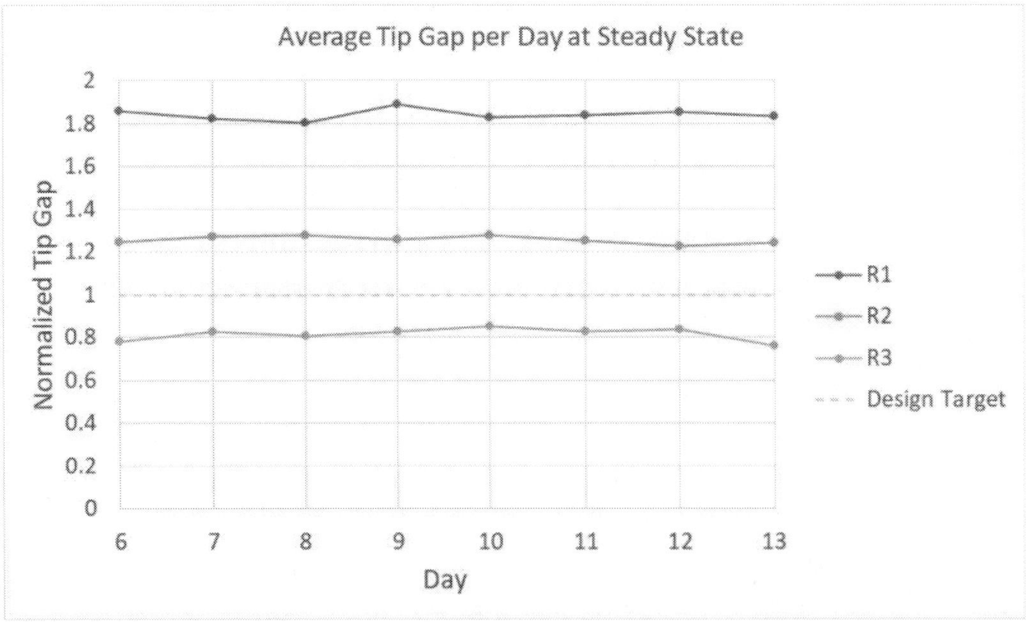

Figure 15. Average normalized rotor tip gap measurements for each test day.

As previously outlined, the test rig is instrumented with surface mounted thermocouples, which are used to measure the temperature of various test article components. At design point, differences between calculated and measured temperatures for the diffuser range between 2 °C and 7 °C. Measured temperature scatter at various circumferential locations is of the same magnitude as the difference between measured and predicted values. These differences are smaller at the leading edge, compared to the trailing edge, where the modelling of flow conditions is more challenging. The close match at design point between predicted and measured values provides confidence in the thermal modelling of the turbomachine, and more specifically, in the approach used to calculate Heat Transfer Coefficients (HTC) in the supersonic portion of the flow path.

The discrepancy between calculated and measured temperatures increases with higher purge flows and at off design conditions indicating that the modelling of HTCs needs to be improved.

Critical stationary components of the main flow path are also instrumented using strain gauges to monitor vibrations of the components and assure that no resonances are observed. During steady state operation, the data collected by the strain gauges indicate that no resonant vibrations occur, which are of concern for the life of the part.

Conclusions

To address the need to decarbonize the olefin production process, Siemens Energy (SE) and Technip Energies (T.EN) are jointly developing the Rotating Olefins Cracker (ROC). The ROC technology employs shock wave gas dynamics to supply the energy input for steam cracking via direct fluid heating of the process flow in place of fired heating in conventional furnaces, thus eliminating the reliance on combustion of fossil fuels for the manufacturing of light olefins.

The ROC technology provides several advantages compared to the current state of the art, namely: GHG emissions reduction via the use of green electricity of hydrogen fired gas turbines, increase in product yield due to the high temperature and short residence times, reduced propensity to coking since the wall temperatures are lower than the fluid temperature, and reduced equipment size and superior scalability which are an inherent characteristic of turbomachinery.

The ROC technology development plan has followed a systematic, step-by-step approach starting from shock tube, laboratory testing to verify the capability of existing chemistry models to accurately simulate the pyrolysis reactions, and proceeding with the design, build, and testing of a reduced-scale, low-temperature aerodynamic demonstrator, operating in similarity to the conditions expected in the actual device. The results of these development efforts have been presented in this paper. The investigation of pyrolysis of pure hydrocarbons and hydrocarbon mixtures when heated through a shock wave show that the current chemistry models are adequate in predicting the products of the pyrolysis process. Moreover, application of these models to the envisioned operation of the ROC provides evidence of improvement in the expected HVC yield. Similarly, testing of the ROC aerodynamic demonstrator has proved that the envisioned turbomachinery architecture can reliably produce the targeted temperature rise required by the cracking process.

Based on these results, Siemens Energy and Technip Energies are proceeding on the next step in the development roadmap, which includes the design and manufacturing of a ROC demonstrator capable of processing and cracking hydrocarbon/steam mixtures. Start of testing of the hydrocarbon demonstrator unit is expected to occur in the 2025 timeframe.

References

[Ref. 1] The Future of Petrochemicals, International Energy Agency (IEA), pg. 50 (2018).

[Ref. 2] The press release can be accessed at https://press.siemens-energy.com/global/en/pressrelease/siemens-energy-and-technip-energies-announce-joint-development-decarbonized-rotating

[Ref. 3] T Ren, M Patel, K Blok. Energy Journal, Issue 31, pg. 425–451 (2006).

[Ref. 4] T Ren, M Patel, K Blok. Energy Journal, Issue 33, pg. 817–833 (2008).

[Ref. 5] Jiankun, S., M., F. A., Rishav, C., J., C. S., F., D. D., & K., H. R. (2020). Combustion and Flame, 221, 364-370 (2020)

Gap in pagination due to formatting issues.

Pages 451-464

61b Electrosynthesis of High-Purity Ethylene Using CO2 and H2O at Near Ambient Conditions

Rohan Sartape, Nishithan Balaji Chidambara Kani, Meenesh R Singh

C_2H_4 is a hydrocarbon of extensive societal, environmental, and industrial importance. Therefore, synthesizing C_2H_4 sustainably via electrochemical CO_2 reduction reaction (CO2RR) is an attractive area to explore. Even though many existing CO2RR systems have reached industrially relevant current densities, almost all use a gas diffusion electrode (GDE)-based electrochemical system with a single-pass conversion of <10%. This leads to a lower concentration of C_2H_4 in the gaseous product stream that mainly comprises CO_2, which contributes significantly to the cost of post-CO2RR separation of products, rendering even processes with high CO2RR current densities unfit for scaling up. Here we develop an aqueous flow-through electrochemical cell to enhance the activity and selectivity of C_2H_4 on a 3D Cu mesh electrode by applying square wave oscillating potentials. We control the oxidation phase of the square-wave oscillating potential for in-situ generation of copper oxide, with some residual sub-surface oxide remaining in the cathodic phase. We hypothesize that the transient sub-surface oxide exerts a significant strain on the metallic copper surface during CO2RR, leading to enhanced C-C bond formation as predicted by DFT. A high C_2H_4 Faradaic efficiency (FE) of ~58%, unprecedented C_2H_4 current density of 306 mA/cm^2 in the aqueous cell, and gaseous C_2H_4 purity of ~52 wt% without CO_2 in the product stream is obtained. Integrating the 3D Cu mesh catalyst in a PV-electrolyzer yields a remarkable solar-to-carbon (STC) efficiency of ~10% with a solar-to-C_2H_4 efficiency of ~4%, almost double the current state-of-the-art solar-driven CO2RR systems. The novel electrochemical cell and catalysts offer several breakthroughs necessary for the sustainable manufacturing of C_2H_4.

61c Intensified Plasma-Driven Methane Coupling to Ethylene through Tailor-Made Catalytic Monoliths

Fabio Cameli, Marco Scapinello, Victor Rosa, Kevin Van Geem, Georgios Stefanidis

The increasing exploitation of shale gas reserves is rapidly changing the energy market, thus the abundant methane (CH_4) streams are gaining the spotlight for chemical valorization.[1] Ethylene (C_2H_4) is the most valuable product of methane upgrade being the most produced organic chemical worldwide due to its role of building block in several chemical processes.

Nonetheless, ethylene is mostly manufactured via steam cracking of naphtha at elevated temperature and with several downstream steps to purify the products mixture. Therefore, extensive research is focusing on alternative methods to exploit abundant methane resources to produce ethylene in a selective and sustainable fashion.[2]

In this context, non-thermal atmospheric plasma (NTAP) represents a promising method to activate stable CH_4 molecules via (renewable) electric energy locally generated next to extraction sites. The highly energetic plasma mixture contains electrons at temperature much higher than the surrounding gas, hence methane conversion beyond the thermodynamic limits can be attained.[3] The most stable products of plasma-assisted methane activation are C_2 species with varying selectivity according to the plasma source.[4]

In this work, we employ a nanosecond pulsed discharge (NPD) plasma reactor to intensify the energy channeled into the system and we couple it with customized catalytic monoliths that drive hydrogenation of acetylene (C_2H_2) to ethylene. High energy efficiency can be attained by regulating the frequency of the nanosecond pulses and the applied voltage to reach single pass CH_4 conversion up to 40% and C_2H_2 selectivity of 80%.[5] A structured electrode in the plate-to-plate configuration is covered with a catalytic layer to shift the selectivity of the C_2 species. The tailor-made, 3D printed monolith is designed upon CFD calculations and assessment of heat integration. Optimization of the (Pd based) catalyst loading and the thickness of the catalytic layer leads to complete conversion of acetylene. Thereby, the lab-scale results of this unique plasma reactor are promising for further development of the technology. The modular nature of the reactor allows numbering up to address large scale production, whilst long term operational stability must be assessed in the context of industrial implementation. This technology could suit decentralized methane valorization into value-added ethylene via an electrified process.

References

(1) Kim, S.; Oh, S. Impact of US Shale Gas on the Vertical and Horizontal Dynamics of Ethylene Price. *Energies* **2020**, *13* (17). DOI: 10.3390/EN13174479.

(2) Dai, Y.; Gao, X.; Wang, Q.; Wan, X.; Zhou, C.; Yang, Y. Recent progress in heterogeneous metal and metal oxide catalysts for direct dehydrogenation of ethane and propane. *Chemical Society Reviews* **2021**, *50* (9), 5590-5630. DOI: 10.1039/d0cs01260b.

(3) Scapinello, M.; Delikonstantis, E.; Stefanidis, G. D. The panorama of plasma-assisted non-oxidative methane reforming. *Chemical Engineering and Processing: Process Intensification* **2017**,

117, 120-140. DOI: https://doi.org/10.1016/j.cep.2017.03.024.

(4) Heijkers, S.; Aghaei, M.; Bogaerts, A. Plasma-Based CH4 Conversion into Higher Hydrocarbons and H2: Modeling to Reveal the Reaction Mechanisms of Different Plasma Sources. *The Journal of Physical Chemistry C* **2020**, *124*(13), 7016-7030. DOI: 10.1021/acs.jpcc.0c00082.

(5) Delikonstantis, E.; Scapinello, M.; Van Geenhoven, O.; Stefanidis, G. D. Nanosecond pulsed discharge-driven non-oxidative methane coupling in a plate-to-plate electrode configuration plasma reactor. *Chemical Engineering Journal* **2020**, *380*, 122477-122477. DOI: https://doi.org/10.1016/j.cej.2019.122477.

61d Catalytic Plastic Pyrolysis and Steam Cracking an Opportunity for a Win-Win Partnership

Leslaw Mleczko, Dennis McCullough, David Sudolsky

Anellotech has spent the last 12 years building a scalable advanced biofuel and chemical recycling suite of technologies that can dramatically reduce greenhouse gas emissions of transportation fuels or plastic manufacturing including packaging and textiles. The company's unique patented catalytic cracking and pyrolysis Plas-TCatTM process approach combines expertise in fluid bed reaction engineering and catalysis with solids handling and pretreatment of waste plastic feedstock materials for direct production of light olefins (C2, C3, C4) and BTX. This paper presents the current stage of Plas-TCat development (TRL-6) with large pilot plant demonstration in the expected commercial-scale olefins production environment.

Floods, droughts, everyday news confirm that climate change happens, and we need to handle. Furthermore, ocean pollution and microplastic issue concern people. The growing market pull for the products with a low CO2 footprint confirm people awareness of this fact. Usually, such developments are
followed by regulatory actions either in form of incentives or taxes. Circular economy is one of the options
to address, above mentioned, changing market and environmental needs but also an opportunity to make
money (develop new business area). Therefore, plastic recycling became in the last few years an increased
interest expressed in the development activities and investments in new recycling plants. However, technologies available on the market can't solve the recycling problem due to their limitations with respect to the voluminal and plastic waste composition.

Chemical recycling is one of the advanced answers to the future market needs. Currently chemical recycling, i.e. conversion of waste plastics to monomers or chemicals, is dominated by the gasification and
thermal pyrolysis processes. Both technology families need multistage processing in order to obtain basic
feedstock olefins and aromatics and to close the recycling loop. The gasification route is an especially long
processing chain and CAPEX intensive technology. In the case of thermal pyrolysis, its product, i.e., pyoil
can be fed either to FCC units or to steam cracker ("SC") furnaces. However, when using it as SC feedstock
it needs to be deeply pretreated in order to remove heteroatoms and contaminates. It is also a low quality
feedstock because of its highly olefinic composition which promotes coking in the cracking furnaces.

Alternatively, Plas-TCatTM 's catalytic pyrolysis of plastic wastes belongs to the second generation technologies for conversion of plastic wastes which overcomes disadvantages of gasification and thermal
pyrolysis. Catalytic pyrolysis integrates thermochemical decomposition of plastics with in-situ catalytic
conversion of pyrolytic intermediates directly to high value chemicals, i.e. aromatic hydrocarbons (mainly
BTX) and light olefins (mainly C2, C3 and C4). Catalytic transformation also eliminates heteroatoms from

the liquid product stream by conversion to components that can be easily separated from the product gas. The ability to remove heteroatoms allows this process to convert mixed plastic waste streams that contain all types of plastics (1-7) with exception of PVC. Combining the expected emission reductions
of the direct conversion process with the scalability of fluid bed processing can deliver large CO2 mass reductions for the world's transportation and chemical manufacturing sectors that are consistent with the olefins industry GHG reduction goals.
Since it is a low pressure and relatively low temperature process it exhibits also competitive CAPEX. As for
most chemical processes it is important to achieve economy of scale. Catalytic pyrolysis offers the possibility to scale the process to large units since it converts plastics in a fluidized bed reactor that utilizes
the same principles as FCC reactors. However, the maximum capacity of a catalytic pyrolysis plant currently is limited by the logistic costs for the PSW supply. This limitation is especially critical for CAPEX
of product separation. One of the possibilities to reduce the downstream costs is an integration with a large-scale units like FCC or Steam Crackers. By this linkage with existing units the catalytic conversion can
profit from the economy of scale. On the other side, by integration of the catalytic pyrolysis with Ethylene
Plant downstream purification trains, the olefin producers can reduce their CO2 footprint and offer plastic
waste-based monomers along with their virgin olefins supply to PE, PP, etc. downstream units without
affecting the steam cracking furnaces except for small paraffin recycling needs.
In the presentation, a catalytic pyrolysis process, developed by Anellotech, Inc., for conversion of mixed
plastics wastes (Plas-TCatTM) will be introduced. This development is at the stage of testing and validation
(TRL-6) in a large fully automatized and continuously operated pilot plant with a nameplate capacity of
0.5 t/d. Furthermore, the potential for integration of Plas-TCat with a steam cracker plant will be discussed.

AIChE Paper Number: 61e

THERMAL CRUDE TO CHEMICALS

Kandasamy M. Sundaram, Distinguished Technologist
Lummus Technology

Ujjal K. Mukherjee, Chief Technology Officer
Lummus Technology

Pedro M. Santos, Technology Director, Chevron
Lummus Global

Ronald M. Venner, Chief Business Officer, Clean
Fuels & Crude to Chemicals,
Lummus Technology

Prepared for Presentation at 2023 Spring National Meeting
Houston, Texas, March 13-16, 2023

Thermal Crude to Chemicals

Kandasamy M. Sundaram, Ujjal K. Mukherjee and Pedro M. Santos, Ronald M. Venner

Lummus Technology, Bloomfield, NJ

Abstract

Steam crackers traditionally use ethane, LPG and naphtha as feedstock. Occasionally, heavier hydrocarbons such as gas oils and unconverted oil from hydrocracking units have been utilized. There are some complexes that process very light crudes and condensates directly in a steam cracker. Lummus Technology, CLG (Chevron Lummus Global) and Saudi Aramco Technologies Company have conducted intense joint research over the last several years to understand the implications of steam cracking crudes heavier than very light crudes and condensates with the bulk of the experiments conducted on Arab Light (AL) crude. The research involved understanding the distribution of molecular species within the whole crude including the very complex structures in asphaltenes. Novel techniques of separation avoiding conventional crude atmospheric and vacuum distillation systems, appropriate catalyst and integrated reactor platforms along with novel heat integration, permits the consistent yield of chemicals to increase from 50 to 60% to between 70 and 80% and higher. Selective hydrogen addition optimizes the molecular species for optimum cracker yields and maximum heater run lengths. One important objective of the research that led to successful commercialization in 2020, was the elimination of low-value stream production such as pyrolysis gas oils even while processing heavier hydrocarbon feeds. Heat integration, operability and reliability were key parameters in the development as was the very significant reduction of overall CAPEX, OPEX, plot space and CO_2 footprint. Some of the salient features of this technology will be discussed in this paper.

Introduction

In 2021 about 214 million metric tons of ethylene were produced worldwide (Fernandez, 2022). Most of the ethylene comes from thermal cracking of hydrocarbons. Primary feeds to produce ethylene are ethane in the Middle East and US Gulf Coast and naphtha in Europe and Asia. In addition to these feeds, LPG, kerosene, diesel and hydroprocessed and virgin vacuum gasoil are used depending upon availability and the local economics. In some parts of the world, ethanol dehydration (bio-ethylene) and methanol to olefins are used but to a lesser extent. In addition, syngas to olefins, oxidative coupling of methane and dehydrogenation or oxydehydrogenation of ethane to ethylene are also under consideration. A small fraction of ethylene produced in the FCC process is also recovered. Steam crackers produce very high purity products (99.95% pure ethylene). As a result, stringent feed specifications are imposed when the feed is destined for steam cracking. Thermal cracking is one of the petrochemical processes that uses very high energy and emits significant quantities of CO_2. A typical naphtha cracker emits from 1.2 to 1.8 kg CO_2/kg C_2H_4; heavier feeds emit more. Often the ethylene plant is integrated with a refinery and the feeds and products and energy, especially in the form of steam, are exchanged between the units. LPG and naphtha feeds used in the cracker are usually obtained from crude by separation using a crude distillation unit (CDU), an energy intensive process. Do we really need a CDU to obtain feeds to produce olefins? This paper will address some of the methods employed to process crude oil to produce olefins bypassing the refinery. This reduces the energy, OPEX, CAPEX and CO_2 emission. Although in principle any crude oil can be used as feed, some are more economically attractive than others. A primary optimization parameter is to maximize ethylene, propylene and other valuable chemicals from crude while producing as low amount of fuel as possible.

Ethylene Production

Most of the ethylene is produced by thermally cracking hydrocarbons mixed with dilution steam in the vapor phase at low pressures and at high temperatures in tubular reactors (known as coils). The effluents are separated in the recovery section consisting of many distillation columns and cryogenic fluids as coolants. Fired heaters are used to supply the energy to the cracking reactions. Super high pressure (SHP) steam (100+ bar) is produced by heat exchange against heater effluent. The SHP is used to drive the compressors in the recovery section. Modern furnaces achieve more than 94% overall thermal efficiency.

Crude cracking has been studied even in the1960's but never commercialized due to poor economics (see Kunii,1980; Corma et.al., 2018). For many decades and even today, the same technology is used to produce ethylene with traditional feeds such as ethane and naphtha. With continued research and development in various disciplines in the last few decades features such as Short Residence Time (SRT®)/low pressure drop coils led to superior olefin selectivity and reduced feed consumption. Ethylene capacity has increased from 25 KTA per furnace in the 1960s to 300 KTA per furnace today and the plant capacity has increased from 300 KTA to over 2000 KTA. The feeds have not changed but the once through yields have improved by more than 10% in the past decades. Therefore, again crude cracking to improve refinery economics is discussed in recent conferences (Masen and Pedro 2020, Mukherjee 2021,2022, Arun 2022, Pedro 2022).

Fundamentally, thermal cracking proceeds via free radical mechanism (Sundaram & Froment,1978; Froment,1992; Dente et.al 1983; Van Geem et.al 2008) and hence lower hydrocarbon partial pressure and short residence time favor higher olefin selectivity (Fernandez-Baujin and Solomon, 1976). However, during this process side reactions leading to the formation of coke occur. Coke, a solid phase, deposits on the walls of the reactor and increases the coil pressure drop, increases the tube metal temperature and limits the ethylene production. Periodically the heater must be steam/air cleaned (decoked). When the frequency of cleaning is increased, it is not economical to crack that feed. Typically, Lummus SRT® heaters run between 30 and 60 days before a cleaning is required. When contaminants are present in the feed, the run length (the time between two decokes) is reduced significantly. Note that the cracking is a homogeneous reaction and hence is independent of surface to volume ratio. On the other hand, coking reactions are surface dependent. Though chemical reaction engineering principles can be used to improve the run length to some extent, feeds or reactor designs resulting in short heater run lengths are not preferred. The difference between the traditional feeds such as naphtha or gasoil and crude oil is the run length of the heater. Typically, when crudes are used as feeds, the run length can be a matter of hours! Therefore, crude oils were not considered as good feeds for olefin production in traditional coil technology. In addition, each type of feed requires optimum processing conditions and hence even with good quality crude oil, the ethylene yield will not be high if optimum conditions are not chosen. How do we achieve maximum performance with crude oil cracking?

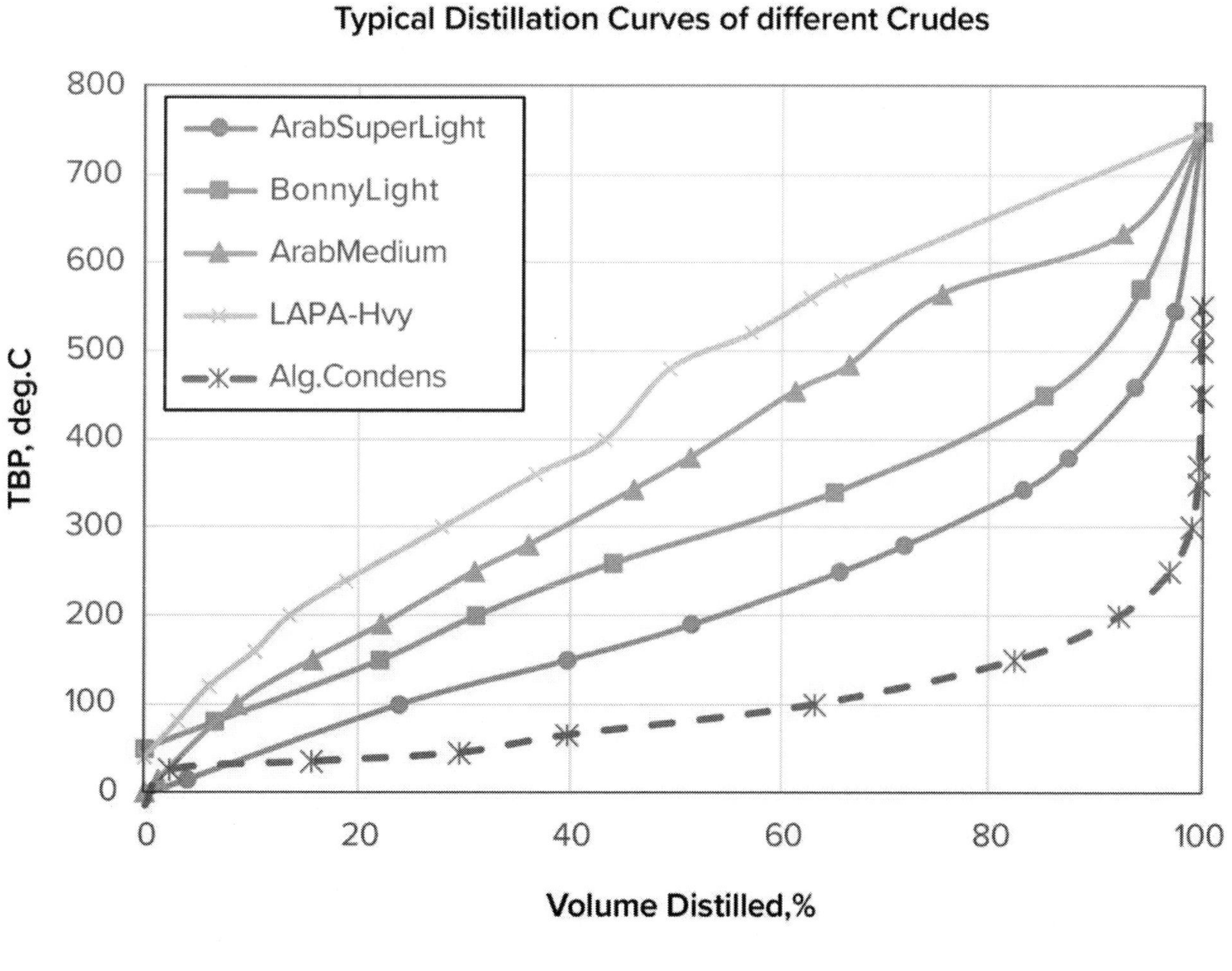

Fig.1 Different types of crudes and Condensate: 1-Arab Superlight (API=49.6, MCR=0.48 wt%), 2-Bonny Light (49.6, 1.24), 3-Arab Medium (30.9, 5.4), LAPA Brazil (22.7,7.0); Dashed line -Algerian Condensate (70.1,0.0009)

Crudes

Unlike ethane or propane, crude is not a single species. It is a mixture of species boiling from very low temperature (~10 °C) to very high temperatures (>700 °C). Typical distillation curves for some crudes are shown in Fig.1. The composition of the crudes also varies significantly. Material boiling above 525 °C is typically called residue. Currently these crudes are sent to a crude distillation unit in a refinery after going through a desalter and separated into different cuts as shown in Fig.2. These cuts are made to some (fuel) specifications (like S.G., distillation range, sulfur, nitrogen, viscosity, etc.) so that these cuts can be used as fuel or processed in downstream units. When the fractions from the crude distillation tower do not meet the required specifications of transportation fuels such as gasoline, aviation turbine fuel, diesel or fuel oil, the fractions have to be processed in other units such as Fluid Catalytic Cracking, Hydrocracking, Residue Hydrocracking, Hydrotreating, and many, many other units. Often these cuts, when not used as fuels are sent to olefin plants as feeds. It is not required that transportation fuel specifications must be met for olefin production. Olefin feeds must meet certain contaminant levels and boiling ranges only. Ethylene plants can tolerate a wide range of feed properties when properly designed. Therefore, some of the refinery operations can be bypassed when used for olefin production. Often field condensates, also called as natural gas liquids, are used as cracker feeds. These are much lighter than crudes but contain some characteristics similar to a crude (Maddock, 1992).

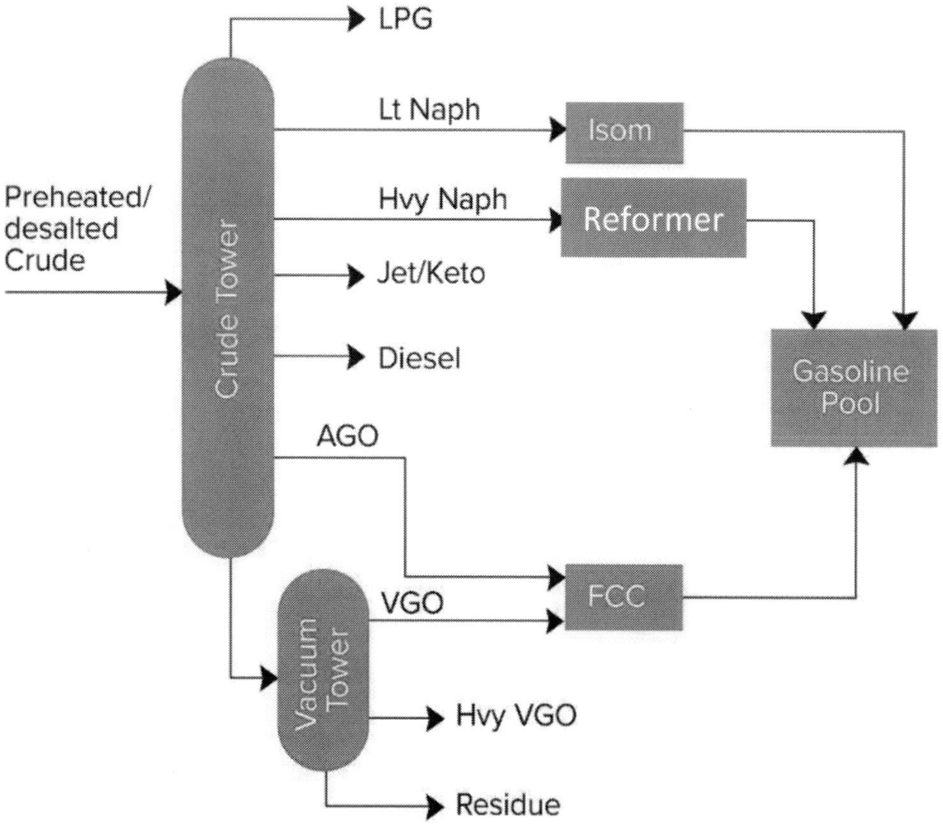

Fig.2 Typical Crude distillation unit showing different cuts

As an example, distillation curve of an Algerian condensate feed is shown in the figure 1. When these condensates are cracked without proper heater design the heater run lengths are also very short. The Algerian condensate shown in Fig.1 is successfully cracked in many plants, but it has some unwanted contaminants (Hg &As) at high levels, which cause problems in the recovery section. If not properly handled, they will damage the recovery section. Therefore, for every crude the effect of the contaminants with respect to olefin plant operation is important. Without a detailed analysis of the crude it cannot be predicted which crude is suitable for steam cracking. Typically, light crudes have low level of contaminants. Based on API value, crudes are classified as superlight (API >45), light (35-45), medium (30-35) and heavy (<30) crudes. Other types of classification also exist. Superlight crudes are almost like field condensates. Superlight and condensates have a high fraction of naphtha boiling range (<200 °C) material (>60%) in the whole crude and only a small fraction of residue (<5%) that boils above 525 °C. Inspection of many crudes shows that many light and superlight crudes are potential candidates as olefin feeds. Figure 3 shows sulfur content of many commercially available crudes as a function of gravity. Sulfur is one of the parameters monitored for refinery operation and for ethylene production.

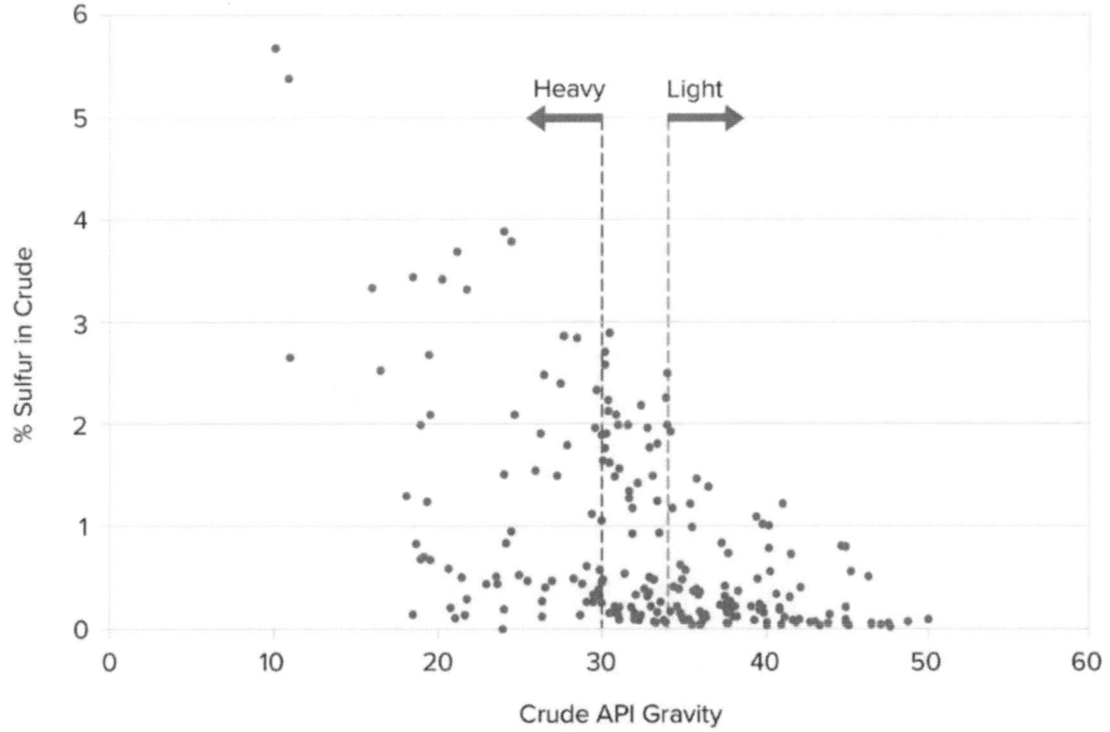

Fig.3 Sulfur content of crudes as a function of gravity

In pyrolysis heaters sulfur is thermally converted to H2S and removed as acid gas.

S in Feed → H2S + Other S-species

For many cracker feeds, more than 90% of feed sulfur is converted to H_2S. The remaining sulfur is converted to mercaptans, thiophenes and other sulfur compounds. When sulfur is present in high level, a large acid gas removal system must be used in the ethylene plant. In addition, other sulfur species such as mercaptans and thiophene act as catalyst poisons for many hydrogenation reactors used in the ethylene plant making the ethylene plant more expensive. Often such high sulfur crudes contain other contaminants such as high MCR (micro carbon residue) and high level of basic nitrogen species. Though high sulfur is not an issue in the ethylene plant, it may not be worthwhile to spend extra CAPEX in the ethylene plant for a large acid gas removal system. It is better to treat (hydroprocess) the feed in the refinery and reduce the sulfur content. When there is no refinery near the petrochemical complex, it is preferable to consider crudes that have less than 0.5 wt% sulfur for ethylene production for economic reasons. The basic thermal crude to chemicals scheme outlined here can be used for any type of crude, but it is less capital intensive and better suited for light, low-sulfur (sweet) crudes and condensates.

Basic Thermal Crude to Chemicals Scheme

Fig.3 shows that there are many light crudes that have low sulfur that can be economically processed to produce ethylene. In our study, we considered three crudes for the initial evaluation.

1) Agbami crude from Africa (very low sulfur)
2) Arab Light from Middle East (high sulfur)
3) Permian crude from USA (very low sulfur)

All these three crudes can be processed to produce ethylene using thermal crude to chemical technology described here. The crude properties are shown in Table 1.

Table 1. Different Crudes Considered in Initial Evaluation

Crude	**Agbami**	**Arab Light**	**Permian**
Source	Nigeria	Saudi Arabia	USA
API	42.3	33.9	48.6
Sulfur, wt%	0.04	1.94	0.01
Nitrogen,wppm	384	873	26
Metals(Ni+V),wppm	2.1	11.1	0.5
MCR	0.67	4.25	0.07
Naphtha (200°C-)	43.5	25.45	46.1
Gasoil (200-350°C)	36.1	26.15	28.9
VGO (350-525°C)	17.28	24.80	20.3
Residue(525°C+)	3.12	23.60	4.7
H- content,wt%	14.2	12.9	14.5
C7 Asphaltene,wt%	0.11	0.76	0.01
Acid Number,mg KOH/kg	0.05	0.09	1.13
K-Factor	12.2	11.3	12.3
Viscosity @50C, cst	1.33	5	1.5

Agbami and Permian crudes are better than Arab Light crude since they contain less residue (>525 °C). High residue crudes cause short heater run length. What are good feeds to produce olefins? Normal paraffins are the best feeds to produce olefins, especially ethylene. Ethylene plants are designed to produce a certain (fixed) quantity of ethylene from all feeds they process. The quantity of other byproducts (propylene, butadiene, etc.) will vary depending upon the quality of the feed. Normal paraffins produce extremely high ethylene and propylene yields. Isoparaffins produce lower ethylene yields and comparable propylene yields to normal paraffins. Naphthenes produce less ethylene and propylene yields than isoparaffins. Aromatics produce very low ethylene and propylene yields. Multi-ring aromatics produce almost no olefins and high amounts of fuel oil that forms coke. The hydrogen content of olefins (C_2H_4, C_3H_6) is 14.3 wt%. The hydrogen content of aromatics is exceptionally low and multi-ring aromatics have even lower hydrogen content (often <8 wt% H). Therefore, a feed that has higher hydrogen content than mono olefin (14.3%) can easily dehydrogenate (lose hydrogen) and produce an olefin. A feed that has lower hydrogen content than an olefin must reject carbon to produce an olefin for carbon-hydrogen balance. Carbon is a solid and hence carbon deficient feeds will shorten the heater run length. This is a simple rule applicable in all types of pyrolysis technologies. Rarely in practice elemental carbon is found in pyrolysis coils. Hydrogen deficient products such as multi-ring aromatics, the principal components of fuel oil, are produced in place of solid carbon. Fuel oil components are so called coke precursors that will produce solid coke eventually. Thermodynamically the ultimate products of thermal cracking are carbon and hydrogen. By manipulating the reactor operating conditions, we maximize intermediate products. One of the intermediate products is ethylene. Thus, the reactor design and operations play a vital role in maximizing ethylene for any feed. Hence, for high ethylene production it is preferable to crack feeds that have reasonable or high hydrogen content in feed. In that respect, light crudes have high paraffin content and they have a higher ethylene potential than heavy crudes; the selection of crude is important for good economic return.

Once the crude is selected then the optimum process configuration must be decided. The type of compounds present in a crude can be used to illustrate the salient features of this technology. Altgelt and Boduszynski (1994) show the type of compounds present in a typical crude as a function of atmospheric equivalent boiling point. A similar type of figure is shown in Fig.4. In every crude and condensate feed all these classes of species are present. The quantity (or concentration) of each species is different in each crude. The figure is explained below in detail.

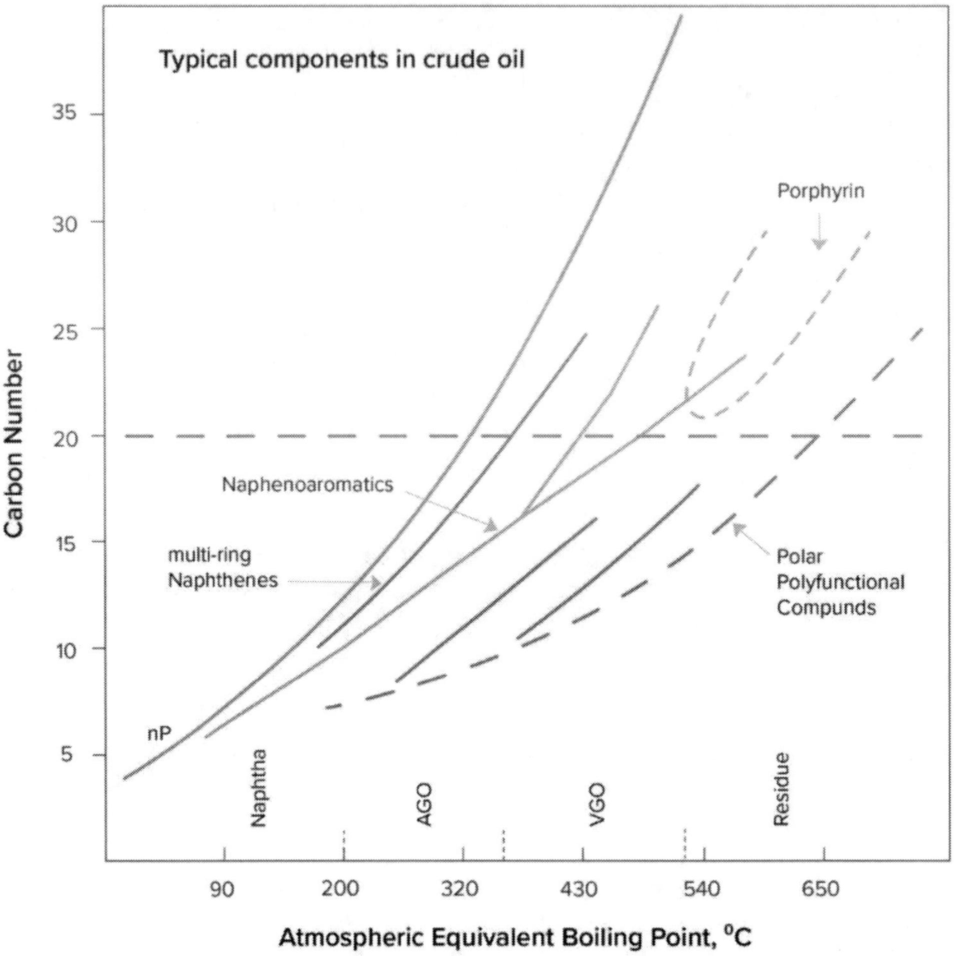

Fig.4 Carbon number versus Atmospheric equivalent boiling point for different classes of species.

Every crude contains normal and isoparaffins, mono and multi-ring naphthenes, mono and multi-ring aromatics, hydrocarbons with sulfur, nitrogen, oxygen and metals (heteroatoms) containing straight chain and/or ring compounds. The species that contain heteroatoms are generally called polar species. For any given carbon number, normal paraffins have the lowest boiling point and polynuclear polar and nonpolar compounds have the highest boiling point. For example, Fig.5 shows C20 species.

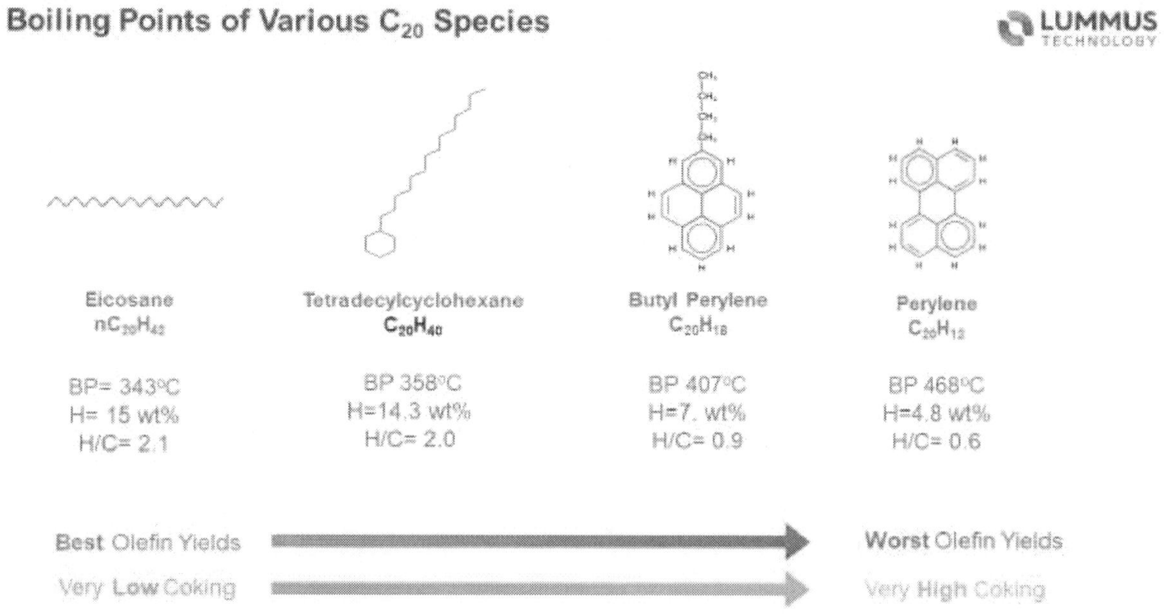

Fig.5 Boiling point of different C20 species

n. $C_{20}H_{42}$ boils at 343 °C (eicosane density 0.787 and MP 37°C) while $C_{20}H_{12}$ boils at 468 °C (perylene density is 1.3 g/cc and MP is 278 °C, brown solid). $C_{20}H_{12}$ is a highly condensed molecule with 4 rings surrounding a benzene ring and has lowest hydrogen content (4.8 wt%). A four-ring aromatic compound with some side chains, $C_{20}H_{18}$ (butyl pyrene 7.0 wt% H) boils slightly lower than $C_{20}H_{12}$ (BP=407 °C, density=1.13g/cc, MP=110°C). $C_{20}H_{12}$ is such a highly hydrogen deficient molecule (H/C=0.6 atomic ratio) compared with n.$C_{20}H_{42}$ (H/C=2.1). Hence $C_{20}H_{12}$ and $C_{20}H_{18}$ (H/C=0.9) will produce mainly coke rather than ethylene. It is the aim of this technology to remove these species selectively and keep the hydrogen rich species in the cracker feed slate. Fig.4 gives the answer how to separate hydrogen rich species from hydrogen deficient species. It shows clearly boiling point is one measure. By limiting the boiling point, the concentration of multi-ring compounds can be reduced in the cracker feed. Thermodynamically we cannot remove them very easily but their concentration can be reduced to acceptable levels. Therefore, cut points are critical parameters. A crude column contains many theoretical stages and hence the overlapping between various fractions is small. That principle is followed in the technology developed by Lummus Technology in heavy oil processing scheme (HOPS), which is in operation for condensate cracking for many years. This is further improved in our patented technology (Dennis et al., 2022) for crude cracking. The concept is explained below in Fig.6 for crude oil cracking. Desalted Crude is preheated to moderate temperatures in the convection section and sent to the HOPS unit. In the HOPS unit, the crude feed is mixed with superheated dilution steam. The HOPS unit contains trays and/or packing. By using dilution steam hydrocarbon partial pressure is reduced, reducing the vaporization temperature. The overhead temperature is controlled and only the vapor (vaporized hydrocarbons+dilution steam) is sent to lower portion (hot flue gas region) of the convection section for further superheating and then to the radiant section for cracking. With the improved HOPS design, the cut point overlap is reduced very significantly. First generation HOPS units are operating since the 1990s for condensate cracking successfully. Note that as shown in Fig.1, condensates are like light crudes with the tail portion boiling well over 700 °C. The bottom fraction of the HOPS unit (unvaporized crude liquid) can be sent to another heater for further preheating and then to another HOPS tower and/or it can be processed further by different technologies such as hydrocracking, fluid catalytic cracking, delayed coking etc. When we use a second HOPS, it is known as two-stage HOPS. One can consider two or three stages and select appropriate boiling (cut) points, depending on the crude properties. One or two stages are adequate for most feeds. The bottom stream of the last HOPS unit (residue) is sent to fuel with or without hydroprocessing and is not processed in the radiant coils. The amount of residue removed depends upon the crude quality. Controlling this residue portion is critical since this material primarily produces coke in the vaporization process in the radiant coil and in the transfer line exchangers. When the crude contains a high residue portion, a lower cut point for the last stage is preferred. High or low residue amount is relative for any crude (for example 4 or 5% residue for many crudes) and hence major portion of the crude (>90% for many crudes) can be used for producing olefins with reasonable run length. For example, when two stages are used, HOPS last stage overhead temperature is at an end point of a typical VGO feed. Therefore, the bottom fraction will contain only the residue portion that cannot be vaporized easily in the convection section. The amount of dilution steam and the level of superheat dictate the maximum cut point possible for the crude. Rejecting the residue from processing in the cracking heaters leads to improved run length and better once-through ethylene yield. Residues have very low hydrogen content and they do not produce any significant amount of ethylene (<5wt%).

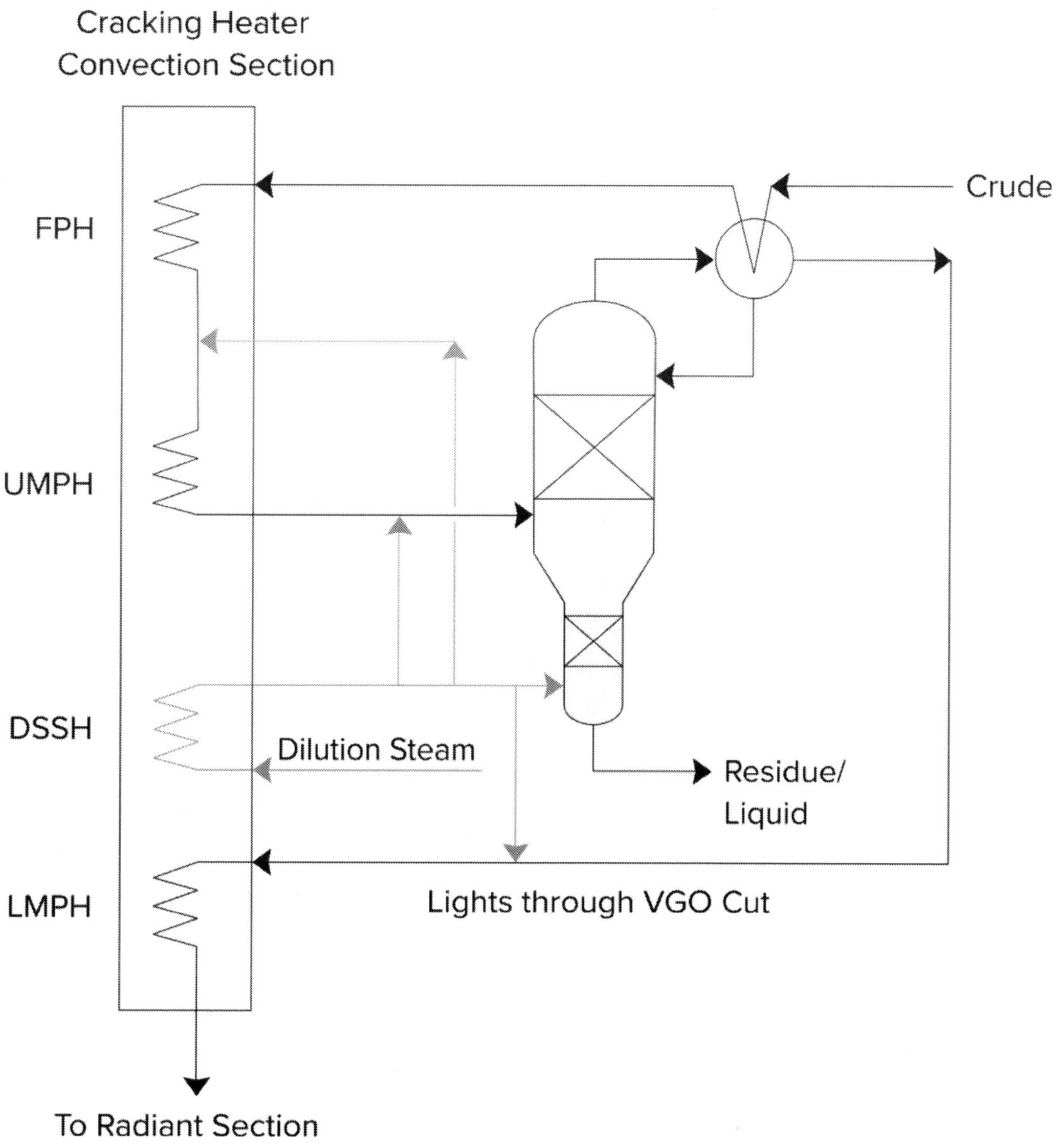

Fig.6 One stage HOPS unit for crude cracking

The residue contains a significant portion of asphaltene molecules. As shown in Table 1, Arab Light crude contains more asphaltene than some other crudes. Typical structure of asphaltene is shown in Fig.7.

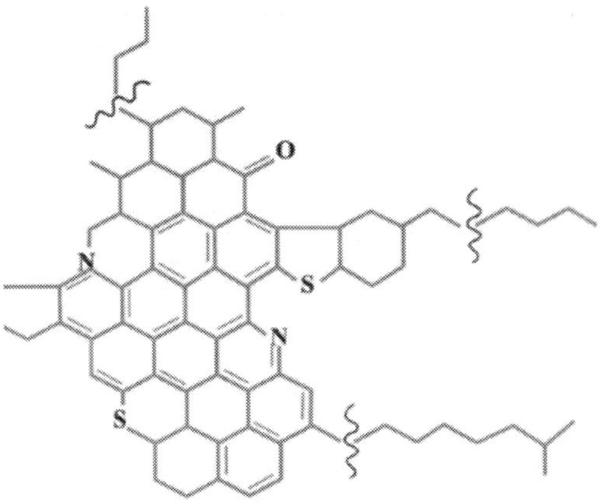

Fig.7 Structure of asphaltene

Asphaltene is a class of high molecular weight compounds containing many heteroatoms (S, N, O & metals). Considerable research has been done in recent years in understanding of these compounds (see Sundaram et al. 2008). Originally it was thought these species have molecular weights in excess of 10,000 but recent studies show they are much less (Badre et al., 2006). Nevertheless, the molecular weight is in the order of thousands. These are multi-ring compounds and are present in peptized form in the crude. Depending upon thermodynamic conditions, they will separate out and cause coking. By removing these species from crude, processing the remaining portion of crude will be easier. These molecules should not enter the cracking heater. Even in traces, they cause enormous problems since the ethylene industry process high volume of feed rate. There is a direct relation between asphaltene content of crude and micro carbon residue (Wiehe , 2007) . Therefore, potential crudes with high MCR should be avoided. Solvent Deasphalting effectively utilizes the thermodynamic principles behind the molecules to remove them. Conversion of asphaltenes rather than rejection through processes such as Solvent Deasphalting, forms the basis of special reactor platforms (liquid circulation or slurry bubble) and catalysts systems all of which form part of the new crude to chemicals scheme.

The choice of one or more stages depends upon the type of crude and the objectives of the project. The fundamentals of thermal cracking dictates that the cracking rate constant increases rapidly with increasing carbon number. Note that (in free radical mechanism) the initiation reaction controls the cracking rate constant. As described in various papers (Dente et al, 1983; Froment, 1992), with higher molecular weight species the number of C-C bonds increases and the probability of breaking that bond increases rapidly. C-C bond is weaker than C-H bond. That is why ethane requires very high temperature (low-rate constant with one C-C bond) while hexadecane requires relatively lower temperature for cracking. For the same reason methane requires extremely high temperature (no C-C bond only C-H bond) higher than 1000 °C for cracking and hence it almost passes as inert in commercial pyrolysis reactors. Typical first order rate constant as a function of liquid feed molecular weight at 800 °C is shown in Fig.8. Feed conversion for a typical coil with 170 millisecond residence time is also shown in the graph. Notice that the high molecular weight species are converted very easily and hence secondary cracking will become dominant if cracking is further continued in the coil. Secondary reactions degrade olefins and produce fuel oil components and coke. Therefore, mixing wide range of molecular weight species results in poorer overall yields. The typical practice is to group narrow boiling cuts and to optimize the cracking.

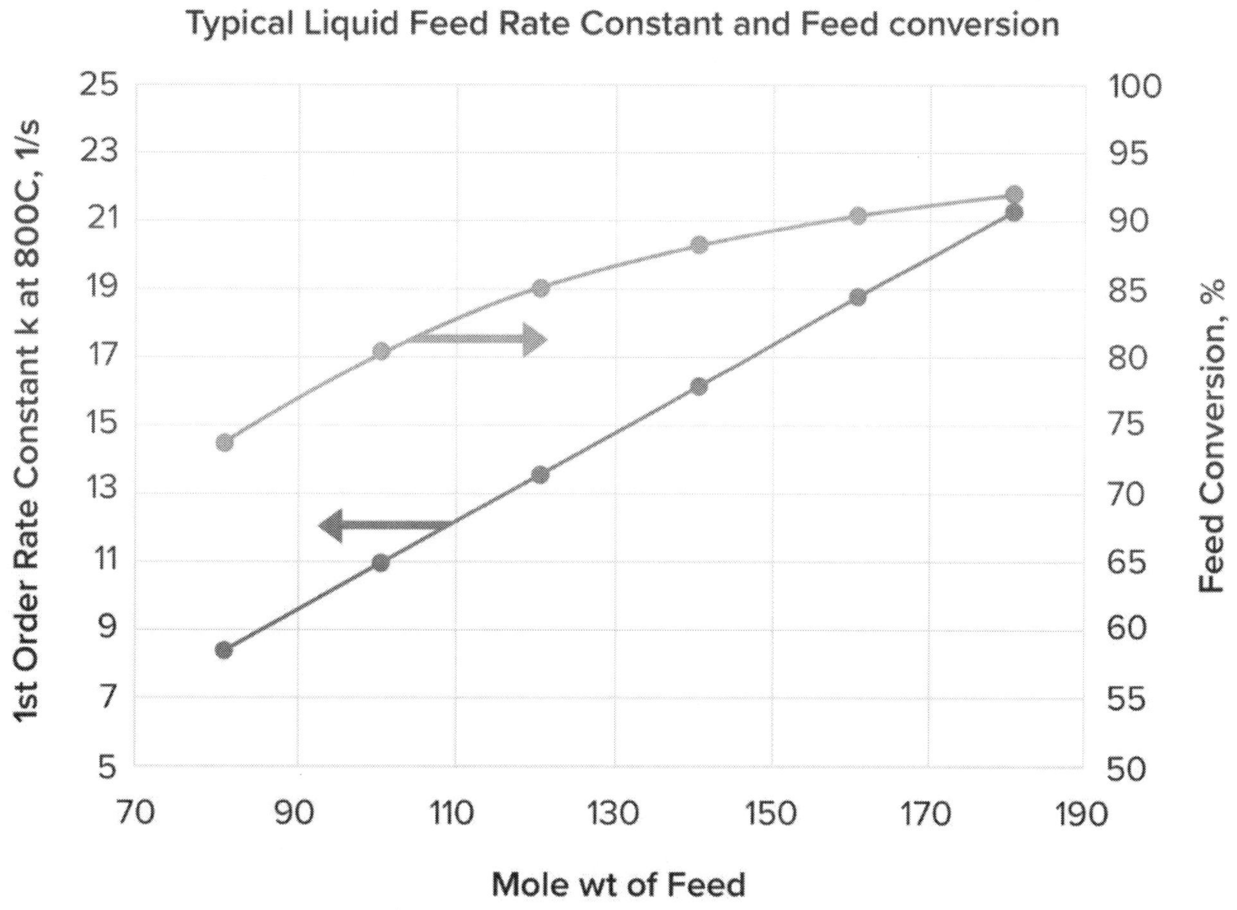

Fig.8 First order rate constants for liquid feed as a function of molecular weight (blue) and feed conversions for typical operating conditions (red)

Typically, it is preferable to keep maximum to minimum molecular weight within a factor of two. For each species there is an optimum cracking condition that will give maximum ethylene yield or maximum ethylene + propylene or other combinations. Theoretically, maximum once through yield increases with increasing coil outlet temperature. At industrial operating conditions with increasing coil outlet temperature ethylene yield always increases. As described above a certain temperature for each feed, secondary reactions, which degrade the olefins and the coking reactions, dominate, drastically reducing the run length. Hence, beyond a certain point it is not worth increasing the coil outlet temperature. Propylene on the other hand reaches a maximum at lower coil outlet temperatures and for most of the commercial operating coil outlet temperature range once-through propylene yield decreases with increasing coil outlet temperature. Typical ethylene and propylene yield behavior for naphtha feed is shown in Fig. 9. In actual operation, instead of monitoring the coil outlet temperature, propylene/ethylene (P/E) ratio is monitored. The P/E ratio reaches maximum value at low temperatures and almost linearly decreases at high coil outlet temperatures. Any destruction of olefin always produces coke. Typically, for most feeds, P/E ratio between 0.45 and 0.5 is the practical limit at high severity or maximum ethylene. Maximum ethylene + propylene is observed at P/E ratio of 0.5 to 0.6. Valuable products are the intermediate products. Therefore, the heater coil is designed to maximize the yield of the valuable intermediate products. It is important to know that at high severities CH_3 radical will enter bimolecular reactions and produce only methane. Hence, methane is considered as fuel component and has to be reduced.

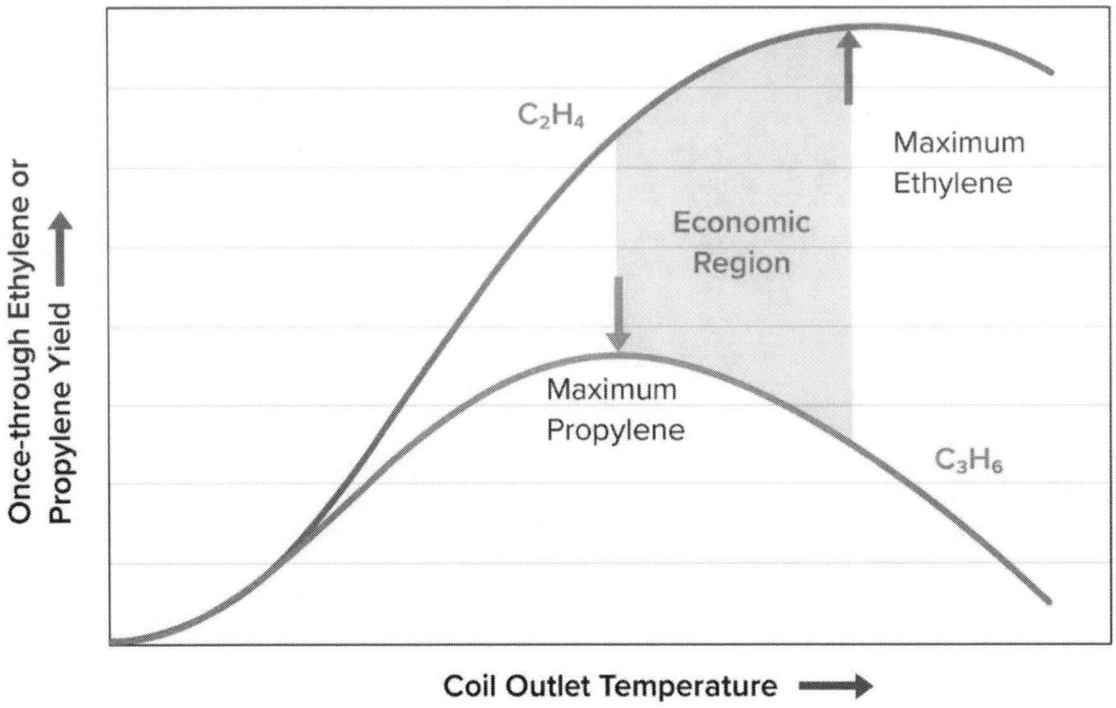

Fig.9 Ethylene and propylene yield as a function of coil outlet temperature for a liquid feed

Note that the reaction proceeds via free radical mechanism and hence the initiation rate controls the overall rate. When the hydrocarbon contains many carbon atoms, simultaneously many propagation reactions are proceeding leading to very rapid conversion. The (primary) products formed also enter into free radical reactions to other (secondary) products. Up to certain temperature, both ethylene and propylene yields increase with increasing coil outlet temperature. The mechanism dictates that the largest olefin reaches the maximum first and then the smallest olefins. In fact, C5/C6 olefins reach maximum at lower temperatures (<775°C) and that is why in the commercial operating conditions these olefins always show decreasing yield with increasing temperature. As mentioned earlier, higher carbon number molecules have more reaction pathways and decompose faster. These can be explained considering formation of ethylene and propylene as examples.

C2H5.-→ C2H4 + H (1)

C2H4+H. → C2H5. (2)

C2H4+CH3→ C3H7. (3)

C3H7.--> C3H6 +H. (4)

C3H7.--> C2H4+CH3. (5)

C3H6+H.--> C3H7 (6)

C3H6+CH3→ C4H9 (7)

There are thousands of free radical reactions taking place even in simple ethane cracking and hence it is not the intention of this paper to develop the reaction scheme for any feed. Ethylene is primarily formed from ethyl radical decomposition and has been extensively analyzed by many workers (Sundaram and Froment 1978, Ranzi et.al, 1983) and this is the main reaction in thermal cracking of all hydrocarbon species. There is only one way ethyl radical can decompose as a first order reaction and all other reactions are addition reactions, which are bimolecular. For propyl radical (there are two isomers and is ignored for this illustration), it can decompose to propylene and hydrogen radical or ethylene and methyl radical. Both routes are equally probable and that is why at very low propane (as an example feed) conversions (where C_3H_7 is dominant), initial molar selectivities of ethylene and propylene are almost the same. Like ethylene, propylene also can enter into addition reactions. There are many addition reactions and only a few are shown for illustration. So, it is clear one cut of crude from C5 to 525 °C boiling range will not produce optimum yields for a light crude. In every plant design, there are practical limits. Our calculation shows two to three HOPS stages are reasonable and often two stages are sufficient. In the first stage the cut point is selected around 160-240 °C and in the second stage 350-525 °C end point is selected depending upon the scope of the project. The first cut is naphtha cut (named for simplicity) and second cut is denoted as a gasoil cut. When three cuts are produced (three stages), naphtha, gasoil and vacuum gasoil cuts are produced. The heaviest portion (residue) is always rejected. Generally, it is mixed with fuel oil produced in the pyrolysis unit and sold as product. No additional processing of cuts is considered in this scheme. The properties of different cuts are shown in Table 2. Typical once through and ultimate yields for a similar naphtha and gasoil feeds for a short residence time coil at high severity are shown in Table 3. Ultimate yields are obtained by recycling ethane and propane only after selectively hydrogenating acetylene and MAPD.

Table 2 Different Ideal Cuts of Arab Light Crude

Cut#	**1**	**2**	**3**	**4**
Name	Naphtha &light	AGO cut	VGO cut	Residue
Cut point, °C	<160C	160-350	350-525	525C+
Yield of crude,wt%	18.83	32.77	24.80	23.61
S.G.	0.7112	0.8322	0.9162	1.0133
Sulfur,wt%	0.04	0.83	2.64	4.16
Nitrogen,wt%	0.00	0.00	0.06	0.25
Metals,ppm	0.00	0.00	1.68	63.50
MCR,ppm	0.00	0.00	0.33	16.55
Paraffins,wt%	74.79			
Naphthenes,wt%	16.78			
Aromatics,wt%	8.43			
Polyaromatics,wt%	0.00	8.76	52.09	>60
C7 Asphaltenes,wt%	0.00	0.00	0.20	3.84
H-Content,wt%	15.4	13.6		

Table 3. Once through and Ultimate Yields

Type	OT	OT	Ult.	Ult.
Feed	Naph	AGO	Naph	AGO
SG	0.7106	0.8118	0.7106	0.8118
P, WT%	71.75		71.75	
N+O, WT%	18.80		18.80	
A, WT%	9.45		9.45	
Yields,wt%				
H2	0.92	0.67	0.96	0.75
CH4	15.65	10.50	16.20	10.97
C2H2	0.91	0.46	0.00	0.00
C2H4	31.20	25.80	35.13	29.10
C2H6	3.00	2.90	0.00	0.00
C3H4	0.89	0.70	0.00	0.00
C3H6	14.00	14.00	14.22	14.19
C3H8	0.23	0.25	0.00	0.00
C4H6	4.75	4.90	4.87	5.00
C4H8	3.50	3.75	3.52	3.77
C4H10	0.20	0.10	0.21	0.11
C5-200°C	20.47	18.55	20.61	18.68
PFO	**4.28**	**17.42**	**4.29**	**17.42**
	------	------	------	------
Total	100.00	100.00	100.00	100.00
C2H4+C3H6	45.20	39.80	49.35	43.29

Notice that naphtha produces reasonable ethylene yield and low fuel oil yield. Gasoil produces lower ethylene yield than naphtha and produces more fuel oil. Vacuum gas oil produces even more fuel oil than the AGO cut. The fuel oil has very low value and hence a significant portion of the crude goes to fuel oil and it does not serve the objective of the project. Therefore, it is necessary to develop a process that will minimize fuel oil leaving the complex. This is discussed next.

A Novel Scheme to Convert Any Crude to Chemicals – TC2C™

Saudi Aramco Technologies Company, Lummus Technology and Chevron Lummus Global (CLG) conducted several years of research to arrive at an improved technology to produce high yields of chemicals from crudes, extending the range beyond just very light crudes or condensates typically considered. The research and subsequent commercialization involved the following vital steps:

- Very detailed componential analysis of crude oils and heavy oils in general
- Development of specialized separation devices to separate the crude into fractions for optimized processing without having to utilize energy and CAPEX intensive crude and vacuum distillation

- Utilization of commercially proven fixed bed, ebullated bed and slurry reactor systems in crude conditioning such that the products from the crude conditioning section could directly be routed to the mixed feed steam cracking reactor section. The steam cracking reactors and recovery systems are an integral part of TC2C™ technology.
- Development of unique catalyst systems for the fixed bed and ebullated reactors that would provide the right amount of hydrogenation without overcracking to naphtha, LPG and light ends.
- Rigorous testing of the impact of varying amounts of pyrolysis fuel oil recycled from the recovery section. TC2C™ successfully upgrades the pyrolysis fuel oil to steam cracker reactor feed

Throughout the development, particular attention was paid to minimizing of equipment count (CAPEX), energy input, carbon footprint, emissions, catalyst deactivation rate and the reactor fouling rates for various feeds.

Crude Analysis and Conditioning

Crude oils contain millions of molecules. There have been many attempts to characterize crude through detailed compositional analysis (Altgelt and Boduszynski, 1994; Boduszynski, 2001; Speight, 2016). In the lower carbon numbers, the total number of n-paraffins, i-paraffins, naphthenes and aromatics are reasonable and easily identified. For example, from C3 through C8, there are 78 P, N, A compounds. With increasing carbon numbers, the number of possible compounds increases exponentially, 66 just for C9 for example. As discussed earlier it is well-known that n-paraffins yield the highest amount of ethylene in steam cracking whereas i-paraffins tend to produce more propylene. Aromatics have nearly zero yield of ethylene and foul the heater tubes. Higher molecular weight multi-ring aromatics and especially, heavy polynuclear aromatics rapidly foul the transfer line exchanger (TLE) in the steam cracker. Dr. M.M. Boduszynski of Chevron started evaluating crude oil using detailed compositional analysis. It is well-known that diverse compounds with similar molecular weight cover a broad boiling range and conversely, a narrow boiling range cut can contain a wide molecular weight range. Boduszynski evaluated compounds with a hydrogen deficiency or "Z" value. He showed that complete hydrogenation and ring opening with no change in carbon number could consume vastly different amounts of hydrogen depending on the "Z" value. General Formula:

C_nH_{2n+Z} where, Z=2-2*(R+DB); n= number of carbon atoms, R=number of rings, DB= number of double bonds, Z=hydrogen deficiency.

In all crudes, the hydrogen deficiency increases with boiling point with higher and higher concentration of condensed rings. Typically, the highest boiling fractions in crude oil containing what is broadly termed, asphaltenes, were the most difficult to convert to transportation fuels or petrochemical feedstocks. Indeed, the analysis of asphaltenes using advanced techniques formed part of the research. The work of Schuller et. al (2017) indicated the incredibly challenging nature of the molecular species within asphaltenes. These authors used atomic force microscopy (AFM) and studied heavy oil molecules in detail.

In TC2C™, a naturally abundant n-paraffin rich light stream is separated with a novel separation device such as that discussed in thermal crude to chemicals section that eliminates heavier molecules from the vapor stream that is routed to the steam cracker. Bottoms from the separation device is routed through another separation device that utilizes hydorgen to separate out a heart cut with carbon number varying between 23 and 35 depending on the TC2C™ variant. The heart cut is sent for fixed bed hydroprocessing to remove nitrogen and sulfur, hydrogenation of aromatics, ring opening and hydrocracking. The catalyst systems are carefully selected to optimize the molecular profile for subsequent processing.

The heaviest fraction of the crude with carbon number exceeding 35, is routed to a liquid circulation reactor with either extrudate or slurry catalyst. These reactors have small on-line catalyst addition and

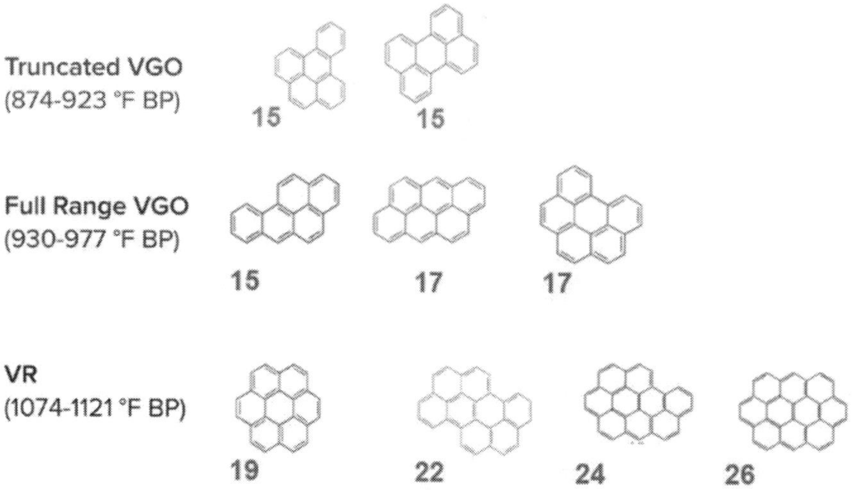

Fig.10 DBE of some species (DBE=C+1-H/2-X/2+N/2; X is halogen, DBE=#rings+ # double bonds for HC molecule)

withdrawal and can run continuously for over five years. Additional LC reactor information can be found elsewhere (Sundaram et.al 2008). The liquid circulation reactors convert the asphaltene and recycle pyrolysis oil to lighter components that are then upgraded to steam cracker feed specification utilizing an integrated fixed bed hydrotreating and hydrocracking system that ensures that no HPNA (heavy polynuclear aromatics) reaches the steam cracker. Some of the known structures that impede full conversion of residue hydrocracked VGO are shown in Fig 10. Through extensive analysis of commercial data from residue hydrocracking and tailored pilot plant tests, it was shown that residue hydrocrackate has an increased concentration of DBE (Double bond equivalent) 15+ compounds. For pure hydrocarbons DBE=C+1-H/2 and it represents the level of unsaturation or hydrogen deficiency. DBE of various class of compounds are shown in Fig.11 and that measured in a typical ethylene plant pyrolysis fuel oil is shown in Fig.12. Within the integrated hydrocracking system, the catalysts system and operating conditions are carefully controlled such that DBE is restricted to 15 or lower in the effluent and that too in trace amounts as shown in Fig.12. Recently, DBE has been used in analyzing the effects of crude blending at Chevron (Rogel et.al, 2023). Slurry Hydrocracking utilizing a very special catalyst can increase the conversion of residue to over 97%. The remaining unconverted oil is filtered and sent over a fixed bed reactor system to meet very low sulfur fuel oil specifications (IMO compliant). Thus, TC2C™ ensures that no part of the converted crude is wasted while maximizing the yield of chemicals. Changes in DBE before and after a LC technology is shown in Fig.13. High DBE value species are almost reduced to zero.

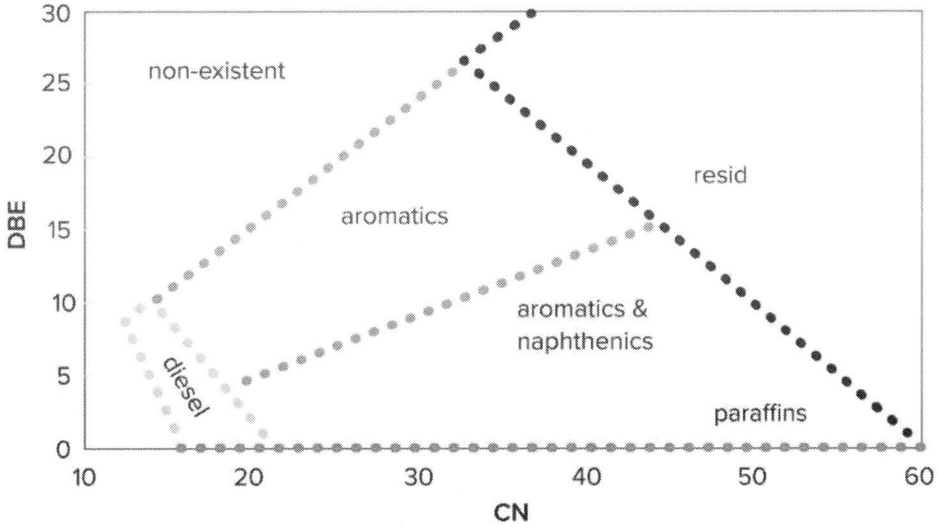

Fig.11 DBE for various molecular structures versus carbon number

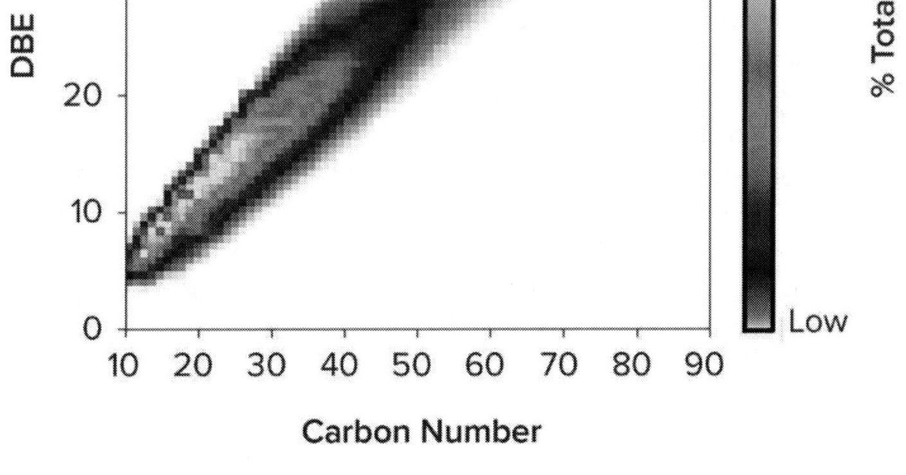

Fig.12 Double Bond Equivalency versus Carbon number for fuel oil

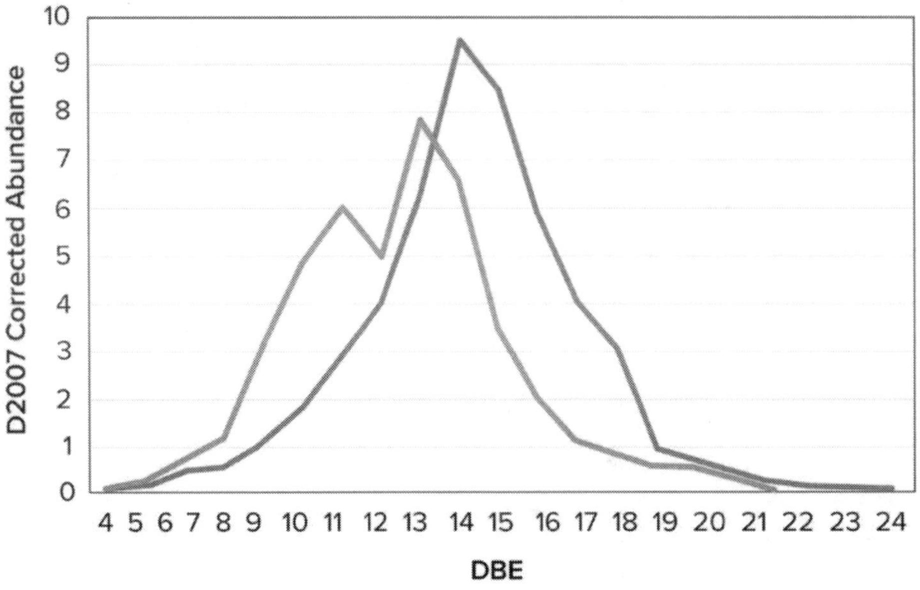

Fig.13 Changes in DBE before (blue) and after (red) hydrocracking

As discussed earlier with a low hydrogen content feed, high ethylene yield cannot be achieved in thermal cracking. When a crude is primarily used to produce chemicals only, then it is important to know whether it is worth to upgrade it or not. Upgrading typically requires either carbon rejection or hydrogen addition. Hydrogen for the refinery requirements mainly comes from steam methane reformer (SMR). This is known to produce high level of CO_2 and nowadays CO_2 emission is an important parameter. There are many potential schemes to produce green hydrogen and blue hydrogen. Within the decade it is expected that at least blue hydrogen will be available at a cost comparable to SMR (grey) hydrogen. For the purposes of this paper, only the technological aspects of upgrading with hydrogen irrespective of the color of hydrogen are considered.

Inspection of the Table 3 clearly shows upgrading naphtha may not improve the yields significantly. Naphtha is widely cracked worldwide without any problem and it produces only a small quantity of fuel oil as shown above. The quality of gasoil and /or vacuum gasoil is reasonable but it is not that great. Moreover, as shown in Table 3, gas oil cracking produces significant amount of fuel oil. While fuel oil may be valuable in certain locations, the intent of crude to chemicals project is to maximize chemicals production from crude and hence the target was to minimize fuel oil. Hydroprocessing will upgrade the quality of the feeds. In the previous section, various cuts are obtained by vaporizing the crude at lower temperature by adding steam in the HOPS tower. Dilution steam is also required for thermal cracking to reduce hydrocarbon partial pressure to increase olefin yields and to suppress coking. However, though a small quantity of saturated water in oil is not harmful for hydrocracking catalysts, the levels considered for thermal cracking are detrimental to the catalyst. TC2C™ utilizes a novel separation device and targeted hydroprocessing of fractions to provide the right amount of hydrogenation of feeds to the steam cracker reactor for maximum crude conversion. The simplified flow scheme is shown in Fig.14.

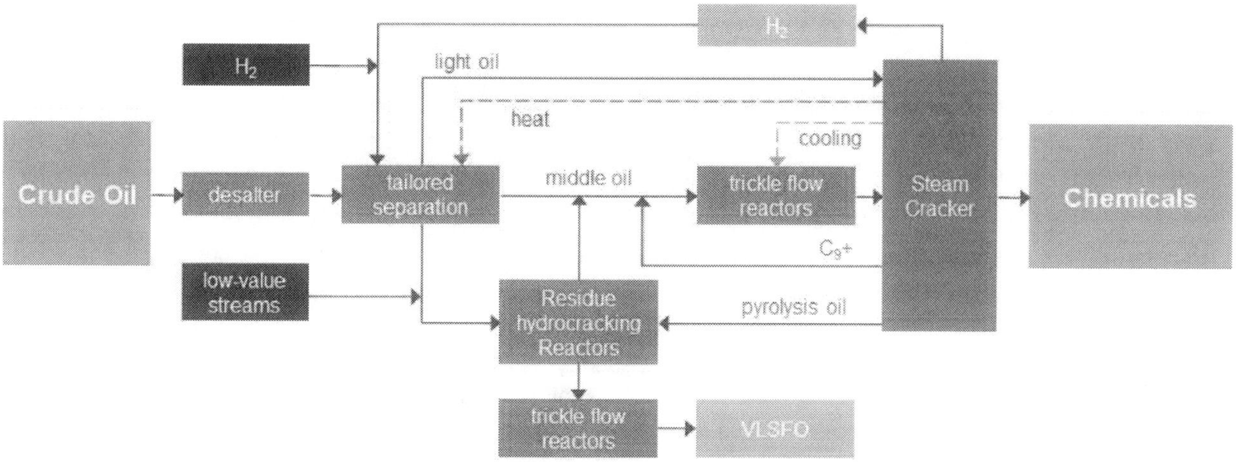

Fig.14 TC2C™ Flow Scheme with Liquid Circulation Reactor Configuration. The total yield of chemicals ranges from 70 to over 85 wt.% depending on the nature of the crude.

The importance of desalter must be stressed at this time. It is present in the previous schemes also (thermal processing without any upgrading). Crude can come from different sources and different transport methods. This will contain some debris and salts. PPM level of chlorides cause havoc on material selection requiring equipment to be alloyed up for resistance against chloride corrosion. PPM level of sodium is sufficient to cause significant damage to catalysts especially when the target run lengths exceed five years. Hence, the investment in desalters and feed filter systems is worthwhile and justified. The fixed bed hydrocracking reactor and liquid circulation reactor technologies explained above are used to condition the feed. Typically, an ethylene plant turnaround is about 5 years and hence all sections are also designed to last that period. A small quantity is taken out of the system as IMO compliant very low sulfur fuel oil (VLSFO) product. By doing so, circulation of multi-ring aromatic rings causing potential catalyst coking problems is also reduced.

An important feature of the TC2C™ is the ability to utilize the pyrolysis oil generated in the steam cracker reactor in the liquid circulation reaction section. Pyrolysis fuel oil is a highly hydrogen deficient molecule rich in multi-ring aromatics (Djokic et al.,2018). Addition of pyrolysis oil to the feed to the liquid circulation reaction section where asphaltene conversion occurs, permits higher conversion of asphaltenes, while maintaining product stability. During the course of the research and development, several different types of pyrolysis oils were tested including very high proportions of pyrolysis oil in the feed mix to ensure that the concept was robust. Extensive pilot plant results show that at low catalyst consumption, heavy boiling material can be converted easily to high quality feeds for olefins production. This is one of the major benefits of the integrated steam cracker and LC-reactor system. Low value fuel oil is upgraded to high value chemicals. More than 90% feed conversions can be achieved at very low catalyst consumption in the liquid circulation reactor system. CLG has designed and the clients have operated many integrated LC-FINING/Hydrocracker plants over the last 25 years. These were mainly designed to produce jet and diesel quality fuels and some plants also use hydrocracked products as feed to an ethylene plant. The difference in fuel and chemical mode is the operation. When certain catalysts and operating conditions are used in the hydrocracker, certain types of species are produced in the hydrocracker, which are not present in the feed. As mentioned earlier, a lot of research was conducted to limit the high DBE HPNA from reaching the cracker such that there would be no highly condensed molecules such as coronenes (6 ring) and ovalenes (10 ring) in the feed to the steam cracker. These molecules deposit as solids fouling the transfer-line exchanger. This has been documented in an earlier article (Fernandez-Baujin et al. 1990). The resulting run length could be a few hours instead of few months. Therefore, CLG and Lummus have strict monitoring procedures for HPNA of the feeds to the cracker. Optimum hydrocracker reactor operation and conversion is essential for high ethylene yield and long run length. For simplicity, we have not shown heat integration. Not only materials (chemicals and refrigerants) are exchanged between ethylene plant and hydroprocessing (feed preparation) section, heat integration (steam and fuel system) is also essential for energy conservation. A Lummus Digitalization system enabled with artificial intelligence and machine learning capabilities provides continuous optimization and improvements based on actual plant data analytics.

Only thermal cracking produces highest level of ethylene from any hydrocarbon feed. Hydroprocessing and FCC share a common reaction mechanism. Both follow Olaf's carbenium ion chemistry while thermal cracking proceeds via free radical mechanism. Many theoretical and experimental studies were conducted in-house and in other laboratories to generate C2H5+ (ethyl carbonium ion, primary ion). Without ethyl cation, ethylene cannot be produced catalytically with high selectivity. Ethylene formed by catalytic process mainly comes from other (thermal) routes. At low FCC operating temperature very low ethylene yield (~5wt%) with very high propylene yield (>30wt%) is observed for many feeds. Stable simplest carbenium ion is secondary C3H7+ and not ethyl cation. Primary ion formed is quickly isomerized to secondary ion leading to the formation of propylene (Froment, 1993).

$s\text{-}C3H7^{+} \rightarrow C3H6 + H^{+}$

and hence FCC is a good way to produce propylene. Lummus has introduced both INDMAX and SRDC technology to maximize FCC propylene. This TC2C™ Case Studies intended to make ethylene as the primary product and hence the whole article is aimed at maximizing ethylene. FCC route is also acceptable to make chemicals, but it will not produce high yield of ethylene. All our experimental data and model calculations clearly show the maximum amount of total valuable chemicals is achieved only through the hydroprocessing/steam cracker route, which also produces the highest amount of ethylene with lowest crude consumption.

Ethylene plant

The previous sections discuss how the feeds to the ethylene pyrolysis reactors are prepared from crude. The contaminants, especially the coke precursors are removed by different methods before cracking in the radiant coil. Since the molecular structure is altered through different processing methods the ethylene and byproduct distributions will be different for different cases. Therefore, in this section, illustrative overall material balances for producing 1500 KTA ethylene at high severity are given in Table 4 for the ethylene plant only. Very short residence time, high selectivity radiant coils are used for this study. All balances are based on Arab Light crude except case 1 which is a typical Middle Eastern full range naphtha for comparison.

Table 4. Overall Material Balance Summary

Case	1	2	3
Feed	Naphtha	Crude	Crude
Mode		Thermal(HOPS)	HCRK(TC2C™)
Feed, KMTA			
Naphtha from Crude+TC2C Crude Conditioning	4,132	1,083	1524
Lt.& Heavy Gasoil from Crude		3829	
Gasoil From TC2C Crude Conditioning			3,045
LPG from Crude+TC2C Crude Conditioning		131	233
Steam Reacted	4	5	5
Total	4,136	5,047	4,807
Products,KMTA			
H2 Product	46	40	43
Methane rich off gas	722	634	671
PG ethylene	1,500	1,500	1,500
PG propylene	649	770	750
Butadiene	195	242	245
Other C4's	167	222	230
BTX	569	567	603
Heavy gasoline	90	137	126
PGO	66	295	195
PFO	126	638	441
Acid gases	4	3	3
Total	4,136	5,047	4,807
High value Chemicals	3119	3325	3351
Amount of crude,KMTA		6,601	4,444
Crude, BBL/day		138,653	93,344
% HV chemicals to Crude		50.4	75.4

Currently there are many plants operating with naphtha feeds. Though the crude contains different ratio of naphtha and gas oil fractions, a comparison is included with 100% naphtha feed. For all the cases shown, acetylene and MAPD are hydrogenated with appropriate selectivity to their respective olefins. Ethane and propane are recycled to extinction. Butene and butadienes are considered as valuable products. Different valuable products such as isoprene, cyclopentadiene are also produced in thermal cracking. However, the quantities will vary with severity and types of feeds. Only very few plants have the capability to separate them as products. Lummus Technology has a technology to separate these species if required. For this study, all C5 species are fully hydrogenated to n.pentane, isopentane or cyclopentane and recycled to extinction. The economics will change if the C5 olefins and diolefins are separated and sold as products. C6-C8 is heart cut gasoline (pygas). Benzene is the most valuable in most places. Generally, BTX (benzene, toluene and xylenes) are extracted and are considered as valuable products. Hence BTX is separated. For BTX extraction, C6-C8 must be hydrogenated and hence two-stage DPG (pyrolysis gasoline hydrotreater) is used. After BTX extraction C6-C8 paraffins and naphthenes (known as raffinate) are recycled to cracker to extinction. These (and, C5s) are usually recycled with naphtha feed. C9-400 °F is a heavy gasoline containing high concentration of C9-C10 aromatics. There are technologies to separate valuable chemicals from this stream, but the quantity is very small. Hence, it is sold as product or processed in the feed preparation section and recycled to heaters for some cases as described below. Pyrolysis gas oil (400F-550 °F) and pyrolysis fuel oil (550 °F+) are generally sold as products. These are highly hydrogen deficient molecules (<8 wt% H content) and rich in multi-ring aromatics. This has low value in the market. The salient feature of TC2C™ technology is to upgrade this material to high value products. Generally, and in this paper combined PGO and PFO is denoted as PFO and no distinction is considered here. For case #3 in Table 4 these are upgraded in the feed preparation section. A small amount of residue is always purged to minimize the coke precursors circulating in the system. This is taken out as IMO compliant very low sulfur fuel oil (VLSFO) for case 3 as shown in Fig.14. Different scenarios considered in Table 4 are discussed below.

Case 1 is standard naphtha cracker. Only C2/C3 recycle, fully hydrogenated C5s and hydrogenated C6-C8 raffinate are recycled to extinction.

Case 2 is two-stage thermal crude to chemicals without any processing of PGO and PFO using HOPS only for separating different cuts. All recycles mentioned in case 1 are applicable here. All unused crude portion (residue) is considered as fuel.

Case 3 TC2C™ for the configuration shown in Fig.14. All recycles of case 1 and recycles with C9-400 °F (C9-204°C) and pyrolysis fuel oil after treatment.

Valuable chemicals include hydrogen, ethylene, propylene, butadiene, butene and BTX. Though VLSFO is also valuable, for this study it is not included since it is considered as fuel. The objective of the project is to produce no fuel. Upgrading the feed (case 3) by hydroprocessing produces the highest amount of valuable chemicals per unit crude. Naphtha cracker is the simplest liquid cracker. When only thermal cracking (no hydroprocessing) is considered, significant amount of residue must be rejected. In addition, vacuum gasoil range molecules must be cracked and higher steam to oil ratio must be used. They consume more crude and more energy. With hydrocracking option residue is also used to some extent. Almost all PGO and PFO produced in the cracker are recycled after upgrading. Therefore, crude consumption is significantly reduced. The amount of high value chemicals is also increased significantly. However, there is a cost associated with that. Thermal mode without residue processing does not require a hydrocracker, residue hydrocracking and hydrogen plant. Once hydrocracking is used, these plants must be added. Considering the reduction in crude and the increase in high value chemicals, the increase in CAPEX is justified. CAPEX depends upon the specific location. The payout is less than two years for most locations. With light crudes such as Permian as high as 70% high value chemicals instead of 50% for Arab Lght crude can be obtained even in thermal mode (Biswas et.al 2022).

Ethylene plant is designed to meet the design ethylene production with differences in feed properties as much as possible. In that process feedrate, severity and other operating conditions can be adjusted as required. The byproduct flowrates will always vary. Though for the design feed rate the design run length can be achieved, it may not be possible for other cases. For some (good quality) feeds, the run length will exceed and for poor quality cases the run length will be less than the design case. For simplicity, some heater parameters are shown for only major liquid cracking furnaces for the case 3 in Table 5. Lummus typically designs a furnace operating between 45 and 60 days, which is considered as economical optimum. There are instances where as low as 25 days and as high as 100 days have been used. Extremely long run length requires longer decoking time. With short run length, it is a strain to the operation when multiple heaters must be decoked one after the other. A 2016 survey of Lummus customers (>50 plants) showed average on-stream time of 98%. Hence, for crude cracking also we maintained similar run length cycle. Using shorter run length some CAPEX can be saved, but it may not be worth at the end.

Table 5 Heater Operating Summary

Feed	**Naphtha**	**Lt.Gasoil**
Flow Rate,% of Naphtha	100	105.4
S/O, w/w	0.50	0.80
Coil Outlet Temp., °C	826	815
TLE outlet Temp., °C	401	410
Fired Duty, MMKcal/h	150.3	166.5
SHP Steam, T/h	95.4	109.0
Overall Th.efficiency, %	94.2	93.6
Total Flue gas, T/h	253	280
C2H4 Prod., kMTA	245	250

In addition to cracking heater, the recovery section plays a vital role in the ethylene plant. The major cost is in the recovery section. Fortunately, after the hot section the recovery section configuration is nearly independent of feed. Since for this study, a naphtha plant is chosen as a reference, relative factors can be used to prorate the capacity of the plant. Fig.15 shows specific energy of the two options relative to a naphtha cracker. In addition, major capacity factors are shown in Fig.16 relative to naphtha cracker. Area factor represents the quantity of feed with that quality relative to reference feed cracking. These are based on simplified mass transfer models and sometimes based on two-component distillation models. Over the years they found to be very reliable and almost the capacity of the unit can be accurately predicted within +/-2% and hence it is a standard design tool for preliminary design. With every newer design the tool is constantly updated based on operating data. Most units have less than 5% over the reference plant except charge gas compression and quench section. As mentioned earlier, naphtha feeds are much better feeds and have high ultimate ethylene yields. Generally crude and in this case Arab Light crude has high residue and the ultimate yield is not that high. However, as shown in Table 4, hydroprocessing has significantly improved the yields.

The specific energy is one of the key parameters of the ethylene plant. It is defined as the energy required to produce per unit weight of ethylene after accounting for energy generated in the plant. Note that in the cracking heater SHP (Super High Pressure) steam is generated, which is primarily used to drive major compressors in the recovery section. Therefore, that amount of steam need not be imported. Methane off gas is produced in the plant and that is used as fuel in the cracking heater and hence that amount of fuel need not be imported. Only excess amounts of steam and fuel are exported. Industry standard energy equivalents are used for import/export of steam, fuel and electricity. Cooling water is another parameter. After accounting for all credits and debits, energy to produce per unit of ethylene is calculated as specific energy. The lowest one is desirable. In the industry pure ethane cracking has the lowest value (~3000 kcal/kg C_2H_4). Using gas turbine and other methods not directly related to the chemical reactions that produce ethylene, specific energy can be reduced by 5 to 20%. In this exercise, no additional schemes are considered and hence direct comparison with reference plant gives a good indication of energy consumption for crude to chemicals. In the traditional way, a refinery is used to produce feeds (naphtha and gasoil etc.) for olefin plant. The refinery costs some CAPEX and some energy. An efficient refinery can be built, but it is beyond the scope of this study. Therefore, additional CAPEX and energy must be included for naphtha plant in principle for proper comparison. That is included based on some in-house data in Fig. 15. Though ethylene plant only shows higher specific energy than naphtha plant, including the refinery to produce that naphtha clearly shows TC2C™ is much superior. When the specific energy is reduced, that also reduces the CO_2 emission, which is more important nowadays.

This clearly shows bypassing the refinery not only reduces the CAPEX, but also reduces the energy consumption. More than 10 patents have been granted and/or pending for different feed treatment/ethylene plant configurations and different technologies covering crude to chemicals. Though only complete crude to chemicals is discussed here, partial transportation fuel production depending upon the local market is also included in some cases. Therefore, TC2C™ not only reduces CAPEX, but also reduces energy consumption and CO_2 emission. Extensive bench scale, pilot scale and demo scale unit (1 bbl/day) were operated over 2 years and all relevant data were collected and modelled. A digital twin of the flow scheme was also constructed. After many simulations, a commercial scale operation is proposed and designed to process Arab Light crude to chemicals. There is currently one plant under construction and is expected to start-up in 2026. There are other projects under various stages of design.

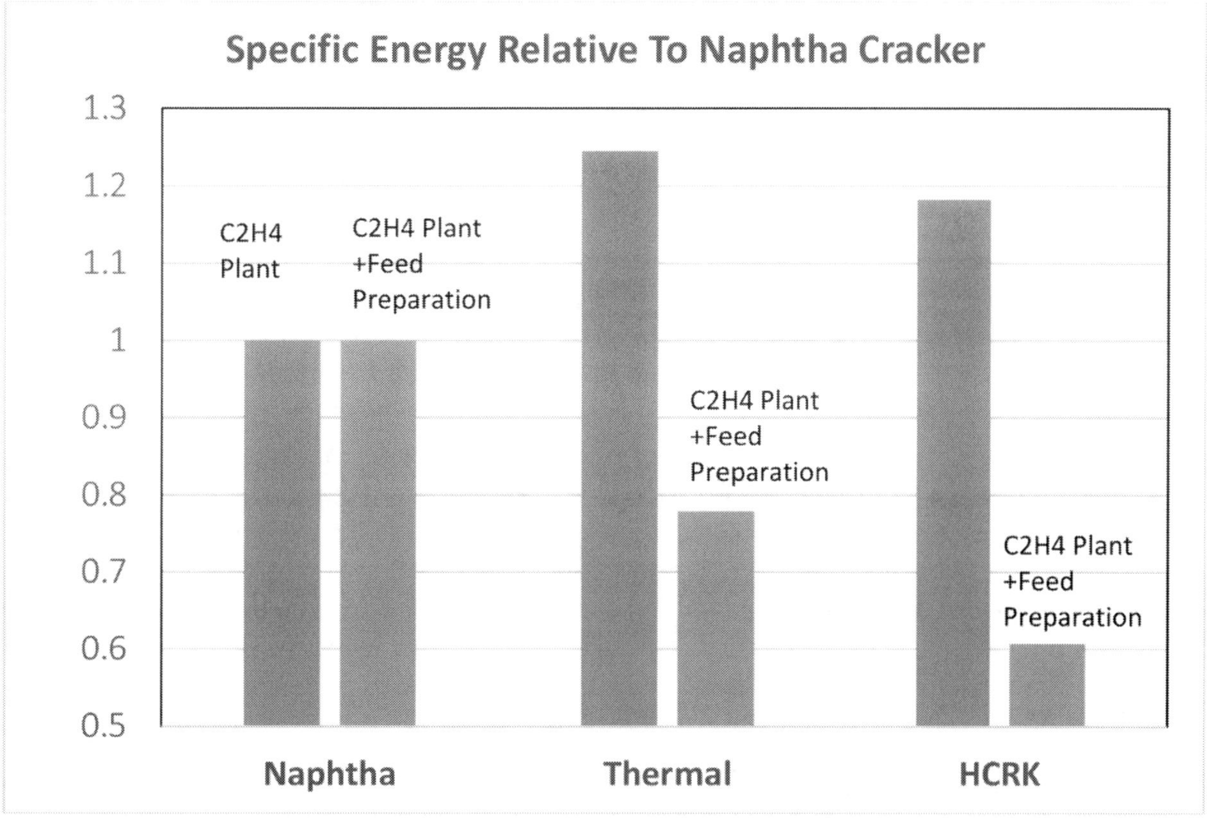

Fig.15. Comparison of specific energy to a naphtha cracker (Thermal has no hydrocracker; HCRK= TC2C™ with hydrocracker).

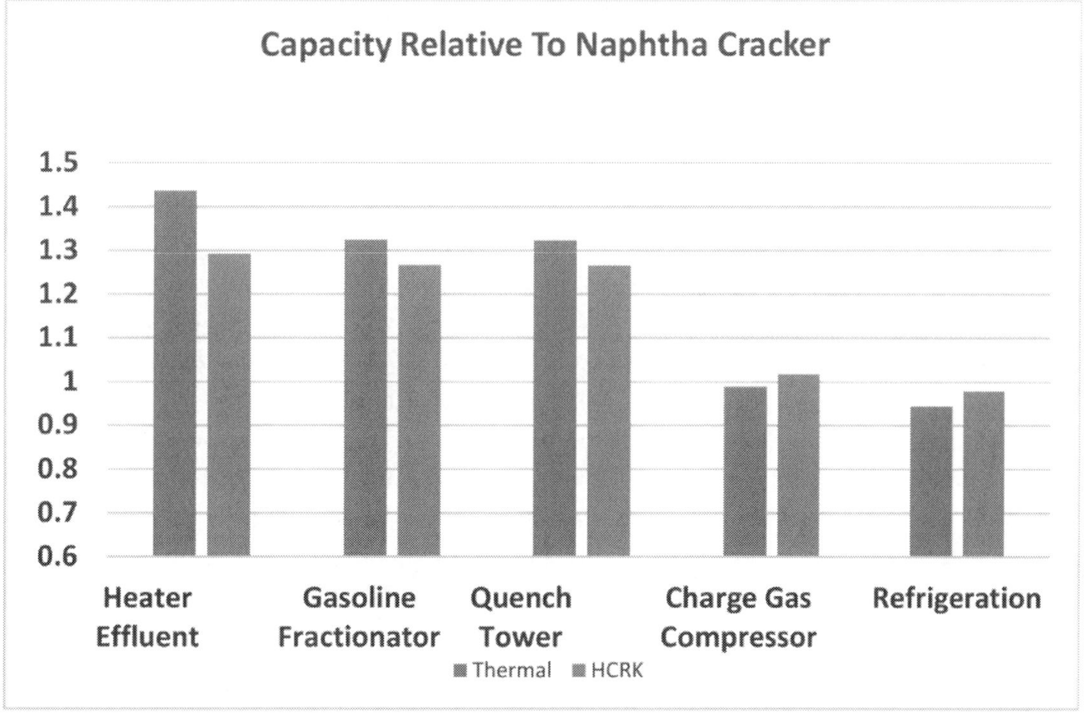

Fig.16 Relative area factors for major ethylene plant recovery section

Summary and Conclusion

Ethylene is produced by thermally cracking hydrocarbons mixed with dilution steam in tubular reactors at high temperatures in short residence time coils. During this process, coke, a solid deposits in the reactor and hence the cracking coils must be cleaned periodically using steam/air. The coking tendency of heavy molecules such as gasoil and vacuum gas oils is high and the relative ethylene yield from these feeds is typically low compared with gaseous feeds and naphtha feeds. Hence, traditionally ethane to naphtha feeds are used in the industry and sometimes gas oils are used to produce ethylene. Crude is not directly used to produce ethylene due to very short run length. The traditional feeds to the cracker are obtained from crude through distillation in a refinery. These are made to specifications to meet the fuel standard. In this paper, crude is cracked to produce ethylene bypassing the refinery. There is no need to build a crude distillation column to produce various cuts and it can be bypassed. In the TC2C™ technology discussed in detail in this paper all crude molecules ranging from naphtha to residue in the crude are processed to produce light olefins. A detailed analysis and understanding of the crude characterization has resulted in better catalyst and reactor systems to upgrade the crude for olefin production. Coke precursors that shortened the heater run length are reduced significantly, thanks to better characterization and experimental techniques of crude characterization. Traditionally, in the ethylene plant only ethane and propane are recycled to extinction. Modern plants further process and recycle some C5s and C6-C8 non-aromatics after BTX extraction. In the scheme proposed, in addition to the above recycles, pyrolysis gasoil and fuel oil produced in the ethylene plant are sent to LC reactor systems and then to hydrocracker. They are upgraded to meet ethylene plant feed requirements and then recycled to ethylene plant. By doing this, the amount of crude required to meet the design ethylene production is significantly reduced. Almost all cuts are used to make ethylene. Only a small purge is taken out as IMO compliany very low sulfur fuel oil, which is also valuable in the market. Without the hydroprocessing route, the yield of high value chemicals is only 50.4% for Arab Light crude and with hydroprocessing it is increased to >75%. Higher yield is possible. CAPEX for the ethylene plant is also reduced and the st,ecific energy is improved compared to a scheme without feed pretreatment. Ethylene plant CAPEX is only 7% more expensive than the typical naphtha cracker. Relative to a 100% naphtha cracker, specific energy of a crude cracking olefin plant increases only slightly with respect to a naphtha cracker and it is less than that of raw AGO/VGO flexible plant obtained from crude. When the energy required to produce naphtha cut from crude in the refinery is included, the TC2C™ route considered here has lower specific energy and hence will emit lower CO_2 for the complex. With hydroprocessing the complete refinery is bypassed. Therefore, CAPEX savings and energy savings are more compared with an integrated ethylene plant/refinery. Extensive bench scale and pilot plant for various sections of TC2C™ technology were run to collect data for design. A plant using this scheme is under construction and is expected to start-up in 2026. Though the scheme is shown with Arab Light crude, the technology is applicable for all types of crudes. Lighter crudes such as Permian or Agbami will reduce the CAPEX and improve crude utilization.

References

1) Fernandez L., "Global production capacities of ethylene 2018-2021", in Statista, June 20,2022
2) Kunii D., "Chemical Reaction Engineering and Research and Development of gas solid systems", Chem.Eng.Sci. 35(9), 1887-1911 (1980). (Kunugi)
3) Corma A., Sauvanaud L, Mathieu Y, Al-Bogami S, Bourane A. and Al-Ghram M, "Direct crude oil cracking for producing chemicals: Thermal cracking modeling", Fuel, 211(1), 726-736 (2018)
4) Masen T., Santos P., "TC2C^{TM} –Thermal Crude to Chemicals- A unique technology maximizing crude to chemicals", ERTC, Madrid , Nov17-18, 2020
5) Mukherjee U., "Thermal Crude to Chemicals", EGYPS, Cairo, Egypt , Feb14-16, 2022
6) Arora A., "TC2C^{TM} –Highest Conversion route from crude to chemicals for Grassroots or existing refineries", ETM, Mubai, India, Sep.15-17, 2022
7) Mukherjee U., "Production flexibility as a key to unlock refinery competitiveness", IDW 2022, Croatia, Sep.2022.
8) Santos P.,"Crude to chemicals", AFPM summit, San Antonio, Texas, USA, Oct.18-20, 2022.
9) Meenakshi Sundaram K. and Froment G.F, IEC Fund., 17,174 (1978)
10) Froment G.F., "Kinetics and Reactor design in the thermal cracking for olefins production", ChemEng.Sci.47, 2163-2177(1992).
11) Ranzi S., Dente M., Pierucci S., and Biardi G. "Initial product Distributions from Pyrolysis of Normal and Branched Paraffins", Ind.Eng.Chem.Fundam. 22, 132-139 (1983).
12) Fernandez-Baujin J.M. and Solomon S.,"An industrial Application of Pyrolysis Technology: Lummus SRT III Module", paper presented at the First Chemical congress of the North American Continent, Mexico city, Nov.30-dec.5, 1975., In Industrial and laboratory Pyrolyses, ACS symposium series, edited L.F.Albright, 1976
13) Altgelt K.H and Boduszynski M.M.,"Composition and Analysis of heavy Petroleum Fractions", Marcel Deckker, NY 1994.
14) Sundaram K.M, Mukherjee U. and Baldassari M.,"Thermodynamic Model of sediment deposition in the LC-FINING Process", Energy & Fuels, 22, 3226-3236 (2008).
15) Wiehe, I.,"Tutorial on the phase behavior of asphaltenes and heavy oils". 2007
16) Badre S, ConcalvesCarla C.C, Norinaga K., Gustavson G. and Mullins O.C.,"Molecular size and weight of asphlatene and asphaltene solubility fractions from coal, crude oils and bitumen", Fuel 85, 1-11 (2006).
17) Boduszynski M.M.,"Composition based Crude oil evaluation", Haverly Systems Asian technical Conference, June 2001.
18) Speight J.J., "Feed Composition" in Handbook of Petroleum Refining, CRC press, 2016
19) Schuller et al., Energy & Fuels, "Heavy oil-based mixtures of different origins and treatments studied by AFM",31, 6856-6861 (2017).
20) Rogel E., Lezcano M., Yee N., and Witt M.,"Effect of aging on deposit characteristics obtained by crude oil blending", Energy &B Fuels, 37, 1848-1856 (2023).
21) Maloney D., Sundaram K.M., Narayanan R and Zafer Akhras A.R (US1407950)
22) Fernandez-Baujin J.M., Maddock M.J. and Sundaram K.M., "A case study based on commercial experience: Vacuum gasoils to Petrochemicals", paper presented at AIChE Spring national meeting held at Orlando,Fl, USA, March 18-20, 1990.

23) Djokic M.R., Muller H., Ristic N.D., Akhras A.R., Symoens S.H., Marin G.B. and Van Geem K.M., "Combined characterization using HT-GC x GC-FID and FT-ICR MS: A pyrolysis fuel oil case study", Fuel Processing Tech. 182,15-25 (2018).
24) Froment G.F."Kinetic modeling of complex processes", Plenum, NY 1993, p409
25) Biswas G., Maesen T. and Sundaram K.M., "Unlocking Permian value", Hydrocarbon Eng., p29, Apr.2022
26) Van Geem K.M., Reyniers M.F., Marin G.B., "Challenges of Modeling Steam Cracking of heavy Feedstocks", Oil & Gas Sci.& Tech. 63, 79 (2008).
27) Maddock M.J."Improved Field Condensate cracking Trends in Industrial Application", paper presented at Refining, LNG and petrochem Asia 92, Dec.1-4, Singapore.

61f THE CHALLENGES OF INTEGRATING PLASTIC WASTE WITHIN A STEAM CRACKER

Bram Van Der Drift, Veronique Reich

Considerable work is being carried out by Technip Energies and others on technologies to reduce the Scope 1 and 2 steam cracker CO2 emissions (from direct firing and imported energy). Reduction of the Scope 3 CO2 emissions, associated with the preparation of feed stock, manufacture of finished products and end of life of the plastics, is significantly more challenging for the industry, particularly for gas cracking plants, for which oils from recycling are not suitable feeds. The reduction of the scope 3 emissions of steam crackers is significantly dependent on its feedstock composition and its associated environmental footprint. Being successful in the transformation from fossil-based to alternative feedstock is one of the levers to achieve the overall reduction of CO2 emissions from ethylene plants. Pyrolysis of plastic waste can help tackle this objective, whilst simultaneously providing for a circular economy as the associated products can be successfully processed in steam crackers. The presentation will showcase a differentiated plastic waste cracking technology; one plastic waste cracking technology that produces an olefin-rich gas as the main product. We will look at how the challenges associated with steam cracker integration can be unlocked through adequate purification and fractionation, all of which has been achieved through the successful collaboration between Synova and Technip Energies. Synova, as a technology provider will discuss the characteristics of its MILENA/OLGA process, including the main differentiators compared to conventional pyrolysis oil processes and Technip Energies, as the purification technology provider, will explain how it's Pure.rGasTM train can bridge the gap towards cracker integration in the recovery section of a gas or liquid cracker plant. The benefits of the combined technology allowing the recovery of high value compounds (ethylene, propylene, light paraffins, butadiene and BTX) from plastic waste while mitigating the presence of contaminants (NOx, CO, metals, chlorides, HCN, acid gases, oxygenates, sulfur compounds, etc.) will be demonstrated and, as an illustration, some typical overall yields will be showcased along with the results of a Life Cycle Assessment (LCA). The LCA shows the expected CO2 reduction provided by the combined technology offering.

4a Reducing Energy Consumption in Cracking Furnaces Using Multivariate Modeling

Williane Carneiro, Helena Amélia F de C Oliviere, Marlene Cardin

Cracking furnaces are a significant energy consumer and consequently there are many long-term sustainability projects focusing on physical design changes that will result in energy reduction. However, the goal of this work was to quickly minimize energy consumption in Braskem's existing furnaces without incurring capex to make physical plant changes. Multivariate analysis can be used to analyze historical operating data to identify opportunities to reduce energy consumption while maintaining throughput and adhering to other manufacturing constraints. This project was achieved through a combination of offline modeling, real-time multivariate process monitoring, and real-time optimization. This presentation will provide an introduction to multivariate analysis, an overview of the Braskem's approach using AspenTech's ProMV suite, and discuss some results achieved.

4b Key Features for Successful AI Applications

Sameer Thorat, Nishant Jaiswal, Rohit Nair

Producing Ethylene has many challenges due to process complexities caused by the large temperature and pressure differentials required to crack and recover products. These complexities create tremendous opportunities for AI/ML to address these challenges efficiently. AI/ML models develop a mathematical model of the relationship between measured variables in the manufacturing process that are unique to the particular unit and can rapidly adjust the unit to optimized operation. When accurately applied AI/ML can drive significant and highly valuable process improvement.

The challenge is to utilize technology appropriately to ensure success and capture the desired benefits. Effective utilization of AI can achieve greater operational efficiencies, maximized capacity utilization, lower production costs, improved quality, reduced emissions, enhanced safety and increased reliability. The uses are numerous, but success is highly dependent on necessary key elements for AI applications that are often ignored.

A smart analytical AI engine is expected to generate **actionable intelligence.** Information must be generated that can be easily interpreted and applied either directly or indirectly by effectively combining human intelligence and control systems. Dashboards from an AI solution must be easy to use and accessible to enhance decision making across all levels of plant operation. Considering the complexities in the ethylene process, having these "**Smart**" dashboards with **interactive levels** to provide the requisite insights to enhance decision making and drive operations to an optimal level is particularly important.

A critical process change in an Ethylene plant will impact operations at multiple points. Optimal response requires a **multi-dimensional** solution covering the wide spectrum of operations to achieve an overall synergistic response. To ensure this synergy and make the most effective solution, a single modeling technique may not be adequate. A **hybrid modeling** approach which utilizes the power of AI/ML and 1st principal models together, can often address the intricacies of the ethylene process more effectively and with the highest accuracy.

AI solutions create maximum impact due to their ability to deliver insight on a **real time** basis. The model predictions must be **quick and accurate**. Compromising on accuracy is not an option as each action has its own consequences, especially in ethylene process. To ensure accuracy, AI solutions need to be **widely compatible** to interact with various data sources for connecting the dots precisely. Development and utilization of **soft sensors** to generate the critical and accurate data insights and confirm accuracy can be critical.

Checking **financial impact** of process change and **customizing** the AI solution to the address the manufacturing problem is critical to capture value. Ensuring the solution is addressing the most financial impactful areas is key.

This paper will present the various improvement opportunities possible from AI/ML solutions application in the ethylene process. The critical features of an AI/ML solution necessary to ensure success and maximize benefits will be explained using ethylene case examples. The focus will be on highlighting critical aspects necessary for successful development and deployment of real time AI applications in ethylene.

4c SPYRO® for Asset Management (SAM®) : T.EN's Cloud Service for Ethylene Plant Optimization

Marco Van Goethem

Digitalization allows the ethylene plant operators to enhance its day-to-day production, availability and reliability. This is achieved by learning from the past and respond to future changes in market dynamics and operation challenges to realize production and margins.

By applying Technip Energies' know-how and experience as an ethylene technology licensor and EPC contractor, as well as Arundo's cloud computation, Artificial Intelligence (AI) and Machine Learning (ML) capabilities, we have developed a valuable digital service that facilitates the improvement of operations, availability and reliability in ethylene plants. SAM enhances the optimization and simulation capabilities of SPYRO by using actual operational data and offering extended visualization, alerts and recommendations.

SAM provides accurate predictions of yield (production numbers versus planned production numbers and expected remaining run length), decoke scheduling, and coil carburization, as well as benchmarking of furnace runs and furnaces.

The environmental module gives insight an GHG emissions and energy consumptions. It also identifies possible GHG reduction options. This is achieved by real-time simulation of furnace operations and a continuous understanding of the furnace conditions, operating parameters, and feedstock. This paper will show the current state of the art of Technip Energies' digital service. The first experiences with SPYRO for Asset Management will be shared.

SAM® part of BeyondTM by T.EN Offering

4d Artificial Intelligence for Cogeneration and Steam System Optimization

Denis Prokofiev, Timothy Holtan

Steam system management and electricity co-generation are vital components to any large, integrated petrochemical facility and are an important utility generally associated with steam cracking operations. In this talk, C3 AI will present the results of BHC3 Process Optimization for steam system allocation and power cogeneration at a large fertilizer facility. The combination of first principles, data science, and a global optimization algorithm allowed C3 AI to fully model high pressure and medium pressure steam systems and produce continuous recommendations on setpoints for three turbogenerators. In production, this solution has been demonstrated to increase internal power generation by up to 10% and reduce Scope 2 emissions through the reduced purchases of externally generated power.

Effective Implementation of AI in Ethylene

James Brigman

Application of data analytics and simulation to address problems in petrochemical and refining operations has been an option for over 40 years. Practically all chemical engineers graduating after 1980 were exposed to first principles simulation while getting their degree. Use of steady state simulation to understand the theoretical capabilities of the processing unit where they are working, the operation of distillation columns, or the sizing of heat exchangers has been a common application of simulation for engineers for the last 40 years and continues to be. Likewise the application of statistical analysis such as using statistical process control techniques to optimize unit performance or applying statistical analysis to equipment reliability to increase overall unit availability has been common practice over this period of time and continues to be. However despite the ready availability of these tools, use for day to day use by engineers for the sake of fine tuning operations and recognizing operational opportunities is limited.

Simulation software vendors have continued to improve their products to make them easier to utilize with significantly better visualization and processing speed. With the increase in processing speed and the ease of accessing years of operational data, data analytics tools such as artificial intelligence and machine learning techniques can be applied. The insights from this analysis can be impressive. However still the use of these tools remains limited despite the fanfare of digitalization and the wealth of tools developed.

Advanced Process Control represents the most successful application of data analytics that is widely used. APC applies a broad range of techniques to provide multi-variable control. Within a set of related process variables, relationships are established to automate the task of pushing performance of certain variables to defined optimal points within the defined constraints of the system. This is common practice. However it should be noted that rarely are a company's engineers supposed to build the APC system for the plant. Typically an internal group of company engineers specializing on the particular APC system provide the upkeep on these systems and do so often with the support of that specific APC vendor.

Digitalization and the application of AI/ML tools are in vogue and where properly applied are generating enormous value. Getting the right solutions in place is seen as necessary to compete. Yet despite this belief, successful implementation of AI/ML tools is estimated to be occurring less than 30% of the time and even then not performing up to expectations. Based on looking at the use simulation and other statistical tools, this should not be surprising. Attempting to treat digitalization, its implementation, and its use in the same manner as simulation will result in the same limited use and applicability despite its tremendous potential. Treating AI/ML digitalization implementation and use more similarly to APC implementation can significantly increase successful application and use.

This paper will go into the details of how successful solutions were built for Ethylene operations that address the major issues found in Ethylene plant operation. These solutions are not competitive to APC but significantly additive to APC in terms of opportunity identification for profitability, reliability and sustainability. The comparison to APC is solely for showing how the application and implementation of these solutions should mirror the application and implementation of APC systems in order to drive the successful use to capture value.

Examples will be given of AI/ML tools implemented in Ethylene facilities as solutions that are able to effectively address:

- Optimizing furnace runlength, yield, and coil life for increased production
- Improving the reliability and performance of key equipment including the quench tower, cracked gas compressor, and various distillation columns
- Equipment and instrumentation reliability
- Establishing awareness of current operational performance in terms of profitability and sustainability versus theoretical or best ever performance

The ongoing use of these solutions to capture tens of millions in increased production and these solution's ongoing maintenance will be described.

While Ingenero does provide an AI/ML tool set, the paper is not concentrating on the tools but the technology implementation and support process necessary to deliver solutions that creates ongoing value in Ethylene facilities that are applicable regardless of selected vendors technology.

Emissions Reduction with Lummus External Air Preheat and Compact SCR

Baozhong Zhou, Alejandro S. Carrillo, Xueping Li

Pyrolysis heaters in ethylene plants are a major source of CO_2 and NOx emissions. Minimizing these emissions is an important target for both, the operating plants and the engineering companies in the petrochemical industry. The biggest challenge for an existing ethylene plant is to maintain the same process operating conditions and, at the same time, reduce the total fuel firing without major heating-surface changes / structure steel modifications.

This paper will discuss Lummus External Air Preheat (LEAP) technology, and will cover various scenarios for retrofitting existing pyrolysis heaters to achieve lower CO_2 emissions and higher fuel efficiency. LEAP performance is based on heating-surface optimization and heat absorption maximization by the process streams. For an equivalent or improved fuel efficiency, LEAP is able to reduce the CO_2 emissions by 10-30%, depending upon the existing heater design. When LEAP is applied together with Lummus' Compact SCR system, the NOx emissions can be controlled at or below the baseline NOx emission level, hence mitigating the hot air combustion impact on the NOx emissions. Lummus' Compact SCR can also be applied by itself to a heater, for lowering the NOx in order to meet tighter emission control regulations.

Lummus' Compact SCR and LEAP are ideal solutions that can be implemented utilizing the existing structure steel and foundation, and they can be accommodated in the existing heater island plot space.

AIChE Paper Number: 27a

Hydrogen and Ammonia: Zero Carbon Fuels for Steam Crackers

Daniel McKenzie

Technology Manager

Lummus Technology

Christoph Strupp

Global VP of Combustion Technology

John Zink Hamworthy Combustion

Mike Henneke

Global CFD Director

John Zink Hamworthy Combustion

Michel Haag

Process and Test Engineer

John Zink Hamworthy Combustion

Prepared for Presentation at the 2023 Spring National Meeting

Houston, TX, March 12, 2023

Hydrogen and Ammonia: Zero Carbon Fuels for Steam Crackers

Daniel McKenzie

Technology Manager

Lummus Technology

Christoph Strupp

Global VP of Combustion Technology

John Zink Hamworthy Combustion

Mike Henneke

Global CFD Director

John Zink Hamworthy Combustion

Michel Haag

Process and Test Engineer

John Zink Hamworthy Combustion

Abstract:

The majority of the emissions from steam crackers are from combustion of hydrocarbon fuel that is obtained from the process as methane-rich offgas. While the offgas contains some byproduct hydrogen, for most crackers the combustion of methane provides most of the required energy input. One of the methods to eliminate the carbon dioxide (CO_2) emissions from steam cracking heaters is to use an alternate fuel source that does not contain any hydrocarbons. If this fuel source is obtainable, either now or in the near future, then making provisions to fire this fuel in the cracking heaters is an attractive option for ethylene producers to lower or potentially eliminate their CO_2 emissions. This option offers a drastic reduction in CO_2 emissions with relatively minor modifications in the ethylene plant, which should be weighed against other CO_2 reduction options that involve fundamental changes to the recovery section and require a significant import of electricity.

Two carbon-free fuel sources that are currently being considered by the industry are hydrogen and ammonia. Both fuels produce zero CO_2 but have vastly different characteristics that present unique challenges for the cracking heater design. This paper will discuss the impact of switching to hydrogen or ammonia on burner design and heater performance as well as the current status of this technology.

Introduction

The primary source of direct CO_2 emissions in steam crackers is the cracking heaters that give the steam cracker its name. The combustion of methane-rich fuel gas in the cracking heaters provides the energy for the steam cracking reactions, producing CO_2 as a byproduct. There are several technical approaches to reducing CO_2 emissions from the cracking heaters, each with its own costs, benefits and level of technical maturity. The ethylene industry is considering various options, each having its own merits and drawbacks:

- Combustion air preheating can significantly reduce fuel firing in the heaters and hence CO_2 emissions, but at the expense of super high pressure (SHP) steam production. Significant modifications to the cracking heater are required, and the decrease in steam production must be compensated for by importing electricity. In order to achieve an overall CO_2 emission reduction, this electricity should be from non-fossil fuel sources. The compressor turbines in the recovery section must be partially changed to electrical motors, with the associated import of electricity having its own availability, cost and carbon footprint to consider.
- Post-Combustion Carbon Capture has relatively high CAPEX and OPEX, which are difficult to justify based on current economics and regulations.
- Electrification of cracking heaters is still under development, requires a significant amount of electricity (approximately 400 MW for the heaters of a 1000 KTA ethylene plant) and produces much less SHP steam, so the compressor drivers must also be electrical motors.

With the increased requirement for green electricity, the availability of the electricity is a key consideration and the required investment in the distribution network could add significant costs to a project. Another approach is to use a fuel source in the cracking heaters that does not contain any hydrocarbons, such as hydrogen or ammonia. Compared to the other approaches listed above, burning zero-carbon fuels can be implemented with minor changes to the cracking heaters and no change to the recovery section.

Hydrogen Fuel

Ethane crackers produce a relatively high yield of hydrogen as a byproduct compared to other feedstocks, so the fuel gas in these plants already contains a high percentage (70-85 mol%) of hydrogen. This hydrogen-rich fuel gas is not new to the industry, and the impact of this level of hydrogen fuel gas on the design of the cracking heaters is well understood. Because of this high hydrogen fuel gas and lower energy consumption per ton of ethylene (i.e., specific energy), ethane crackers have a much lower carbon footprint than heavier feed crackers.

Nonetheless, ethylene producers of all feed types are looking to further reduce CO_2 emissions, and one simple way to achieve that is to increase the hydrogen content of the fuel gas. Figure 1 shows the relative impact on CO_2 emissions as a function of hydrogen content in the fuel gas. Figure 1 also shows the effect of hydrogen content on SHP steam production, which will be discussed later.

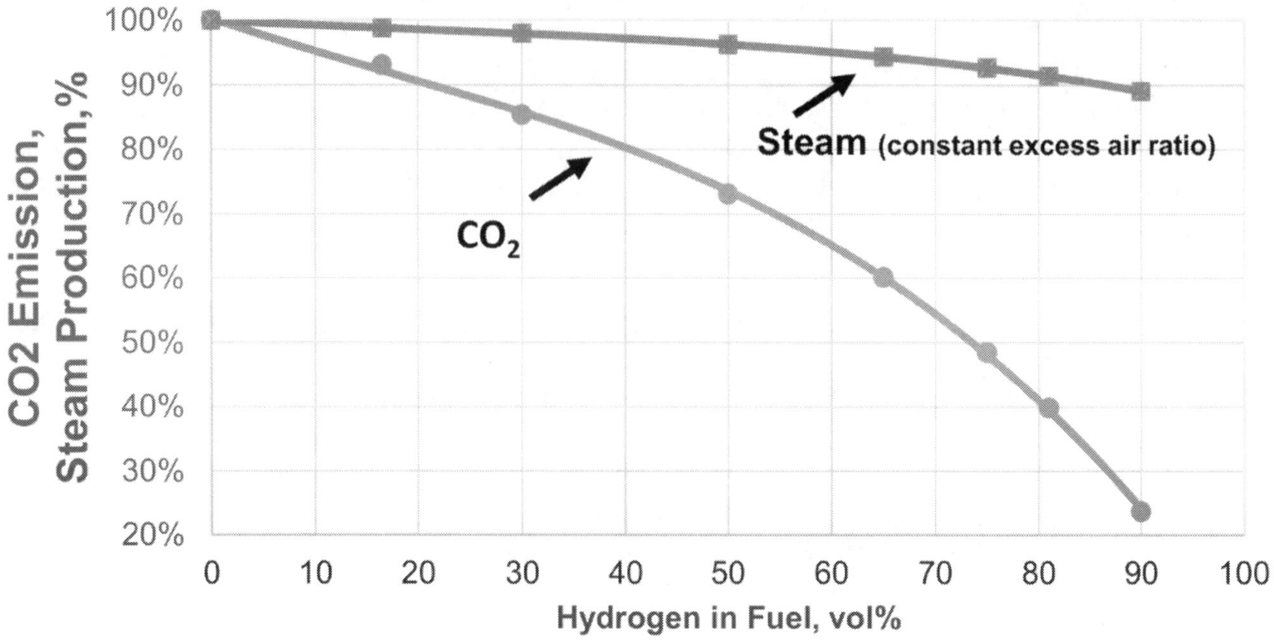

Figure 1. Relative CO_2 emissions and steam production with hydrogen content in fuel gas.

To take credit for a reduction in overall CO_2 emissions from a steam cracker, the carbon footprint of the imported hydrogen itself must be considered. Almost all (>99%) of the hydrogen that is currently produced worldwide is from steam reforming of natural gas and coal gasification, and less than 1% of that hydrogen is produced with carbon capture⍰⍰. Most of the hydrogen produced is used in refineries for hydrodesulfurization, ammonia synthesis, other chemicals and steel production⍰⍰. When the hydrogen is produced from fossil fuels without carbon capture, it is considered gray hydrogen and this process emits a lot of CO_2. When the carbon is captured, it is considered blue hydrogen and therefore has lower indirect or "Scope 2" emissions. Green hydrogen is produced from electrolysis using renewable electricity and therefore has zero indirect emissions. However, electrolysis consumes a lot of electricity, approximately 50 kWh/kg of hydrogen⍰⍰. To put this electrical demand into the context of an ethylene plant:

- A 1000 KTA ethane cracker would need approximately 155 MMkcal/hr or 5400 kg/hr of imported hydrogen to replace the methane in the fuel gas and eliminate CO_2 emissions. This import is approximately 40% of the total fuel required. Production of this amount of green hydrogen by electrolysis would require approximately 270 MW of renewable electricity.

Directionally, cracking heavier feeds would require greater amounts of green hydrogen to fully decarbonize the process. A few cases studies follow that indicate the amount of imported hydrogen that would be needed for world-scale crackers.

Although neither blue hydrogen nor green hydrogen is widely adopted, production of hydrogen from these low-emission routes is expected to grow. About 350 electrolysis projects are currently under development, and even more are in early stages of development, which in total have the potential to bring the total green hydrogen production to 8 million tons by 2030, or about 10% of the demand [4]. For blue hydrogen, 16 plants are currently operational, producing about 0.7 million tons, and another 50 projects are under development, which could bring the total production to 9 million tons by 2030 [4]. With increasing production of blue and green hydrogen, and with increased policy incentives, the cost is projected to decrease and may even be competitive with gray hydrogen by 2030 [2].

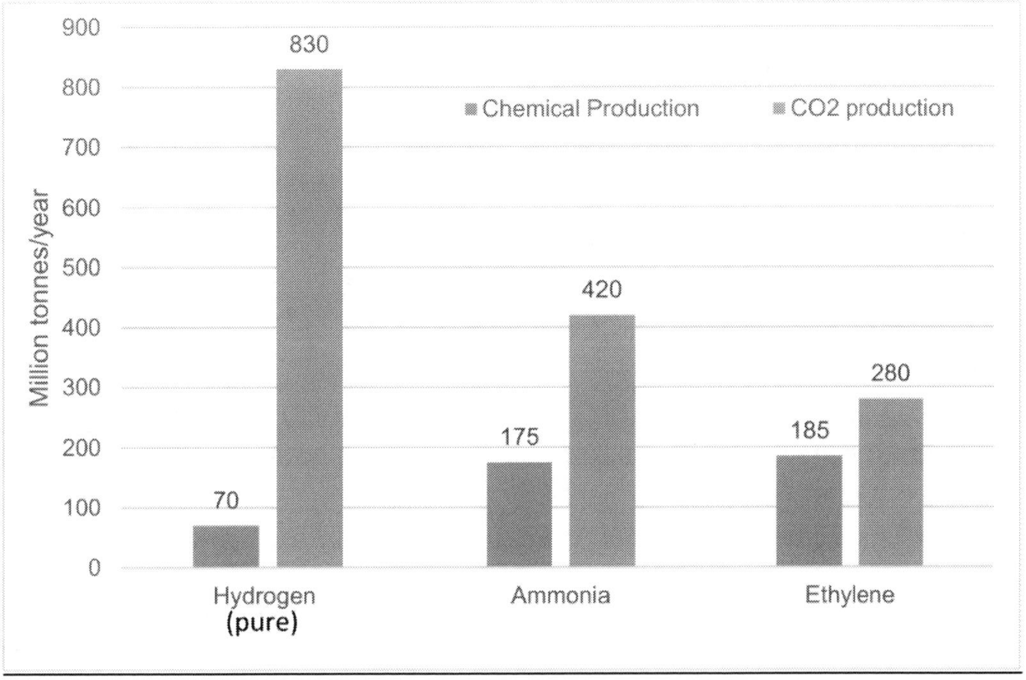

Figure 2. Annual hydrogen, ammonia and ethylene production and their corresponding CO_2 emissions in 2019. [2][3][5]

Ammonia Fuel

Ammonia has a strong potential to be used as a carbon-free energy carrier and fuel. It is one of the largest volume chemicals produced globally, and a commonly distributed chemical to industrial as well as agricultural and commercial end users. Compared to hydrogen, ammonia has higher molecular weight and volumetric energy density, is easier to ship and store, and has more commercial, industrial and end-user experience with already existing widespread infrastructure. Most of the renewable ammonia activity is centered on the production of renewable/green hydrogen, while the ammonia synthesis is from the conventional Haber-Bosch process, in which hydrogen and nitrogen are combined at high temperature and pressure using a catalyst.

While research and development are active on the use of ammonia directly in gas turbines, internal combustion engines and fired heaters, the use of ammonia as fuel in steam cracking heaters has not been demonstrated. The main challenge in using green ammonia as fuel is the limited worldwide supply of ammonia from non-fossil fuel sources. Like hydrogen, ammonia is considered blue if the CO_2 is captured in the reforming process and if renewable electricity is used in the air separation plant to make the nitrogen. Green ammonia is also possible if the hydrogen is produced through electrolysis.

Storage and Transport

Storage and transport facilities for both hydrogen and ammonia exist and are well understood. As such, little to no development of engineering solutions is required in these areas. Hydrogen is an extremely light gas and therefore is difficult to store and transport unless pipelines are available. The hydrogen molecule is very small and therefore difficult to contain without leakage, and can diffuse into certain materials. Due to these challenges with transportation, its use as imported fuel in steam crackers is limited to those with nearby hydrogen pipelines or petrochemical complexes with reformer units. Hydrogen is highly flammable and has an invisible flame in daylight, which makes leak detection more difficult.

Ammonia can be easily stored and transported, so it has higher potential for import as fuel where hydrogen infrastructure does not exist. Ammonia has similar properties to LPG (boiling point is -33°C versus propane -42°C). Ammonia is corrosive and toxic but is easily detected by smell.

Both fuels have safety and environmental concerns that must be considered, although both are well-known chemicals in steam crackers so protocols for safe handling have been established.

Characteristics

Regardless of the color label associated with the hydrogen or ammonia, the chemical composition of the product, high purity H_2 or NH_3 is considered identical. Therefore, the source of the fuel is not a consideration in the combustion process. However, these zero carbon fuels have vastly different physical and combustion properties from methane and therefore present challenges for both the burners and radiant section as well as the convection section.

Impact on Burners and Radiant Section

Over the history of ethylene production, different cracking heater designs have been developed. Depending on the heater design, several types of burners are used. Many heater designs in ethylene cracking application have burners that fire against a refractory wall. The combustion heat is radiated from the wall toward the coils, where the heat is transferred to the process fluid. Other heater designs use free-standing burners in between the coils.

The first cracking heaters were designed with only premixed burners mounted on the side wall of the heater. Later, diffusion-type burners were installed on the heater floor and fired upward along the wall. In some heaters, a combination of floor burners and wall burners are used to increase the flexibility of the heat flux profile. Another heater type, the balcony heater, has two rows of floor burners at different elevations firing along the wall.

Burners used in steam cracking heaters must have a wide degree of fuel flexibility. Typical fuel gases range from heavy fuels, such as butane, to natural gas up to fuel gas with 85 vol% hydrogen. However, as we will see, extending this fuel flexibility to include firing 100% hydrogen and 100% ammonia is a technological challenge that requires additional R&D effort.

When compared to conventional hydrocarbon fuels, hydrogen and ammonia present challenges. Table 1 provides properties for hydrogen, ammonia and methane. Both hydrogen and ammonia in the gas phase have substantially lower energy density when compared to hydrocarbon fuels. Also note the substantial differences in the laminar flame speed. These differences in laminar flame speed are an important factor in burner design.

Table 1: Combustion properties of methane, hydrogen, and ammonia.

	Methane (CH_4)	**Hydrogen (H_2)**	**Ammonia (NH_3)**
Lower Heating Value (LHV), kcal/kg kcal/Nm^3	 11,946 8,556	 28,651 2,581	 4,443 3,394
Adiabatic flame temperature, °C	1,950	2,110	1,800
Laminar flame speed @ 0% excess air, cm/s	35	225	7
Flammability limits in air	5%-15%	4%-75%	15%-28%
Boiling temperature, °C	-161	-253	-33

Hydrogen

Combustion performance

As indicated in Table 1, hydrogen has a laminar flame speed that is roughly seven times that of methane. For diffusion type (non-premixed) burners this flame speed does not present a challenge–in fact, hydrogen flames are much easier to stabilize, even under conditions such as low excess air, high fuel gas pressure and high air velocity. However, for premixed burners that are typically applied as wall burners in steam cracking heaters, this high flame speed causes a phenomenon called flashback. Flashback occurs when the flame propagates upstream into the fuel/air mixture and burns inside the burner. Flashback presents a safety and operation risk and will lead to the complete destruction of the burner.

Modern burner technology uses a diffusion-type wall burner to address the need to fire high hydrogen fuel without any risk of flashback. The John Zink Hamworthy Combustion (JZHC) Walfire™ burner (Figure 4) is a diffusion-type wall burner that has been successfully applied in petrochemical applications. In contrast to a premix burner, the air flows through the burner independent from the fuel gas pressure. The Walfire™ uses a single point radial fuel injection that injects the fuel over an air disk along the wall. The air disk induces stabilizing vortices to optimize the flame shape. Low NO_x performance is achieved with a combination of air staging and internal flue gas recirculation. Due to the diffusion type combustion, a larger turndown of 10:1 and significantly lower noise emissions can be achieved compared to premix burners.

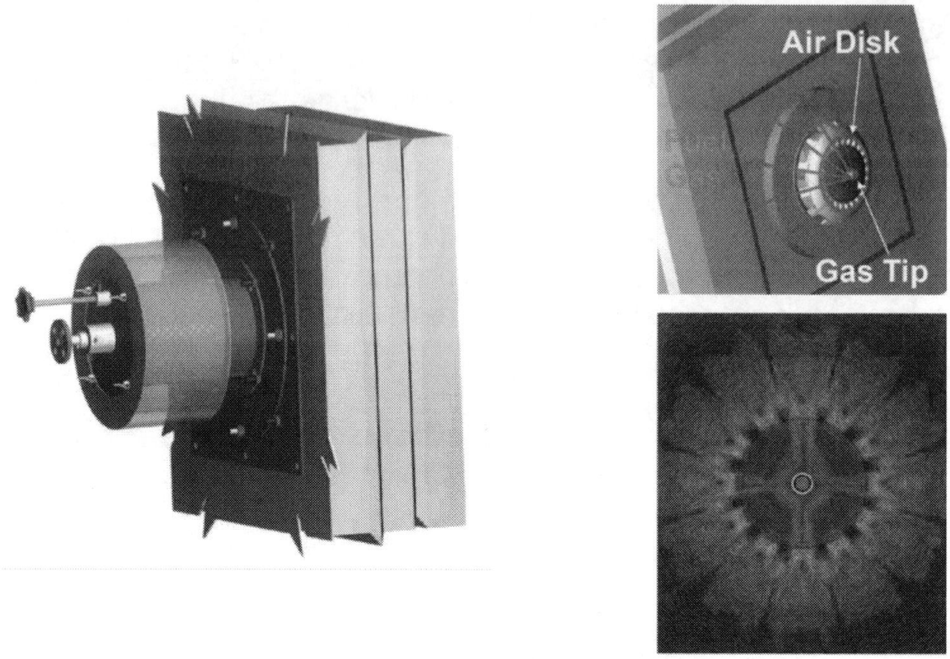

Figure 4: Walfire™ diffusion flame burner for application as a wall burner.

NO_x emissions

NO_x emissions in hydrocarbon flames are produced by two mechanisms: thermal NO_x and prompt NO_x. Thermal NO_x depends strongly on the flame temperature. By increasing the hydrogen fraction, the flame temperature and thermal NO_x increases. Prompt NO_x is formed through a chemical pathway involving cyanic radicals. Consequently, the fewer hydrocarbon radicals that are present in the combustion zone, the less prompt NO_x is formed.

Many installed floor burners (diffusion flame) are based on a first-generation fuel staged NO_x reduction technology. For these burners, NO_x emissions show an increasing trend with hydrogen percentage, with an accelerated increase at elevated hydrogen percentages. However for Ultra-low NO_x burners with highly flue gas diluted flames, NO_x emissions will drop as the hydrogen fraction in the fuel increases above about 80%. For these burners the amount of thermal NO_x is significantly lower and the reduction of the prompt NO_x at higher hydrogen levels will overcompensate the thermal NO_x production. This behavior is shown in Figure 5.

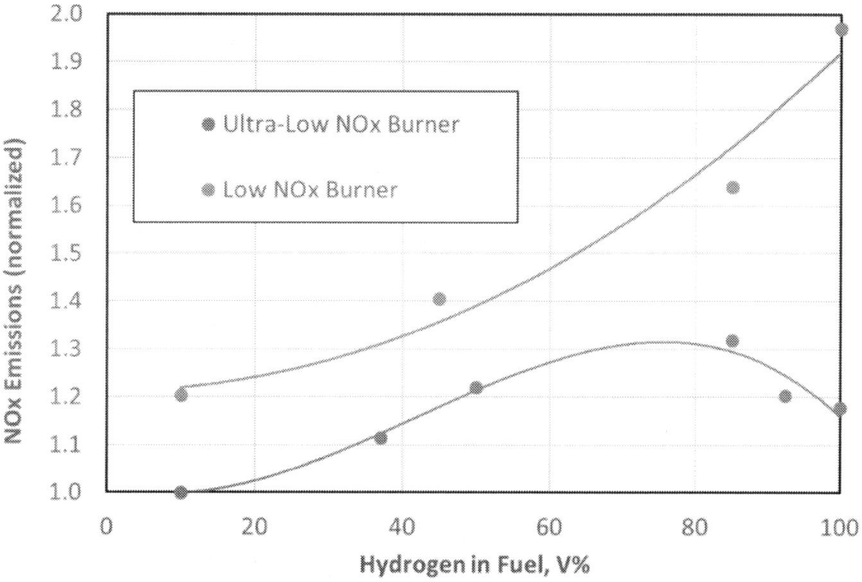

Figure 5: NO_x emissions versus hydrogen content in a natural gas mixture. Comparison of a first generation staged fuel burner with a modern ultra-low NO_x burner.

Premix burners typically show an increasing NO_x level with an increasing H_2 content up to their flashback limit.

Cracking heater heat transfer

When changing the fuel gas composition, it is important to evaluate the impact on the heat flux distribution. A 3-D CFD simulation has been performed to investigate this impact. A JZHC test furnace with two natural draft, diffusion type, ultra-low NO_x COOLStar®-18 burners was used as the modeling case. This model represents a typical steam cracking application with floor burners fired against the wall. The geometry comprised the entire furnace including the cooling tubes and a detailed model of both burners (see Figure 6). The computational mesh consisted of approximately 13 x 10^6 finite volume cells. Turbulence was modeled using the k-ε model with two-layer all y+ wall treatment. Combustion was modeled with a hybrid eddy breakup combustion model.

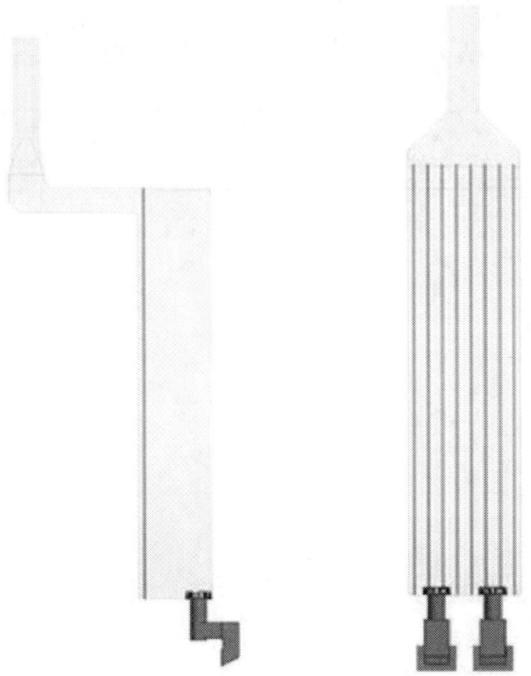
(a) (b)

Figure 6: CFD furnace geometry with 2 COOLStar®-18 floor burners: (a) side view and (b) front view.

The graphs in Figure 7 show the local normalized and absolute heat fluxes over the normalized height of the furnace. Figure 7a shows that the location of the peak heat flux is shifted closer to the burner with increasing hydrogen in the fuel gas. The incident heat flux, shown in Figure 7b, indicates the total amount of heat transferred to the coils increased with an increasing amount of hydrogen in the fuel gas. Both effects are related to the higher adiabatic flame temperature and higher combustion velocity of hydrogen, which lead to shorter flames. So, more heat is released in closer proximity to the burner which leads to higher thermal radiation. Table 2 shows that the efficiency of the radiant section increases by 4.4% when going to 100% hydrogen and keeping the fired duty constant. Also, the change in top furnace temperature and the position of the peak heat flux can be seen in Table 2.

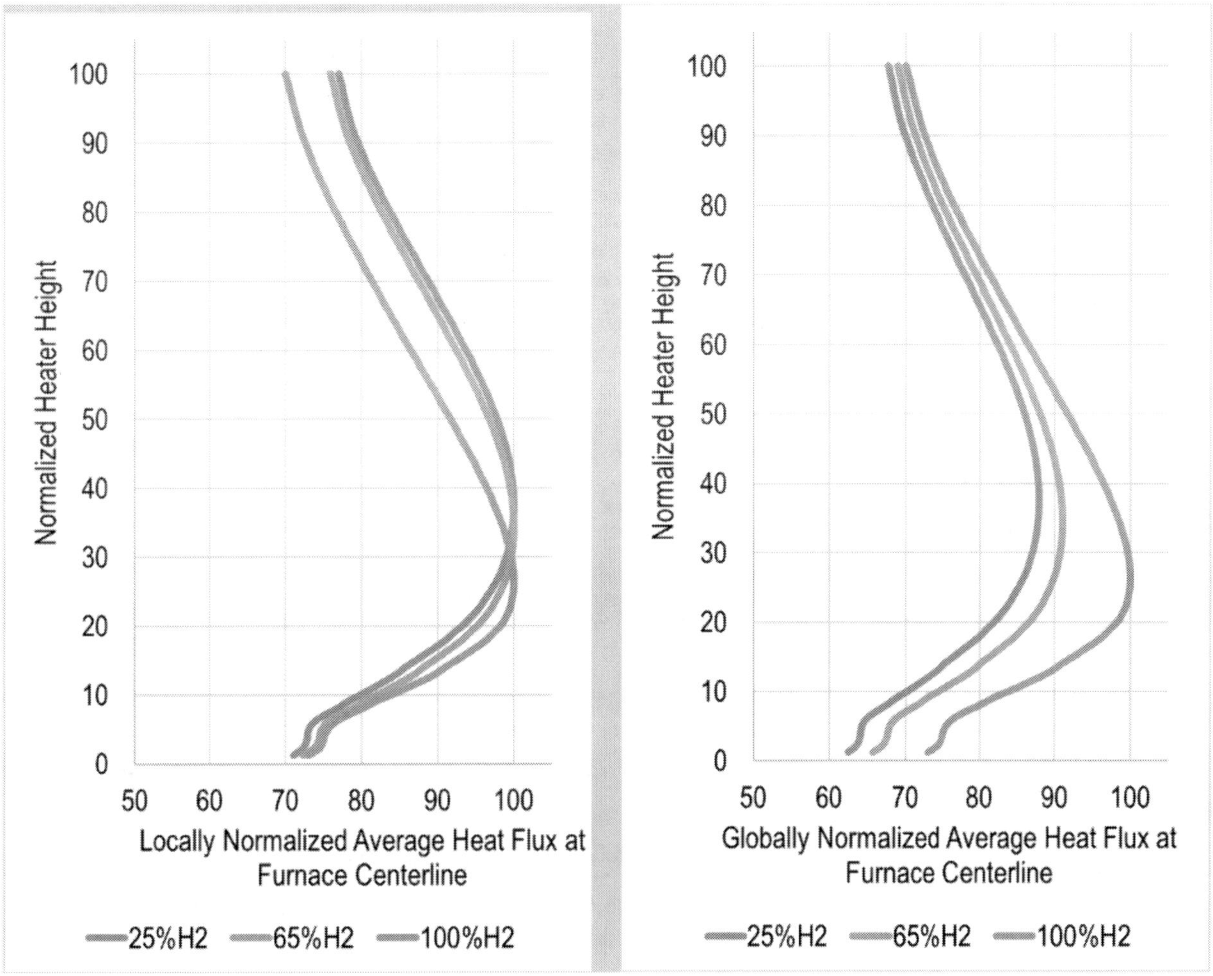

(a) (b)

Figure 7: CFD results as a function of normalized heater height for (a) locally normalized average heat flux at the heater centerline and (b) local heat flux at the heater centerline.

Table 2: Selected CFD boundary conditions and result.

	Base case (28% H_2, 69% CH_4, 2% C_2+, 1% N_2)	**65% H_2, 35% CH_4**	**100% H_2**
Chemical heat release, MMkcal/hr	5.72	5.72	5.75
Relative radiant section duty (radiant efficiency), %	39.8	41.0	44.2
Top furnace temperature, °C	1123	1125	1142
Peak heat flux elevation, %	37.8	35.0	25.9

Material selection

High hydrogen fractions in the fuel gas may impact material selection. The impact of hydrogen on refractory ceramic materials is a subject of current research [6]. Typical burner refractory consists of a mixture of Al_2O_3 and SiO_2. Studies have shown that the high hydrogen concentrations will lead to a degradation of the SiO_2 inside the tile material [7]. For high hydrogen firing, it is important to select the appropriate tile material, as well as the insulating brick next to the burners, refractory near the wall burners, and decoking nozzles (if present). More development work is ongoing by tile material suppliers for full hydrogen firing in process burners.

In addition, some burners use steel components as flame holders. With its higher adiabatic flame temperature and faster flame speed, hydrogen flames produce more thermal radiation incident on these metal components, which leads to reduced service life. It may be necessary to upgrade the metallurgy used for these components to increase their service life.

Ammonia

Ammonia is a novel fuel gas in steam cracking applications. The combustion of ammonia has two major challenges: very high NO_x emissions and slow reaction kinetics which are unfavorable for the burner stability.

Combustion performance

In terms of burner stability, the worst fuel used in today's steam cracking applications is natural gas. Table 1 shows that ammonia has a lower flame temperature than methane and a flame speed five times lower than that of methane. Lower flame speed and temperature increase the potential for an unstable flame. Therefore, the burner technology needs to be properly chosen to ensure stable combustion for firing ammonia. In case the burner technology cannot be changed on an existing heater, the combustion performance can be improved by adding hydrogen to ammonia. In fact, the hydrogen content can be increased in the fuel gas by the upstream cracking of ammonia to hydrogen and nitrogen.

NO$_x$ emissions

A significant fraction of the nitrogen from the ammonia is oxidized to produce an oxide of nitrogen such as NO, NO_2 or N_2O. In contrary to NO or NO_2, is N_2O not toxic but it is a potent greenhouse gas with a CO_2 equivalent of 298 t CO_2 / t N_2O [8]. During low temperature operations such as startup, hot steam standby, and decoking, an increased production of N_2O and even toxic ammonia slip could be observed during burner tests. Those operation scenarios should be carefully evaluated.

Large-scale burner tests have shown that NO$_x$ production is strongly impacted by the combustion type (premix or diffusion) and the type of molecules in the fuel gas. In standard premixed burners firing hydrogen/ammonia blends, NO$_x$ emissions can exceed 20,000 ppm. Optimized radiant wall burner technology can reduce NO$_x$ by up to 90%, which brings the emissions down to a range that can be handled by selective catalytic reduction (SCR). Full-scale burner tests on burners used for cracking applications showed that NO$_x$ emissions increase with increasing amount of ammonia in the fuel gas up to a specific maximum. After reaching this maximum, NO$_x$ emissions decrease with increasing ammonia. The location of this maximum depends on the burner technology and also on the fuel gas mixture. Figure 8 shows such a typical trend for a premix burner. Similar NO$_x$ trends are confirmed by various laboratory scale experiments [12]. Newer burner models like the Walfire^{TM} can achieve significant NO$_x$ reductions compared to the values above, and additional burner technology development is ongoing for ammonia applications.

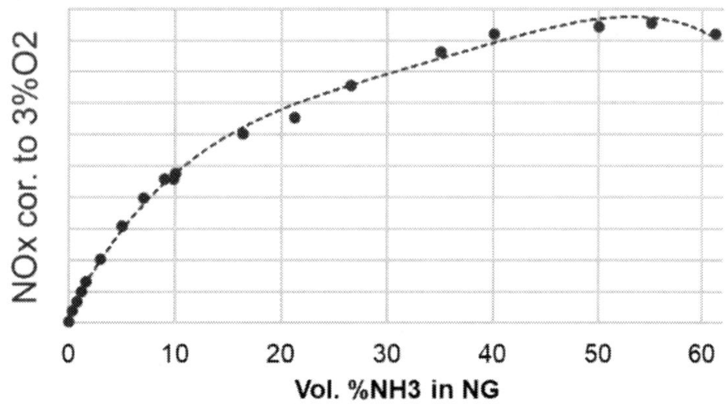

Figure 8: NO$_x$ emissions as a function of ammonia percentage in fuel gas for a premixed burner.

Impact on Convection Section

The impact of zero-carbon fuels on the convection section is primarily due to the difference in heating value, combustion air requirements of the fuel and the resulting volume of flue gas. This is a function of the stoichiometry of the primary combustion reactions:

$$CH_4 + 2O_2 \rightarrow 2H_2O + CO_2$$

$$H_2 + \frac{1}{2}O_2 \rightarrow H_2O$$

$$NH_3 + \frac{3}{4}O_2 \rightarrow \frac{1}{2}N_2 + \frac{3}{2}H_2O$$

Compared to methane, both hydrogen and ammonia have much lower molar energy density; i.e., per unit of duty, the required molar rate of hydrogen is 3.3x that of methane and ammonia is 2.5x that of methane. However, on a molar basis, hydrogen and ammonia generate fewer combustion products and less oxygen (combustion air) is required per mole of fuel. Therefore, with a fixed excess air ratio and heat liberation, when compared to 100% methane fuel, the net result is that hydrogen has a *lower* flue gas rate (~20%) whereas ammonia has a *slightly higher* flue gas rate (~4%). The resulting flue gas composition is similar except that hydrogen and ammonia produce zero CO_2 and about twice the concentration of water compared to that from methane combustion. The heat capacity of the flue gas also plays a factor since it also increases with higher water content. The net result of these factors is a decrease in convection section duty for hydrogen and an increase in convection section duty for ammonia.

Table 3. Flue gas production from methane, hydrogen and ammonia fuels (at 10% excess air).

	100% CH_4	**100% H_2**	**100% NH_3**
Molar heating value, kcal/Nm^3	8,556	2,581	3,394
Moles of combustion products per mole of fuel	3	1	2
Moles of O_2 per mole of fuel	2	0.5	0.75
Flue gas rate per fired duty, kg/MMkcal (normalized)	**1660 (1.0)**	**1342 (0.81)**	**1722 (1.04)**
Flue gas composition, mol%			
O_2	1.7%	1.6%	1.4%
N_2	72.1%	66.3%	69.6%
CO_2	8.7%	0%	0%
H_2O	17.4%	32.1%	29.0%

This different flue gas duty will affect the heat recovery and temperature profile in the various banks of the convection section, most importantly the crossover temperature and SHP steam production. Crossover temperature is the process temperature between the convection section outlet and the radiant coil inlet. This is selected for a given feedstock to maximize heat absorption without reaching temperatures high enough to initiate cracking reactions and subsequent fouling in the convection tubes, which are not easy to clean. As the flue gas rate decreases for a given heating surface, the crossover temperature will also decrease. This leads to an increase in fuel firing to achieve the coil outlet temperature in the radiant section, which increases heat flux and decreases run length.

Table 3 shows the extreme differences in flue gas generated from pure methane, hydrogen or ammonia, but in an actual plant the fuel gas will be a blend of these components, as well as minor amounts of CO, C2s and other components. As such, the performance of a given heater with some amount of zero-carbon fuel will be within the range of the extremes in the table. This needs to be evaluated on a case-by-case basis. However, the following general rules apply:

- For an existing heater design with increasing amount of hydrogen in the fuel, the decrease in flue gas heat load and resulting impact on the convection section coils can be mitigated without mechanical modifications simply by increasing the amount of excess air. Consequently, the fired duty will increase slightly due to the additional duty required to heat the extra combustion air from ambient to firebox temperature, and NO_x generation will increase, which needs to be addressed.
  For an existing heater, this represents a viable solution to utilize zero-carbon fuels without significant impact on the heater performance while maintaining the level of SHP steam production that is necessary to drive the compressor turbines in the recovery section.
- For a case with hydrogen firing and increased excess air, the absolute combustion air requirement is still less than for the normal plant fuel gas so modification of the ID fan is not necessary.
- For a case with ammonia firing, the flue gas rate may increase so the ID fan needs to be evaluated.
- For a new heater design, if minimum fuel firing is desired to reduce the imported fuel cost, then the convection heating surface can be modified to achieve the desired crossover temperature at the normal excess air rate, although steam production will be lower.

Case studies

Several case studies are presented below to illustrate the impact of adding hydrogen or ammonia fuel to an ethane cracker and a naphtha cracker.

Table 4. Ethane cracking heater performance with increased hydrogen fuel.

	Base (Normal plant fuel)	**100% H_2, Normal excess air**	**100% H_2, Increased excess air**	**100% H_2, Modified convection surface**
HC feed rate, kg/hr	46,902	46,902	46,902	46,902
Fuel composition, mol%				
H_2	**82.7**	**100**	**100**	**100**
CH_4	16.8	0	0	0
Other	0.5	0	0	0
% Excess air	10	10	17	10
Fired duty, MMkcal/hr	86.8	83.5	86.8	82.2
Fuel flow rate, kg/hr	4,847	2,909	3,022	2,868
Imported H_2, kg/hr	0	1,111	1,224	1,071
Overall thermal efficiency, %	93.9	94.3	93.9	94.3
Crossover temperature, °C	Base	-12	-1	-3
Run length	Base	< Base	~Base	~Base
SHP steam production, kg/hr	69,890	65,796	69,632	63,824
% of base	100.0	94.1	99.6	91.3
Flue gas rate, kg/hr	131,756	115,384	127,250	113,711
CO_2 emission, kg/hr	8,393	0	0	0
% of base	100	0	0	0

Table 4 shows the performance of a single 200 KTA ethane cracking heater and various options that can be considered with 100% hydrogen firing. In a typical ethane cracker, the plant fuel gas contains 70-85 mol% hydrogen, with the balance mostly methane. In order to satisfy the overall fuel demand of the heaters, some makeup fuel is usually required, which is often methane rich (e.g. natural gas). To get to 100% hydrogen firing, the supplemental methane-rich fuel gas is considered to be replaced by imported pure hydrogen and the hydrogen produced by the plant is purified such that the resulting fuel composition is essentially 100% hydrogen. Moving to 100% hydrogen firing while maintaining the normal excess air therefore requires a hydrogen import of approximately 5.4-6.1 t/h for a world scale 1000 KTA ethane cracker in order to achieve (essentially) zero CO_2 emissions from the heaters.

For the same convection section design, increasing excess air from 10% to 17% results in almost identical performance with 100% hydrogen firing when comparing to the base case. For a grassroots heater, the convection section could be modified to use 10% excess air and achieve the same desired crossover temperature, but with reduced firing / hydrogen import and with less SHP steam production, as shown in the fourth column in the table.

Table 5. Ethane cracking heater performance with increased ammonia fuel.

	Base (Normal plant fuel)	**Partial NH_3 firing**
HC feed rate, kg/hr	46,902	46,902
Fuel composition, mol%		
H_2	82.7	59.5
CH_4	16.8	0
NH_3	**0**	**40**
Other	0.5	0.5
% Excess air	10	10
Fired duty, MMkcal/hr	86.8	91.2
Fuel flow rate, kg/hr	4,847	11,266
Imported NH_3, kg/hr	0	9,420
Overall thermal efficiency, %	93.9	93.3
Crossover temperature, °C	Base	+13
Run length	Base	> Base
SHP steam production, kg/hr	68,890	75,144
% of base	100	103
Flue gas rate, kg/hr	131,756	142,546
CO_2 emission, kg/hr	8,393	0
% of base	100	0

Table 5 shows the performance of a single 200 KTA ethane heater with partial ammonia firing. The base case is the same as in Table 4. Rather than replacing makeup fuel with hydrogen, for this scenario the makeup fuel is ammonia to achieve an overall fuel balance. This results in a fuel composition of approximately 60 mol% H_2 / 40 mol% NH_3. At this condition, the heater performance is almost identical to the base case but with a slight increase in fired duty, an increase in SHP steam production and a slight increase in run length. Extrapolating to a world-scale 1000 KTA plant, the required ammonia to achieve essentially zero CO_2 emissions from the heaters is approximately 47 t/h.

Table 6. Naphtha cracking heater performance with partial and full hydrogen fuel.

	Base (Normal plant fuel)	**60% H_2**	**89% H_2**	**100% H_2**
HC feed rate, kg/hr	86,433	86,433	86,433	86,433
Fuel composition, mol%				
H_2	**18**	**60**	**89**	**100**
CH_4	81	39	10	0
Other	1	1	1	0
% Excess air	10	14	21	25
Fired duty, MMkcal/hr	138.6	138.6	138.6	137.8
Fuel flow rate, kg/hr	11,292	9,489	6,797	4,803
Imported H_2, kg/hr	0	1,190	3,058	4,501
Overall thermal efficiency, %	94.0	94.0	94.0	94.1
Crossover temperature, °C	Base	+1	+0	-3
Run length	Base	~Base	~Base	~Base
SHP steam production, kg/hr	98,198	98,197	98,020	97,053
% of base	100	100	100	99
Flue gas rate, kg/hr	233,486	229,781	223,546	214,620
CO_2 emission, kg/hr	30,143	22,013	9,277	0
% of base	100	73	31	0

For naphtha crackers, the plant fuel gas contains much less hydrogen so a substantial amount of hydrogen import would be required to achieve a significant level of CO_2 reduction. Table 6 shows the performance of a single 200 KTA naphtha cracking heater. As the amount of hydrogen in the fuel is increased, excess air must be increased as high as 25% to compensate for the associated large change in flue gas rate and to achieve the same performance in terms of heater run length and SHP steam production. Such a large shift in hydrogen content of the fuel requires an enormous amount of imported pure hydrogen (approximately 23 t/h for a 1000 KTA plant) which may not be practical, and thus incremental cases are shown at 60% hydrogen and 89%.

As discussed above for ethane cracking, for a grassroots naphtha heater design the convection section could be modified to use 10% excess air and achieve the same desired crossover temperature, but with reduced firing / hydrogen import and with less SHP steam production.

Table 7. Naphtha cracking heater performance with partial and full ammonia fuel.

	Base (Normal plant fuel)	**Partial NH_3 firing**	**Full NH_3 firing**
HC feed rate, kg/hr	86,433	86,433	86,433
Fuel composition mol%			
H_2	18	18	0
CH_4	81	41	0
NH_3	**0**	**40**	**100**
Other	1	1	0
% Excess air	10	10	10
Fired duty, MMkcal/hr	138.6	140.4	146.8
Fuel flow rate, kg/hr	11,292	16,755	33,025
Imported NH_3, kg/hr	0	8,380	33,025
Overall thermal efficiency, %	94.0	93.7	93.7
Crossover temperature, °C	Base	+11	+39
Run length	Base	≥Base	≥Base
SHP steam production, kg/hr	98,198	100,171	106,672
% of base	100	102	109
Flue gas rate, kg/hr	233,486	238,904	257,450
CO_2 emission, kg/hr	30,143	22,218	0
% of base	100	74	0

Table 7 shows the performance for a 200 KTA naphtha cracking heater if the methane is partially replaced with ammonia. With 40% ammonia in the fuel gas, the increased flue gas duty results in 2% higher steam production and higher crossover temperature/slightly longer run length, but with a slight reduction in thermal efficiency. For illustration purposes, a case is also given where the total fuel gas is replaced by ammonia to achieve essentially zero CO_2 emissions from the heaters. A detailed evaluation would be required for a specific cracking heater to ensure that the crossover temperature does not exceed the recommended range for the given feedstock, or else some convection surface modifications might be required.

<u>Other considerations</u>

The type of fuel in the firebox does not affect cracking reactions in the radiant tubes, so the amount of methane generated by the process does not change. If hydrogen or ammonia is used to replace the methane in the fuel gas that is produced from the steam cracker, then that methane must be exported or used elsewhere in the petrochemical complex. Exporting the methane to be burned as fuel gas in another process unit defeats the purpose of reducing overall carbon emissions, so ethylene producers must find another destination for this methane. One option is to send it to a reformer with CO_2 capture to produce blue hydrogen, but the captured CO_2 must also have a destination. Another option is to use oxidative coupling of methane (OCM) technology to convert the methane into valuable hydrocarbons like ethylene and propylene, which is discussed in a separate paper [13].

Combustion air preheating can be combined with hydrogen or ammonia firing to reduce the fired duty requirement by up to about 30%. This will minimize the quantity of imported zero--carbon fuel required to make the transition to zero CO_2 emission cracking heaters. However, the associated reduction in SHP steam production must be studied along with the recovery section compressor drivers.

NO_x reduction

As discussed, increasing hydrogen and ammonia (up to a certain point) will increase NO_x production. NO_x emissions limits are typically defined on a dry basis, and since the flue gas produced from hydrogen or ammonia has a higher fraction of water vapor than flue gas produced from hydrocarbons, when adding these fuels the NO_x *concentration* may actually increase more than the absolute increase in *mass* of NO_x. Depending on the fuel selection and environmental permit, it may be necessary to use SCR to meet emission requirements. If the cracking heater has already been designed with SCR or space has been left for future SCR addition, this will not present any difficulty when retrofitting an existing cracking heater for hydrogen firing. For existing heaters with limited space, compact SCR designs can be considered. New heater designs with natural gas combustion can be made hydrogen-friendly by leaving space for future SCR addition. Even limited ammonia firing will require SCR addition to meet tolerable NO_x emissions. For high ammonia concentrations in the fuel, further burner development is required to produce NO_x levels that can be handled by SCR.

Control system and instrumentation

The fuel supply and control system must be carefully evaluated before increasing the hydrogen content or using ammonia as fuel gas. Since the heating value of these fuels is so different than methane's, the control systems should measure the heating value of the fuel mixture (which is typically done with a Wobbe meter) and automatically manipulate the fuel flow to maintain a steady firing rate, i.e., do not keep the pressure constant. The system must be operable under both startup, normal and upset scenarios. If an upset scenario is plausible where the hydrogen rate could be suddenly reduced then the control system may need to provide automated control to increase the combustion air to avoid incomplete combustion and a fuel rich environment. Due to the increased molar flowrate of hydrogen or ammonia fuel compared to methane for a given duty, the fuel gas pressure needs to increase for a given burner design and orifice size. Therefore, the fuel gas piping system and control valves need to be evaluated for the anticipated range of composition. Depending on the amount of ammonia in the mixture, additional preheating may also be required to ensure no condensation at the burner inlet.

Conclusion

Ethylene producers are looking at various ways to reduce CO_2 from their plants. Some CO_2 reduction strategies that involve a reduction or elimination of fuel firing unfortunately also result in a reduction in SHP steam production. This shortfall in energy must be compensated for by green electricity to drive the major compressors, along with the CAPEX associated with converting drivers to electric. By substituting some or all of the hydrocarbon fuel with fuels that contain zero carbon, the CO_2 emissions can be reduced or eliminated from the cracking heaters while maintaining the original SHP steam production. This strategy has the benefit that only minor modifications are required to the cracking heaters and no modifications are required to the recovery section compressor drivers. The feasibility of firing these fuels, however, depends on their availability and economics which are expected to improve in the future as blue and green production become more prevalent.

The two available zero-carbon fuels, hydrogen and ammonia, have very different combustion properties than typical methane-rich fuel gas, and therefore require special consideration on both the burner and convection section design. Burner designs have already been developed and proven for use in high hydrogen applications, and burner designs capable of operating on ammonia/hydrogen/hydrocarbon mixtures have successfully been tested. Further burner development is needed to mitigate the very high NO_x production at high ammonia concentrations in the fuel. The impact on convection section can be easily addressed by adjusting excess air or modifying the heating surface to achieve the desired performance.

References:

[1] IEA (2022), Hydrogen Supply, IEA, Paris https://www.iea.org/reports/hydrogen-supply, License: CC BY 4.0

[2] The Royal Society (2022), The role of hydrogen and ammonia in meeting the net zero challenge, See https://royalsociety.org/-/media/policy/projects/climate-change-science-solutions/climate-science-solutions-hydrogen-ammonia.pdf (accessed 1 Feb 2023)

[3] IEA (2019), The Future of Hydrogen, IEA, Paris https://www.iea.org/reports/the-future-of-hydrogen, License: CC BY 4.0

[4] IEA (2021), Global Hydrogen Review 2021, IEA, Paris https://www.iea.org/reports/global-hydrogen-review-2021, License: CC BY 4.0

[5] Z. Zhao et al. "An economic analysis of twenty light olefin production pathways" J. Energy Chem. (2021)

[6] Hemrick, James, "Refractory issues related to the use of Hydrogen as an Alternative Fuel". American Ceramic Society Bulletin, March 2022

[7] Palmer, G, "Volatilization of Refractory Silica in Hydrogen Water Vapour Gas Streams" Refractories WORLDFORUM 4 (2012)

[8] B. Metz et al (eds) "Contribution of Working Group III to the Fourth Assessment Report of the Intergovernmental Panel on Climate Change" Cambridge University Press 2007 section 7.4.3.1.

[9] Ren et al "Olefins from conventional and heavy feedstocks: Energy use in steam cracking and alternative processes" Energy, 2006, vol. 31, issue 4, 425-451

[10] Baukal, C., et al "High Hydrogen Fuels in Fired Heaters" Proceedings of the 2021 American Flame Research Committee fall meeting

[11] Londerville, S. et al "Ammonia For Industrial Combustion" Proceedings of the ASME 2022 International Mechanical Engineering Congress and Exposition IMECE2022-96499

[12] Zhang et al "Emission analysis of the CH4/NH3/air co-firing fuels in a model combustor" Fuel, 2021, vol 291, ISSN 0016-2361.

[13] Laurin, Marc "OCM & Steam Cracker Integration", AIChE paper number 155d, prepared for presentation at the 2023 Spring National Meeting, March 2023.

AIChE Paper Number: 27b

Blue Hydrogen Firing or Flue Gas Carbon Capture? Tailored Options for Steam Cracker CO_2 Elimination

Torben Hoefel

Product Manager Sustainable Olefin Technologies
Sustainable Hydrocarbon Solutions
Linde Engineering

Thomas Haselsteiner

Head of Process Engineering
Hydrogen and Synthesis Gas Plants
Linde Engineering

David Maher

Sr Business Development Manager - Technology
Americas Large Projects
Linde

Prepared for Presentation at the
2023 Spring Meeting and 19th Global Congress on Process Safety
Houston, Texas, March 13, 2022

Blue Hydrogen Firing or Flue Gas Carbon Capture? Tailored Options for Steam Cracker CO_2 Elimination

Torben Hoefel
Product Manager Sustainable Olefin Technologies
Sustainable Hydrocarbon Solutions
Linde Engineering

Thomas Haselsteiner
Head of Process Engineering
Hydrogen and Synthesis Gas Plants
Linde Engineering

David Maher
Sr Business Development Manager - Technology
Americas Large Projects
Linde

Abstract

Near-zero emission olefin plants can be designed today by applying approaches characterized by elimination/capture of the generated carbon dioxide resulting from combusting or converting carbon-containing fuels. In the context of steam crackers, commercially ready solutions are available for "post-combustion CO_2 capture" in the form of flue gas washing units, and "pre-combustion CO_2 capture" in the form of "blue" hydrogen firing.

This paper will give an introduction into these fundamentally different solution approaches for petrochemical applications. It is outlined how new or existing steam crackers can be equipped or retrofitted with flue gas carbon capture units and it is described how CO_2 emission sources from one or multiple petrochemical sites can be decarbonized by setting up a blue hydrogen plant and/or supply network. The importance of integrated solutions is described, highlighting required design changes to the cracking furnaces to accommodate hydrogen firing or flue gas carbon capture as well as possible design features for the cracker separation section.

Profound experience across multiple technology fields and industries, including petrochemical, hydrogen and CO_2 processing plants, is required to evaluate the wide range of decision criteria influencing the concept selection for given site-, project- and customer-specific constraints. This conference contribution introduces and comments the most important rationales for concept selection.

Finally, examples of decarbonizing world-scale petrochemical sites with different individual constrains will be presented. The paper will highlight how a holistic approach by an experienced licensor can support the solution definition in such a multi-criteria decision process.

1. Introduction

Being a highly energy-intensive sector, the petrochemical industry accounts for about 5% of global CO_2 emissions [1]. Steam cracker plants are the largest greenhouse gas emission source within this sector and are estimated to account for approximately 25% of the overall greenhouse gas (GHG) emissions in the European chemical industry [2].

Including direct and indirect emissions, an average European steam cracker generates about 0.9-1 ton of CO_2 per ton of generated high value chemicals. This means that an average European large-scale steam cracker generating >1,200 kta high value chemicals may likewise generate about 1,200 kta of CO_2 emissions. The major share of these emissions (about 90%) is contributed to the direct emissions originating from the furnaces and auxiliary boiler systems.

One possible approach for the reduction of these emissions to near-zero is carbon capture, utilization and/or storage (CCUS). CCUS is characterized by elimination/capture of the generated carbon dioxide resulting from combusting or converting carbon-containing fuels. Such approaches are typically distinguished into three possibilities: "Post-combustion CO_2 capture", "Pre-combustion CO_2 capture" and "Oxy-fuel combustion CO_2 capture" [3,4,5].

While "Oxy-fuel combustion CO_2 capture" is not yet commercially available for cracking furnaces, commercially ready solutions can today be implemented for "Post-combustion CO_2 capture" in the form of flue gas washing units, and "Pre-combustion CO_2 capture" in the form of "blue" hydrogen firing. In the context of steam cracking plants these two solutions may possess fundamentally different characteristics. The two different approaches are described on a general level within the next chapters and various decision criteria for the solution finding process are further outlined.

2. Hydrogen generation from cracker methane fraction

The first option of steam cracker direct CO_2 emission elimination (by carbon capture) is often referred to as "pre-combustion CO_2 capture", which involves the "decarbonization" of a carbonaceous fuel. In the most relevant scenario, a methane rich stream, which could originate from the cracker tail gas and might be supplemented with a natural gas import, is sent to a unit, where it is converted to hydrogen and CO_2. A generic flow diagram of such a "blue hydrogen" plant is shown in **Figure 1**. The methane stream(s) are first sent to a tail gas processing unit. If the methane stream still contains increased amounts of hydrogen (as it might be for example the case for ethane cracker tail gases), a hydrogen separation could be carried out. Furthermore, if required, methane is compressed to increased pressures and purified from contaminants (if any). The central step is then the methane conversion to synthesis gas (H_2, CO, CO_2), which can be carried out by different reaction systems as further outlined below. Some reaction systems like autothermal reformer (ATR) or partial oxidizer (POX) will require supplemental oxygen. Other systems like steam methane reformer (SMR) require the supply of significant amounts of heat from a combustion chamber. In any case the first reaction step is followed by a CO shift, where the CO shall be converted to CO_2. After the two reaction steps, some amounts of methane and/or CO typically remain in the synthesis gas.

In the remaining sections of the plant, hydrogen, CO_2 and water are basically separated and brought to the required specifications. Again, depending on the specific targets and boundary constraints, the process design could drastically change. For example, the CO_2 removal could be ideally carried out within the synthesis gas at elevated pressure and CO_2 concentration, as indicated in **Figure 1**. In some setups, where a SMR is utilized in the reaction step, flue gas carbon capture could be applied as well (as further described below).

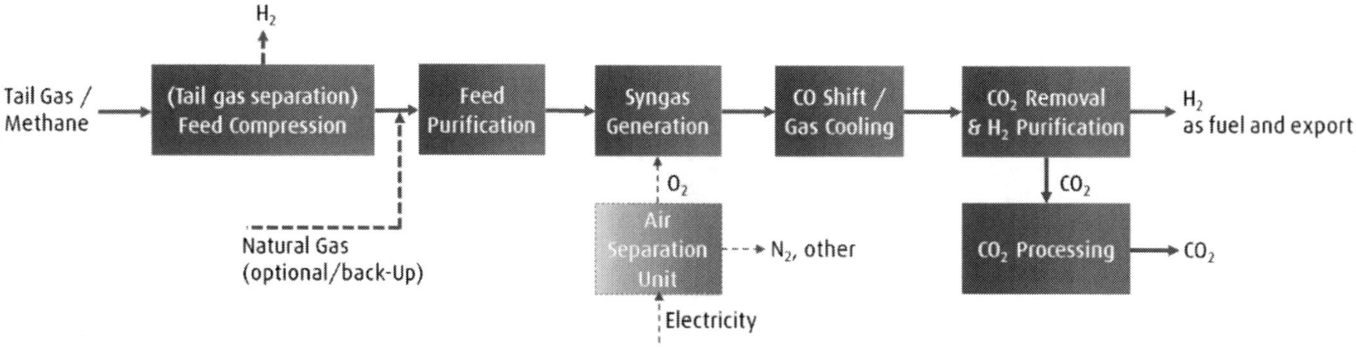

Figure 1: *Generic process flow scheme for a "blue" hydrogen plant*

As outlined above, the process design of such blue hydrogen plant can drastically change depending on various boundary constraints. Some exemplary technology elements are further described in the following examples.

In a first example shown in **Figure 2**, the tail gas processing unit design is highlighted, which might be required for hydrogen-rich cracker tail gases. Typically, such cracker tail gases are discharged at fuel gas pressure conditions and it may be the task of the tail gas processing unit to generate separate hydrogen and methane fractions, ideally at high purities and elevated pressure(s) (if required). Depending on the design, such separation could consume large amounts of power for compression steps. One possible solution is indicated in **Figure 2** on the left-hand side, where a separation scheme utilizing pressure swing adsorption (PSA) in combination with membrane technology is utilized. Using a combination of both technologies, a hydrogen yield of 98 % can be achieved. Such a unit can be furthermore characterized as a "bolt-on" design, which is independently operated from the upstream cracker plant. As a result, it can be later added to a plant and it can also be designed as a central unit for processing of more than one tail gases.

An alternative design is shown in **Figure 2** on the right-hand side, where the actual hydrogen and methane separation is carried out within the cracker itself, leveraging on the cold separation equipment anyway required for ethylene purification. The task of hydrogen/methane separation can be conducted in a more CAPEX and OPEX optimized design and even higher yields close to 99-100 % are reached. However, such unit is integrated into the cracker design and might not be suitable under brownfield conditions.

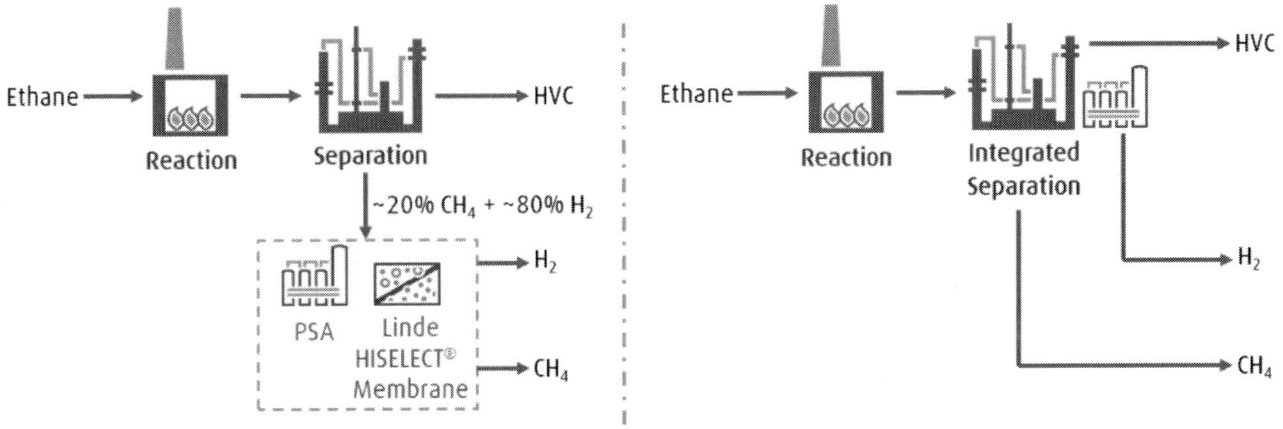

Figure 2: *Technology example 1: H_2/CH_4 separation (for ethane cracker). The separation could be carried out in a separate and fully electrified unit ("Bolt-on" design, left) or integrated into the cracker separation scheme (right).*

The second example shown in **Figure 3** evaluates on selection criteria relevant for the central reaction step. Conventional hydrogen generation from methane is frequently carried out with SMR technology, as shown in the upper part of **Figure 3**. The heat duty required for the reaction is provided by sending all carbon-containing molecules from the product gas to the combustion chamber of the SMR. The process scheme has a simple hydrogen purification step and does not require any conditioning of the oxidant for the combustion. If a decarbonization in the range of 95% is however targeted, the most straight-forward solution in these brown field conditions is the application of flue gas carbon capture, although the CO_2 must be captured at low pressure conditions here.

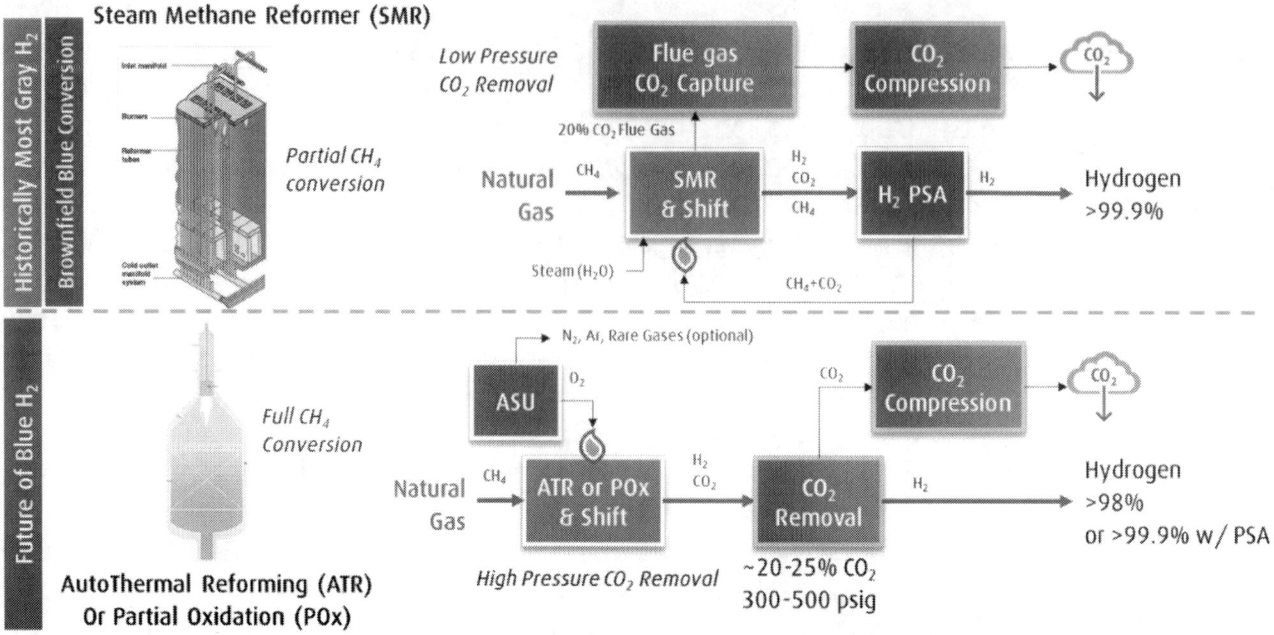

Figure 3: *Technology example 2: Reforming step and selection criteria. A first scenario (upper section) shows an existing steam methane reformer (SMR) generating grey hydrogen and it's conversion to a decarbonized unit using flue gas carbon capture. For flue gas carbon capture, amine wash units are the most familiar solution, but other alternatives like Linde's PSA based HISORP® CC solution are commercially available. SMR technology for blue hydrogen generation might also be a consideration for green field under certain boundary conditions (not shown here). A second scenario shows a greenfield solution of a decarbonized hydrogen plant using an autothermal reformer (ATR) or partial oxidation (POX), converting and shifting nearly all carbon to CO_2. The CO_2 is removed from the synthesis gas at elevated concentration and pressure, again either using amine wash or PSA based technology.*

Under certain boundary conditions SMR technology for blue hydrogen generation might also be a consideration for green field, for example by designing a hydrogen fired SMR. However, the typically reduced methane conversion within the SMR compared to an ATR proves to be a difficult aspect within this concept, when it comes to decarbonization targets around 95%. Either the application of flue gas carbon capture or the recycle of methane and/or CO to upstream of the SMR would be required.

Alternatively, for newly built large-scale blue hydrogen plants with a decarbonization target of 95%, ATR technology is applied, as shown in the lower part of **Figure 3**. Using pure oxygen from an air separation unit, the main heat generation is carried out within the reactor itself. Due to the high conversion of CH_4 and CO, about 95 % of the carbon can be shifted to CO_2, which is then fully removed within an efficient and compact CO_2 removal unit in the synthesis gas.

A third technology example shown in **Figure 4** is indicating different technology options for the removal of CO_2 from synthesis gas. A rather conventional technology is the utilization of amine wash units, as shown on the left-hand side. Such technology requires thermal energy for the CO_2 removal. If the boundary constraints of the project favor however the export of steam from the blue hydrogen plant, such CO_2 removal can also be designed by based on a PSA scheme (**Figure 4**, right-hand side), leveraging for example on low-carbon electrical power.

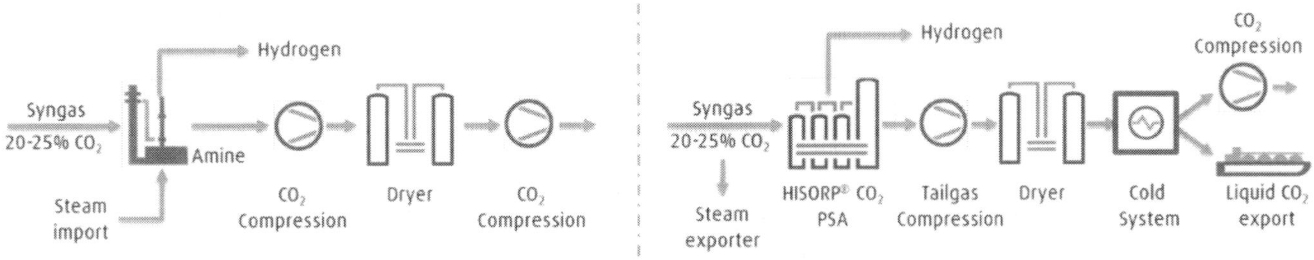

Figure 4: *Technology example 3: CO_2 removal from syngas. A conventional solution is shown on the left-hand side, where an amine wash unit removes CO_2 using thermal heat, which is available at surplus. An alternative solution is shown on the right-hand side, where CO_2 is removed by a PSA. Surplus thermal heat can be exported from the synthesis gas plant. The PSA tail gas is compressed, dried, and sent to a liquefaction, where remaining light fractions are separated and recycled. Depending on the final specification of the CO_2, it can be either further compressed or sent to BL as liquid.*

In principle, both CO_2 removal methods indicated in **Figure 4** are even suitable for the capture of CO_2 from flue gas, where CO_2 partial pressure is significantly lower and other difficult conditions may apply. However, special amines are utilized in case of amine wash units, as described in the next chapter. For somewhat increased CO_2 concentrations in flue gas (like for example applicable for SMR firing as shown and explained in **Figure 3** as well as certain oxyfuel processes in other industry applications), the PSA based CO_2 recovery is an alternative, which should be carefully assessed in early project stages.

3. Hydrogen fuel switching of cracking furnaces

In the context of steam cracker decarbonization, hydrogen fuel switching of course requires design changes to the cracking furnaces' fueling system. Although the heat value of the generated hydrogen might be sufficient for the endothermal cracking reaction under most ideal conditions (ethane cracking), it will not be sufficient considering an actual furnace design, where only a fraction of the introduced heat can be transferred to the reaction itself (e.g. about one third). Most of the remaining heat is still utilized e.g. in the form of feed preheating and/or steam generation, but as consequence, import hydrogen fuel will be required. Such import can for example be delivered by "pre-combustion CO_2 capture" in the form of blue hydrogen plants as described in the previous chapter. As any import hydrogen is however expected to contribute significantly to the operating costs, it should be the task to minimize the fuel demand of the hydrogen-fired furnaces by deprioritizing on secondary tasks like steam generation. The reduction of such secondary heat carrier can be efficiently compensated for example by the electrification of large machinery in the separation section [6].

Summarizing the above considerations, an ideal setup of a hydrogen fired furnace also includes measures like air preheating, which results in significantly reduced fuel demand at the expense of reduced steam generation. The exemplary results shown in **Figure 5** give an idea on how such a furnace could perform.

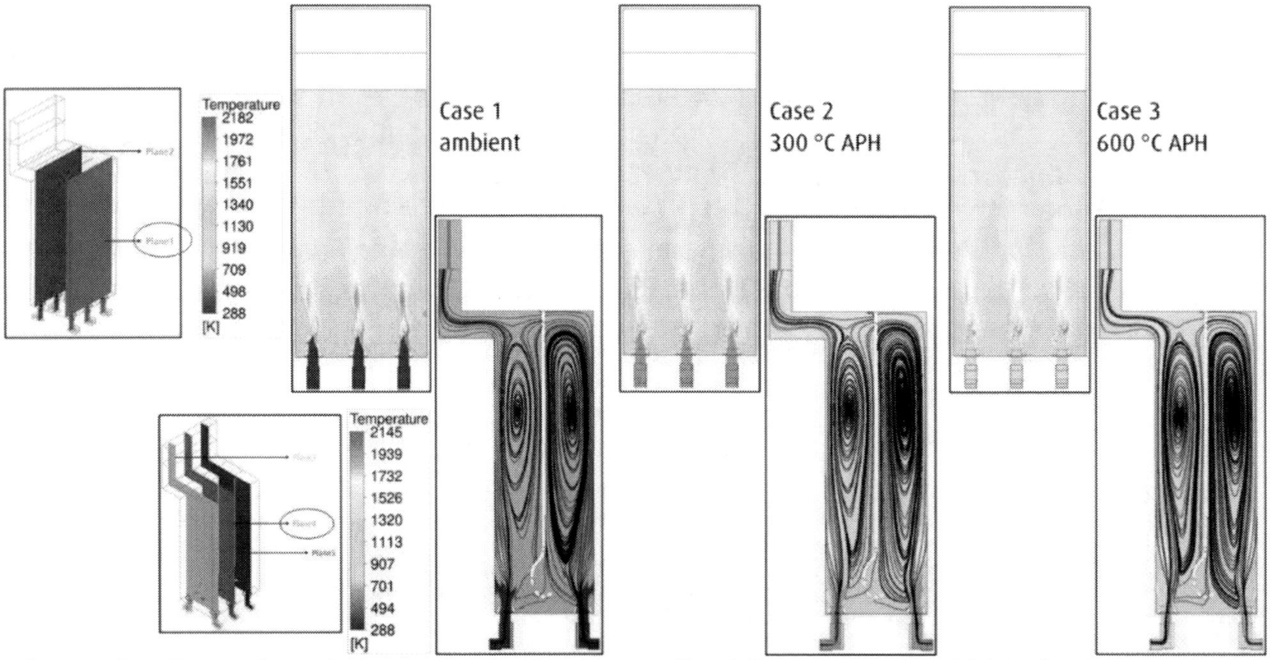

Figure 5: *Exemplary CFD results for hydrogen fired furnaces with 100% hydrogen fuel. Upper section shows a side view of a furnace chamber and lower section shows a cut view from the front of a single radiation chamber. Three cases are shown. Case 1, left, 100% H_2 firing, without air preheat. Case 2, middle 100% H_2 firing, with 300 °C preheated combustion air. Case 3, right, 100% H_2 firing, with 600 °C preheated combustion air. A diffusion type burner was modelled, which was designed for operation with preheated air.*

The CFD results, which incorporated appropriate diffusion type burner designs, show that stable operation conditions are achieved with comparatively moderate flame temperatures for all cases. Even for very high air preheating temperatures, acceptable maximum flame temperatures are observed. Furthermore, significantly increased fuel efficiencies can be expected. The "fuel efficiency" shall be described herein as heat absorbed in the radiant section (reaction + temperature increase) divided by the introduced fired duty. For a fully hydrogen fired furnace (case 1), an already high fuel efficiency of for example 45.5 % can be expected. If the heat in the flue gas is now "recycled" to the radiant section, for example by flue gas air preheating, and an air preheating temperature of 300 °C is targeted (case 2), the fuel efficiency rises to approximately 55.4 %, which represents a fuel saving of 19 % compared to the already efficient case 1. Further maximization of the air preheating temperature (e.g. 600 °C, case 3) results in even higher fuel efficiency of 66.6% and accordingly a fuel consumption reduced by 33% compared to case 1.

Considering the exemplary results shown in **Figure 5** and several further design aspects not described herein in detail, it is concluded that 100% hydrogen firing can be implemented in new designs today without major technical obstacles, with or without (high) air preheating.

For existing furnace designs however, switching to hydrogen fuel might be more challenging, which very much depends on the given constraints. Various aspects including the existing burner and burning chamber design, the heat balance of the furnace convection section, as well as requirements regarding NOx emissions must be carefully taken into account. In some instances, such fuel switching might be possible with reasonable effort, like for example for rather modern gas cracking furnaces, which already fire hydrogen-rich fuel and already provide space for flue gas denitrification. For other furnace designs, such fuel switching might be associated with unreasonable effort.

4. Flue gas CO_2 capture and integration into steam cracker

The second option of steam cracker direct CO_2 emission elimination (by carbon capture) is flue gas carbon capture, also referred to as "post-combustion CO_2 capture". The most mature and commercially available solution for flue gas carbon capture at steam crackers utilizes chemical absorption of CO_2 [7]. As shown in **Figure 6** at the example of the BASF OASE® blue flue gas carbon capture technology, the respective flue gases are collected and cooled down to ambient conditions, for example by a direct contact cooler. If required, a booster fan is installed downstream of the cooling unit. The flue gas is further routed through a column with circulating liquid amine solvent, which absorbs the CO_2 molecules. The loaded solvent is sent to a regeneration column, where a heating medium such as low-pressure steam enables the release of highly concentrated CO_2 from the solvent. The CO_2 is then sent to a processing unit for further upgrade (e.g. drying, compression, liquefaction) for storage or utilization. The performance of such unit, including key characteristics like thermal demand and solvent replacement costs, very much depends on the properties of the applied solvent as well as the detail process design. Typically these systems can reach a carbon capture rate of 95 %.

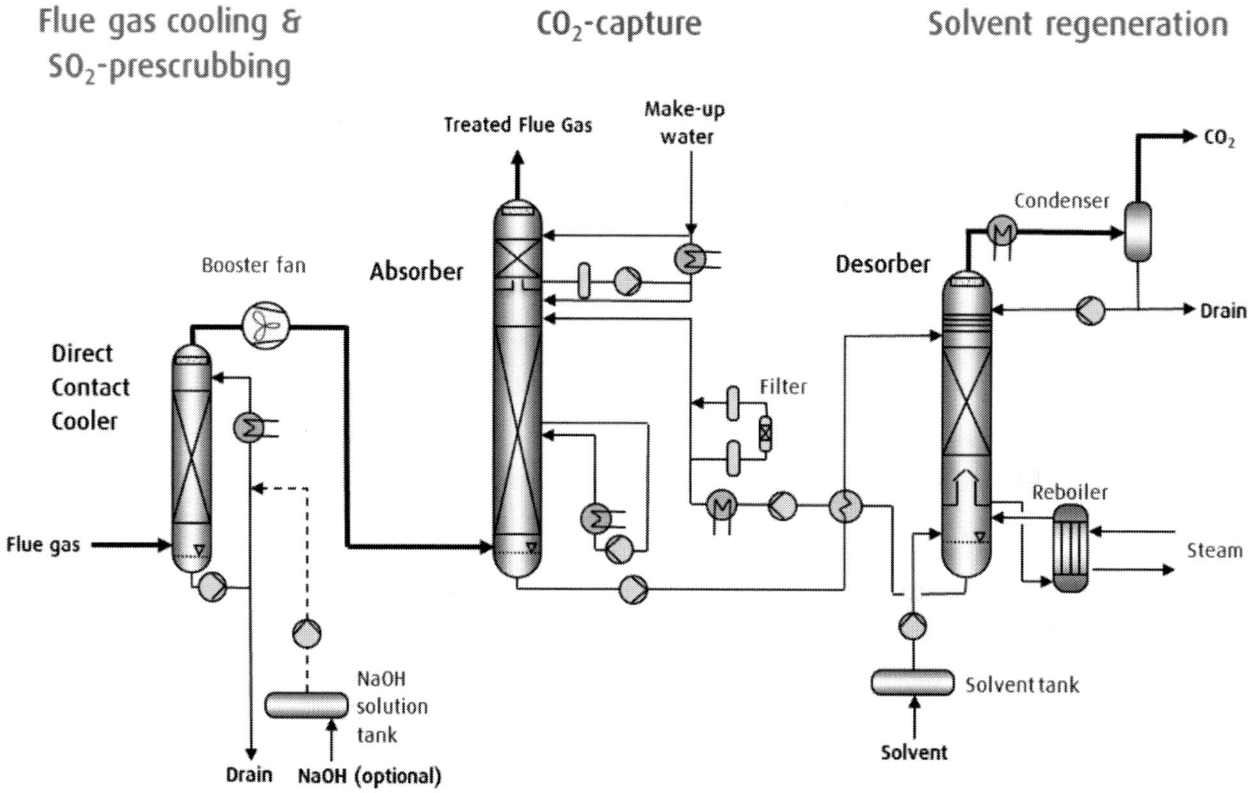

Figure 6: *Simplified flowsheet of a BASF OASE® blue flue gas carbon capture plant.*

Generally, the flue gas carbon capture unit can be considered as an elegant "add-on" solution with relatively low impact to the steam cracker and its fired capital equipment. It is however vital that sufficient plot space is available in close proximity to the primary CO_2 emitters to avoid long-distance ducting for the volumetrically large flue gas flows. Furthermore, a sound engineering design regarding the required flue gas ducting is of high relevance, ensuring a careful consideration of the mechanical and static design, as well as operation and safety aspects, especially in brownfield conditions [7].

Moreover, the thermal energy demand for the capture unit itself may cause "extra" CO_2 generation (e.g. from generating "on-purpose" steam by additional auxiliary boilers), which may result in additional CO_2 sent to storage or utilization in the range of 10 to even 30 % of the total CO_2. This unfortunate "windup" effect in the CO_2 generation can however be mitigated or even avoided if the thermal energy supply is smartly integrated into the cracker steam system [7]. Not considering such energy-integration opportunities may cause a misjudging in early phase studies, as these integration opportunities may have a major impact on the economic assessment.

5. Decarbonization scenarios for petrochemical sites

The different technology characteristics described in the previous chapters and their suitability to actual project boundary constraints are exemplarily described at two generic cases.

In a first example as shown in **Figure 7** a chemical complex is considered. A possible conventional setup is shown in the top. An existing steam cracker converts ethane to ethylene and generates a hydrogen-rich tail gas, which is used for firing in the cracking furnaces, resulting in a dilute CO_2 stream with concentrations around approximately 4-6 mol% in the wet flue gas. A new cracker facility is planned also cracking ethane and generating hydrogen-rich tail gas. The complex further requires the generation of electrical power inside battery limits, which shall be implemented by a gas turbine, generating oxygen-rich flue gas also with dilute CO_2 concentrations. Furthermore, hydrogen is required on-site and/or may be a valuable export for existing OSBL infrastructure. An on-purpose hydrogen plant converting methane could be considered for hydrogen generation, also generating flue gas containing CO_2. The described CO_2 emitters might be scattered, and restricted plot conditions might apply around some of these emitters.

The described conditions are likely unsuitable for decarbonization by flue gas carbon capture, due to the mainly dilute and scattered CO_2 emission sources, requiring several large flue gas capture units, which might not fit into the existing plot conditions. Instead, a large hydrogen extraction and generation unit could be a suitable solution, as shown in in the possible decarbonized setup at the bottom of **Figure 7**. In the shown example, any hydrogen-rich streams are collected centrally for a hydrogen extraction. The remaining methane and natural gas import – if required – are sent to a large hydrogen generation unit, which could be composed of an autothermal reformer. The setup only requires a single CO_2 extraction at elevated pressure and concentration (within the synthesis gas) inside the hydrogen generation unit. The generated hydrogen is used for the export requirement as well as any fuel consumer on-site. The new cracker can be designed greenfield for full hydrogen firing and the existing cracker can be likely retrofitted rather easily to full hydrogen firing, as high hydrogen concentrations were already fired in the existing design. Also, efficient gas turbines will likely become available in the near future, which are suitable for operation with 100% hydrogen, generating no direct CO_2 emissions. The hydrogen generation unit is an asset for itself and can pivot to next generation uses, independent of steam cracker development and its requirement with respect to fueling. Further, if necessary and/or preferred, such central hydrogen unit could be built, owned and operated by a 3rd party.

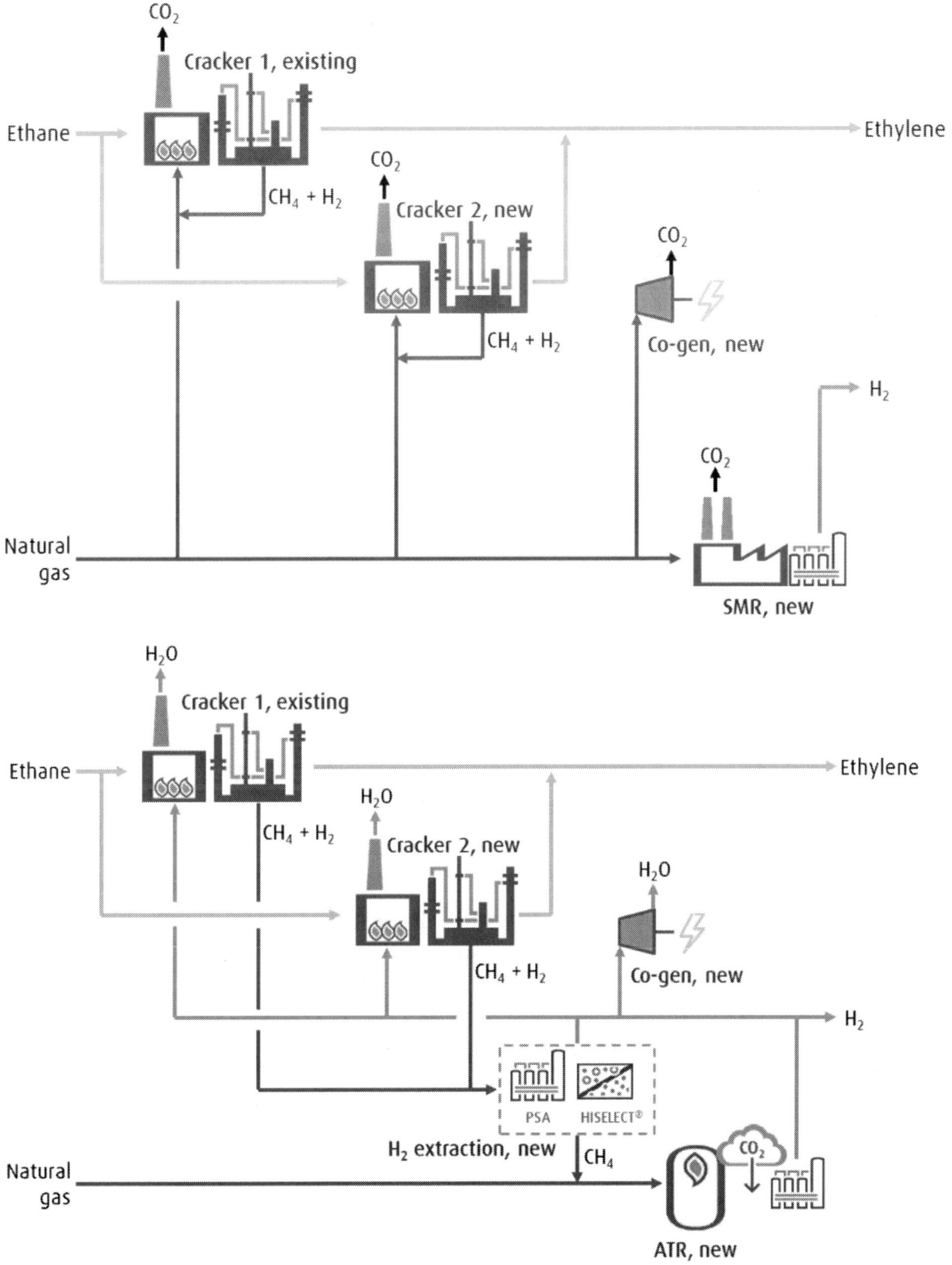

Figure 7: *A generic example suitable for hydrogen fuel switching: Conventional setup (top) and possible decarbonized setup (bottom).*

A second example is shown in **Figure 8**, where the direct CO_2 emissions of a single large-scale liquid cracker generating ethylene, propylene and other high value chemicals (HVC) shall be eliminated. Furnaces and possible auxiliary boiler systems are co-located and these systems are fired with methane-rich fuel, generating flue gas with CO_2 concentrations in the wet flue gas around 7-9 mol%. It is further assumed that sufficient plot is available in proximity of the emitters.

The described conditions are likely unsuitable for decarbonization by hydrogen fueling, as the existing assets are designed for CH_4-rich firing and a modification to 100% hydrogen fueling is likely associated with major efforts. Despite its comparably large size compared to a syngas amine wash unit, a flue gas amine wash unit appears to be a reasonable choice in this scenario, as it would not require the several other units associated with hydrogen generation, including feed compression/purification as well as synthesis gas generation and shift. Instead, the flue gas capture unit also benefits from the economy of scale in this scenario. Moreover, comparably simple energy integration opportunities can be identified for flue gas capture units and steam crackers, mitigating or even avoiding the extra thermal energy demand of a flue gas capture unit [7].

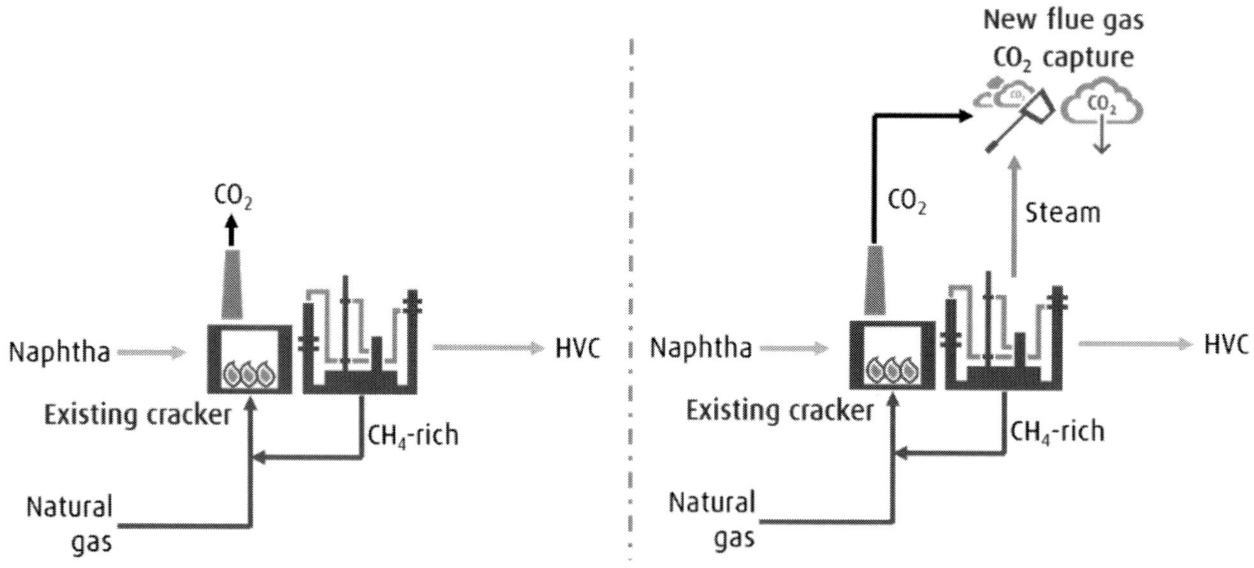

Figure 8: *A generic example suitable for flue gas CO_2 capture: Conventional setup (left) and possible decarbonized setup (right).*

6. Summary

The generic scenarios described in the previous chapters are inspired by real ongoing projects and shall highlight how the solution finding process can be influenced by site- and client-specific as well as economic constraints. In the following table, some key features and key limitations of each technical solution are again summarized.

Table 1: *Some selected technology features and limitations regarding hydrogen fuel switching and flue gas carbon capture in the context of steam cracker decarbonization.*

Hydrogen Fuel Switching	Flue gas CO_2 capture
↑ Most suitable for decarbonizing regionally distributed assets, which may have dilute CO_2 flue gas streams and are already capable of operating with near 100% H_2. ↑ Additional (future) renewable H_2 sources can be faded in ↑ Surplus and/or multiple methane sources can be collected and processed centrally ↓ Dedicated large space required ↓ Loss of energy in the reforming of methane fraction into H_2 and CO_2, resulting in additional CO_2 generation. ↓ Existing combustion processes/furnaces must be suitable or retrofitted for hydrogen firing. In case of retrofitting temporarily production loss of individual units to be expected.	↑ Most suitable for large emission sources in proximity to each other with decently high CO_2 concentrations ↑ Heat demand (LPS) can be integrated into steam cracker, buffering the additional energy demand ↑ Implementation of cracker ISBL modifications in regular plant turnaround feasible ↓ Space requirement in proximity of the emission source. ↓ Low concentration and low pressure of CO_2 to be captured make the actual capture process CAPEX and OPEX intensive (if unintegrated) ↓ Additional CO_2 produced due to thermal energy demand leading to higher fuel import (if unintegrated). ↓ Existing CO_2 emission sources need to be retrofitted with flue gas ducting.

In summary, hydrogen fuel switching is most suitable for decarbonizing regionally distributed assets already compatible with hydrogen firing, whereas flue gas CO_2 capture is most suitable for large emission sources in proximity to each other and with decently high CO_2 concentrations in the flue gas.

However, a multi-criteria decision process is required to identify the most promising solution(s) and the following aspects should be considered already in the initial project phase:

- Individual properties of CO_2 emitters, like volume and CO_2 concentration
- Required carbon capture ratio / decarbonization ratio
- Emitter distribution and access, plot availability
- Impact on & suitability to existing assets as well as integration opportunities
- Possible Leverage on existing adjacent infrastructure (like existing H_2 infrastructure)
- Aspects on economy of scale
- Capital allocation and 3rd party involvement for operation of CO_2 elimination units
- Impacts on the operation and dynamics as well as the reliability of the olefin plant
- General flexibility towards combination with other CO_2 reduction methods and various possible long-term strategies intended for the given site
- Possibility for staged implementation scenarios

Especially integration aspects can be easily overlooked in early project phases, which might lead to misjudgment of certain routes in early (economic) assessments. Selection of suitable technology elements and integration of petrochemical units and H_2/CO_2 processing facilities play a vital role for the selection of the most promising solution(s).

7. References

1. Lange, J. P., "Towards circular carbo-chemicals – the metamorphosis of petrochemicals", *Energy & Environmental Science*, 2021
2. European Commission, JRC Science for Policy Report, "Energy efficiency and GHG emissions: Prospective scenarios for the Chemical and Petrochemical Industry", 2017, available under https://publications.jrc.ec.europa.eu/repository/bitstream/JRC105767/kj-na-28471-enn.pdf, p. 14, Table 3.
3. Santos, M. P. S. and D. P. Hanak, "Carbon capture for decarbonisation of energy-intensive industries: a comparative review of techno-economic feasibility of solid looping cycles", *Frontiers of Chemical Science and Engineering*, 2022
4. Jha, A., V. Ravuru, M. Yadav, S. Mandal and A. K. Das, "A critical analysis of CO_2 capture technologies", *Hydrocarbon Processing*, 2021
5. Buehler, J., "Overview of decarbonization pathways for the oil and gas and petrochemical industries—Part 2", *Hydrocarbon Processing*, 2022
6. Reyneke, H., M. Whitney, M. Kamann and P. Brehm, "Full machine electrification of a grass roots ethylene plant", 2023 Spring Meeting and 19th Global Congress on Process Safety, Houston, Texas, Paper Number 155e, 2023
7. Hoefel, T., A. Laugwitz and M. Anas, "Practical flue gas carbon capture at steam crackers", 2023 Spring Meeting and 19th Global Congress on Process Safety, Houston, Texas, Paper Number 155f, 2023

Gap in pagination due to formatting issues.

Pages 547-571

AICHE PAGE NUMBER: 27C

BENEFITS AND OPTIMIZATIONS OF OXYFUEL COMBUSTION

Galip H. Guvelioglu

Director of Gas Processing

Lummus Technology

Melanie O'Sullivan

Technology Manager for Carbon Capture & Energy Storage

Lummus Technology

Prepared for Presentation at the Year (2023) Spring National Meeting

Houston, TX, March 13, 2023

BENEFITS AND OPTIMIZATIONS OF OXYFUEL COMBUSTION

Galip H. Guvelioglu

Director of Gas Processing

Lummus Technology

Melanie O'Sullivan

Technology Manager for Carbon Capture & Energy Storage

Lummus Technology

Abstract:

Oxyfuel combustion provides an attractive alternative for decarbonization of fired process heaters used in steam cracking units. In Oxyfuel, fuel is combusted with oxygen from an air separation unit rather than combustion air to significantly reduce inert nitrogen in the resultant flue gas. In Oxyfuel combustion, a portion of the flue gas is recycled back to the fired heater to reduce the oxygen concentration and obtain combustion and heat transfer characteristics like conventional combustion air to minimize the impact on furnace design. The significant reduction of nitrogen in flue gas from oxyfuel increases the CO_2 concentration approximately ten times and reduces the volume of flue gas product by an order of magnitude. The higher concentration of CO_2 in the flue gas requires purification to remove water, excess oxygen required for efficient combustion, and trace inerts nitrogen and argon with a CO_2 purification unit. Within the CO_2 purification unit, either gaseous CO_2 or liquid CO_2 can be produced. With liquid CO_2, pumping to supercritical pressures can be achieved with lower overall power requirements compared to CO_2 compression. In this paper the application of Oxyfuel combustion for steam cracking unit heaters, oxygen requirements from the ASU, and considerations for the CO_2 purification unit will be discussed.

Introduction

In Oxyfuel combustion, fuel is combusted with oxygen from an Air Separation Unit (ASU) compared to conventional combustion, which uses air. Using oxygen for combustion nearly eliminates the inert nitrogen in the resultant flue gas and the flue gas consists of primarily carbon dioxide. An Oxyfuel combustion block flow diagram for a Steam Cracking Unit (SCU) heater is shown below in Figure 1. To minimize heater modifications by retaining similar combustion and heat transfer characteristics to conventional combustion with air, a portion of the flue gas is recycled back to the cracking heater. The gross flue gas is compressed with a blower and cooled with circulating water in a Direct Contact Cooler (DCC). A purge stream of condensed water from the flue gas exits from the circulating water to maintain water balance. The flue gas recycle is taken from the downstream of the Direct Contact Cooler.

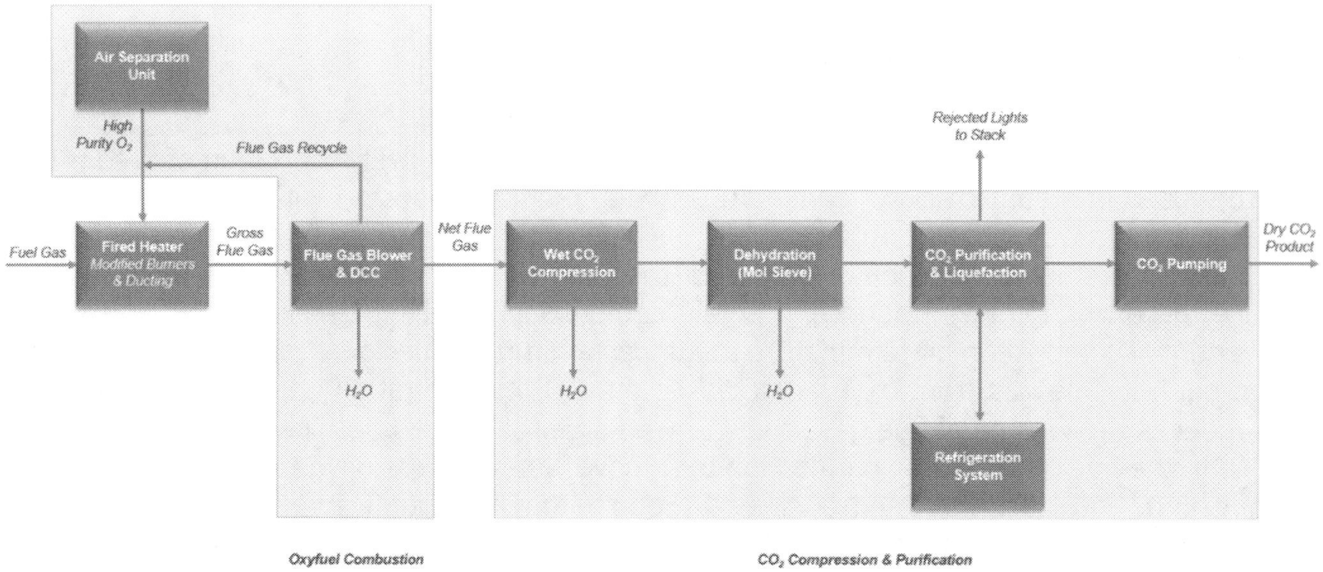

Figure 1: Oxyfuel Combustion Block Flow Diagram

In Oxyfuel combustion the required amount of oxygen is determined by the stoichiometric requirement of fuel and some excess oxygen to ensure complete combustion of the fuel, like air combustion. The amount of flue gas that needs to be recycled is determined by the process and heat recovery section duties. In a Steam Cracking Unit (SCU) application approximately 85-90% of the total flue gas is recycled. For a generic 1080 KTA naphtha cracker, the resultant flue gas composition and flow considering Oxyfuel combustion and air combustion are shown in Table 1.

Table **1**: Air Combustion and Oxyfuel Combustion Flue Gas Data

	Air Combustion	**Oxyfuel Combustion 97% O_2**
Gross Flue Gas		
Flow, Nm^3/h	1,076,993	1,181,656 (+10%)
Temperature, °C	106.0	106.0
Pressure, barg	0.00	0.00
Combustion Air, Nm^3/h	978,612	-
Oxygen from ASU, Nm^3/h	-	189,891
Combustion Fluid Composition, mol%		
CO_2	-	69.6%
H_2O	-	6.8%
N_2	78.1%	1.4%
O_2	20.9%	18.4%
Ar	0.9%	3.8%
Net Flue Gas Flow, Nm^3/h	1,076,993	101,988 (-91%)
Net Flue Gas Composition, mol%		
CO_2	8.2%	84.4%
H_2O	18.3%	8.2%
N_2	71.0%	1.5%
O_2	1.8%	1.8%
Ar	0.9%	4.1%

The net flue gas, primarily carbon dioxide with lighter impurities oxygen, nitrogen and argon, is water saturated leaving the DCC. The net flue gas is then sent to the CO_2 Purification Unit (CPU) where the flue gas is dehydrated and light impurities such as oxygen, nitrogen, and argon can be removed.

Oxyfuel CO_2 Purification Unit

A process schematic of Lummus' proprietary CPU developed for oxyfuel applications is shown in Figure 2. The net flue gas saturated with water from the DCC is compressed in a multi-stage centrifugal compressor with interstage cooling where some of the water is removed during compression. The compressor discharge pressure is optimized depending on ASU oxygen purity, resulting flue gas composition and CO_2 recovery and purity targets.

The compressed CO_2 is precooled with high pressure refrigerant to 20°C to reduce the water content in the stream before feeding a dehydration unit. The CO_2 must be partially condensed at a temperature below -20°C to remove light impurities. The water content needs to be reduced to less than 1 ppmv to prevent ice and CO_2 hydrate formation which requires a molecular sieve.

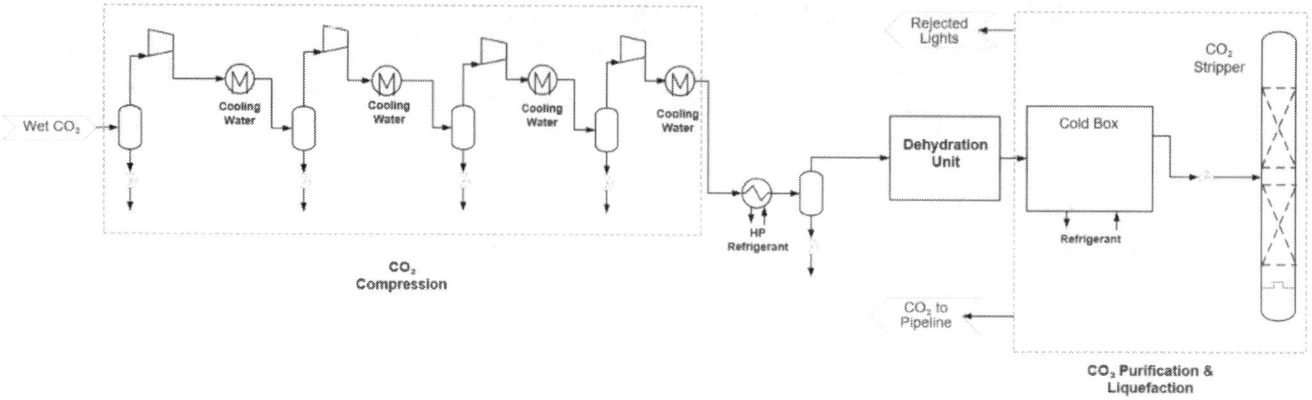

Figure 2: CO_2 Purification Unit Configuration

The dry flue gas stream is condensed in a brazed aluminum heat exchanger against high pressure and medium pressure refrigerant, the CO_2 Stripper Reboiler, and offgas streams before feeding the CO_2 Stripper. The CO_2 Stripper is condensed against vaporizing refrigerant and offgas streams. The CO_2 Stripper is operated at slightly lower pressure than feed chilling to account for frictional and static head losses between the two systems as well as to ensure a sufficient temperature approach between the feed condensing and reboiler vaporizing heat curves.

The CO_2 Stripper bottoms will consist of high-purity CO_2, which may be pumped and heated against dry flue gas before reaching final battery limit conditions. The CO_2 Stripper net overhead consists of some CO_2, excess oxygen, and any trace impurities (i.e., nitrogen and argon) and is reheated in the Cold Box. The CO_2 Stripper condenser operating temperature along with the light impurities content in the flue gas determine the maximum CO_2 recovery achievable in the CPU, which is typically over 95%.

The refrigeration system is a 3-stage conventional refrigeration system that utilizes propylene as refrigerant.

Comparison to Post Combustion Carbon Capture Unit

Another method for carbon capture in SCU heaters is utilizing Post Combustion Carbon (PCC) technology. The PCC unit is located downstream of conventional SCU heaters as shown in Figure 3 and utilizes amine solvents such as monoethanolamine (MEA) to absorb CO_2 from the flue gas. The flue gas pressure is increased with a blower to overcome the pressure drop of additional ducting, DCC and CO_2 Absorber towers, and then sent to the DCC. The DCC cools the flue gas to below 50°C prior to feeding the CO_2 Absorber.

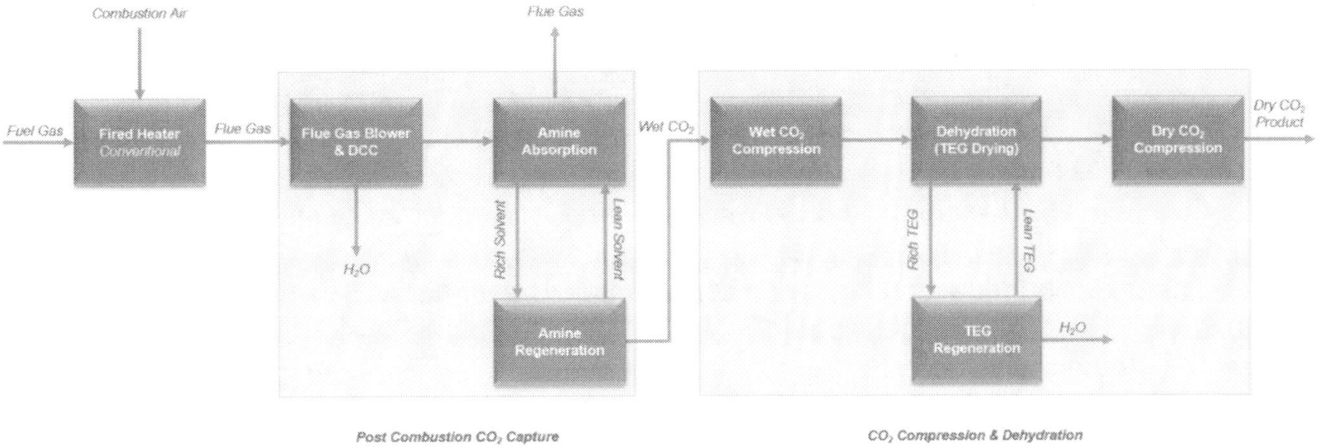

Figure 3: Post Combustion CO_2 Capture Block Flow Diagram

As shown in Table 1, the gross flue gas flowrate to the Blower and DCC for both carbon reduction configurations are similar and large. It is anticipated that for large ethylene units, each individual heater will need to be equipped with its own dedicated blower and DCC in both carbon capture approaches.

Compared to PCC, Oxyfuel combustion offers some key benefits:

- The main advantage of using Oxyfuel combustion is the high concentration of CO_2, greater than 90% mol on dry basis, in the net flue gas. Because inert nitrogen is not present in the system, the net flue gas volume is reduced by more than 90%. This allows the use of a common CPU for multiple large capacity heaters, providing significant CAPEX savings and economy of scale benefit. In contrast for PCC, tower diameters, even when considering high capacity/high efficiency internals, become very large and all heaters must be equipped with dedicated PCC units.
- Oxyfuel combustion can achieve over 95% CO_2 capture whereas PCC is typically limited to 90% CO_2 capture, as additional capture increases solvent circulation and regeneration duties significantly.
- No additional chemicals are needed in oxyfuel combustion. In post combustion carbon capture, a solvent, traditionally an amine, monoethanolamine (MEA), is required to absorb the CO_2. The system is corrosive in nature and requires upgraded metallurgy due to the high temperatures and chemicals present (typically carbon steel with post weld heat treatment and stainless-steel cladding).
- Significant steam consumption is required for solvent regeneration in a post combustion unit. Minimal steam is required for the dryer regeneration heater in an Oxyfuel combustion unit.
- The Oxyfuel refrigerant system could be integrated with the cracker refrigeration system to reduce equipment count, cost, and footprint.
- Oxyfuel has lower NOx emissions as the primary source of NOx, i.e., nitrogen, is at much lower concentrations in the Oxyfuel burners.

Compared to PCC, Oxyfuel combustion has some additional requirements.

- A post combustion carbon capture unit requires no change to current heater operation except incorporating ducting to direct flue gas to the PCC. Oxyfuel combustion typically requires some burner and flue gas duct modifications.
- An Air Separation Unit for oxygen supply is required for Oxyfuel combustion.
- The CO_2 dehydration is typically accomplished via a lower CAPEX triethylene glycol (TEG) unit in post combustion CO_2 capture to meet typical pipeline specifications. In Oxyfuel combustion, a mole sieve dryer is required to meet the lower allowable moisture content due to the colder operating temperatures of the CPU.

Either carbon reduction approach can be applied to either a grass-roots design or retrofit design. A PCC retrofit design may be more challenging to implement based on available plot space near the existing heaters. As discussed above, a PCC design may require individual PCC units per heater depending on capacity. These units should be located near the existing heaters.

Alternately, an Oxyfuel retrofit design can have different sections of the process located in different areas. An Oxyfuel design may require individual blowers and DCC towers per heater depending on capacity. This equipment should be located near the existing heaters. However, the CPU and ASU can be located further away from the existing heaters. Locating the CPU further from the heaters will increase blower power consumption to account for the longer piping to reach the CO_2 Compressor suction.

Impact on Utilities and Overall Emissions

A breakdown of the utility requirements and emissions summary for each carbon reduction method is provided in Table 2 for a generic 1080 KTA naphtha cracker. The CO_2 emissions, Scope 1 and Scope 2 CO_2 equivalent (CO_2-e) are calculated with Green House Gas Protocol global warming potential values based on fourth assessment report (AR4) [1] [2]. Scope 1 emissions are direct greenhouse (GHG) emissions that occur from sources that are controlled or owned by an organization (e.g., emissions associated with fuel combustion in boilers, furnaces). Scope 2 emissions are indirect GHG emissions associated with the purchase of electricity, steam, heat, or cooling.

Table **2**: Utilities and Emissions Summary

	Base Case No CO_2 Capture	**Post Combustion CO_2 Capture**	**Oxyfuel Combustion CO_2 Capture**
Fuel Fired	BASE	BASE	98% BASE
Scope 1 CO_2-e Generated, kg/h	172,733	172,733	169,028
% Scope 1 CO_2-e Captured	0%	90%	95%
Scope 1 CO_2-e Emitted, kg/h	172,733	17,273	8,451
Electrical Consumption, MW	BASE	BASE + 24 (CCSU)	BASE + 28 (CPU) + 46 (ASU) (1)
Steam Consumption, MW (3)	BASE	BASE + 236	BASE + 1
Cooling Water Consumption, ton/h	BASE	BASE + 10,130	BASE + 13,974
Scope 2 CO_2-e Generated, kg/h	54,819	123,517	84,975
Scope 1+2 CO_2-e Generated, kg/h (2)	227,552	296,249 +30%	254,002 +12%
Scope 1+2 CO_2-e Emitted, kg/h	227,552	140,790 -38%	93,426 -59%
Scope 1+2 (Green) CO_2-e Emitted, kg/h	221,682	125,317 -44%	58,067 -74%

Notes
1. ASU Specific Power of 170 kWh/ton O_2 contained (97% mol O_2)
2. Electricity emission factor of 396 kg CO_2-e/MWh
3. PCC steam consumption considers 30 wt% MEA solvent solution with specific energy of 4.4 GJ/ton CO_2

While the addition of any form of carbon reduction increases the CO_2 generated from increased utility demand, once carbon reduction is accounted for the net CO_2 emitted is appreciably lower than the Base Case (no carbon capture considered). As discussed above, Oxyfuel can achieve higher CO_2 recovery than PCC, resulting in lower Scope 1 CO_2-e emitted. While Oxyfuel has a much higher electrical demand than PCC, PCC results in higher Scope 2 CO_2-e emissions due to the high steam demand. These two effects result in a significant reduction in CO_2-e emissions for Oxyfuel compared to PCC. Furthermore, most of the Oxyfuel Scope 2 emissions are from electrical consumption. If a green source of electricity is considered, Scope 1 and Scope 2 emissions can be reduced by 74% compared to the Base Case.

Impact on Equipment

One significant benefit of oxyfuel combustion is the reduction in the quantity of flue gas that must be processed. This is especially significant when considering large capacity ethylene plants. If post combustion carbon capture is considered for large capacity plants, tower diameters, even when considering high capacity/high efficiency internals, become very large.

As cracker capacities continue to increase, multiple post combustion carbon capture units are required to process the full flue gas from all heaters.

As shown in Table 3, the total Blower and Direct Contact Cooler capacity for both carbon reduction configurations is very high. It is anticipated that for large units, multiple pieces of equipment operating in parallel will be required for both carbon reduction configurations.

In post combustion capture, the CO_2 Absorber capacity is also very high and it is anticipated that multiple columns would need to operate in parallel as well. Considering high efficiency internals to limit the required bed heights, the Direct Contact Cooler and CO_2 Absorber may be housed in the same shell to save equipment cost and plot space. While the Regenerator capacity is significantly lower than the other post combustion towers, it is anticipated that a separate Regenerator is required for each CO_2 Absorber for ease of operation. By comparison, a single CO_2 Stripper may be used for all heaters.

While the blower capacity is nearly the same between post combustion and oxyfuel combustion, the power consumption is higher for the post combustion unit. This is largely a result of the higher molecular weight of the flue gas for Oxyfuel combustion. There is also a minor difference in compression ratios due to the additional pressure drop associating with packing in the CO_2 Absorber for post combustion carbon capture.

The CO_2 Compressor power consumption is higher for Oxyfuel combustion because the compressor processes the full CO_2 generated in the heaters (recovered and unrecovered CO_2) as well as impurities, whereas the post combustion unit only processes the recovered CO_2 and remaining moisture.

Table **3**: PCC v. Oxyfuel Combustion Equipment Capacities

	Post Combustion	Oxyfuel Combustion
Blower		
Capacity, Nm^3/h	1,076,993	1,181,656 (+10%)
Power, kW	6,276	6,058 (-3%)
Compression Ratio	1.10	1.09
Molecular Weight	27.66	37.62
Direct Contact Cooler		
ID x T/T, m	-	7.2 x 15.5
Direct Contact Cooler/CO_2 Absorber		
ID x T/T, m	6.9 x 35.0	-
Regenerator		
ID x T/T, m	5.0 x 32.2	-
CO_2 Stripper		
ID x T/T, m	-	2.3 x 26.0
CO_2 Compression		
Power, kW	15,298	17,362
CO_2 Product Pump Power, kW	-	563
CO_2 Dehydration	TEG System	Mole Sieve Unit
Capacity	0.9 m^3/h circulation	93,630 Nm^3/h
Heating Duty, MMkcal/h	0.19	0.53 (1)
Refrigerant Compressor		
Power, kW	-	3,423
ASU		
Contained O_2 (97% mol purity)	-	6,311 TPD
Notes: 1. Intermittent operation		

Variations of CO_2 Purification Unit Design

The CPU design depends on several key design parameters that are interconnected to one another:

- CO_2 product specification
- Impurities content in the oxyfuel flue gas / oxygen supply purity
- CO_2 recovery target
- Refrigerant

CO_2 Product Specification

The CO_2 product's final use, whether for Carbon Capture and Sequestration (CCS), for Enhance Oil Recovery (EOR), or to be sent to an existing CO_2 pipeline, determines the target specifications. A typical existing CO_2 pipeline specifications is shown in Table [3].

Table **4**: Typical CO_2 Pipeline Specifications

CO_2	> 95% mol
Water	< 30 lbs/MMSCF
H_2S	< 20 ppmv
Total Sulfur	< 35 ppmw
Nitrogen	< 4% mol
Hydrocarbons	< 5 % mol Dew Point < -20°F
Oxygen	< 10 ppmw
Other	< 0.3 gallons /MMSCF

The oxyfuel flue gas can have more than 95% mol CO_2 on dry basis, however the oxygen content is approximately 2% due to excess oxygen required for combustion. The oxygen specification of existing pipelines is maximum 10 pmmw due to corrosion concerns, potential for other undesired reactions and its byproducts if used for EOR. To achieve less than 10 ppmw oxygen in the CO_2 product, a stripping column is required. However, for CCS applications where higher oxygen content in the CO_2 is acceptable, a separator vessel can be used instead of a stripping tower, reducing the plot space, CAPEX, and OPEX. This change in configuration replaces a distillation tower with a flash drum and eliminates the tower reboiler, reflux drum and pump, and all associated instrumentation and controls. This can also reduce the Refrigerant Compressor power by approximately 25% considering the same feed chilling pressure and 97 mol% oxygen concentration.

Impurities in Oxyfuel Flue Gas – Oxygen Supply Purity

The main impurities in the oxyfuel flue gas are excess oxygen, which needs to be minimized as much as practical while ensuring complete combustion and nitrogen and argon from the oxygen supply. For typical SCU oxygen supply a large-scale cryogenic Air Separation Unit (ASU) will be required.

A key design parameter that may be optimized is the oxygen purity from the ASU. Reducing the oxygen purity will result in a more impurities in the net flue gas as well as a higher volume of flue gas. In turn, all compressor power consumption and equipment capacities will increase within the CO_2 purification unit. Furthermore, the increased nitrogen and argon content will reduce the CO_2 Stripper Condenser temperature required to meet 95% CO_2 recovery. The feed condensing curve will also shift to a lower temperature due to the change in composition.

At 97% oxygen purity the CO_2 Stripper pressure needs to be increased to achieve 95% CO_2 recovery with propylene refrigerant. At oxygen purities less than 97%, the CO_2 Stripper Condenser temperature is too low to utilize propylene refrigerant and achieve 95% CO_2 recovery with the discussed configuration and CO_2 recovery needs to be reduced.

The ASU power consumption depends on the oxygen purity, oxygen delivery pressure, and whether any other products such as nitrogen and argon are recovered. For Oxyfuel applications, the oxygen pressure required is relatively low (0.1 barg) and lower purity (<97%) oxygen can be used. The ASU specific power consumption decreases with reducing oxygen purity required and above 97% the specific power consumption increases significantly [4]. For Oxyfuel applications, ASU process configuration optimizations for 95% oxygen purity have been developed with a specific energy of 158 kW/ton of oxygen contained, which can be further reduced with integration [5].

The sensitivity results of the ASU oxygen purity to CPU and overall oxyfuel combustion system are summarized in Table ***3***. For this analysis ASU specific energy considered is as follows:

- 95.0% O_2 purity = 160 kWh/ton O_2 product
- 97.0% O_2 purity = 170 kWh/ton O_2 product
- 99.5% O_2 purity = 190 kWh/ton O_2 product

As the oxygen purity increases the CPU power requirements decreases, however in this study the optimal oxygen purity in terms of energy efficiency/emissions is found to be 97%, above which the increase in ASU power consumption most outweighs the decrease in CPU power consumption.

Table **3**: ASU Oxygen Purity Sensitivity Analysis

ASU Oxygen Purity, mol%	**95.0%**	**97.0%**	**99.5%**
Flue Gas Composition, mol%			
CO_2	80.9%	84.4%	89.0%
H_2O	8.2%	8.2%	8.2%
N_2	3.7%	1.5%	0.0%
O_2	1.8%	1.8%	1.8%
Ar	5.5%	4.1%	1.0%
CO_2 Recovery	95%	95%	95%
CPU Power Consumption, MW	28.9	26.8	24.5
ASU Power Consumption, MW	44.5	46.3	50.2
Total Power Consumption, MW	73.4	73.1	74.7
CO_2 Captured, kg/h	160,582	160,582	160,582
Specific Energy, kWh/T CO_2	456.8	455.4	465.5
Scope 1+2 CO_2-e Emitted (1), kg/h	37,481	37,393	38,030
Notes 1. Electricity emission factor of 396 kg CO_2-e/MWh. Only considers major emissions sources.			

An SCU and ASU can provide several heat integration opportunities where heat of compression in ASU could be recovered in the steam or boiler feed water heater, further reducing the specific energy consumption of oxygen.

Refrigerant

Refrigerants such as ammonia, propane, and propylene can be used in the CPU. Since many SCUs already utilize a propylene refrigeration system and propylene allows for lower condenser temperatures and higher CO_2 recoveries, propylene could be more advantageous for SCU applications.

Conclusion

In conclusion, Oxyfuel combustion is an attractive option for CO_2 mitigation. Compared to post combustion carbon capture, the net flue gas that must be processed is an order of magnitude lower flow and an order of magnitude higher CO_2 purity. This makes Oxyfuel combustion an especially attractive option for large capacity ethylene plants, where multiple post combustion carbon capture units would be needed in parallel. Oxyfuel is also able to reduce Scope 1 and Scope 2 CO_2-e emission by 33% compared to a PCC unit due to the higher recovery and lower emissions related to utility demand. Furthermore, there are optimizations that may be made within the Oxyfuel combustion unit to reduce cost and improve energy efficiency, such as modifying the CO_2 separation process in exchange for a lower CO_2 product purity, as well as optimizing the ASU oxygen purity, refrigerant selection, and CO_2 recovery target.

References

[1] "Global Warming Potential Values," Greenhouse Gas Protocol, [Online]. Available: https://www.ghgprotocol.org/sites/default/files/ghgp/Global-Warming-Potential-Values%20%28Feb%2016%202016%29_1.pdf. [Accessed 9 2 2023].

[2] "Scope 1 and Scope 2 Inventory Guidance," EPA, [Online]. Available: https://www.epa.gov/climateleadership/scope-1-and-scope-2-inventory-guidance. [Accessed 9 2 2023].

[3] P. A. Brownsort, "Briefing on carbon dioxide specifications for transport," EU CCSUS Projects Network, 2019.

[4] A. Darde, R. Prabhakar, J.-P. Trnier and N. Perrin, "Air sepration and flue gas compression and purification units for oxy-coal combustion systems," *Energy Procedia,* vol. 1, pp. 527-534, 2009.

[5] P. Higginbotham, V. White, K. Fogash and G. Guvelioglu, "Oxygen Supply for Oxycoal CO2 Capture," *Energy Procedia,* vol. 4, pp. 884-891, 2011.

AIChE Paper Number 27d

Development of a Commercial Scale Ethylene Furnace Hydrogen Burner

David B. Spicer
Chief Engineer – Olefins
ExxonMobil Technology & Engineering Company

Melissa J. Bishop
Pyrolysis Process Engineer
ExxonMobil Technology & Engineering Company

Taemin Eom
Pyrolysis Process Engineer
ExxonMobil Technology & Engineering Company

Prepared for Presentation at the 2023 Spring National Meeting
Houston, TX, March 13 – 16, 2023

Development of a Commercial Scale Ethylene Furnace Hydrogen Burner

David B. Spicer
Chief Engineer – Olefins
ExxonMobil Technology & Engineering Company
Melissa J. Bishop
Pyrolysis Process Engineer
ExxonMobil Technology & Engineering Company
Taemin Eom
Pyrolysis Process Engineer
ExxonMobil Technology & Engineering Company

Abstract

As the ethylene industry considers potential pathways to reduced CO_2 emissions, partial or full fuel switching from methane-rich tail-gas to hydrogen has been proposed in numerous papers and presentations.

While the direct CO_2 emission benefits of hydrogen combustion are straight-forward to calculate, the design of a hydrogen burner compatible with modern ethylene cracking furnaces must address several technical challenges. Included in those challenges are:

1. Maintaining adequate heat distribution and avoidance of localized high coking rates.
2. Managing NOx emissions as the fuel-gas hydrogen content increases.
3. Providing acceptable burner stability across the full fuel-gas range from natural-gas to hydrogen.

This paper reviews progress made in developing a full-scale hydrogen burner for 100% floor-fired ethylene furnaces, including test-furnace performance trends.

Contents

1. Introduction

Ethylene and propylene are petrochemicals commercially produced in large quantities. They are used in the production of a wide variety of chemical products, including polyethylene and polypropylene. These chemical products in turn are used in numerous applications considered essential to modern standards of living.

The production of ethylene and propylene is an inherently energy-intensive process. Steam-cracking, the process by which the majority of the world's ethylene is produced, involves free radical reactions that occur at very high temperatures. Steam cracking furnaces, also referred to as pyrolysis reactors, generate the high temperatures required for olefins production by the combustion of fuel-gas.

The fuel-gas typically consumed is primarily a methane/hydrogen mixture, with the combustion of the methane component leading to CO_2 emissions in the cracking furnace flue-gas.

As the ethylene industry seeks commercially applicable methods to reduce the associated GHG emissions, the potential for fuel-switching to higher-hydrogen content is generating significant interest.

However, the use of very high hydrogen content fuel-gas as a steam cracking furnace fuel brings technical challenges:

a) Generation of hydrogen with minimal GHG emissions and acceptable energy consumption.
b) Design of the steam cracking furnace to be fueled by the very high hydrogen content fuel-gas.
c) Design of the burners to combust the very high hydrogen content fuel-gas while providing acceptable heat distribution within the furnace/reactor, and acceptable combustion emissions...with oxides of nitrogen (NOx) of particular interest.

Other papers in this and other industry conferences address the first two challenges above. This paper focuses on the burner design challenges for using very high hydrogen content fuel gas in a steam cracking furnace.

2. CO_2 Emission Trends from Hydrogen Rich Fuel Gas Combustion

Steam cracking furnaces typically fire plant "tail-gas", the mixture of hydrogen and methane byproducts from the cracking reactions.

Tail-gas from naphtha crackers is typically methane rich...being in the range 15 – 25 vol. % hydrogen. Variation within that range is a function of the feed quality, cracking furnace selectivity and the level of hydrogen recovered from tail-gas for other purposes (e.g.; for use in hydrogenation reactors). In contrast, tail-gas from an ethane cracker with limited (or no) hydrogen recovery may be as high as 85 vol. % hydrogen.

Since the industry typically characterizes fuel-gas by "volume % hydrogen" it is interesting to plot the CO_2 emissions that result from varying tail-gas quality.

While commercial steam-cracker tail-gas can contain numerous minor components (including CO, N_2, C_2H_4, etc.) for the purposes of this study we have considered a binary component fuel-gas consisting only of methane and hydrogen.

Figure 1 plots the CO_2 emissions from a hypothetical steam cracking furnace firing 350 MBtu/hr (LHV) heat release, and operating 8,400 hrs./year.

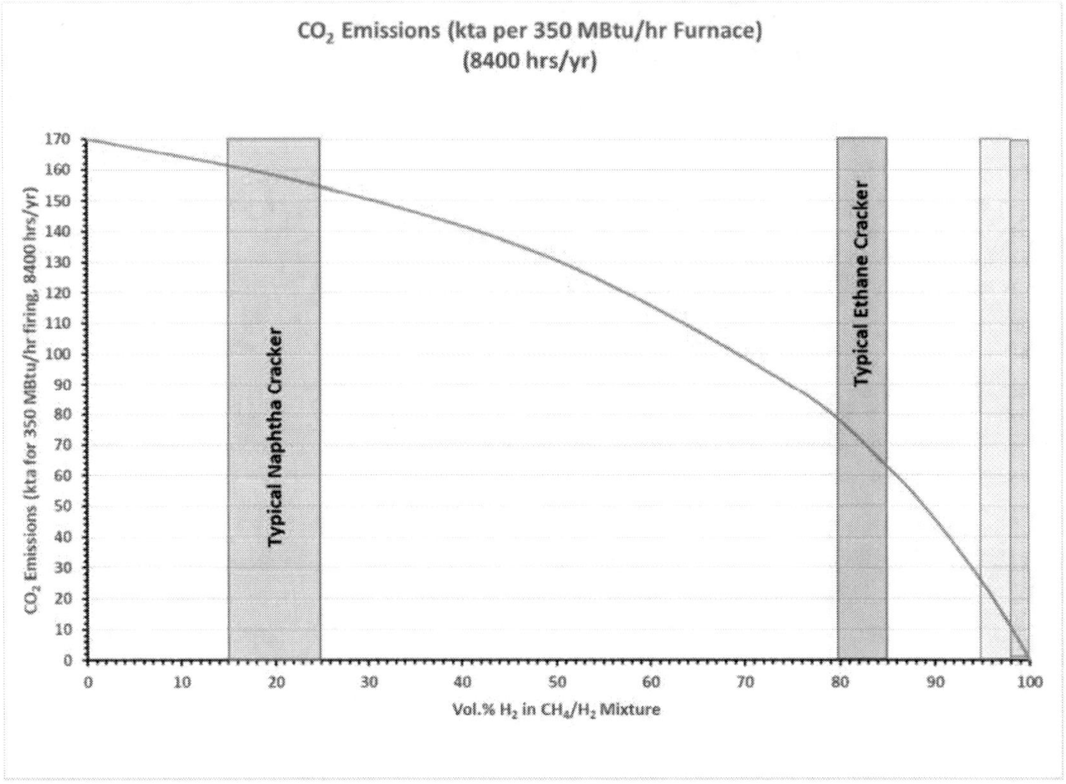

Figure 1: CO2 Emissions vs. Fuel-Gas Vol, % Hydrogen

GHG estimates for Scope 1 emissions only, based on CH_4/H_2 fuel-gas composition

Figure 1 illustrates the non-linear nature of CO_2 emissions when plotted against the commonly used "vol. % hydrogen" parameter. While CO_2 emissions do fall as vol.% hydrogen in fuel increases, a furnace firing 80-85 vol.% hydrogen can have CO_2 emissions as high as 50% of the emissions of a furnace firing a 15-25 vol.% hydrogen fuel gas.

Figure 2 provides additional insight into this behavior.

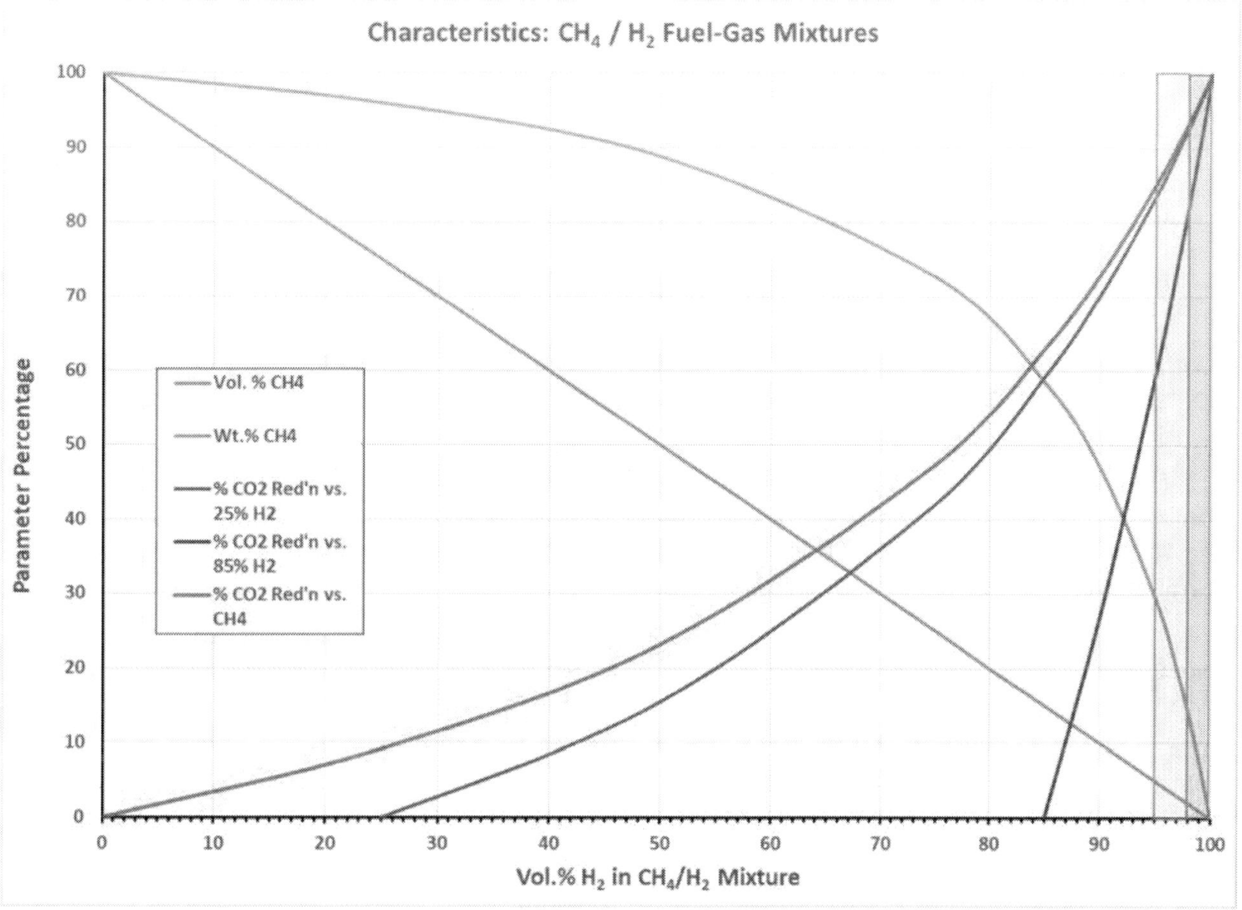

<u>Figure 2: Characteristics of Methane/Hydrogen Fuel-Gas Mixtures</u>

GHG reduction estimates for Scope 1 emissions only, based on CH_4/H_2 fuel-gas composition

Figure 2 illustrates the non-linear relationship between wt. % methane and vol. % hydrogen in the binary fuel-gas mixture.

Also plotted are curves showing the percentage CO_2 emission reduction achieved as hydrogen concentration increases compared to three base cases.

- 100% methane
- 25 vol. % hydrogen (representative of a naphtha cracker)
- 85 vol. % hydrogen (representative of an ethane cracker with no H_2 product recovery)

The curves indicate that to achieve a 90% reduction in CO_2 emissions from cracking furnaces in a typical naphtha cracker one may need a fuel-gas with greater than 95 vol. % hydrogen. To achieve an 80% reduction in the lower CO_2 emissions from cracking furnaces in an ethane cracker, one may need a fuel-gas with 98 or higher vol. % hydrogen.

The challenge to burner designers for ethylene furnaces becomes clear from this representation. It will be desirable to demonstrate a burner that can acceptably fuel 98 - 100% hydrogen in a commercial cracking furnace environment.

3. Challenges Firing High Hydrogen Content Fuel-Gas in Ethylene Furnaces

Firing very high hydrogen content (98 - 100%) fuel-gas in a steam cracking furnace poses several technical challenges.

These challenges can be attributed to the very high flame speed and high flame temperature of hydrogen relative to natural gas, and even to the methane/hydrogen mixtures currently fired in conventionally fueled cracking furnaces.

A review of the literature reveals much discussion about diffusion flames and pre-mixed turbulent flames, but all references are in agreement that hydrogen has a much higher flame speed than methane, and that flame speed in hydrogen/methane mixtures increases with increasing hydrogen concentration.

Some designs and vintages of cracking furnace employs burners which use a pre-mix architecture. In such burners fuel and air are mixed, typically in a venturi, prior to exiting the burner tip. If the flame speed sufficiently exceeds the velocity at which the mixed gas exits the tip, the flame can propagate back into the venturi, a phenomena known as flashback. Over time such operation can cause significant damage to the burner.

When a raw-gas burner is used, the issue of flashback is eliminated. However the high flame speed of hydrogen makes it difficult to generate a sufficiently tall flame with such a burner to provide acceptable heat distribution in a commercial, 100% floor-fired cracking furnace.

Floor firing has several advantages. It makes combustion air-preheat systems (with their associated combustion air ductwork) more practical and less costly. All burners in a floor-fired furnace operate with the same draft, enabling improved air:fuel ratio consistency across a natural-draft or induced-draft furnace. This is not the case for furnaces with wall-mounted burners located at various elevations.

As modern cracking furnaces have increased in capacity, radiant fireboxes have tended to increase in height. Recent designs have featured radiant sections as tall as 44 feet (top-of-floor to bottom-of-arch). Figure 3 illustrates a recent example. This trend increases the challenge of using very-high hydrogen fuel-gas.

Figure 3: Commercial Steam Cracker Radiant Section Height

The third challenge of using hydrogen fuel in a steam cracking furnace is managing NOx emissions. Steam cracking furnaces operate at very high radiant section temperatures, and high temperatures are known to favor the generation of high NOx emissions. Over time burner technology has evolved to reduce NOx emissions using such techniques as combustion-staging (staged-fuel or staged-air) and flue-gas-recirculation or flue-gas-entrainment. The recent generations of ethylene plants in the United States have incorporated SCR (Selective Catalytic Reduction) to further reduce NOx emissions.

There is a large body of technical literature that suggests than NOx emissions from a given burner rise as the fuel hydrogen content rises.

Figure 6.19 (page 208) in Reference 1 depicts a trend of NOx emissions rising as fuel hydrogen increases.

While installing a higher capacity SCR is always an option for a new, hydrogen-fueled cracking furnace, there is a large population of cracking furnaces in operation that do not have SCR systems installed. For such furnaces the historical expectation of higher NOx emissions as fuel hydrogen increases may pose a challenge for fuel-switching.

4. ExxonMobil's Existing Steam-Cracking Furnace Burner

ExxonMobil's current low NOx commercial steam-cracking furnace burner design is described US Patent 6,877,980 B2. The burner is a staged-air, internal flue-gas-recirculation burner, with optional steam-injection. A schematic illustration is included below as Figure 4.

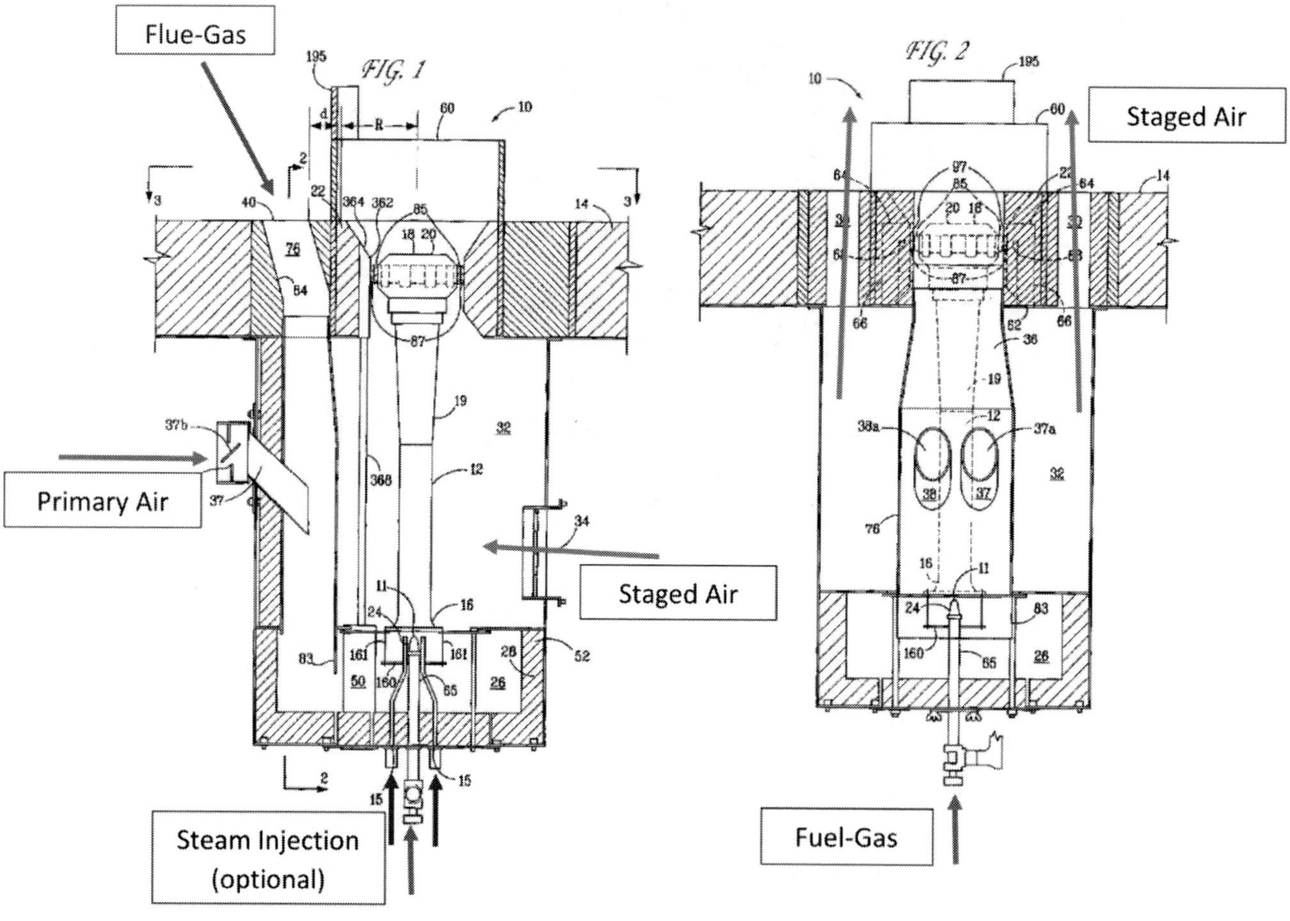

Figure 4: Existing Generation Ethylene Furnace Burner

In the burner the fuel-gas enters through a single fuel spud. A high-performance venturi inspirates both primary-air and flue-gas from the furnace floor to mix with the fuel-gas. The level of flue-gas recirculation can be adjusted using the position of the primary-air damper.

A larger air-stream enters the burner through the staged-air plenum. The majority of this air enters the furnace through a series of staged-air ports spaced radially around the fuel/primary-air/FGR stream.

Steam, where employed, is injected directly into the venturi entrance. Thus the steam mixes well with the fuel-gas prior to combustion, maximizing NOx suppression effectiveness.

The flue-gas recirculation port draws flue-gas from the furnace floor in response to the fuel-gas/venturi system. A small volume of staged-air is bled into the FGR system to moderate temperatures and prevent degradation of the metallic FGR duct.

A full or partial ring provides additional physical separation between the primary combustion zone and the staged-air injection and FGR ports.

This burner has been employed, with some variations, in commercial cracking furnaces since 2004. Embodiments with maximum firing rates between 6.0 and 8.5 MBtu/hr (LHV) have been used. The population of installed equipment covers more than 30 furnaces and more than 1,800 burners.

Applications include ambient-air and air-preheat furnaces, and fuel compositions as high as 85 vol. % hydrogen.

Figure 5 is a photograph of the radiant section of a furnace firing 80-85 vol. % hydrogen fuel-gas with ambient combustion air.

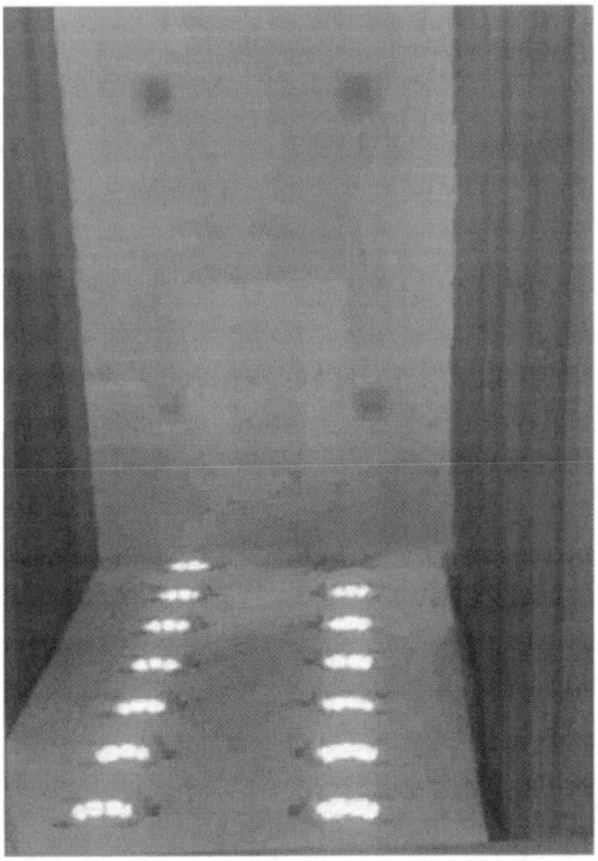

Figure 5: Current Generation Burner Firing 80-85 vol. % Hydrogen

A full-scale burner of this geometry was recently tested over a range of hydrogen content fuels. Testing was conducted at John Zink's test center, located in Tulsa, Oklahoma.

For the test program, the burner was operated at approximately 6.0 MBtu/hr (LHV) heat release. The test furnace was generally operated at 2050 – 2100 °F operating temperature. Excess oxygen was held in the range of 1.5 – 2.0 wet vol. %. To provide the best possible data trends, data points were adjusted to a consistent basis of 2.0 wet vol. % excess O_2 and 2100 °F furnace operating temperature.

Figure 6 shows the NOx emissions achieved in the test furnace during this test-firing. ***Note that test-furnace NOx emissions do not exactly match commercial furnace NOx emissions.*** However, trends observed in the test furnace are generally a good indicator of trends observed in commercial furnaces.

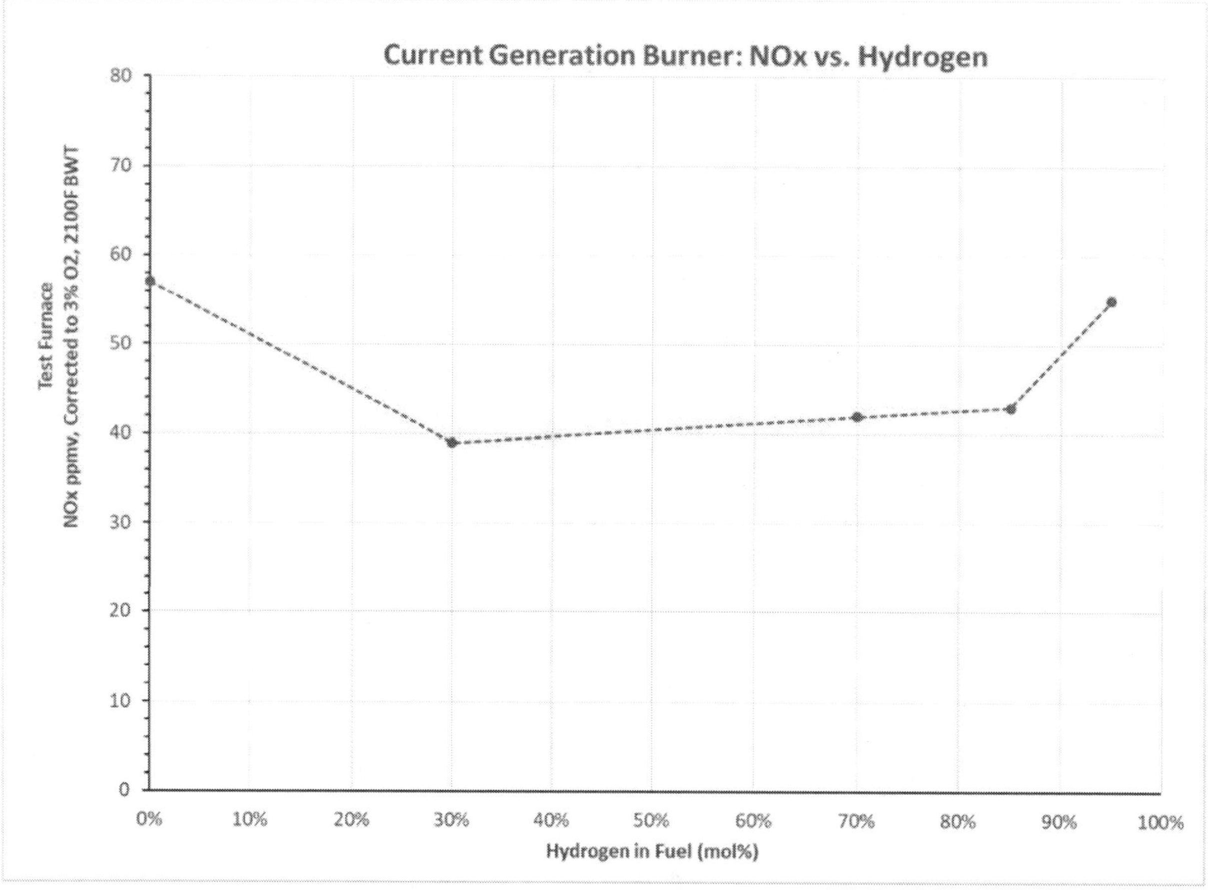

Figure 6: NOx vs. Fuel Hydrogen for Current Generation Burner

The data shows slowly increasing NOx emissions as fuel hydrogen increases from 30 to 85 vol. %, but then increasing more rapidly as fuel hydrogen increases to 95 – 97 vol. %.

At hydrogen concentrations above 95 vol. % the burner began to exhibit a tendency to flashback.

5. Development Program: "Proof of Concept" Burner

While the existing commercial burner provides competitive NOx emissions and moderate NOx increases as fuel hydrogen increases, additional broadening of the operating window is desirable for a commercial very-high hydrogen fuel burner.

The internal flue-gas recirculation of the existing burner dilutes the fuel/primary-air mixture with inerts, thus reducing NOx. The same phenomena also reduces flame speed in the primary combustion zone, enabling the burner to operate at quite high hydrogen levels in fuel before flashback is encountered.

However, recirculating hot flue-gas from the floor of the furnace increases the temperature of the mixture in the venturi, partially negating the positive effects of the recirculated flue-gas.

Systems exist to draw cool flue-gas from the furnace stack and mix it with the combustion air to limit the increase in mixture temperature in the primary combustion zone. However these systems require extensive duct-work and a means of drawing the flue-gas down to the required combustion air ducts. In addition, such systems increase the fuel-consumption of the furnace as the cooled flue-gas must be reheated to the radiant firebox temperature.

Instead, a development of the current commercial burner was envisaged where a compact heat-exchanger inside the burner cooled the recirculated flue-gas by heating the staged-air flowing into the burner.

In this manner the primary combustion zone mixture is cooled, reducing both NOx generation and flame speed in the primary combustion zone. By heating staged-air, no energy is lost from the system, so no change to the furnace fuel-consumption occurs.
The preheated staged-air makes little difference to NOx emissions as the secondary combustion zone is not a region of high NOx generation.

The compact heat exchanger design requires adequate heat transfer capability while at the same time operating within the available hydraulic margins of the furnace. As in the current generation burner, a small flow of staged-air is fed into the FGR duct upstream of the heat exchanger to avoid exceeding its allowable operating temperature.

Such a burner is described in US Patent 6,869,277 B2.

Figure 7 provides a schematic view of the burner.

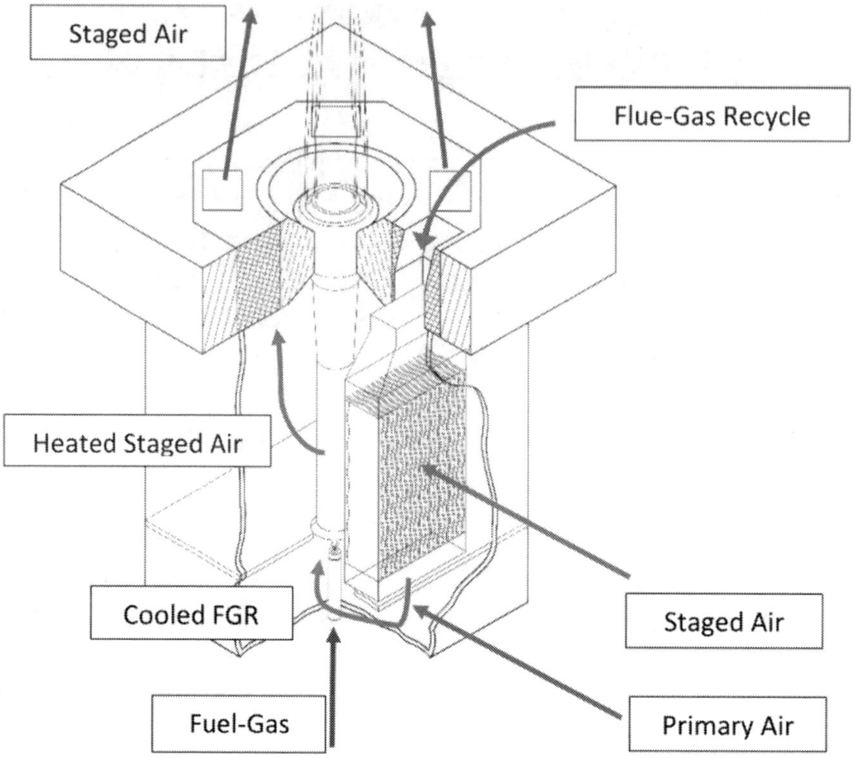

Figure 7: Internal FGR Burner with Integral Heat Exchanger

With this concept we aimed to develop a burner with the following capabilities:

- Able to light-off and fire at or near design rates on 100% natural-gas.
- Provide lower NOx emissions than the existing burner over all hydrogen/methane mixture ranges.
- Capable of firing hydrogen/natural-gas mixtures of greater than 95 vol. % hydrogen.
- Capable of transitioning from high-hydrogen content fuel-gas to natural-gas without flame lift-off.

6. "Proof of Concept" Burner Testing Summary

In 4Q2021 "proof of concept" burners of the new design were tested at John Zink's test facilities in Tulsa, Oklahoma.

Burners were commercial scale (6.0 – 6.5 MBtu/hr (LHV) heat release) and the test furnace was configured to operate at 2050 – 2100 °F temperature.

A number of heat exchanger geometries and burner tip geometries were tested on both ambient combustion air and preheated combustion air.

Burner light-off on natural gas was found to be satisfactory.

NOx emissions were found to be lower than the current commercial burner, and NOx emissions were observed to reduce as the fuel hydrogen content was increased.

Figure 8 illustrates the NOx performance of the test burner.

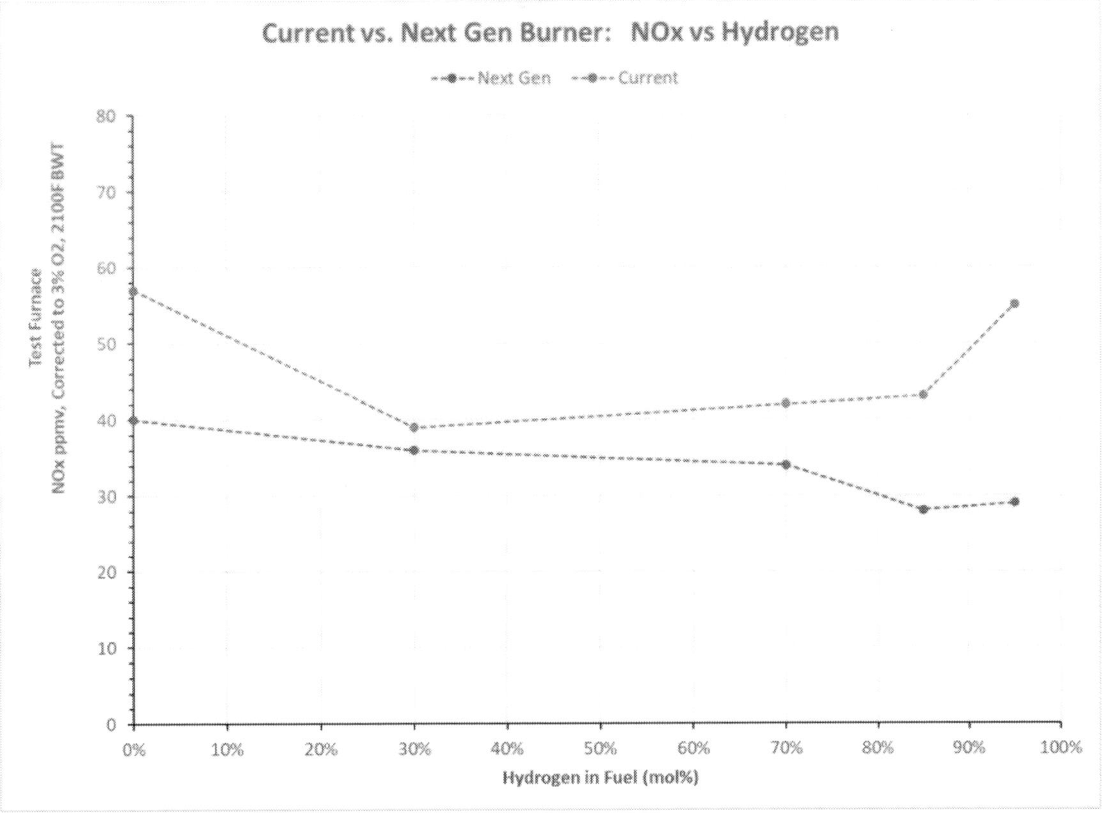

Figure 8: NOx Performance of the "Proof of Concept" Burner

NOx emissions were reduced versus the current burner, with the greatest reductions apparent at very-high hydrogen content fuels. NOx emissions reduced as fuel hydrogen content increased across the full fuel-range tested (with NOx values at 85% and 95% hydrogen being within the expected data scatter).

The new burner exhibited improved flashback resistance at very high hydrogen levels (> 95 vol. %).

The performance characteristics (NOx versus Fuel Hydrogen) of the "Proof of Concept" burner appears to defy conventional wisdom in that NOx emissions decreased as fuel hydrogen increased.

The authors hypothesize that this may be due to two characteristics of the burner architecture.

First, as fuel hydrogen increases in a given burner at a given firing rate, the fuel spud pressure rises. This generates a higher energy fuel-jet into the venturi, and can draw in additional flue-gas recycle. The additional FGR further dilutes the mixture entering the primary combustion zone. As the flammability range of the mixture increases as fuel hydrogen content increases, this poses no concerns for burner stability. Hence the burner architecture effectively increases FGR rates and therefore increases thermal-NOx suppression effectiveness as the fuel-hydrogen increases.

Figure 9 shows the trend of fuel spud pressure for a given burner (fixed spud geometry) firing at a constant 6 MBtu/hr (LHV).

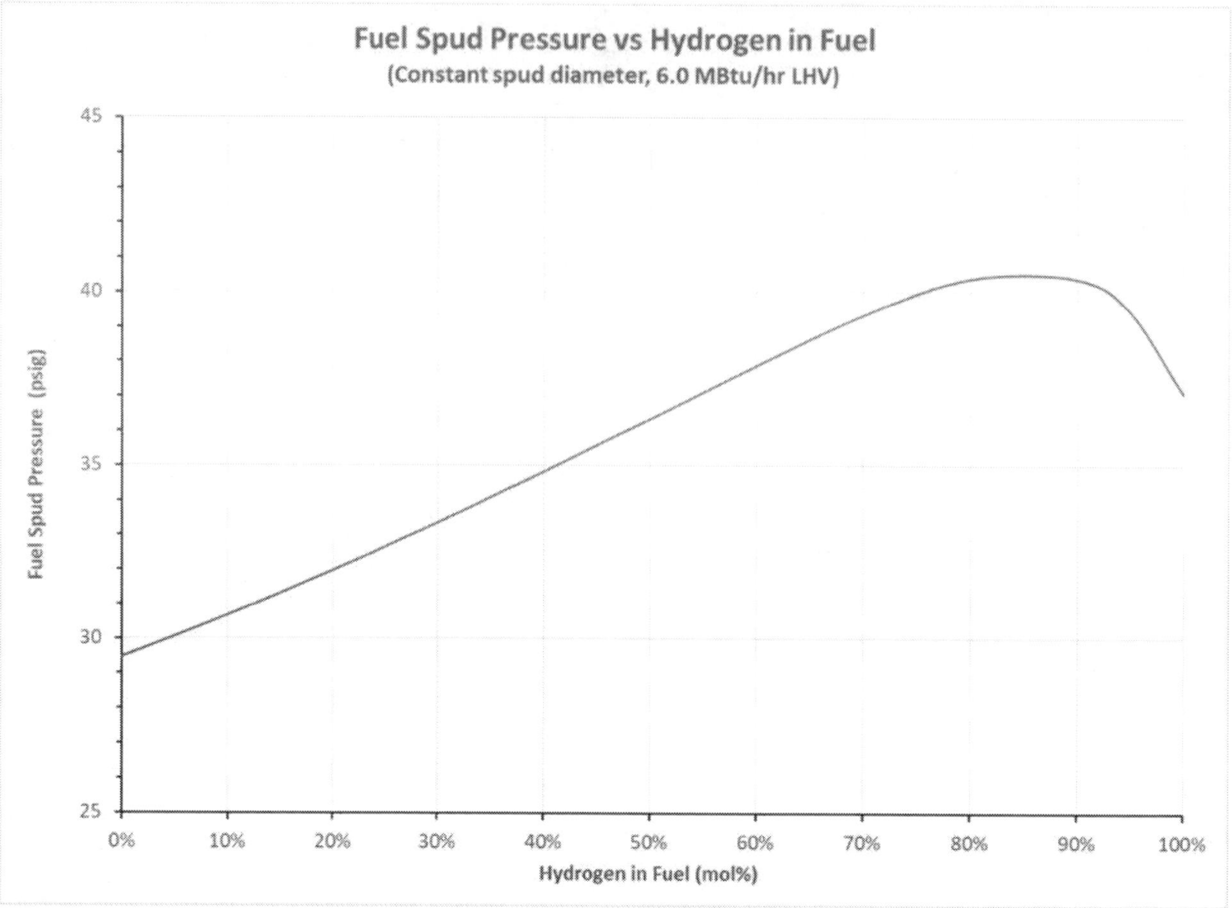

Figure 9: Fuel Spud pressure vs. Fuel Hydrogen (in a Methane/Hydrogen Mixture)

Second, as fuel hydrogen increases, prompt-NOx generation decreases due to the declining availability of carbon atoms in the combustion mixture.

Since the burner is highly effective at suppressing thermal-NOx, the reduction in prompt-NOx is hypothesized to more than offset any increase in thermal NOx as the fuel hydrogen increases.

As with the current generation of burner, injecting moderate quantities of steam into the venturi proved effective where very low NOx emission levels are required.

Figure 10 shows the trend of NOx emissions with steam injection rate for burners firing at 6.0 MBtu/hr (LHV).

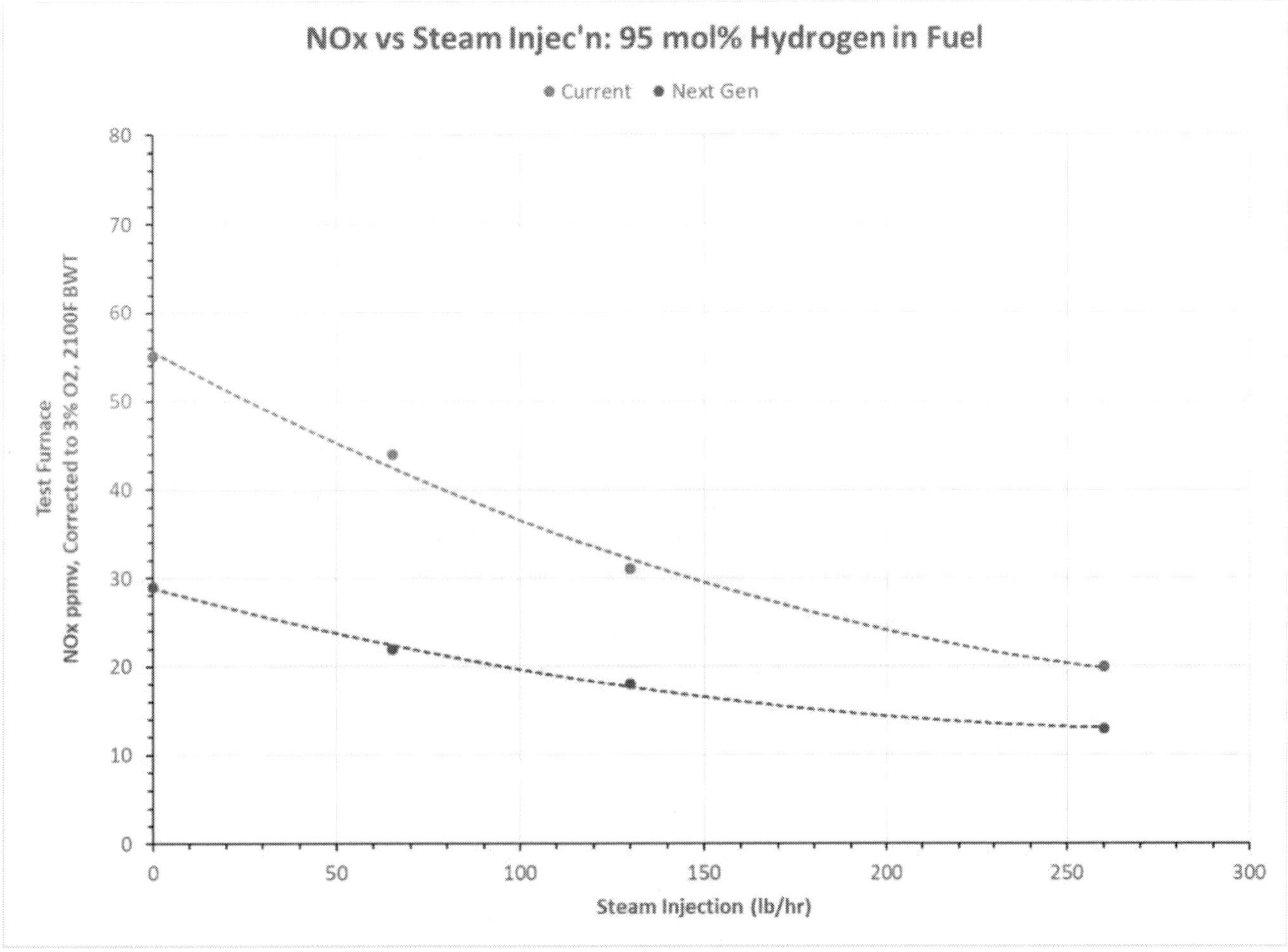

Figure 10: NOx Emissions vs. Steam Injection Rate

Figure 11 shows the "proof of concept" burner firing 6.0 MBtu/hr (LHV) heat release of 100% hydrogen fuel, operating with steam injection.

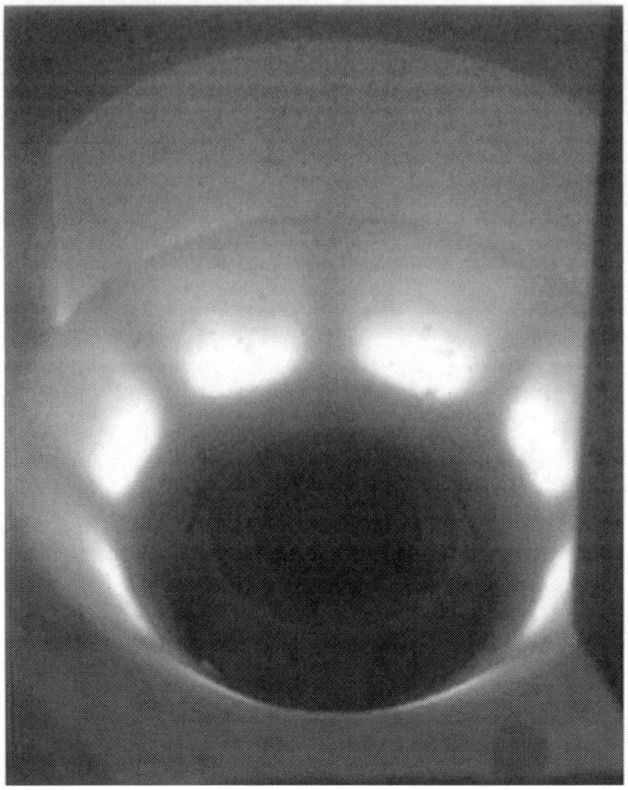

Figure 11

"Proof of Concept" Burner
6.0 MBtu/hr. (LHV) Heat Release
100% H2, 130lb/hr. Steam Inj.

7. Conclusions

A "proof of concept" burner has demonstrated very-high hydrogen fuel capability at commercial scale, in 2050 – 2100 °F furnace environment.

The burner characteristics were consistent with 100% floor-firing in a commercial steam-cracking furnace.

The burner was demonstrated on both ambient and pre-heated combustion air, across the full range from methane-rich tail-gas to very-high hydrogen.

The burner was consistently lit off on 100% natural-gas with no issues.

Initial testing suggests that transitions from very-high hydrogen to natural-gas fuels can be safely managed.

The burner NOx emissions declined as fuel-hydrogen increased.

The "proof of concept" burner provides a solid engineering foundation upon which to design a commercial hydrogen fuel burner for a 100% floor-fired steam-cracking furnace application.

References:

1. The John Zink Combustion Handbook, C. Baukal & R. Schwartz, 2001 Figure 6.19, p.208. CRC Press LLC.

AIChE Paper Number: 27e

ELECTRIFICATION OF STEAM CRACKING – SRT-e™ ELECTRIC CRACKING HEATER

Authors:

Richard Jibb

Technology Director

Lummus Technology LLC, Bloomfield NJ, USA

Baozhong Zhao

Director of Engineering

Lummus Technology LLC, Bloomfield NJ, USA

Alejandro Carrillo

VP Lummus Heat Transfer USA

Lummus Technology LLC, Bloomfield NJ, USA

Kishan Thakurdin

Thermal Engineer,

Lummus Technology LLC, Bloomfield NJ, USA

Charles Kawalski

Thermal Engineer,

Lummus Technology LLC, Bloomfield NJ, USA

Prepared for Presentation at the 2023 AIChE Spring National Meeting Houston, Texas, USA, March 12-14, 2023

35th Ethylene Producers Conference: Heaters/H2 Firing - Decarbonization Options Session

ELECTRIFICATION OF STEAM CRACKING – SRT-e™ ELECTRIC CRACKING HEATER

Authors:

Richard Jibb

Baozhong Zhao

Alejandro Carrillo

Kishan Thakurdin

Charles Kawalski

Lummus Technology LLC, Bloomfield NJ, USA

Abstract:

Societal demand for reductions in greenhouse gas (GHG) emissions is leading to a major shift in the way energy is used within an ethylene plant. This paper discusses options for electrification of new and existing ethylene plants to reduce emissions by replacing heat from burning fuel with electrical sources. Whilst electric drives are currently available to replace steam turbine driven machinery, electrification of the steam cracking heater is in demonstration phase [1]. Any device for providing large heating duty to a steam cracking heater using electricity as an energy source needs to consider three aspects – (i) Conversion of electric power to heat at high temperature and heat flux; (ii) heater structure, layout and safety considerations (iii) electrical supply and related interfaces. This paper describes how Lummus Technology's patent pending SRT-e technology addresses each aspect using essentially proven components. The SRT-e system can be deployed initially at smaller scale such as capacity expansion or expanded to a full-scale cracking heater subject to the available decarbonized electrical power supply.

Introduction Electrification of Ethylene Production

Electrification of services in the transportation, buildings, and industrial sectors coupled with decarbonization of electricity generation is a proposed pathway to achieving greenhouse gas reduction targets set by governments and private sector. A recent survey reported in Chemical Week magazine [2] indicates that 65% of companies surveyed in the Americas and 97% in EMEA have issued a 2050 net-zero commitment.

The term "electrification" generally refers to the process of making a machine or system operate using electricity when it did not before – e.g. the conversion of a machine or system that uses fossil fuels to the use of electrical power. In the context of a machine or system for producing ethylene and other important bulk chemicals "electrification" may include:

i. Conversion of steam turbines to electrical motors
ii. Direct electrical heating (or "Power to Heat")
iii. Use of electrical energy to generate hydrogen fuel
iv. Electrochemical production routes
v. or combinations thereof.

Conversion of steam drives to electric drives [3] and use of green or blue hydrogen as a fuel source [4, 5] have been considered in previous papers. In this paper we will focus on the direct use of electrical energy as a heat source for steam crackers, specifically conversion of the Lummus SRT® platform from a fuel fired system to an electrically heated system – the SRT-e™ electrically heated steam cracker - using readily available proven technologies.

Certainly, lowering the carbon intensity of electricity generation and substituting electricity for higher-emissions fossil fuels in energy intensive end-use sectors in the chemical industry can realize significant reductions in carbon dioxide emissions. Electrical heating reduces Scope 1 emissions from owned and controlled sources, however indirect Scope 2 emissions from generation of purchased electricity must also be considered when analyzing greenhouse gas emissions resulting from ethylene production. Although it is not the primary topic of this paper, the generation of electricity from renewable sources, or from fossil fuels with carbon capture, at the scale required for significant adoption is clearly a related issue. The use of any electric heater technology may be limited at a given location by the amount of electricity available, the Scope 2 emissions resulting from generating that electricity, and effects on the grid due to start-up and shutdown of the heater. The CO2 footprint for electricity generation is expected to decrease over time, with the phasing out of fossil fuels, carbon capture technologies and increased use of renewables. Nonetheless, it will be necessary to source significant amounts of electricity with a low or negligible CO2 footprint. Hence, the demand for electrically heated steam crackers must go hand in hand with the decarbonization of the electrical supply infrastructure.

For this reason and others, the SRT-e system has been designed to be highly modular at scales starting at ~20-30 KTA ethylene, but capable of expanding to much higher capacity including that of a world-class commercial facility. Electrical heating can be phased into operations, initially as capacity expansion, but later expanded to a full-scale cracker as a decarbonized electricity infrastructure develops.

SRT-e System Design Philosophy

In this section, we describe the design philosophy that was followed for electrification of a steam cracking heater. As with any significant design problem, it was first necessary to create a "Design Brief" outlining the key features of the product. The SRT-e system Design Brief identified the following key objectives:

- Leverage the existing SRT heater safety, reliability, and operational experience.
- Deployable at scales from 20-30 KTA - but capable to expand to any world scale capacity
- Use best available proven technology from best-in-class technology partners
- Maintain 6-8 years run length between turnarounds
- Optimize electrical system cost based on available components such as transformers and controls
- Be ready for commercialization in 2023

Figure 1 shows a three-dimensional rendering of an SRT-e heater with a capacity of ~ 200 KTA ethylene production and a radiant heating capacity of 60-80 MW. In the following sections we will cover the various aspects of the design in three areas:

(i) Conversion of electrical energy to heat
(ii) Heater structure, layout and safety
(iii) Electrical supply and related interfaces

Figure 1 – Lummus SRT-e electrically heated steam cracking heater

Conversion of electrical energy to heat - Electrical heating solution

The energy input required for steam cracking represents a problem of scale and material selection. The high endothermicity of the cracking process, high reaction temperature (800 °C - 900 °C) and short residence time (below 1 s) require heat flux at the reactor coil in the range 50 – 110 kW/m^2. The ideal heating device should be capable of providing the same energy input as a fired heater but should be reliable enough to operate continuously over the interval between coil replacements – typically 6-8 years.

An initial question in the design process was, whether to electrically heat the same coils from the outside in (i.e., in a similar fashion to conventional heaters) — or to develop new heating concepts (such as heating from the inside, directly passing a current/inducing a current in the cracking coils themselves, or internally heating the reaction coil). The latter approach would likely result in a significant change to the reactor design. Hence, outside-in heating was selected based on the objectives in the Design Brief.

Various (external) electrical heating technologies were considered for the SRT-e system, and there are several possible configurations as shown in Figure 2. Induction heating is a possibility for future development, however there is a need to address the lower efficiency of induction heating due to cooling of the electromagnetic coils used to induce current flow in the cracking coil or susceptor. Likewise direct (Joule) heating of a cracking coil can achieve essentially similar conditions to external resistance heating, avoiding the need to consider radiative heat transfer entirely; however, the lower resistivity of the cracking coils leads to very high currents, which in turn raises the cost of the electrical supply infrastructure. The need to maintain a constant cross section for the flow of current compromises the coil design and may require additional materials development and validation.

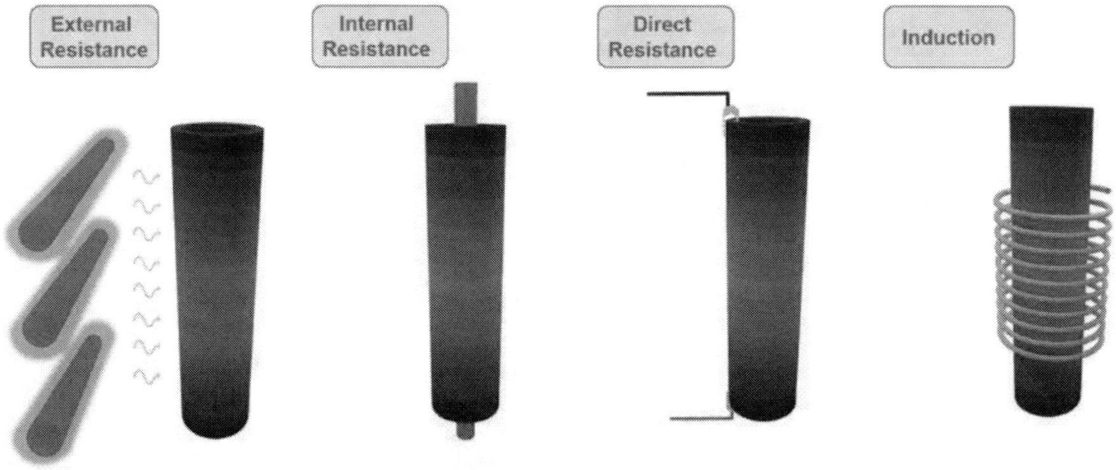

Figure 2 – Electrical heating methods – from left – external resistance heating, internal resistance heating, direct resistance heating and induction heating.

While all these options might eventually be considered for the SRT-e platform, initial development has focused on the outside-in heating using resistance heating elements to replace the burners used in conventional heaters. This is the most straightforward development path and provides the quickest route to commercialization. The subsequent section focuses on options for resistance heating at conditions required for an ethylene heater.

Electric Resistance Heating

Electric resistance heating is an established method of converting electrical power into heat. High temperature (> 1000 °C) resistance heating is already practiced in many other industrial applications including glass and aluminum manufacturing, heat-treating, annealing and for laboratory furnaces. In an isolated system, such as a cracking heater, converting power to heat by means of resistance heating, is highly efficient. Energy losses due to heat leak through insulation and transmission (so called I^2R losses) are small compared to the total heat supplied, and efficiencies around 98% are attainable. Experience in the glass industry includes heaters that operate at ~6 MW at a temperature of 1300 °C and heat fluxes as high as 180 kW/m^2

Resistance heating takes place by means of the "Joule effect". Joule's first law states that the power of heating (P) generated by an electrical conductor is proportional to the product of its resistance and the square of the current ($P = I^2R$, wherein I is the current and R is the resistance). Materials with relatively low resistance will require a higher current for a given power output. Most high temperature heating elements will operate under low voltage conditions, below 1000V. Although medium voltage heating elements (4160 V) are available, they are limited to lower temperatures by virtue of the need to use a dielectric to prevent arcing from the element.

The selection of a heating element needs to consider several factors, including heater temperature, atmosphere, service life, and power required. In particular, the element temperature (°C) and loading (power applied per unit surface, kW/m^2) dictate how a heating element can be used in each application.

The **maximum element operating temperature** (MET, °C) is the maximum temperature that the element can reach before failure. MET is a property of the element material, although the furnace atmosphere becomes important at high temperatures, since materials may react differently to various chemical compounds. A system that works very well at one temperature in air, may fail quickly if exposed to a moisture rich or reducing atmosphere at the same temperature. In a similar manner to the materials used for radiant coils, many of the elements rely on the formation and maintenance of a protective oxide layer.

By contrast, the **maximum element loading** (MEL, kW/m^2) is determined by the inter-relationship of operating temperature and the ability of the element to radiate the energy to a load – in this case the radiant coils - whilst remaining below the MET. MEL is affected by the element placement relative to the wall, and view factor to the load. An energized element will always be operating at a higher temperature than its surroundings. At the limit of operation, as the operating temperature increases, so does the element temperature and the heat flux must decrease to prevent the element overheating.

Metallic heating elements

Mineral insulated sheathed tubular heating elements are widely available in a large variety of sizes and configurations, but are limited to temperatures below 650 °C. These elements are not suitable for use in an ethylene heater.

Bare wire Nickel-Chromium (NiCr) heating elements are widely used in industrial furnaces and electric household appliances. NiCr elements are robust and repairable but their use is limited to a MET of ~1100 °C. and a corresponding MEL below ~30 kW/m^2. FeCrAl (Fecralloy) elements offer an improvement in performance allowing MET of 1300 °C at ~30-45 kW/m^2. Fecralloy heating elements require an oxidizing atmosphere (>200 ppm O2) to maintain an Al2O3 protective layer on the elements.

The combined temperature and heat flux required for the cracking heater are beyond the capability of sheathed elements and are at the edge of the materials limitations of the freely radiating bare wire metallic elements (NiCr and FeCrAl). Just as cracking coils materials have been developed for higher maximum tube metal temperature and longer life, resistance element materials have been developed for longer life and higher watt density at high temperatures. However, with currently available materials and product forms it seems likely that metallic elements will have short life and need to be replaced regularly, possibly while the heater is operating, or else the heat flux will have to be reduced so much that the resulting heater capacity (or size) becomes uneconomic.

High temperature heating elements

Other heating element types are available that have higher MET and MEL under the conditions required for steam crackers. These are discussed in more detail below.

Silicon carbide (SiC) heating elements exploit the fact that this material has a relatively high electrical conductivity and can be formed by extrusion or molding into a variety of shapes. SiC heating elements can achieve temperatures up to ~1600 °C and have MEL values in the range 80 – 140 kW/m^2 or higher. The elements are available as rigid tubular devices at relatively large diameters (commercially available up to 70 mm) allowing a high heating duty per element. The rigidity of the SiC elements means that they are self-supporting and can be installed across a span of a heater. Ceramic element types such as SiC can achieve higher performance than metallic elements at high temperatures. However, unlike metallic elements they are porous, forming intergranular oxides, which manifests as a change in resistivity over time - aging. The aging is managed by using a voltage reserve and by means of thyristor control, typically using silicon-controlled rectifiers (SCR) for each heating circuit.

Molybdenum disilicide (MoSi2) heating elements are available in various grades, with operating temperatures up to 1850 °C. MoSi2 elements can withstand oxidation at high temperatures due to the formation of a thin layer of SiO2 on the surface provided that a slightly oxidizing atmosphere (>200 ppm O2) is provided to maintain the protective layer. This property has led to the use of MoSi2 heating elements in various high temperature gas conversion processes. At temperatures ~1200 °C the elements become ductile while being brittle below this temperature. After a period in operation the elements become very brittle in cold conditions and thus are easily damaged. The electric resistivity of the elements is a function of temperature, but the resistance does not change due to aging.

Heater structure and layout

The SRT-e heater system is a patent pending design developed by Lummus Technology. The principle of the design is to use a single heated enclosure for each increment of ethylene capacity, which is considered a "unit cell" of production. Each "unit cell" consists of cracking coils and associated transfer line heat exchangers. It was found that a capacity of ~20-30 KTA and heat input of 8-10 MW was a particularly convenient size based on the electrical components required. The yields and performance of the system will be equivalent to, or better than, a fired system since control of heat flux profile with electrical heating can be much more uniform than the heat flux profile inherent to fired heaters.

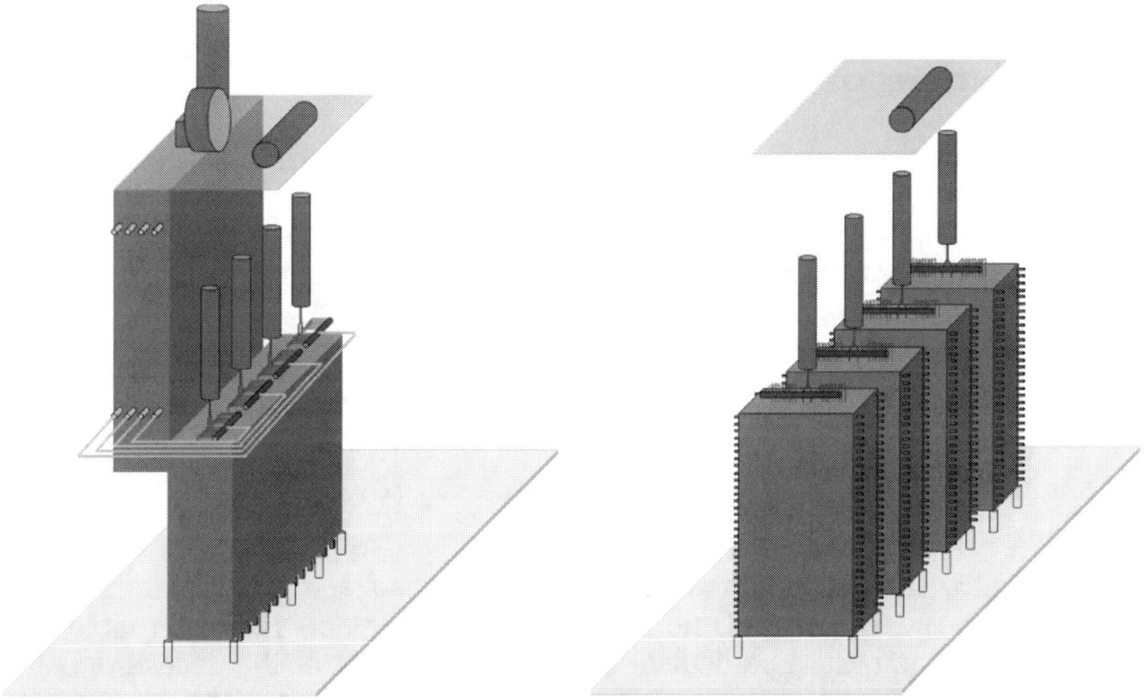

Figure 3 –Conventional SRT with major components and equivalent SRT-e

Grey: Radiant Box & Convection Section, Purple: Radiant SRT coil, Red: Burners (left) or heating elements (right), Blue: TLEs & steam drum, Orange: Convection Coil & cross-overs, Green: ID Fan and Stack

Figure 3 shows a schematic of a conventional SRT heater on the left side, and on the right side, an equivalent SRT-e "heater" configured as a group of multiple unit cells. The unit cells can be grouped in sets of 2-16 or more and arranged such that both coil removal and element replacement can be readily accommodated.

For a fully electric heater there is no flue gas for waste heat recovery and hence the convection section is not required, so heat recovery from the cracked gas product as well as preheating of feed streams also needs to be addressed in the design. Finally, but importantly, since there is normally no combustion occurring, the implications of a tube leak or rupture need to be carefully considered.

"Unit Cell" design – features and advantages

The reasoning to use a unit cell design rather than to retain a conventional heater casing and size requires some further explanation. Some key advantages of the SRT-e unit cell design are as follows:

- As previously discussed, the SRT-e unit cell design enables a modular approach, which was a key part of the original Design Brief. The SRT-e design is suitable for initially smaller scale capacity expansions and can later be scaled to larger heaters.

Additionally,

- The modular system eliminates the need to maintain an entire "spare" heater. Decoking can be carried out on a partial-heater (cell) basis.
- Heat flux distribution is not dependent on variables such as flame height, etc. and is expected to be both more uniform and controllable, which can positively impact radiant coil run length and life.
- The surface area available for mounting the elements can be maximized which in turn reduces the element operating temperature.
- The components required to deliver electrical power to the unit cell will have a capacity set by the maximum current flow at the voltage of the electrical load to the heater. As an example, a typical transformer to step down from a supply voltage to ~600V might be approximately 5-10 MVA based on a 6000A breaker limit. This lines up with the capacity of a single coil to produce ~20-30 KTA ethylene at 8-10 MW per coil.
- Each individual unit cell is of a similar power output to existing experience in other industries.

Any SRT coil design can be used in a unit cell – for instance SRT I, SRT III, SRT VI or SRT VII configurations. The preferred modular capacity of ~20-30 KTA is primarily based on electrical requirements and transformers/controls for each unit cell.

The electrical system is designed to consider decoking and high steam standby (HSSB) modes of operation. Since there is no firebox decoking, effluents will be routed to a decoke drum (with cyclone) to reduce particulate emissions. As a part of the decoke operation, HSSB (High steam standby) is also included. Both decoke and HSSB operations are like conventional fired heaters, except that the energy source is electricity.

Quenching and Heat recovery

The radiant duty, the heat input required for initiating and maintaining the endothermic cracking reactions, is provided by electrical resistance heating in the SRT-e system. Electrical heating replaces the combustion of fuel gas in a conventional SRT cracking heater. The crossover temperatures for the process gas entering the cracking coils in the SRT-e heater are in the same range as SRT fired heaters - between 550 and 750 °C, and likewise, outlet temperatures are in the range between 800 and 900 °C. An important part of the SRT-e design is how to heat the feed gas to the highest practical crossover temperature and how to quench, further cool and recover heat from the product gas.

In a fuel fired system, 35 to 50% of the total firing duty is used for the radiant section whilst 40 to 60% of the heat input into a heater (i.e., of the firing duty) is recovered in the convection section. The flue gas is cooled down to between 80-120 °C before being released to the atmosphere via the stack. Taken together, the radiant and convection zone of a modern Lummus SRT steam cracking heater can make use of more than 95% of the overall fired duty (lower heating value or net calorific value). In the SRT-e system there is no convection section, but a high crossover temperature is still desirable to minimize the electrical consumption to that required to provide the heat of reaction.

In the absence of a convection section, heat input to the vaporized feed mixture can be maximized through heat recovery from the cracked gas. In later discussion, we will refer to the equipment for quenching of cracked gas using boiler feed water to generate super high pressure steam as "Primary Transfer Line Exchanger" – PTLE, and the equipment for secondary cooling as "Secondary Transfer Line Exchanger" – STLE. In effect, the transfer line exchangers are two exchangers in series – the first (PTLE) to quench the reaction by generating steam, the second (STLE) to preheat the reaction mixture. The amount of heat that is recovered to generate steam in the PTLE can be minimized to increase the temperature of the reaction mixture entering the cracking coils, which in turn reduces the electrical heating required in the radiant section.

Primary Transfer Line Exchanger (PTLE)

Cracked gases leaving the radiant coil of the ethylene heater at 800-900 C should be cooled with minimum residence time in the adiabatic section between the coil outlet and the quench system cooling zone. Quenching is achieved in a smaller PTLE than that typically used for fuel fired SRT heaters, but the major design features are retained – i.e., minimum residence time in the adiabatic zone, uniform flow to prevent eddies in the entry zone to the PTLE and low pressure drop. Since the adiabatic interface between cracking coil and PTLE matches the current SRT best practice, residence time and reactor yield are not affected. The PTLE cools the cracked gas product from the reaction coil outlet against high pressure boiler feed water to generate saturated super high-pressure steam. As shown in Figure 4, the steam can be used to superheat a vaporized feed (in the case of naphtha cracking), although other heat integration schemes for other feed types can also be accommodated.

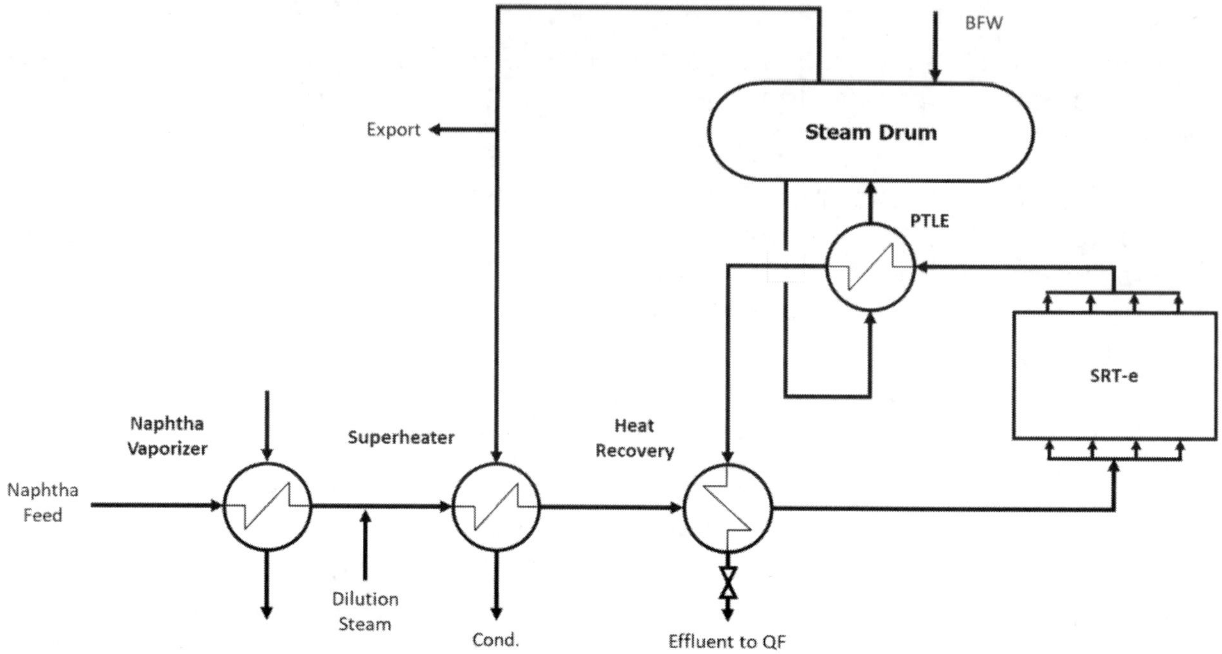

Figure 4 – Potential flow scheme and heat recovery for SRT-e system with a naphtha feedstock

Secondary Transfer Line Exchanger (STLE)

The STLE is essentially a feed effluent type heat exchanger. Lummus has extensive experience over the last 40 years of preheating feed mixtures (hydrocarbon feed or hydrocarbon plus dilution steam) using STLE and the same concept is applied in the SRT-e electric cracking heater.

The cracked effluent is cooled on the tube side of a dedicated set of shell and tube heat exchangers against a mixture of hydrocarbon feed and dilution steam on the shell side. The extent of the available heat recovery from the cracked gas is limited at the upper end by the temperature of the quenched effluent stream and at the lower end by the lowest temperature for the cracked effluent, which in turn depends on the type of feed. The lowest outlet temperature is to avoid fouling due to condensation of heavy components, which rapidly increases outlet temperature and pressure drop to a point where the resulting run length or time between decoking cycles becomes controlled by the STLE operation.

Venting of the SRT-e unit cells

Protection of the SRT-e heater is necessary in the event of damage to the reaction tubes. The scenarios considered range from minor leakages to one or more reaction tubes being severed completely – a coil rupture event; during which, there is an escape of combustible gas into the unit cell. Whilst such events are rare, they must still be considered in the design of the system.

For electrically heated reactors, there is normally no combustion occurring, so the atmosphere must be specified as part of the design of the system. The SRT-e system operates with a nominally neutral pressure in each of the multiple individual unit cells. The atmosphere in each unit cell is in communication with a duct system leading to a common flare stack for each group. It is possible to operate at atmospheric or reduced oxygen concentration, with provision to monitor for CO, CO2, or hydrocarbon to detect any leakages.

Any input of flammable gases could have the following consequences:

i. Accumulated hydrocarbons in the reactor vessel have the potential to reach the explosion or detonation limit with the normal oxygen content of the air and temperatures above the auto-ignition temperature.

The amount of oxygen in each unit cell is relatively small in comparison to the leakage of hydrocarbons. For ignition to occur the mixing of cracked gas with the oxygen-containing gas must pass through the flammable envelope of the cracked gas and a suitable source of ignition must be present. Since open flame and spark ignition sources do not exist in the SRT-e heater system, the remaining ignition source possibilities are the elevated temperature of the cracked gas and the high temperature of the interior surfaces, particularly the electric heating elements. The SRT-e system is designed so that individual unit cells feed to a common vent system to ensure that ignition of remaining hydrocarbon occurs in a controlled location.

ii. The energy input due to combustion could lead to overheating.

The energy potential of the cracked gas is of course very large, although as previously noted, the available oxygen is quite limited, and in addition there is no fan or natural draft drawing air into the heater. If the available oxygen is completely consumed then the energy input will result in a temperature excursion at the front envelope of the hydrocarbon leak stream with the air inside the heater. The duration of combustion will be within a few seconds before the oxygen is exhausted, however, the hydrocarbon will continue endothermic cracking which leads to a temperature drop. Therefore, the overheating of heater components, if any, will be limited.

iii. Complete or incomplete combustion, together with the volume of gas flowing out of the reaction tubes, can lead to a rapid increase in pressure.

The relief area from each unit cell to the common duct is selected to avoid over pressurization due to the combination of gas flowing into the enclosure, and the pressure rise due to heat from combustion of the hydrocarbon material with the oxygen contained.

Deployment of the SRT-e system at smaller scale

For an initial, smaller scale, deployment the SRT-e system can be integrated with a facility operating conventional fired heaters in parallel. For instance, if a capacity expansion is 20-30 KTA with one unit cell, it is likely that the overall steam balance will not be affected by the fact that the electric heater generates less SHP steam than a conventional cracking heater, and this steam is also not superheated. For a larger expansion in a modern plant, the overall steam balance can be closed by putting one (or more, as needed) of the compressors in electrical drive so that loss in SHP steam is compensated. In the scheme shown in Figure 4, saturated SHP steam is used as a superheating medium.

Deployment of the SRT-e system at larger scale

At a larger scale, multiple unit cells will be grouped together. The concept of an individual "heater" is slightly different from the conventional system. There will be a small amount of SHP steam available, which can be used for preheating of the feed, but more of the steam turbine drives will likely be replaced by electrical motor drives. Multiple configurations are possible depending on the level of isolation between unit cells (i.e., the number of transfer line valves). The capital cost associated with the electrical supply is somewhat offset by the reduction in structural steel since there is no longer a convection section.

Electrical System

The SRT-e heater system includes the electrical infrastructure to deliver 5-10 MVA to the heating elements. As previously noted, this approach to the electrical system design follows much of the already established practices in other industries for large high temperature furnaces operating at similar conditions to a steam cracking heater.

Unlike supply of power to a motor drive where inductive loads must be considered; the electrical heating load is entirely resistive. Additionally, the load is required at a low voltage level so we must supply a relatively high current.

Figure 6 illustrates the electrical system in the form of a one-line diagram, indicating the major components required. All of the components illustrated have been sourced based on commercially available and proven equipment suppliers. It is noteworthy that the available capacity of some of these components is similar scale to the energy input required by each unit cell. As such, the natural grouping is to use 2 x 5MVA transformers per unit cell with corresponding fuses and breakers.

In the example shown in Figure 6, the high voltage supply from the switchyard is stepped down from high voltage to low voltage via 2 x 5 MVA transformers per unit cell. Each transformer is paired to a control panel which houses a set of contactors, breakers and a control unit feeding a set number of heating circuits.

The power to each circuit is separately controlled using silicon-controlled rectifier units (SCRs) which are solid state switching devices capable of fast, highly variable proportional control of electric power. The SCR control option is preferred over other control methods, based on the availability of relatively low cost, well proven SCR technology already in use.

The transformers and control panels will be housed in a dedicated electrical building. Standard low temperature cables will run a three-phase electrical supply from the electrical building to a junction box located adjacent to each face of the unit cell. From the junction box, high temperature cables run up the side of the heater casing, through a special gland in the casing to a bus bar inside a weather-protected enclosure. Each heating element connects to a bus bar inside the enclosure. The electrical/heater interface of the SRT-e heater has been carefully designed to ensure that the element connections are both robust and accessible for installation and maintenance. The cracking coils and heating elements can be accessed independent of one another.

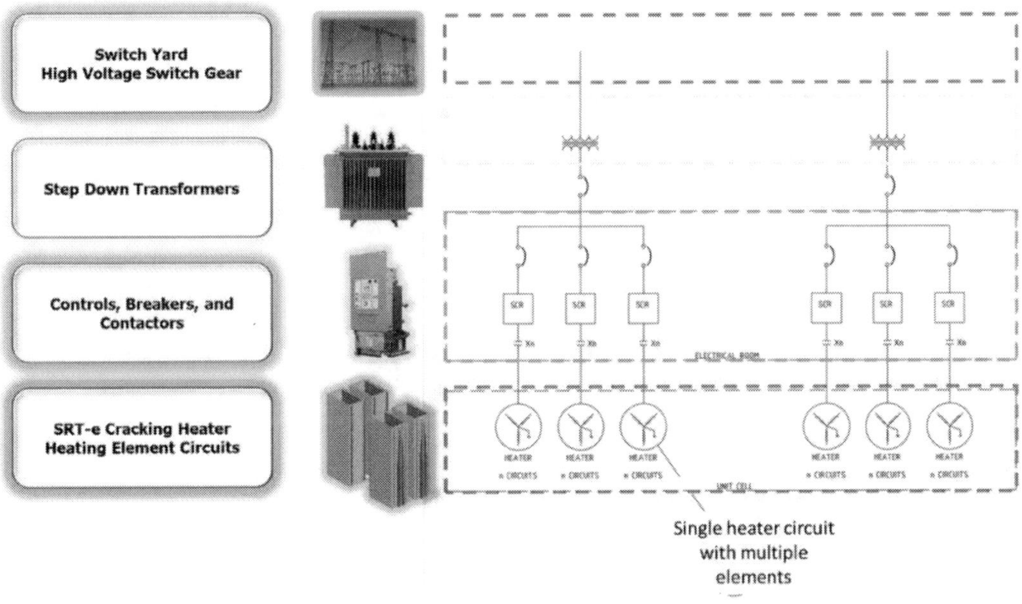

Figure 5 – Major electrical components required for each SRT-e Unit Cell

Based on a preferred option to use high temperature resistance heaters, the SCR controllers will allow control of electrical power to each heating circuit, as well as providing the ability to adjust voltage over time to compensate for any changes in the resistance of the heating elements with either time or temperature. The heat flux profile for the SRT-e electrical heater can be controlled by setting the power input to each element heating circuit. Additionally, flow control can be applied by using techniques already in operation for fuel fired SRT heater technology.

Summary

The Lummus SRT-e platform has been developed to extend electrification of ethylene plant operations to the radiant section of cracking heaters. The preferred heating method is electrical resistance heating, using commercially proven high temperature resistance heating elements installed on the heater walls. In the future the same platform may be used to develop other electrical heating methods as the technology matures and becomes commercially available. The SRT-e platform is modular at the scale of a single unit cell of ~20-30 KTA ethylene capacity and maintains as far as possible the best features of Lummus SRT technology from fuel fired systems. The technology can be applied at smaller scale for capacity expansion type projects, as well as for new grassroots commercial facilities.

References

[1] Chemical Engineer online "Greener chemicals: steam cracking could go electric by 2023" March 24th 2021

[2] Chemical Week Magazine 2/9 January 2023 p20 "Sustainability remains an imperative"

[3] AIChE technical paper reference number: 6c "Mechanical and Electrical considerations of electric motor/Drivers for large centrifugal compressors in Ethylene cracker applications" Malnedi K., Harvey S., Rauber A.

[4] AIChE technical paper reference number: 118f "The reforming of cracker fuel gas", Rogers N. 2022

[5] AIChE technical paper reference number: 27a "Hydrogen and Ammonia: Zero Carbon Fuels for Steam Crackers" McKenzie D., Strupp C., Henneke M., Haag M. 2023

Review Technology and Operational Issues for Utilizing High Hydrogen Containing Fuels in Ethylene Furnaces and Process Heaters

Raghu Narayan, Shara Narayan

The objective of this paper is to provide a 'deeper dive' and determine the technical and operational issues of converting fired heaters, such as Steam-Methane Reformers, Ethylene Cracker Furnaces and Process heaters to using hydrogen as a fuel, for the purpose of reducing greenhouse gas emissions. The focus of this overview paper will be on the following areas:

- Heater performance, influenced by thermodynamic and combustion characteristics of the fuel
- Required Burner performance and robustness
- Fuel gas system retrofit requirements
- Impact of GHGs and Retrofit Needs for Emission Control System
- Mechanical Materials related issues and requirements

AIChE Paper Number: 27f

Electrification of steam cracker furnaces – Technology demonstration and the route to commercialization

Mathieu Zellhuber
Process Design Engineer
Sustainable Hydrocarbon Solutions - Linde Engineering

Martin Hofstaetter
Process Design Engineer
Sustainable Hydrocarbon Solutions - Linde Engineering

Rainer Kemper
Technology Advisor
Sustainable Hydrocarbon Solutions - Linde Engineering

Gunther Kracker
Head of Product Management & Cost Estimation
Sustainable Hydrocarbon Solutions - Linde Engineering

Prepared for Presentation at the 2023 AIChE Spring National Meeting
EPC Technology Session: Heaters/H2 Firing - Decarbonization Options
Houston, TX., March 12-16, 2023

Electrification of steam cracker furnaces – Technology demonstration and the route to commercialization

Mathieu Zellhuber
Process Design Engineer
Sustainable Hydrocarbon Solutions - Linde Engineering
Martin Hofstaetter
Process Design Engineer
Sustainable Hydrocarbon Solutions - Linde Engineering
Rainer Kemper
Technology Advisor
Sustainable Hydrocarbon Solutions - Linde Engineering
Gunther Kracker
Head of Product Management & Cost Estimation
Sustainable Hydrocarbon Solutions - Linde Engineering

Abstract

To keep up with the targets of the 2015 Paris Climate Agreement and limit global warming to preferably below 1.5 degrees, all energy- and emission-intensive industries must undergo a profound and fast transformation of their production assets and applied technologies. Olefin technology licensors and operating companies are thus urged to provide viable solutions in the form of carbon emission neutral plants, which ensure the required production in a safe, reliable, and sustainable manner.

Cracking furnaces are the largest greenhouse gas emission source in steam cracker plants and cause a significant share of the overall emissions in the petrochemical industry. The paper and associated presentation focusses on the electrification of cracking furnaces. By converting the heat generation from fossil fuel firing to electric heating using renewable power, flue gas emissions from the cracking furnaces are completely avoided. The use of electric energy at such high-power duties for such a high-temperature process brings a disruptive innovation to the industry. The electric cracking furnace is a key enabler to reduce the total CO2 emissions from steam cracker plants by more than 95% and hence to achieve the sustainability targets of the petrochemical industry.

In 2021, the consortium partners BASF, SABIC and LINDE announced a collaboration to jointly develop electric cracking furnaces. In 2022, construction works began for a multi-megawatt demonstration plant at BASF in Ludwigshafen (Germany), targeting the demonstration of two different electric heating technologies at industrial scale. With start-up scheduled for 2023, the two additional furnaces will be completely integrated into the commercial production of ethylene and other high value chemicals. The subsequent commercialization of electric furnaces is planned by the end of 2024.

The conference contribution provides information on the two electric heating technologies developed and now ready for demonstration. Characteristics of the demonstration plant are presented, underlining that its operation will be highly representative for the operation of future commercial-size electric furnaces.

Finally, an outlook on integration scenarios of commercial-size electric furnaces into grassroot and existing plants is given. It shows the potential and benefits of considering electric furnaces timely in the development of long-term sustainability roadmaps of olefin production sites.

Introduction and motivation

The petrochemical industry accounts for approx. 5% of the global CO2 emissions [1]. In particular the upstream conversion from fossil feedstocks (e.g. naphtha) to base chemical building blocks (e.g. olefins) is generally a highly energy-intensive step and therefore accounts for a significant share in overall CO_2 emissions.

The steam cracking industry, as one of the main process routes, was estimated to account for approximately 25% of the overall greenhouse gas (GHG) emissions in the European chemical industry [2]. At a global scale, estimations for the total CO_2 emissions from steam cracking are in the range of several hundred millions of tons per annum [3, 4, 5]. For instance, a single state-of-the art naphtha steam cracker produces approx. 1,200 kta of CO_2 considering a name plate capacity of 1,500 kta ethylene.

This underlines the importance of developing novel technologies quickly and making them accessible to the industry, in order to counteract climate-damaging emissions and to take an important first step towards greenhouse gas neutrality.

More than 90% of the CO_2 emissions of a modern steam cracker plant can be attributed to the high energy demand of the process [3]. It is required to provide the reaction heat for the endothermic conversion in the cracking furnaces, which moreover needs to be supplied at high temperatures (e.g. gas temperature of around 850 °C at the coil or reaction tube outlet and up to 1100 °C maximum tube metal temperature). Significant amounts of energy are also needed to run the work-up section, which separates the raw cracked gas into the commercial "high value chemicals (HVCs)" like ethylene, propylene, the crude C4 stream containing butadiene, and the aromatic pyrolysis gasoline containing benzene and other aromatics.

The main primary energy supply for the overall plant is currently ensured by burning the "methane fraction" (which is a by-product from the cracking reaction) and natural gas in the cracking furnaces, to provide the necessary high-temperature reaction heat, but at the same time also generate from waste heat the necessary utility steam for the subsequent process steps. Consequently, a very large proportion of scope 1 GHG emissions of a steam cracker plant are released to atmosphere via the furnace stacks. Therefore, the cracking furnaces have been identified as an optimal starting point for a significant reduction of the GHG footprint of steam cracking.

The electrification of cracking furnaces eliminates large flue gas streams emanating from the furnaces and thus completely avoids the related direct (scope 1) CO_2 emissions. Indirect (scope 2) emissions linked to electric power supply are ideally avoided by supplying 100% renewable power to electrified furnaces. As will further be explained in this paper, electrically heated cracking furnaces will be preferably combined with the electrification of main compressor drives of the steam cracker, to compensate for the reduced utility steam production, compared to a state-of-the-art fired furnace. The overall CO_2 balance of base chemical producers can thus be significantly improved.

The CO2 avoidance character differentiates the furnace electrification from other steam cracker decarbonization options such as blue hydrogen firing (pre-CCS) or flue gas CO_2 capture (post-CCS), which act via sequestration of CO_2 and require CO_2 outlets plus subsequent handling for storage and/or utilization. As a further distinction, electric furnaces are particularly suited to locations with high availability of renewable power supply, potentially enabling in the long-term low energy import costs without additional operating expenses for CO_2 capture and handling.

Hence, there is a great incentive to develop technically and commercially mature solutions for the electrification of cracking furnaces, as also pointed out by Amghizar et al. [3].

This paper presents the corresponding technology demonstration activities and the roadmap to future commercialization laid out by the joint collaboration partners BASF, SABIC and LINDE. In the development process, the technical teams touched new ground. Two alternative electrical heating concepts for steam cracker furnaces have been developed and are being implemented in a large-scale demonstration plant. Furthermore, case study examples will shed light on the integration possibilities of electric furnaces in existing and in grassroot plants.

Scope and impact of transition from fired to electric furnaces

Conventional fired furnaces form the heart of a steam cracking plant, not only because they host the cracking reaction in their radiant section or firebox, but also because of their central role in the plant's energy system. Large quantities of waste heat are recovered from the flue gas and process gas streams leaving the radiant section. They are used to preheat furnace feed streams and to generate high pressure utility steam, which is subsequently used in turbines and heat exchangers for mechanical work and heat supply in the downstream separation section.

Transitioning from fired to electric furnaces therefore requires solutions at three different scales of unit/plant design:

1. At the core of an electric furnace, in the coilbox, a reliable, temperature-resistant electric heating technology that supplies the required heat for the reaction, with high power control accuracy and well-suited heat flux density profiles for maximum selectivity and runtimes between decoking cycles.
2. At the periphery of the coilbox, a furnace heat integration system, which efficiently recovers waste heat from the hot cracked gas stream, supplies energy to the feed preheaters of the furnace, and exports potentially residual balance heat to other process units, e.g. in the form of steam.
3. At an overall plant level, energy flows and in particular steam balances are bound to change if one or more electric furnaces are included in a plant. Consequently, power and heat supply for normal operation, turndown, start-up/shutdown and upset scenarios need to be ensured by alternative means compared to conventional cracker plants with fired furnaces and boilers.

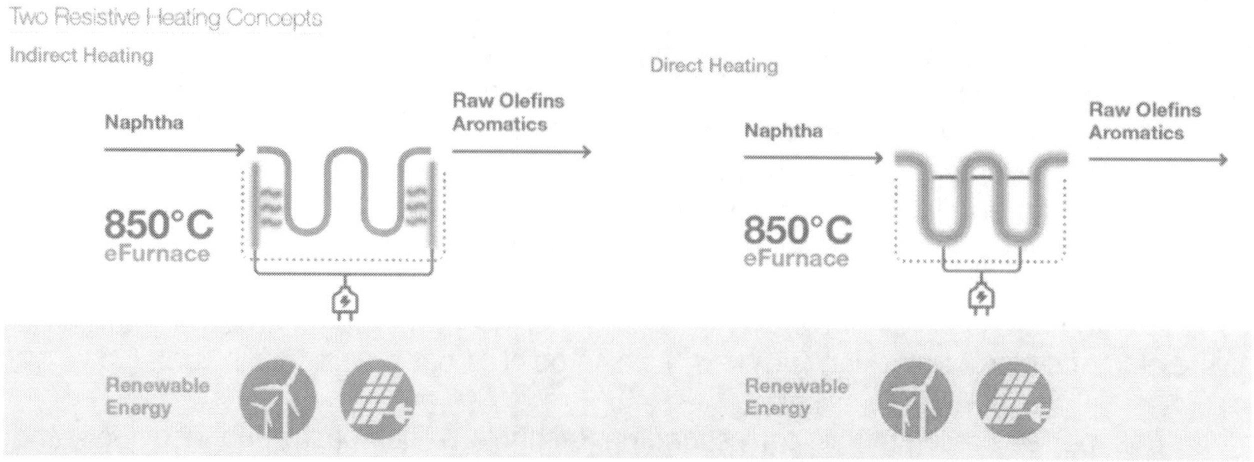

Figure 1: Principle sketches of the developed indirect (left) and direct heating concepts for electrified steam cracker furnaces (from [6])

For the core part of the electric furnace, the partners BASF, SABIC and LINDE have developed two alternative technical solutions, whose principles are sketched in Figure 1.

Both heating concepts are based on resistive heating i.e. with electric power supply to the resistive heating material via solid electrical conductors (unlike inductive heating). The difference between the two concepts lies in the location of the resistive heating material within the furnace.

In a furnace with "direct heating", an electric current is applied directly to the process tubes/cracking coils inside the reactor. Joule heating inside the tube wall, caused by the electric resistivity of the cracking coil material, provides the required thermal energy for the chemical reaction.

In a furnace with "indirect heating", an electric current is applied to distinct heating elements that are brought to the necessary temperature by Joule heating. The supply of the thermal energy required in the process tubes is subsequently achieved by mostly radiative heat transfer from the heating elements and the insulated walls of the coilbox.

Extensive pilot testing activities and numerical simulations with different calculation approaches were undertaken to confirm the feasibility of both heating technologies and to establish a sound knowledge and data basis for the design of a demonstration plant and future commercial plants. The techno-economic analysis conducted by the partners showed that both heating concepts have their own advantages depending on the specific use case, while offering comparable energy efficiency/power consumption. This is the reason why both approaches will be demonstrated alongside each other in one demonstration plant, see corresponding section below. Further technical details about the heating technologies and performance data will be made publicly available after start-up and initial operation of the plant.

Both heating concepts in the furnace core are compatible with the wide range of feedstocks typically processed in existing cracking furnaces. This also includes emergent, sustainable feedstock alternatives such as bio-feedstock and plastic waste pyrolysis oil, which enable reductions in scope 3 GHG emissions. This high feedstock flexibility results from the fact that either type of electrically heated coilbox, i.e. with direct or indirect heating, can be combined unrestrictedly with multiple layouts/topologies of furnace heat integration. The customization of an electric furnace to a given feedstock scenario therefore mostly takes place at the periphery of the coilbox, i.e. in the heat integration, see Figure 2. This is analogous to fired furnaces, which do also show most significant differences in convection section and cracked gas cooling section designs, for example between liquid and gas feed furnaces. As a supplement, for both electric heating concepts, also the cracking coil design and/or the power control of the coilbox can be tailored to optimize the reactor performance for a preferred feed and operation scenario.

As will be explained more in detail in a separate section below, the integration of an electric furnace into the overall plant needs to be considered with a long-term perspective. The furnace heat integration scheme must be aligned with the overall steam and utility system of the plant. In case of grassroot projects, these central systems are likely to be very different than in existing plants, with high degrees of electrification throughout the plant and reduced utility steam circulation. In case of plant revamps, the installation of an electric furnace as add-on or replacement of a fired furnace may also induce the need for simultaneous modifications in the separation section of the plant. As for the feed flexibility, both electric heating technologies are equally suitable for grassroot and revamp applications as add-ons or replacements.

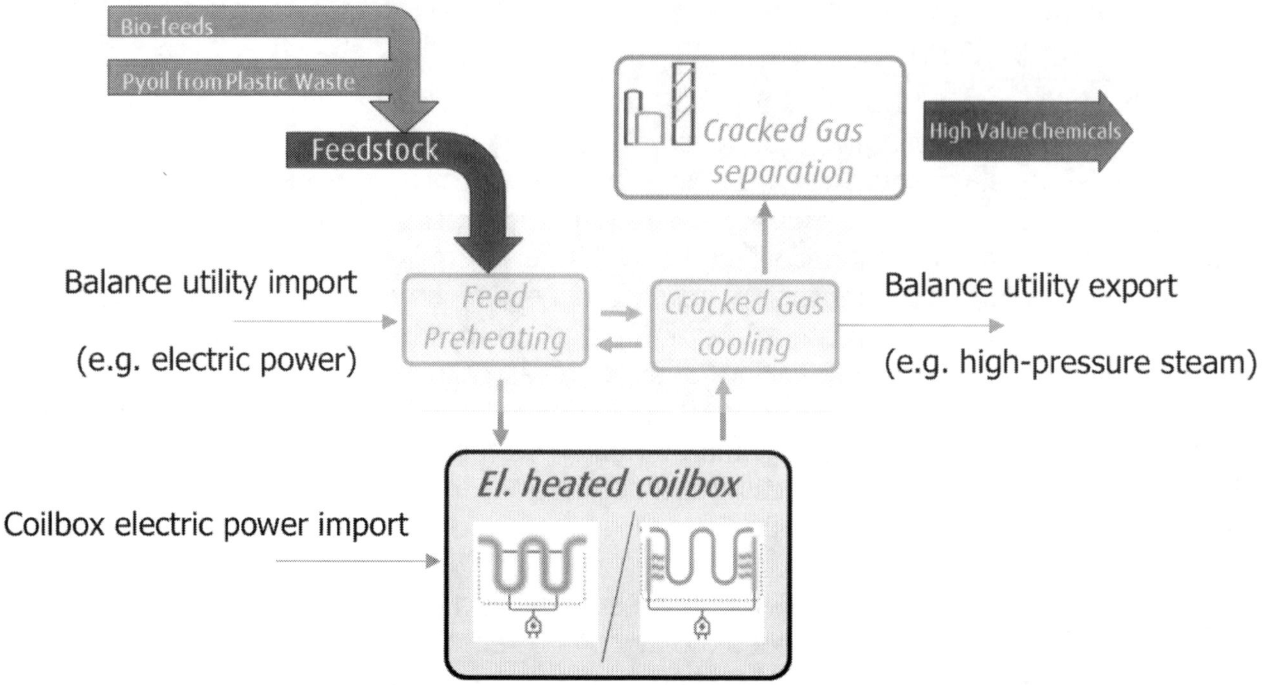

Figure 2: Sketch of overall structure of an electrified furnace system with heat integration. Both heating technologies are suitable for the entire range of relevant feedstocks.

Technology demonstration, scale-up & commercialization

Developing, demonstrating and scaling up two alternative electrical heating concepts for steam cracking furnaces is a major undertaking spanning over several years. It involved the three named collaboration partners, thereby bringing together two major operating companies and an ethylene technology licensor, as well as a number of selected partners supplying key components for the implementation of the new technologies. The overall development roadmap and setup is illustrated in Figure 3.

The fundamental technology development by the collaboration partners seamlessly transitioned into parallel, but strongly interweaved workstreams focussing on:

1. the execution & operation of a highly versatile demonstration plant, and
2. the upscaling of the developed technologies to full-size commercial furnaces.

The combined knowledge, design and operation experience resulting from both workstreams will serve to ensure the maturity of the technology and of the execution workflows for the commercialization of electric furnace technology, which is scheduled for end of 2024.

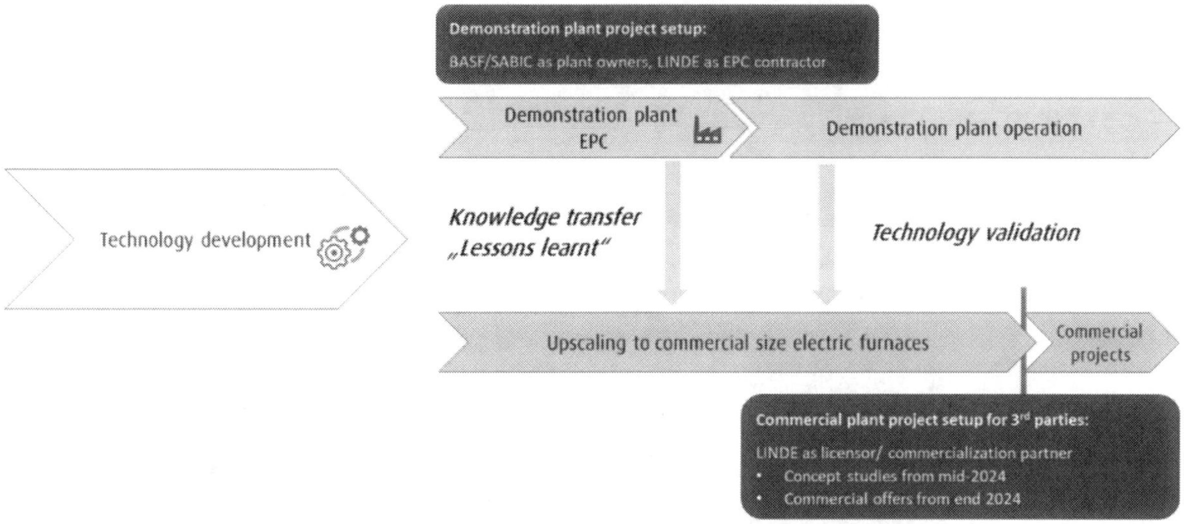

Figure 3: Overall technology development roadmap and setup

The demonstration plant at the BASF Verbund site in Ludwigshafen (Germany) takes a key role on this path. Its main goal is to demonstrate continuous olefin production using electric energy as a heat source and to serve as a reference for an industrial-scale plant. The plant is currently under construction and will be operated under industrial conditions with the purpose to demonstrate the reliability of key components, like materials of construction and custom-made components, for use in this type of high-temperature reactors. Its start-up is scheduled for the second half of 2023. The test operation will last for several years and the plant will be accessible for reference visits. The demonstration project has been awarded funding by the German Federal Ministry for Economic Affairs and Climate Action [6, 7].

The demonstration plant has the following key features:

- Both heating technologies (direct and indirect) will be demonstrated in separate coilboxes.
- In total, up to 6 MW of renewable power can be drawn from the 6kV grid and used for preheating the feed and sustaining the cracking reaction.
- In total 4 tons per hour of naphtha feed can be processed, and the cooled cracked gas is being transferred to the existing BASF steam cracker for the commercial production of HVCs.
- Cracking coils in both coilboxes are of commercial dimensions.

The demonstration plant is designed for maximum flexibility, allowing parallel operation of the two radiant boxes, but also cracking operation of only one furnace, while the second furnace is either in decoking, standby or shutdown. The decoking procedure will be similar as for conventional fired cracking furnaces (introduction of steam/air mixture, burning of coke).

Figure 4: 3D model snapshot of electric furnace demonstration plant at BASF Ludwigshafen

A 3D model snapshot is shown in Figure 4, indicating also with faded colours the surrounding elements of the existing steam cracker into which the demonstration plant is integrated. A functional sketch of the demonstration plant architecture is given in Figure 5.

The integration of the demonstration plant into the infrastructure of one of the existing BASF Ludwigshafen steam crackers is at the same time a key feature and a prominent challenge. It means that the demonstration plant EPC execution as well as the plant operation must comply with the strict regulations and permitting processes of a commercial production facility. Consequently, all common industry-standard design reviews have been conducted, documented, and followed up, based on best practice from conventional process plant execution projects. In addition, the standard procedures were supplemented by a newly devised multi-disciplinary review format, namely an "electric HAZOP", which was conducted with support from notified body experts. This review specifically focussed on the novel aspects of the implemented technologies and their innate, high share of electric components. Furthermore, the demonstration plant will undergo the complete certification process as for a commercial unit.

The demonstration unit serves to prove the full maturity of an electrified furnace and as valuable basis for the future implementation at larger commercial scale. The indicated capacity and the layout of the demo unit were selected to enable a reliable upscaling. Major equipment is already close to commercial-size dimensions or the scaling factor is below 10, which is considered in the industry as factor not implying significant upscaling risks. Partially numbering up will be applied, thereby making use of industrially proven equipment.

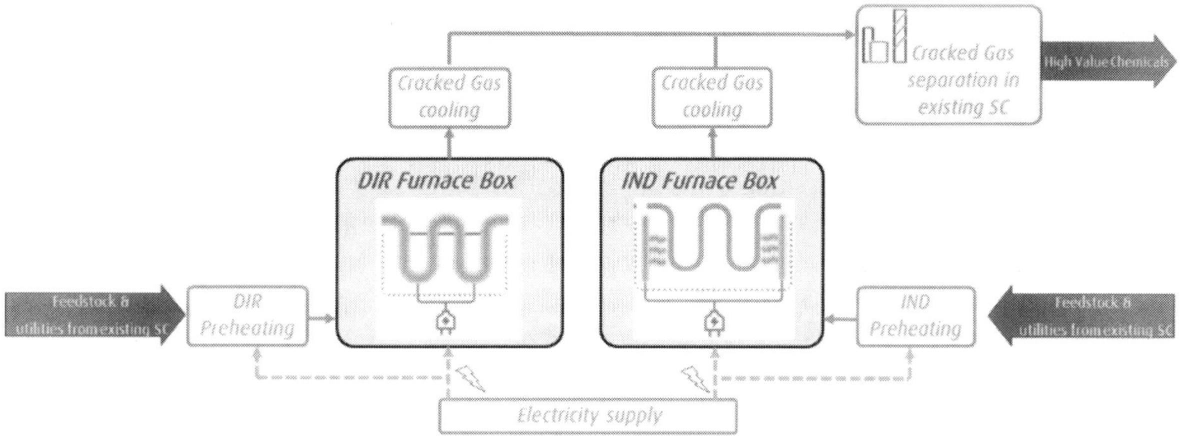

Figure 5: Functional sketch of demonstration plant architecture

Case study examples for future integration of electric furnaces into steam cracker plants

It is expected that the first electrified furnace projects at commercial scale will be add-on furnaces or replacement furnaces to existing sites. Capacities will likely be in the range of 40 to 80 kta ethylene to fit to the capacity of existing steam crackers for example in Europe and to stay in line with existing single furnace capacities of a site.

In this context, it can be of great interest for operating companies to already include the electrification of individual furnaces into their long-term sustainability and decarbonization roadmaps for existing steam cracker plants. As mentioned earlier, the stepwise integration of electric furnaces is preferably synchronized and aligned with modifications in the separation train, in particular with stepwise electrification of large rotating machinery to balance out the reduced steam production in the furnaces. Moreover, it can as well be advantageous to combine the partial furnace electrification with alternative decarbonization measures in the furnace area, for example by adding air preheating, flue gas carbon capture and/or by switching to blue hydrogen firing in remaining fired furnaces.

An exemplary illustration of possible long-term decarbonization pathways resulting from a revamp case study is shown in Figure 6, starting from a conventional plant setup, based on naphtha feedstock with 7 methane fired furnaces. A first moderate CO_2 emission reduction step may for example be realized by equipping a pair of furnaces with air preheating and electrifying the C3 machine. A second, more significant CO_2 reduction step could be done by electrifying three furnaces as well as the main cracked gas compressor. Alternatively, a flue gas carbon capture system may be installed for four of the existing furnaces. The final step towards zero scope 1 CO_2 emissions can in this case study be reached by combining three electric furnaces with either 4 air-preheated furnaces powered by blue hydrogen or with the 4 methane-fired furnaces previously equipped with flue gas carbon capture. Alternatively, a full electrification of furnaces and large machinery may also be executed as a final step for complete avoidance of

CO_2 emissions. In all scenarios, electric furnaces will be the key element in the decarbonization concept in this case study.

As indicated in the figure, multiple bifurcations and reunifications can be envisaged on the way when transforming an existing plant into a zero-emission one. Early roadmap development including all applicable decarbonization approaches facilitates the identification of pathways with "no-regret character", which avoid early dismantling of previous modifications and keep multiple pathways accessible for later steps. In the example shown, it would for instance be advantageous to electrify three furnaces already at transformation step 2. This permits to still choose between pre- or post-carbon capture for the transformation step 3, as well as to have the option of a full electrification of all furnaces as step 3.

In particular for plant revamp applications, it shall be noted that for the integration of electric furnaces additional plot space is required for accommodating the surrounding electrical infrastructure (e.g. substations, transformers, cabinets, switchgears, cables/busbars).

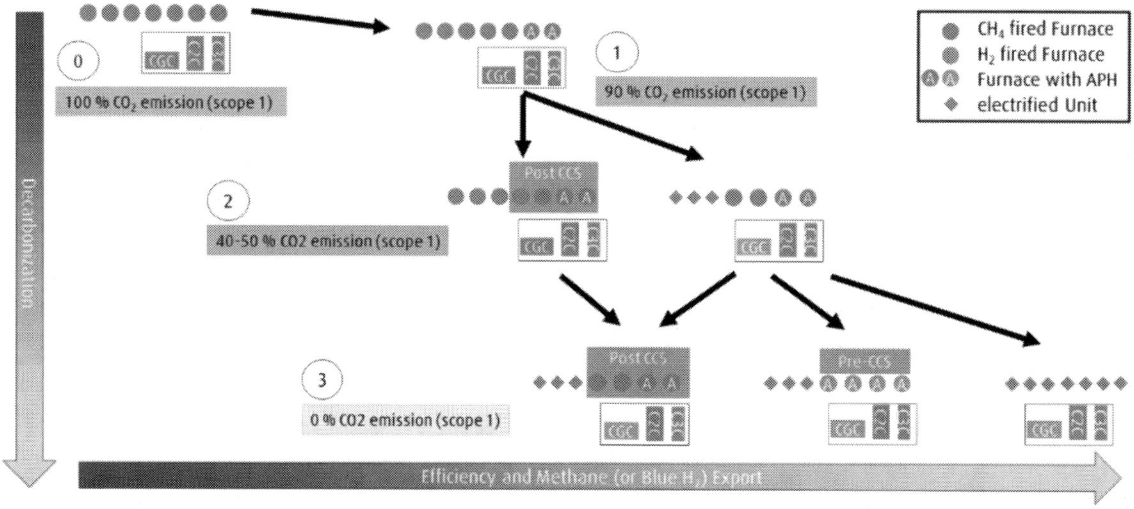

Figure 6: Potential decarbonization roadmap for exemplary plant using furnace electrification in conjunction with further decarbonization approaches.

Future greenfield plant layouts may also benefit from the integration of electric furnaces, which are a key enabler to lower the overall energy demand for the steam cracking process. Projection study results for exemplary gas and naphtha feed steam crackers with zero scope 1 CO2 emissions in normal operation are shown in Figure 7 and Figure 8, respectively.

The considered zero-emission plant concepts include either full or partial electrification of the furnace sections, combined with full electrification of the large rotating machinery. The high degree of electrification leads for all shown zero-emission concepts to a significant improvement of the energy efficiency of the plant, compared to conventional reference plants with significant CO_2 emissions. This is an inherent advantage of highly electrified steam cracker plants, see also [8].

For ethane feed steam crackers, see Figure 7, the overall energy demand of a plant can be reduced by almost a third (compared to the indicated study reference) when using only electric furnaces or a mix of electric and high-efficiency, hydrogen fired furnaces. The latter ones are thereby consuming the hydrogen separated from the cracked gas. In a setup with electric furnaces only, significant amounts of hydrogen are exported (equivalent to "turquoise" hydrogen), at the expense of a higher total electric power import from renewable sources. In this regard, it must be noted that for all feed types the electric power needed for electric coilbox heating is more than 50% lower than the fired duty of a conventional fired furnace with the same production capacity. Therefore, it may very well be an attractive scenario to continuously export large quantities of hydrogen with zero carbon footprint (in terms of scope 1 & 2) from a fully electrified ethane feed steam cracker.

In naphtha feed steam crackers, see Figure 8, the produced hydrogen quantities are lower, making a layout with solely electric furnaces more likely. As for ethane feed steam crackers, the total energy demand of a zero-emission plant concept is reduced compared to conventional cracker layouts. Nevertheless, in the example shown, the reduction is limited by the significant amounts of electrical power additionally required by process heaters in the separation section. Indeed, in such a setup, a "heat gap" appears, as steam export from the fully electrified furnace train is insufficient to supply heat to all consumers in the backend of the plant, even with full machine electrification. This opens the door for additional efficiency optimization measures in the separation section, by using e.g. heat pumps to minimize this additional power import for the supply of low-temperature process heat.

The case study examples illustrate that the development of successful decarbonization roadmaps ideally includes a wide range of applicable low- to zero-emission technology solutions. Close collaboration between operating companies and licensors is advised to identify the most advantageous combination for a specific project site, depending on relevant factors like feedstock scenarios, plot availability, access to renewable power supply, surrounding infrastructure and others.

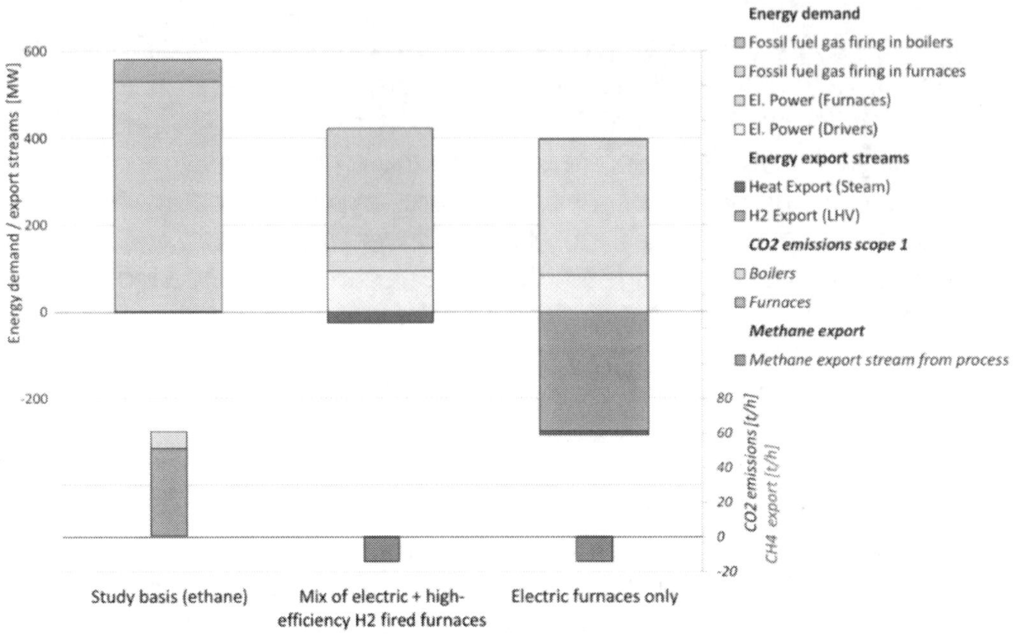

Figure 7: Projection study results for exemplary gas feed cracker setups with zero scope 1 CO_2 emissions in normal operation (based on 1000 kta ethylene production). Energy demand and CO_2 emissions are plotted as positive values, while exported energy and methane balance quantities are plotted as negative values.

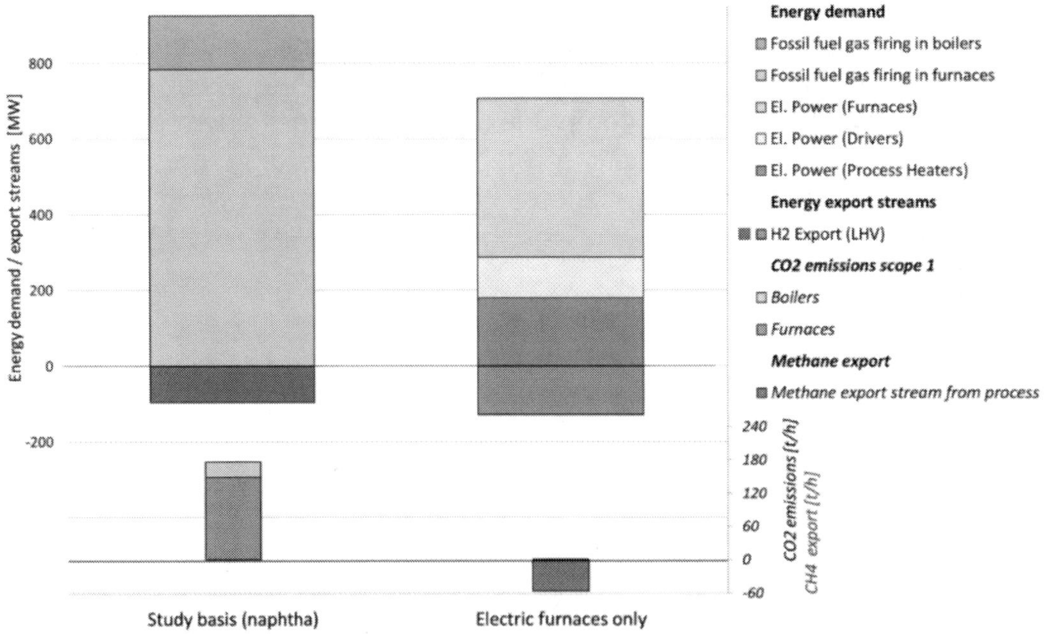

Figure 8: Projection study results for exemplary liquid feed cracker setups with zero scope 1 CO_2 emissions in normal operation (based on 1000 kta ethylene production). Energy demand and CO_2 emissions are plotted as positive values, while exported energy and methane balance quantities are plotted as negative values.

Conclusion & outlook

The electrification of cracking furnaces is set to play a key role in the transformation of the steam cracking industry towards net-zero GHG emissions. The demonstration of two applicable electrical heating technologies in an industrial plant setting will be a significant milestone on this path. A successful demonstration will prove that renewably produced electricity can be used in the future as main energy source in steam crackers while still ensuring reliable long-term furnace operation for base chemical production.

The large scale of the demonstration plant and its integration into a commercial production facility will pave the way for the future implementation of commercial-size electric furnaces, with reduced upscaling risk in all relevant project phases, spanning from engineering until commercial operation.

The paper further illustrates the manifold possibilities to integrate electrified furnaces into steam cracker plants, thereby giving the recommendation to timely develop long-term decarbonization roadmaps for individual plants based on their specific boundary conditions. The integration of electric furnaces can be effectively combined with other decarbonization approaches to enable the transition to net-zero emissions for both existing and future grassroot plants.

References

1. European Commission, JRC Science for Policy Report, "Energy efficiency and GHG emissions: Prospective scenarios for the Chemical and Petrochemical Industry", available under https://publications.jrc.ec.europa.eu/repository/bitstream/JRC105767/kj-na-28471-enn.pdf, p. 14, Table 3, 2017.

2. Lange, J. P., "Towards circular carbo-chemicals – the metamorphosis of petrochemicals", *Energy & Environmental Science*, 2021.

3. Amghizar, I., Dedeyne, J. N., Brown, D. J., Marin, G. B., Van Geem, K. M., "Sustainable innovations in steam cracking: CO_2 neutral olefin production", *Reaction Chemistry and Engineering*, 2020.

4. Middleton, J., "Decarbonisation of steam crackers", *Decarbonisation Technology*, 2021.

5. Mynko, O., Amghizar, I., Brown, D. J., Chen, L., Marin, G. B., Freitas de Alvarenga, R., Civancik Uslu, D., Dewulf, J. Van Geem, K. M., "Reducing CO2 emissions of existing ethylene plants: Evaluation of different revamp strategies to reduce global CO2 emission by 100 million tonnes", *Journal of Cleaner Production*, 2022.

6. Joint News Release, "BASF, SABIC and Linde start construction of the world's first demonstration plant for large-scale electrically heated steam cracker furnaces", available under https://www.basf.com/global/en/media/news-releases/2022/09/p-22-326.html, 2022.

7. Kompetenzzentrum Klimaschutz in energieintensiven Industrien (KEI), available under https://www.klimaschutz-industrie.de/foerderung/projekte/basf-efurnace, 2022.

8. Kracker, G., Becker, J., Zellhuber, M., "Electrification of Steam Cracking and Related Challenges", *33rd Ethylene Producers Conference, 2021 AIChE Virtual Spring Meeting and 17th Global Congress on Process Safety,* 2021.

Operational Based Petrochemicals Decarbonization System

Abduljelil Iliyas, Manuel Dias

Developing optimum strategies for CO2 emission reduction and circular economy are some of the key challenges faced by the chemical industry today. Chemical companies need to estimate, track, report, and ultimately reduce their environmental footprint, for example from direct and indirect GHG emissions, or from the use and end-of-life disposition of final products such as plastics.

Petrochemical plants, especially the olefin plants, are energy-intensive with significant direct CO_2 emissions. However, for most plants, the focus is mostly on production margin optimization with less attention to the carbon footprint associated with such optimization efforts. Significant development efforts are ongoing within the industry to reduce the carbon footprint of petrochemical complexes. For example, furnace electrification, renewable feedstocks, green/blue H2, and carbon capture; however, many of these will take several years to mature and will be capital intensive. While these efforts are ongoing, companies must focus on what they can do with existing assets to reduce their CO_2 footprint based on available degrees of freedom.

This presentation will discuss and present examples of how SABIC is utilizing key day-to-day operational decision variables to reduce CO_2 emission levels in our assets. For example,

- Feed slate (e.g. ethane vs LPG vs naphthas)
- Co-cracking and cracking severity
- Fuel mix and steam boiler fuel make-up (e.g. natural gas vs. fuel oil)
- Hydrogen disposition

11a EPC Keynote:Developing a Roadmap to Net Zero

Sarah Marshall

How can the petrochemicals industry meet global ambitions to be net-zero by 2050? What are short- and long-term milestones on the journey to industrial decarbonization? How can manufacturers reduce Scope 1, 2 and 3 emissions?

Sarah Marshall, VP of Sustainability at NOVA Chemicals, will share the company's plans to drive decarbonization and align to global ambitions to be net-zero by 2050. In addition, she'll share NOVA's decarbonization strategy and roadmap for GHG reductions that includes efficiency & electrification projects, renewable and low carbon power, hydrogen and carbon capture, utilization, and sequestration (CCUS) strategies as well as new and emerging technologies.

AUTHOR INDEX